Lecture Notes in Computer Science 8986

Commenced Publication in 1973
Founding and Former Series Editors:
Gerhard Goos, Juris Hartmanis, and Jan van Leeuwen

More information about this series at http://www.springer.com/series/7407

Jan Kratochvíl · Mirka Miller
Dalibor Froncek (Eds.)

Combinatorial Algorithms

25th International Workshop, IWOCA 2014
Duluth, MN, USA, October 15–17, 2014
Revised Selected Papers

 Springer

Editors
Jan Kratochvíl
Faculty of Mathematics and Physics
Charles University
Praha
Czech Republic

Dalibor Froncek
Department of Mathematics
University of Minnesota, Duluth
Duluth, MN
USA

Mirka Miller
University of Newcastle
Newcastle, NSW
Australia

ISSN 0302-9743   ISSN 1611-3349 (electronic)
Lecture Notes in Computer Science
ISBN 978-3-319-19314-4   ISBN 978-3-319-19315-1 (eBook)
DOI 10.1007/978-3-319-19315-1

Library of Congress Control Number: 2015940420

Springer Cham Heidelberg New York Dordrecht London

Printed on acid-free paper

Springer International Publishing AG Switzerland is part of Springer Science+Business Media
(www.springer.com)

Preface

The 25th International Workshop on Combinatorial Algorithms (IWOCA) was held during October 15–17, 2014, in the picturesque harbor town Duluth, located in the south-west corner of Lake Superior in Minnesota, USA. Autumn is a favorite time of the year for visiting Duluth, owing to the amazing range of colors of tree and shrub foliage on display at this time of the year. The IWOCA 2014 Organizing Committee timed the event perfectly!

IWOCA – the workshop that originated 25 years ago as the (Australasian) AWOCA – has over the years established itself as a truly international conference. The name change (to IWOCA) reflected the expanse of the conference beyond local boundaries, motivated by the growing global interest in the conference. The first IWOCA events were still held in Australia in 2007, and the subsequent years brought it to Japan (2008), the Czech Republic (2009), the UK (2010), Canada (2011), India (2012), France (2013), and to the USA this year. During the last six years the proceedings have been published by Springer in the LNCS series.

IWOCA 2014 received 68 submissions, most of them of very high quality. The Program Committee was faced with hard work and sometimes difficult decisions and we regretted that some good papers had to be rejected because of the limited capacity of the conference schedule. In the end, 32 contributed talks were presented during the conference.

We would like to thank all who have sent their submissions and to congratulate all the authors of the accepted papers. We extend special thanks to the distinguished invited speakers Josep Domingo-Ferrer, Pinar Heggernes, Saketh Saurab, and Xuding Zhu. We also thank all the authors who submitted posters for the poster session (which are, however, not included in these proceedings).

Finally, we thank all the members of the Program Committee, all external reviewers, and all the members of the Organizing Committee for all the hard work they have done. While all committee members worked well as a team, some names must be singled out: Special thanks go to Sergei Bezrukov for tirelessly updating the website and running the technology support during the workshop, and to Xiaofeng Gu for handling technical issues of papers included in both the pre-workshop proceedings and this volume.

March 2015

Dalibor Froncek
Jan Kratochvíl
Mirka Miller

Organization

Program Committee

Jemal Abawajy	Deakin University, Australia
Hideo Bannai	Kyushu University, Japan
Ljiljana Brankovic	University of Newcastle, Australia
Stolting Brodal	Aarhus University, Denmark
Pino Caballero-Gil	University of La Laguna, Spain
Charlie Colbourn	Arizona State University, USA
Maxime Crochemore	King's College London, UK
Pinar Dundar	Ege University, Turkey
Jiri Fiala	Charles University in Prague, Czech Republic
Dalibor Froncek	University of Minnesota Duluth, USA
Roberto Grossi	Università di Pisa, Italy
Jan Holub	Czech Technical University in Prague, Czech Republic
Costas Iliopoulos	King's College London, UK
Ralf Klasing	LaBRI and CNRS, France
Christian Komusiewicz	TU Berlin, Germany
Jan Kratochvíl	Charles University in Prague, Czech Republic
Dieter Kratsch	Universite de Lorraine-Metz, France
Gregory Kucherov	University Paris-Est Marne-la-Vallée and CNRS, France
Thierry Lecroq	Université de Rouen, France
Zsuzsanna Lipták	Università di Verona, Italy
Paul Manuel	Kuwait University, Kuwait
Mirka Miller	University of Newcastle, Australia
Ian Munro	University of Waterloo, Canada
Kunsoo Park	Seoul National University, South Korea
Solon Pissis	King's College London, UK
Hebert Pérez-Rosés	University of Lleida, Spain
Sohel Rahman	Bangladesh University of Engineering and Technology, Bangladesh
Vojta Rodl	Emory University, USA
Frank Ruskey	University of Victoria, Canada
Bill Smyth	McMaster University, Canada
Lynette Van Zijl	Stellenbosch University, South Africa

Program Committee Co-chairs

Jan Kratochvíl   Charles University in Prague, Czech Republic
Mirka Miller   University of Newcastle, Australia

Problem Session Co-chairs

Uwe Leck   University of Wisconsin-Superior, USA
Zsuzsanna Lipták   Università di Verona, Italy

Proceedings Technical Editor

Xiaofeng Gu   University of Wisconsin-Superior, USA

Steering Committee

Costas Iliopoulos   King's College London, UK
Mirka Miller   University of Newcastle, Australia
Bill Smyth   McMaster University, Canada

Organizing Committee

Sergei Bezrukov   University of Wisconsin-Superior, USA
Dalibor Froncek (Chair)   University of Minnesota Duluth, USA
Xiaofeng Gu   University of Wisconsin-Superior, USA
Steven Rosenberg   University of Wisconsin-Superior, USA
Uwe Leck   University of Wisconsin-Superior, USA

Additional Reviewers

Avrachenkov, Konstantin   Dogan, Derya
Baca, Martin   Escoffer, Bruno
Baisya, Dipankar   Feria-Puron, Ramiro
Bari, Md. Faizul   Fernau, Henning
Bevern, René Van   Fertin, Guillaume
Boeckenhauer, Hans-Joachim   Fici, Gabriele
Burcsi, Péter   Firoz, Jesun
Caballero-Gil, Candido   Foucaud, Florent
Caceres Cruz, Jose   Gerbner, Daniel
Cechlarova, Katarina   Grigorious, Cyriac
Chen, Jiehua   Gu, Xiaofeng
Cheng, Eddie   Harju, Tero
Cicalese, Ferdinando   Hartung, Sepp
D'Arco, Paolo   Hocquard, Hervé
Dev, Himel   Hoksza, David

Hossain, Md. Iqbal
Hsieh, Sun-Yuan
Hüffner, Falk
I., Tomohiro
Inenaga, Shunsuke
Irvine, Veronika
Islam, A.S.M. Sohidull
Keil, Mark
Lefebvre, Arnaud
Li, Rao
Liu, Daphne
Lukovszki, Tamas
Martin-Fernandez, Francisco
Mary, Arnaud
Medvedev, Paul
Miltzow, Tillmann
Molina-Gil, Jezabel
Mondal, Debajyoti
Nisse, Nicolas
Oner, Tahsin
Ordin, Burak
Peterlongo, Pierre
Phanalasy, Oudone
Pineda-Villavicencio,
    Guillermo
Prieur-Gaston, Elise

Rajan, Bharati
Rios-Solis, Yasmin
Rivals, Eric
Rosenberg, Steve
Ryan, Joe
Ryjacek, Zdenek
Saba, Sahand
Salikhov, Kamil
Scholtzova, Jirina
Sebe, Francesc
Sgall, Jiri
Sgall, Jiří
Shaw, Dipan Lal
Sorge, Manuel
Stephen, Sudeep
Suchy, Ondrej
Tantau, Till
Tiwary, Hans Raj
Valla, Tomáš
Valtr, Pavel
Vialette, Stéphane
Walen, Tomasz
Weimann, Oren
Xu, Min
Zhu, Xuding
Zohora, Fatema Tuz

Contents

On the Complexity of Various Parameterizations of Common Induced
Subgraph Isomorphism 1
  Faisal N. Abu-Khzam, Édouard Bonnet, and Florian Sikora

Approximation and Hardness Results for the Maximum Edges
in Transitive Closure Problem.. 13
  Anna Adamaszek, Guillaume Blin, and Alexandru Popa

Quantifying Privacy: A Novel Entropy-Based Measure of Disclosure Risk ... 24
  Mousa Alfalayleh and Ljiljana Brankovic

On the Galois Lattice of Bipartite Distance Hereditary Graphs 37
  Nicola Apollonio, Massimiliano Caramia, and Paolo Giulio Franciosa

Fast and Simple Computations Using Prefix Tables Under Hamming
and Edit Distance 49
  Carl Barton, Costas S. Iliopoulos, Solon P. Pissis, and William F. Smyth

Border Correlations, Lattices, and the Subgraph Component Polynomial 62
  Francine Blanchet-Sadri, Michelle Cordier, and Rachel Kirsch

Computing Minimum Length Representations of Sets of Words
of Uniform Length 74
  Francine Blanchet-Sadri and Andrew Lohr

Computing Primitively-Rooted Squares and Runs in Partial Words 86
  Francine Blanchet-Sadri, Jordan Nikkel, J.D. Quigley, and Xufan Zhang

3-Coloring Triangle-Free Planar Graphs with a Precolored 9-Cycle 98
  Ilkyoo Choi, Jan Ekstein, Přemysl Holub, and Bernard Lidický

Computing Heat Kernel Pagerank and a Local Clustering Algorithm 110
  Fan Chung and Olivia Simpson

A Γ-magic Rectangle Set and Group Distance Magic Labeling 122
  Sylwia Cichacz

Solving Matching Problems Efficiently in Bipartite Graphs 128
  Selma Djelloul

A 3-Approximation Algorithm for Guarding Orthogonal Art Galleries
with Sliding Cameras 140
  Stephane Durocher and Saeed Mehrabi

On Decomposing the Complete Graph into the Union
of Two Disjoint Cycles   153
  Saad I. El-Zanati, Uthoomporn Jongthawonwuth, Heather Jordon,
  and Charles Vanden Eynden

Reconfiguration of Vertex Covers in a Graph   164
  Takehiro Ito, Hiroyuki Nooka, and Xiao Zhou

Space Efficient Data Structures for Nearest Larger Neighbor   176
  Varunkumar Jayapaul, Seungbum Jo, Venkatesh Raman,
  and Srinivasa Rao Satti

Playing Several Variants of Mastermind with Constant-Size Memory
is not Harder than with Unbounded Memory   188
  Gerold Jäger and Marcin Peczarski

On Maximum Common Subgraph Problems in Series-Parallel Graphs   200
  Nils Kriege, Florian Kurpicz, and Petra Mutzel

Profile-Based Optimal Matchings in the Student/Project Allocation Problem   213
  Augustine Kwanashie, Robert W. Irving, David F. Manlove,
  and Colin T.S. Sng

The Min-max Edge q-Coloring Problem   226
  Tommi Larjomaa and Alexandru Popa

Speeding up Graph Algorithms via Switching Classes   238
  Nathan Lindzey

Study of $\kappa(D)$ for $D = \{2,3,x,y\}$ 250
  Daniel Collister and Daphne Der-Fen Liu

Some Hamiltonian Properties of One-Conflict Graphs   262
  Christian Laforest and Benjamin Momège

Sequence Covering Arrays and Linear Extensions   274
  Patrick C. Murray and Charles J. Colbourn

Minimum r-Star Cover of Class-3 Orthogonal Polygons   286
  Leonidas Palios and Petros Tzimas

Embedding Circulant Networks into Butterfly and Benes Networks   298
  R. Sundara Rajan, Indra Rajasingh, Paul Manuel, T.M. Rajalaxmi,
  and N. Parthiban

Kinetic Reverse k-Nearest Neighbor Problem   307
  Zahed Rahmati, Valerie King, and Sue Whitesides

Efficiently Listing Bounded Length *st*-Paths   318
    Romeo Rizzi, Gustavo Sacomoto, and Marie-France Sagot

Metric Dimension for Amalgamations of Graphs   330
    Rinovia Simanjuntak, Saladin Uttunggadewa, and Suhadi Wido Saputro

A Suffix Tree Or Not a Suffix Tree?   338
    Tatiana Starikovskaya and Hjalte Wedel Vildhøj

Deterministic Algorithms for the Independent Feedback Vertex
Set Problem   351
    Yuma Tamura, Takehiro Ito, and Xiao Zhou

Lossless Seeds for Searching Short Patterns with High Error Rates   364
    Christophe Vroland, Mikaël Salson, and Hélène Touzet

Author Index   377

On the Complexity of Various Parameterizations of Common Induced Subgraph Isomorphism

Faisal N. Abu-Khzam[1], Édouard Bonnet[2], and Florian Sikora[2(✉)]

[1] Lebanese American University, Beirut, Lebanon
faisal.abukhzam@lau.edu.lb
[2] PSL, Université Paris-Dauphine, LAMSADE, UMR CNRS 7243, Paris, France
{florian.sikora,edouard.bonnet}@dauphine.fr

Abstract. MAXIMUM COMMON INDUCED SUBGRAPH (henceforth MCIS) is among the most studied classical NP-hard problems. MCIS remains NP-hard on many graph classes including bipartite graphs, planar graphs and k-trees. Little is known, however, about the parameterized complexity of the problem. When parameterized by the vertex cover number of the input graphs, the problem was recently shown to be fixed-parameter tractable. Capitalizing on this result, we show that the problem does not have a polynomial kernel when parameterized by vertex cover unless NP ⊆ coNP/*poly*. We also show that MAXIMUM COMMON CONNECTED INDUCED SUBGRAPH (MCCIS), which is a variant where the solution must be connected, is also fixed-parameter tractable when parameterized by the vertex cover number of input graphs. Both problems are shown to be W[1]-complete on bipartite graphs and graphs of girth five and, unless P = NP, they do not belong to the class XP when parameterized by a bound on the size of the minimum feedback vertex sets of the input graphs, that is solving them in polynomial time is very unlikely when this parameter is a constant.

1 Introduction

A common induced subgraph of two graphs G_1 and G_2 is a graph that is isomorphic to induced subgraphs of each. The problem of finding a common induced subgraph of maximum number of vertices (or edges) has many applications in a number of domains including bioinformatics and chemistry [11–13,16,17]. In the decision version of the problem, we are given an integer k and the question is to decide if there is a solution with at least k vertices. We say that k is the natural parameter of the problem, that is the solution size.

Concerning its classical complexity, MAXIMUM COMMON INDUCED SUBGRAPH is NP-complete, and remains so on bipartite graphs and graphs with bounded treewidth. However, the problem is in P for trees [10] and graphs of (both) bounded treewidth and bounded degree [3].

A decision subproblem of MAXIMUM COMMON INDUCED SUBGRAPH is the well known INDUCED SUBGRAPH ISOMORPHISM (ISI) problem, which consists of deciding whether G_1 is isomorphic to an induced subgraph of G_2. In other words,

© Springer International Publishing Switzerland 2015
J. Kratochvíl et al. (Eds.): IWOCA 2014, LNCS 8986, pp. 1–12, 2015.
DOI: 10.1007/978-3-319-19315-1_1

it is equivalent to MAXIMUM COMMON INDUCED SUBGRAPH where $k = |G_1|$. In this case G_1 is called the pattern graph while G_2 is the host graph. ISI is W[1]-hard in general, by a straightforward reduction from k-Clique. Therefore MCIS is also W[1]-hard. On the other hand, if ISI is in FPT on a certain graph class, then so is MCIS. To see this, note that an arbitrary instance (G_1, G_2, k) of MCIS can be reduced in fpt-time to two instances of ISI by enumerating all possible graphs on k vertices and checking whether each is an induced subgraph of each of the two input graphs. This implies that ISI and MCIS have the same parameterized complexity when parameterized by the solution size, which we refer to as the natural parameter in this paper. Of course, the latter reduction takes time $O(2^{k^2})$ (multiplied by the time needed to solve ISI on the given graph class), which makes it prohibitively impractical. We shall provide a simpler reduction that takes $O(c^k)$-time on a class of graphs that includes H-minor free graphs and graphs of bounded degree.

Another way to deal with the hardness of a problem is to study its complexity with respect to auxiliary (or structural) parameters, to better understand the behavior of the problem (see for example [8]). MCIS is already hard on graphs with bounded treewidth, being NP-hard on forests, as we shall observe based on a classical result from Garey and Johnson [10]. Accordingly, the problem is W[1]-hard when parameterized by the treewidth of the input graphs. Therefore we need to look for bigger parameters. We shall study the problem with respect to the size of a (minimum) feedback vertex set that of a (minimum) vertex cover of input graphs. We observe that MCIS is not in XP when parameterized by the feedback vertex set number of the input graphs. This also implies that the problem is not in XP when parameterized by treewidth.

We observe that ISI remains W[1]-hard on graphs where G_1 has a k-vertex cover by a reduction from the W[1]-hard INDUCED BIPARTITE MATCHING problem [14]: if the pattern consists of k disjoint edges, its vertex cover is k. Therefore, MCIS is W[1]-hard when the parameter is the vertex cover of one of the input graphs, even if the other graph is bipartite. However, if the parameter k is the combination of the vertex cover of both input graphs, then the problem is in FPT, with a running time of $O((24k)^k)$ [1]. We shall prove in Sect. 3 that MCIS does not have a polynomial-size kernel in this case unless NP $\subseteq$ coNP/$poly$.

We also consider the MAXIMUM COMMON CONNECTED INDUCED SUBGRAPH problem. We observe that the problem is in FPT on graphs of bounded degree and show it to be W[1]-complete on the class of bipartite graphs, even if the input graph is C_4-free. Consequently, MCCIS is W[1]-complete on graphs of girth five. Finally, we show that MCCIS is fixed-parameter tractable when parameterized by a bound on the minimum vertex covers of the input graphs.

2   Preliminaries

Two finite graphs $G_1 = (V_1, E_1)$ and $G_2 = (V_2, E_2)$ are *isomorphic* if there is a bijection $\pi : V_1 \to V_2$ such that $\forall u, v \in V_1 : uv \in E_1 \Leftrightarrow \pi(u)\pi(v) \in E_2$. Given a graph $G = (V, E)$, a graph $G' = (V', E')$ is an *induced subgraph* of G

if $V' \subseteq V$ and $E' = E(V') = \{uv \in E \mid u, v \in V'\}$, i.e. E' is the edge set with both extremities in V'. We also say that G' is the subgraph of G induced by V'.

The *girth* of a graph G is the length of the shortest cycle contained in G. Contracting an edge uv consists of deleting uv and replacing the vertices u and v by a single vertex w in the incidence relation (edges incident on u or v become incident on w). A graph H is a *topological minor* of graph G if H is obtained from a subgraph of G by applying zero or more edge contractions. Given a fixed graph H, a family $\mathcal{F}$ of graphs is said to be *H-minor free* if H is not a minor of any element of $\mathcal{F}$.

The MAXIMUM COMMON INDUCED SUBGRAPH problem is defined formally as follows.

MAXIMUM COMMON INDUCED SUBGRAPH (MCIS):
- **Input**: Two graphs $G_1 = (V_1, E_1)$ and $G_2 = (V_2, E_2)$.
- **Output**: An induced subgraph G'_1 of G_1 isomorphic to an induced subgraph G'_2 of G_2 with a maximum number of vertices.

MAXIMUM COMMON CONNECTED INDUCED SUBGRAPH (MCCIS) is defined as MCIS with the additional restriction that the solution must be connected. INDUCED SUBGRAPH ISOMORPHISM is defined similarly:

INDUCED SUBGRAPH ISOMORPHISM (ISI):
- **Input**: Two graphs $G_1 = (V_1, E_1)$ and $G_2 = (V_2, E_2)$.
- **Output**: An induced subgraph G'_1 of G_1 isomorphic to G_2.

Parameterized complexity. A parameterized problem (I, k) is said *fixed-parameter tractable* (or in the class FPT) w.r.t. (with respect to) parameter k if it can be solved in $f(k) \cdot |I|^c$ time (i.e. in fpt-time), where f is any computable function and c is a constant (see [7,15] for more details about fixed-parameter tractability). The parameterized complexity hierarchy is composed of the classes FPT $\subseteq$ W[1] $\subseteq$ W[2] $\subseteq \cdots \subseteq$ XP. The class XP contains problems solvable in time $f(k) \cdot |I|^{g(k)}$, where f and g are unrestricted functions. A W[1]-hard problem is not fixed-parameter tractable (unless FPT = W[1]) and one can prove W[1]-hardness by means of a *parameterized reduction* from a W[1]-hard problem. This is a mapping of an instance (I, k) of a problem A_1 in $g(k) \cdot |I|^{O(1)}$ time (for any computable function g) into an instance (I', k') for A_2 such that $(I, k) \in A_1 \Leftrightarrow (I', k') \in A_2$ and $k' \le h(k)$ for some function h.

A powerful technique to design parameterized algorithms is *kernelization*. In short, kernelization is a polynomial-time self-reduction algorithm that takes an instance (I, k) of a parameterized problem P as input and computes an equivalent instance (I', k') of P such that $|I'| \le h(k)$ for some computable function h and $k' \le k$. The instance (I', k') is called a *kernel* in this case. If the function h is polynomial, we say that (I', k') is a polynomial kernel. It is well known that a problem is in FPT iff it has a kernel, but this equivalence yields super-polynomial kernels (in general). To design efficient parameterized algorithms, a kernel of polynomial (or even linear) size in k is important. However, some lower bounds on the size of the kernel can be shown unless some polynomial hierarchy collapses. To show this result, we will use the cross composition technique developed by Bodlaender et al. [4].

Definition 1 (Polynomial Equivalence Relation [4]). *An equivalence relation $\mathcal{R}$ on Σ^* is said to be* polynomial *if the following two conditions hold: (i) There is an algorithm that given two strings $x, y \in \Sigma^*$ decides whether x and y belong to the same equivalence class in time $(|x| + |y|)^{O(1)}$. (ii) For any finite set $S \subseteq \Sigma^*$ the equivalence relation $\mathcal{R}$ partitions the elements of S into at most $(\max_{x \in S} |x|)^{O(1)}$ classes.*

Definition 2 (OR-Cross-Composition [4]). *Let $L \subseteq \Sigma^*$ be a set and let $Q \subseteq \Sigma^* \times \mathbb{N}$ be a parameterized problem. We say that L cross-composes into Q if there is a polynomial equivalence relation $\mathcal{R}$ and an algorithm which, given t strings $x_1, x_2, \ldots, x_t$ belonging to the same equivalence class of $\mathcal{R}$, computes an instance $(x^*, k^*) \in \Sigma^* \times \mathbb{N}$ in time polynomial in $\sum_{i=1}^{t} |x_i|$ such that: (i) $(x^*, k^*) \in Q \Leftrightarrow x_i \in L$ for some $1 \leqslant i \leqslant t$. (ii) k^* is bounded by a polynomial in $\max_{i=1}^{t} |x_i| + \log t$.*

Proposition 1 ([4]). *Let $L \subseteq \Sigma^*$ be a set which is NP-hard under Karp reductions. If L cross-composes into the parameterized problem Q, then Q has no polynomial kernel unless $NP \subseteq coNP/poly$.*

A parameterized problem is said to be *fixed-parameter enumerable* if all feasible solutions can be enumerated in $O(f(k)|I|^c)$ where f is a computable function of the parameter k only, and c is a constant.

3   Structural Parameterization of Maximum Common Induced Subgraph

Let us first recall that $tw(G) \leqslant fvs(G) \leqslant vc(G)$, where $tw(G)$ (resp. $fvs(G)$, $vc(G)$) represents the treewidth (resp. the feedback vertex set number, the vertex cover number) of G [8]. As noted before, if the parameter is the combination of $tw(G_1)$ and $tw(G_2)$ then MCIS is known to be W[1]-hard. Even more, if the parameter is the combination of $fvs(G_1)$ and $fvs(G_2)$ (which is bigger than the combination of the treewidth), then the problem is not even in XP since MAXIMUM COMMON INDUCED SUBGRAPH and INDUCED SUBGRAPH ISOMORPHISM are NP-hard on forests, a case where the parameter is equal to 0. Indeed, one can modify the reduction from 3-PARTITION done by Garey and Johnson in [10] for SUBFOREST ISOMORPHISM to our problem, by building chains of $B + 3$ vertices instead of $B + 1$ in G_2 such that each chain of G_1 is separated by a vertex. The following theorem follows.

Theorem 1. *Unless $P = NP$, MAXIMUM COMMON INDUCED SUBGRAPH is not in XP when parameterized by a bound on the minimum feedback vertex sets of the pair of input graphs.*

The hardness of MCIS on forest also implies the following.

Corollary 1. *Unless* P = NP, MAXIMUM COMMON INDUCED SUBGRAPH *is not in* XP *when parameterized by the treewidth of the input graphs.*

It was shown in [1] that MCIS is in FPT if the parameter is the combination of $vc(G_1)$ and $vc(G_2)$. Accordingly, the problem has a kernel, but no polynomial bound is known on its size. We show that, in this case, the kernel cannot be polynomial unless NP $\subseteq$ coNP/*poly*.

Theorem 2. *Unless* NP $\subseteq$ coNP/*poly*, MAXIMUM COMMON INDUCED SUB-GRAPH *has no polynomial kernel when parameterized by the sum of the sizes of vertex covers in the two input graphs.*

Proof. We will define an OR-cross-composition from the NP-complete CLIQUE, problem, where the given instance is a tuple (G^c, l) and the question is whether the graph G^c contains a clique on l vertices.

Given t instances, $(G_1^c, l_1), (G_2^c, l_2), \ldots, (G_t^c, l_t)$, of CLIQUE, where G_i^c is a graph and $l_i \in \mathbb{N}, \forall 1 \leqslant i \leqslant t$, we define our equivalence relation $\mathcal{R}$ such that any strings that are not encoding valid instances are equivalent, and $(G_i^c, l_i), (G_j^c, l_j)$ are equivalent iff $|V(G_i^c)| = |V(G_j^c)|$, and $l_i = l_j$. Hereafter, we assume that $V(G_i^c) = \{1, \ldots, n\}$ and $l_i = l$, for any $1 \leqslant i \leqslant t$. We will build an instance of MAXIMUM COMMON INDUCED SUBGRAPH parameterized by the vertex cover (G_1, G_2, l', Z) where G_1 and G_2 are two graphs, $l' \in \mathbb{N}$ and $Z \subseteq V(G_2)$ is a vertex cover of G_2 computed in fpt-time, such that there is a solution of size l' for MAXIMUM COMMON INDUCED SUBGRAPH iff there is an $i, 1 \leqslant i \leqslant t$ such that there is a solution of size l in G_i^c. We will now describe how to build G_1 and G_2.

To build G_2 (see also Fig. 1):

- $V(G_2) = \{p, q, r\} \cup \{a_i \mid 1 \leqslant i \leqslant t\} \cup \{e_{uv} \mid 1 \leqslant u < v \leqslant n\} \cup \{v_i \mid 1 \leqslant i \leqslant n\}$,
- $E(G_2)_1 = \{pq, pr, qr\}$,
- $E(G_2)_2 = \{ra_i \mid 1 \leqslant i \leqslant t\}$,
- $E(G_2)_3 = \{a_i e_{uv} \mid uv \in E(G_i^c)\}$,
- $E(G_2)_4 = \{e_{uv}v_u, e_{uv}v_v \mid \forall 1 \leqslant u < v \leqslant n\}$,
- $E(G_2) = E(G_2)_1 \cup E(G_2)_2 \cup E(G_2)_3 \cup E(G_2)_4$.

To build G_1 (see also Fig. 2):

- $V(G_1) = \{p, q, r, a\} \cup \{e_i \mid 1 \leqslant i \leqslant \binom{l}{2}\} \cup \{v_i \mid 1 \leqslant i \leqslant l\}$,
- $E(G_1)_1 = \{pq, pr, qr, ra\}$,
- $E(G_1)_2 = \{ae_i \mid 1 \leqslant i \leqslant \binom{l}{2}\}$,
- $E(G_1)_3 = \{e_i v_u, e_i v_v \mid \forall 1 \leqslant i \leqslant \binom{l}{2}, e_i = uv\}$,
- $E(G_1) = E(G_1)_1 \cup E(G_1)_2 \cup E(G_1)_3$.

We set $l' = |V(G_1)|$, and $Z = \{p, r\} \cup \{e_{uv} | 1 \leqslant u < v \leqslant n\}$. It is easy to see that Z is indeed a vertex cover for G_2 and that its size is equal to $\frac{n(n-1)}{2} + 2$, which

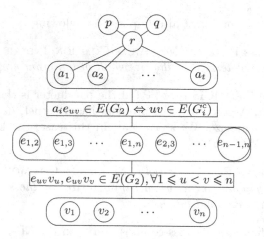

Fig. 1. Illustration of the construction of G_2.

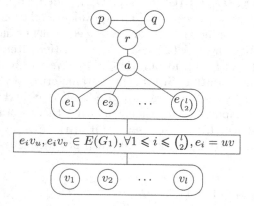

Fig. 2. Illustration of the construction of G_1.

is polynomial in n and hence in the size of the largest instance. Note that the size of the graph G_1 does not depend on t and is polynomial in n, so the size of its vertex cover is also polynomial in n and independent of t.

Let us show that G_1 is an induced subgraph of G_2 iff at least one of the G_i^c's has a clique of size l.

($\Leftarrow$) Suppose that G_i^c has a clique of size l. We denote by $S \subseteq V(G_i^c)$ a clique of size exactly l in G_i^c. We show that there is an induced subgraph S' of G_2 of size l', isomorphic to G_1. We set $V(S') = \{p, q, r\} \cup \{a_i\} \cup \{e_{uv} \mid \forall uv \in E(S)\} \cup \{v_u \mid u \in S\}$. One can easily check that this subgraph is isomorphic to G_1.

($\Rightarrow$) Assume now that G_1 is an induced subgraph of G_2. Denote by S' the subgraph of G_2 isomorphic to G_1. Note that the only triangle in G_2 is pqr.

Indeed, $T(V(G_2) \setminus \{p\})$ is bipartite. The triangle pqr in G_1 has therefore to match pqr in G_2. Moreover, r in G_1 has to match r in G_2 since p and q have no edges besides the clique pqr. The vertex a in G_1 can only match a vertex a_i for some $i \in \{1, \ldots, t\}$. Then, e_1 up to $e_{\binom{l}{2}}$ in G_1 has to match $\binom{l}{2}$ vertices in $\{e_{uv} \mid 1 \leqslant u < v \leqslant n\}$ of G_2 which correspond to actual edges in G_i^c. Finally, v_1 up to v_l in G_1 has to match l vertices among the v_j's in G_2. Note that the number of edges in $E(G_1)_3$ between the e_j's and the v_j's is exactly $2\binom{l}{2} = l(l-1)$. More precisely, each e_j touches 2 edges in $E(G_1)_3$ and each v_j touches $l - 1$ edges in $E(G_1)_3$. In order to get a match in G_2, one should find a set of $\binom{l}{2}$ edges inducing exactly l vertices. So, this set of l vertices is a clique in G_i^c.

Note that the parameter of MCIS in this reduction is exactly the size of G_1. Therefore, this negative result holds for ISI too.

Despite the fact that ISI and MCIS have the same parameterized complexity when parameterized by the natural parameter, they exhibit different complexities with respect to structural parameters. In fact, the latter is not even in XP when parameterized by the vertex cover of only one of the two graphs while ISI is FPT when parameterized by the vertex cover of the second (host) graph. To see this, note that when the host graph has a k-vertex cover, the minimum size of a vertex cover in the pattern graph must be bounded by the parameter k, otherwise we have a no instance. The claim follows from the fixed-parameter tractability of MCIS in this case [1].

Although MCIS is not in XP w.r.t. some structural parameters such as treewidth and feedback vertex set number, it is, together with MCCIS and ISI, in W[1] w.r.t. the natural parameter.

Theorem 3. *MCIS, MCCIS and ISI are* W[1]-*complete w.r.t. the natural parameter.*

Proof. Since ISI, MCIS and MCCIS are W[1]-hard by a straightforward reduction from k-Clique, it suffices to show membership in W[1]. In [5], it is shown that if a problem can be reduced in FPT time to simulating a non-deterministic single-taped Turing Machine halting in at most $f(k)$ steps, for some function f, then it is in W[1]. The Turing Machine can have an alphabet and a set of states of size depending on the size of the input of the initial problem. In our case, we can design a Turing Machine that guesses in $2k$ steps the corresponding right k vertices in G_1 (for ISI this part is not necessary) and the right k vertices in G_2 (our alphabet being isomorphic to an indexing of $V(G_1) \cup V(G_2)$) and then check in time $O(k^2)$ whether the two induced subgraphs are isomorphic (and that they are connected for MCCIS). □

We now turn our attention to the case where the MCIS is parameterized by a combination of the natural parameter and some structural parameter. For example, consider the case where the parameter is the sum of some bound t on the feedback vertex set of the input graphs and the natural parameter k. The problem is FPT in this case since graphs of t-feedback vertex set are H-minor free (let H be the "fixed" graph consisting of a disjoint union of $t + 1$ triangles).

Moreover, we know ISI is FPT in this case due to [9]. However, and as stated in Sect. 1, a solution to an instance of MCIS (in this case) is obtained via an exhaustive enumeration of $O(2^{k^2})$ instances of ISI. This can be improved on classes of graphs that are given with some fixed coloring t, such as H-minor free graphs and graphs of bounded maximum degree. In fact, if the input to MCIS is a pair of t-colored graphs, then a reduction algorithm would first check whether each of the two graphs has an (independent) color class of size k. If so, then both have an edgeless common subgraph of size k. Otherwise, the order of at least one of the two graphs, say G_1, is smaller than tk. In such case, the algorithm proceeds by running a (fixed-parameter) algorithm for ISI on each of the $O(2^{tk})$ induced subgraphs of G_1.

In [14] it was shown that INDUCED MATCHING is W[1]-hard on bipartite graphs. As mentioned earlier, this proves that MCIS is W[1]-hard in this case. We show that MCIS remains W[1]-hard on C_4-free bipartite graphs, which proves its W[1]-hardness on graphs of girth five.

Theorem 4. MAXIMUM COMMON INDUCED SUBGRAPH *is* W[1]-*complete w.r.t. size of the solution, as parameter, even on* C_4-*free bipartite graphs.*

Proof. Membership in W[1] comes from Theorem 3. For the hardness, consider the following reduction from the W[1]-hard problem CLIQUE. Given an instance $(G = (V, E), k)$ of CLIQUE, we build an instance (G_1, G_2, k') of our problem as follows. The graph G_2 is the bipartite incidence graph of G (the bipartition is between vertices representing V and vertices representing E), the graph G_1 is the bipartite incidence graph of K_k, and $k' = k + \binom{k}{2} = |V(G_1)|$.

Note that a bipartite incidence graph is C_4-free since, in a simple graph, no two edges are incident on the same pair of vertices.

It is clear that G_1 occurs as a connected induced subgraph of G_2 iff there is a clique of size k in G, because w.l.o.g. $k > 2$ and the vertices representing edges in G_1 and G_2 are of degree 2. □

Corollary 2. MAXIMUM COMMON INDUCED SUBGRAPH *is* W[1]-*complete w.r.t. size of the solution on graphs of girth five.*

4   Maximum Common Connected Induced Subgraph

MAXIMUM COMMON CONNECTED INDUCED SUBGRAPH is trivially FPT whenever INDUCED SUBGRAPH ISOMORPHISM is FPT, including H-minor free graphs, since the enumeration of all $O(2^{k^2})$ possible induced connected subgraphs can be used as described before. The converse is also true. In fact, an instance (G_1, G_2, k) of ISI can be reduced to an equivalent instance $(G_1', G_2', k + 1)$ of MCCIS by letting G_i' be the graph obtained by adding a single (universal) vertex to G_i that is made adjacent to all other vertices of G_i. It follows that MCIS and MCCIS have the same parameterized complexity with respect to the natural parameter (i.e., solution size).

Note that MCIS is NP-hard on forests while MCCIS is solvable in polynomial-time in this case: given two forests G_1 and G_2, run the polynomial-time MCCIS algorithm of [2] on every pair of trees from G_1 and G_2.

In the following of this section we study the complexity of MCCIS with respect to structural parameters.

Lemma 1. INDUCED CONNECTED SUBGRAPH ISOMORPHISM *is* NP-*hard even when both graphs have feedback vertex set number equal to one.*

Proof. Given an instance of INDUCED SUBGRAPH ISOMORPHISM on forests G_1 and G_2 (each with at least 2 trees), we build an instance of INDUCED CONNECTED SUBGRAPH ISOMORPHISM by adding a universal vertex (connected to every node) in G_1 and in G_2. One can see that these two universal vertices must be matched together since they are the only ones with sufficiently high degree. Then, there is a solution for INDUCED SUBGRAPH ISOMORPHISM iff there is a solution for INDUCED CONNECTED SUBGRAPH ISOMORPHISM. The result of course holds for MCCIS too. □

Corollary 3. *Unless* P $=$ NP, MAXIMUM COMMON CONNECTED INDUCED SUBGRAPH *is not in* XP *when parameterized by a bound of the minimum feedback vertex set number of the input graphs (and hence then when parameterized by a bound on the treewidth of each of the two input graphs).*

Given the above negative result, the next question is whether MCCIS is in FPT w.r.t. the parameter vertex cover. In [1], a parameterized algorithm is presented for MCIS when the parameter is a bound on the minimum vertex cover number of the input graphs. However, that algorithm cannot help us much for solving MCCIS since it relies on the existence of a feasible solution of size at least $\approx n - k$ which consists of mapping the two *big* independent sets of the two graphs onto each other. Of course, this is not a feasible solution for MCCIS. In the following we prove that MCCIS is fixed-parameter tractable w.r.t. $k = \max(vc(G_1), vc(G_2))$.

Theorem 5. MAXIMUM COMMON CONNECTED INDUCED SUBGRAPH *parameterized by a bound on the vertex covers of the input graphs is fixed-parameter tractable.*

Proof. In time $O^*(2^k)$ (even $O^*(1.2738^k)$ [6]), we can find minimum vertex covers C_1 and C_2 in G_1 and G_2 respectively. Let $I^{(j)}$ be the independent set $V(G_j) \setminus C_j$ for $j \in \{1, 2\}$. By assumption, our parameter k is $\max(C_1, C_2)$, so we can enumerate all tripartitions of C_1 and C_2 in time $O^*(9^k)$. We denote by $C_{1,m}$, $C_{1,u}$ and $C_{1,i}$ (respectively $C_{2,m}$, $C_{2,u}$ and $C_{2,i}$) the three sets of a tripartition of C_1 (respectively C_2). For $j \in \{1, 2\}$, $C_{j,u}$ corresponds to the vertices of C_j that are not matched, so they may be deleted. $C_{j,m}$ comprises the vertices matched to $C_{3-j,m}$ (that is, to the vertex cover of the other graph), and $C_{j,i}$ are the vertices matched to $I^{(3-j)}$, the independent set of the other graph. See Fig. 3.

We observe that for $j \in \{1, 2\}$, $I^{(j)}$ can be partitioned into at most 2^k classes of twins: $I_1^{(j)}, I_2^{(j)}, \dots I_{2^k}^{(j)}$. A class of twins in this context is a set of vertices with an identical neighborhood in the vertex cover and there are at most 2^k subsets of C_j. Potentially, some classes can be empty: they correspond to a subset of the vertex cover C_j that is not the (exact) neighborhood of any vertex in $I^{(j)}$.

At this point, we can enumerate the mappings between $C_{1,m}$ and $C_{2,m}$ in time $O^*(k^k)$ and the mappings between $C_{j,i}$ and $I^{(3-j)}$ in time $O^*((2^k)^k) = O^*(2^{k^2})$. Indeed, to match a vertex u with a vertex v or a twin of v is equivalent. Thus, in time $O^*((9k)^k 2^{k^2})$ we can enumerate all the solutions of MCIS where only vertices of $I^{(1)}$ could still be matched to vertices of $I^{(2)}$. The optimal map of the independent sets can be done in linear time by matching the greatest number of vertices in each *equivalent* twin class (which is the size of the smaller of the two equivalent twin classes), where a twin class $I_r^{(j)}$ in $I^{(j)}$ is equivalent to a twin class $I_s^{(3-j)}$ in $I^{(3-j)}$ if the vertices of $N(I_r^{(j)}) \setminus C_{j,u}$ and $N(I_s^{(3-j)}) \setminus C_{3-j,u}$ are in one-to-one correspondence.   □

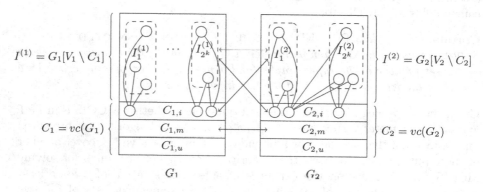

Fig. 3. Illustration of the proof of Theorem 5. Dashed boxes represent the classes inside the independent set. Arrows represent the matching between sets of vertices.

To find a solution for MCCIS, the algorithm described in the above proof enumerates all possible maximal common induced subgraphs in time $O^*((9k)^k 2^{k^2})$. As such, it can be used as an enumeration algorithm for MCIS.

Theorem 6. MAXIMUM COMMON INDUCED SUBGRAPH *parameterized by vertex cover, is fixed-parameter enumerable.*

Finally, the following corollaries follow easily from the proofs of Theorems 2 and 4 since the graphs used in both proofs are connected.

Corollary 4. MAXIMUM COMMON CONNECTED INDUCED SUBGRAPH, *parameterized by a bound on the minimum vertex covers of input graphs, does not have a polynomial-size kernel unless* NP $\subseteq$ coNP/poly.

Corollary 5. MAXIMUM COMMON CONNECTED INDUCED SUBGRAPH *is* W[1]-*complete on bipartite graphs and graphs of girth five.*

Table 1. Summary of different parameterized complexity results of ISI, MCIS and MCCIS for different structural parameters.

	$vc + vc$	$vc + fvs$	$fvs + fvs$	vc
ISI	FPT; no Poly Kernel (Theorem 2)	Open	$\notin$ XP	FPT for $vc(G_2)$, $\notin$ XP for $vc(G_1)$
M(C)CIS	FPT ([1],Theorem 5); no Poly Kernel	Open	$\notin$ XP (Corollary 3)	$\notin$ XP

In the following table we give a summary of some results obtained in this paper along with open questions. Note that for ISI, $vc + fvs$ is not the same parameter as $fvs + vc$. In the latter, the parameter is a bound on the vertex cover of G_2 (as well as the feedback vertex set of G_1) which makes ISI in FPT, while it remains open for $vc + fvs$. We also note that ISI is not in XP w.r.t. $vc(G_1)$ by a simple reduction from INDEPENDENT SET (let G_2 be an edgeless graph on k vertices, then its vertex cover number is 0).

5   Conclusion

We studied the MAXIMUM COMMON INDUCED SUBGRAPH and MAXIMUM COMMON CONNECTED INDUCED SUBGRAPH problems with respect to the solution size as natural parameter on special graphs classes, such as forests, bipartite graphs and graphs of girth five. The two problems are fixed-parameter tractable on H-minor free graphs, which include forests, but they are W[1]-complete on bipartite graphs and graphs of girth five.

We also considered the use of auxiliary parameters, such as a bound on the minimum vertex covers of the input graphs. Although both MCIS and MCCIS are in FPT in this case, we proved that no kernel of polynomial bound can be obtained unless NP $\subseteq$ coNP/*poly*. We noted that MCIS is not even in XP with respect to other (smaller) auxiliary parameters, such as treewidth and feedback vertex set (see Table 1). A few corresponding open problems remain to be addressed. For example, are MCIS/MCCIS in FPT when parameterized by the combination of the vertex cover number and the feedback vertex set number, or by the vertex cover number and the treewidth?

Acknowledgements. Work partially supported by the bilateral research cooperation CEDRE between France and Lebanon (grant number 30885TM).

References

1. Abu-Khzam, F.N.: Maximum common induced subgraph parameterized by vertex cover. Inf. Process. Lett. **114**(3), 99–103 (2014)
2. Akutsu, T.: An RNC algorithm for finding a largest common subtree of two trees. IEICE Trans. Inf. Syst. **75**(1), 95–101 (1992)

3. Akutsu, T.: A polynomial time algorithm for finding a largest common subgraph of almost trees of bounded degree. IEICE Trans. Fundam. Electron. Commun. Comput. Sci. **76**(9), 1488–1493 (1993)
4. Bodlaender, H.L., Jansen, B.M.P., Kratsch, S.: Kernelization lower bounds by cross-composition. SIAM J. Discrete Math. **28**(1), 277–305 (2014)
5. Cesati, M.: The turing way to parameterized complexity. J. Comput. Syst. Sci. **67**(4), 654–685 (2003)
6. Chen, J., Kanj, I.A., Xia, G.: Improved upper bounds for vertex cover. Theor. Comput. Sci. **411**(40–42), 3736–3756 (2010)
7. Downey, R.G., Fellows, M.R.: Fundamentals of Parameterized Complexity. Texts in Computer Science. Springer, London (2013)
8. Fellows, M.R., Jansen, B.M.P., Rosamond, F.A.: Towards fully multivariate algorithmics: parameter ecology and the deconstruction of computational complexity. Eur. J. Comb. **34**(3), 541–566 (2013)
9. Flum, J., Grohe, M.: Fixed-parameter tractability, definability, and model-checking. SIAM J. Comput. **31**(1), 113–145 (2001)
10. Garey, M.R., Johnson, D.S.: Computers and Intractability: A Guide to the Theory of NP-Completeness. W.H. Freeman, San Francisco (1979)
11. Grindley, H.M., Artymiuk, P.J., Rice, D.W., Willett, P.: Identification of tertiary structure resemblance in proteins using a maximal common subgraph isomorphism algorithm. J. Mol. Biol. **229**(3), 707–721 (1993)
12. Koch, I., Lengauer, T., Wanke, E.: An algorithm for finding maximal common subtopologies in a set of protein structures. J. Comput. Biol. **3**(2), 289–306 (1996)
13. McGregor, J., Willett, P.: Use of a maximal common subgraph algorithm in the automatic identification of the ostensible bond changes occurring in chemical reactions. J. Chem. Inf. Comput. Sci **21**, 137–140 (1981)
14. Moser, H., Sikdar, S.: The parameterized complexity of the induced matching problem. Discrete Appl. Math. **157**(4), 715–727 (2009)
15. Niedermeier, R.: Invitation to Fixed Parameter Algorithms. Lecture Series in Mathematics and Its Applications. Oxford University Press, Oxford (2006)
16. Raymond, J.W., Willett, P.: Maximum common subgraph isomorphism algorithms for the matching of chemical structures. J. Comput. Aided Mol. Des. **16**, 521–533 (2002)
17. Yamaguchi, A., Aoki, K.F., Mamitsuka, H.: Finding the maximum common subgraph of a partial k-tree and a graph with a polynomially bounded number of spanning trees. Inf. Process. Lett. **92**(2), 57–63 (2004)

Approximation and Hardness Results for the Maximum Edges in Transitive Closure Problem

Anna Adamaszek[1], Guillaume Blin[2,3], and Alexandru Popa[4]([✉])

[1] Max-Planck-Institut Für Informatik, Saarbrücken, Germany
anna@mpi-inf.mpg.de
[2] LaBRI, UMR 5800, University of Bordeaux, 33400 Talence, France
[3] CNRS, LaBRI, UMR 5800, 33400 Talence, France
guillaume.blin@labri.fr
[4] Faculty of Informatics, Masaryk University, Brno, Czech Republic
popa@fi.muni.cz

Abstract. In this paper we study the following problem, named Maximum Edges in Transitive Closure, which has applications in computational biology. Given a simple, undirected graph $G = (V, E)$ and a coloring of the vertices, remove a collection of edges from the graph such that each connected component is *colorful* (i.e., it does not contain two identically colored vertices) and the number of edges in the transitive closure of the graph is maximized.

The problem is known to be APX-hard, and no approximation algorithms are known for it. We improve the hardness result by showing that the problem is NP-hard to approximate within a factor of $|V|^{1/3-\varepsilon}$, for any constant $\varepsilon > 0$. Additionally, we show that the problem is APX-hard already for the case when the number of vertex colors equals 3. We complement these results by showing the first approximation algorithm for the problem, with approximation factor $\sqrt{2 \cdot \mathrm{OPT}}$.

1 Introduction

The Maximum Edges in Transitive Closure problem we consider in this paper belongs to the framework of colorful components problems.

COLORFUL COMPONENTS FRAMEWORK: Given a simple, undirected graph $G = (V, E)$ and a coloring $\sigma : V \to C$ of the vertices with colors from a given set C, remove a collection of edges $E' \subseteq E$ from G such that each connected component in the resulting graph $G' = (V, E \setminus E')$ is a *colorful component* (i.e., it does not contain two identically colored vertices). We want the graph G' to be optimal according to some fixed *optimization measure*.

In our problem, the optimization measure is the number of edges in the transitive closure. For a graph consisting of k connected components, each containing respectively $a_1, a_2, \ldots, a_k$ vertices, the number of edges in the transitive closure of the graph is

© Springer International Publishing Switzerland 2015
J. Kratochvíl et al. (Eds.): IWOCA 2014, LNCS 8986, pp. 13–23, 2015.
DOI: 10.1007/978-3-319-19315-1_2

$$\sum_{i=1}^{k} \frac{a_i \cdot (a_i - 1)}{2} .$$

Maximum Edges in Transitive Closure (MEC): Given a simple, undirected graph $G = (V, E)$ and a coloring $\sigma : V \to C$ of the vertices, remove a collection of edges $E' \subseteq E$ from G such that each connected component in the resulting graph $G' = (V, E \backslash E')$ is colorful, and the number of edges in the transitive closure of G' is maximum.

Motivation. The colorful components framework is motivated by applications in comparative genomics [8,10], which is a fundamental branch of bioinformatics studying the relationship of the genome structure between different biological species. One of the key problems in this area, the multiple alignment of gene orders, can be captured as a graph theoretical problem, using the colorful components framework, where the colorful graphs represent similarity relationships between genes from different homologous gene families. A partition into colorful components corresponds then to a partition of genes into orthology sets, where any two genes from the same genome belong to different orthology sets. We refer the reader to [10] for a more detailed description of the connection between the multiple alignment of gene orders and the graph theoretic framework considered.

The understanding of orthologous genes of two different genomes as originating from a single gene in the most recent common ancestor of the two species leads to transitivity as a property of the orthology relation. This motivates the study of MEC (see [10] for more details, and for a discussion why MEC yields good results in practice).

Related Work. The Maximum Edges in Transitive Closure problem has been introduced by Zheng et al. [10]. They present heuristic algorithms for the problem, without giving any worst-case approximation guarantee. They also conjecture the problem to be NP-hard. Adamaszek and Popa [1] prove that MEC is APX-hard, even in the case of 4 vertex colors.

The colorful components framework appeared first in the paper by Zheng et al. [10] and has been formally defined by Adamaszek and Popa [1], although problems which fit into this framework have already been studied earlier. We now summarize known results for these problems.

In the problem named either Colorful Components [3,4] or Minimum Orthogonal Partition [5,10], the objective function is to minimize the number of edges removed from G to obtain a graph in which all connected components are colorful. Bruckner et al. [4] show that the problem is NP-hard for three or more colors and they study fixed-parameter algorithms for the problem. Their NP-hardness reduction can be modified slightly to show the APX-hardness of the problem (see [1]). Zheng et al. [10] and Bruckner et al. [3] study heuristic approaches for the problem, and He et al. [5] present an approximation algorithm for some special case of the problem. As the general problem is a special case of the Minimum Multi-Multiway Cut problem, it admits a $O(\log |C|)$ approximation algorithm [2].

Zheng et al. [10] introduce the Minimum Singleton Vertices problem (MSV), where the goal is to minimize the number of isolated vertices in the resulting graph. Zheng et al. [10] present heuristic algorithms for the problem, without giving any worst-case approximation guarantee. They also conjectured that the problem is NP-hard. Tremblay-Savard and Swenson [9] consider a Maximum Orthogonal Edge Cover problem (MAX-OREC), which is a dual problem to MSV. There, the goal is to cover a maximum number of vertices of a graph using vertex-disjoint, non-singleton connected colorful subgraphs. In [9], a 2/3-approximation algorithm for MAX-OREC is presented. Adamaszek and Popa [1] prove that MSV (and therefore also MAX-OREC) can be solved exactly in polynomial time, thus disproving the conjecture in [10].

Adamaszek and Popa [1] introduce another problem, termed Minimum Colorful Components, in which the goal is to delete a subset of edges such that the resulting graph has only colorful components and the number of connected components is minimized. They show that this problem cannot be approximated within a factor of $|V|^{1/14-\varepsilon}$ unless $P = NP$, and within a factor $|V|^{1/2-\varepsilon}$ unless $ZPP = NP$.

Our Results. In this paper we improve the hardness results for the MEC problem, and we present the first approximation algorithm.

First, we show that MEC is APX-hard even for the case when $|C| = 3$. This settles the complexity of the problem when the number of colors is a constant, as for $|C| = 2$ the MEC problem can be solved exactly in polynomial time by using a maximum matching algorithm. Our proof is via a reduction from the Maximum Bounded 3-Dimensional Matching problem (Max 3-DM-3).

For the general case, when the number of colors is arbitrary, we show that MEC is NP-hard to approximate within a factor of $|V|^{1/3-\varepsilon}$ for any constant $\varepsilon > 0$. This result holds even if the input graph is a tree and each color appears at most twice in the graph. We use the same reduction from the Independent Set as Rizzi and Sikora for proving hardness of approximation of the Graph Motif problem [7].

We also show the first polynomial-time approximation algorithm for MEC, which has a ratio of $\sqrt{2 \cdot \text{OPT}}$. We use the exact polynomial time algorithm for the Minimum Singleton Vertices problem [1] to obtain a partition into colorful components and then we show that this partition has a big enough number of edges in the transitive closure.

2   APX-hardness of MEC for $|C| = 3$

In this section, we prove that the MEC problem restricted to instances using only 3 colors is APX-hard. The proof is via a reduction from the Maximum Bounded 3-Dimensional Matching problem. This result strengthens the one presented in [1], which holds for problem instances using 4 colors.

Before we give the reduction, we first state the definition of Max 3-DM-3 and the known hardness result for it.

MAXIMUM BOUNDED 3-DIMENSIONAL MATCHING (MAX 3-DM-3): The input consists of pairwise disjoint sets X, Y, Z and a collection $T \subseteq X \times Y \times Z$ of triples such that each element from X, Y and Z occurs in at least one and at most three triples in T. The aim is to find a feasible subset of triples $T' \subseteq T$ (i.e., no two elements of T' agree on any coordinate) of maximum cardinality.

Theorem 1 (Theorem 4.4 in [6], Rephrased). *There exists a constant $\varepsilon > 0$ such that it is NP-hard to distinguish between the instances of Max 3-DM-3 with the following properties:*

1. *There is a feasible collection of triples $T' \subseteq T$ such that every element of X, Y and Z belongs to some triple in T'.*
2. *For every feasible collection of triples $T' \subseteq T$ less than $(1 - \varepsilon)$ fraction of elements from $X \cup Y \cup Z$ belong to some triple of T'.*

Without loss of generality we can assume that $|X| = |Y| = |Z| = n$, since if $|X|$, $|Y|$ and $|Z|$ are different, then the case 1 of Theorem 1 cannot hold. Also, define $N = |T|$. It holds that $N \leq 3n$, since each element of $X \cup Y \cup Z$ appears in at most three triples. In the rest of the section, we use $\mathrm{OPT_{3DM}}$ to denote the size of an optimal solution of a Max 3-DM-3 instance (the instance we refer to will always be clear from the context), and $\mathrm{OPT_{MEC}}$ to denote the value of an optimal solution (i.e., the number of edges in the transitive closure of the graph) of the MEC instance obtained via the reduction.

Reduction. Given an instance (X, Y, Z, T) of Max 3-DM-3, we create an instance $(G = (V, E), \sigma)$ of the MEC problem in the following way. See Fig. 1 for a partial illustration. We create the set of vertices V as follows.

1. For each triple $t_j \in T$, we add six vertices $\{t_j^X, t_j^Y, t_j^Z, t_j^{XY}, t_j^{XZ}, t_j^{YZ}\}$.
2. For each element $x_i \in X$ (resp. $y_i \in Y$ and $z_i \in Z$), we add a corresponding vertex x_i (resp. y_i and z_i).

We have that $|V| = 6 \cdot |T| + |X| + |Y| + |Z| = 6N + 3n$. Let us now define the coloring $\sigma : V \to C$ of the vertices using the set of colors $C = \{1, 2, 3\}$.

1. For any $1 \leq i \leq n$ and $1 \leq j \leq N$, $\sigma(x_i) = \sigma(t_j^{XY}) = \sigma(t_j^Z) = 1$.
2. For any $1 \leq i \leq n$ and $1 \leq j \leq N$, $\sigma(y_i) = \sigma(t_j^{YZ}) = \sigma(t_j^X) = 2$.
3. For any $1 \leq i \leq n$ and $1 \leq j \leq N$, $\sigma(z_i) = \sigma(t_j^{XZ}) = \sigma(t_j^Y) = 3$.

Finally, let us define the collection of edges E.

1. For each $1 \leq j \leq N$, each of $\{t_j^X, t_j^{XY}, t_j^{XZ}\}$, $\{t_j^Y, t_j^{XY}, t_j^{YZ}\}$, $\{t_j^Z, t_j^{XZ}, t_j^{YZ}\}$ forms a clique of size three.
2. For each $1 \leq i \leq n$ and $1 \leq j \leq N$, if x_i (resp. y_i and z_i) appears in t_j, connect x_i (resp. y_i and z_i) to t_j^X (resp. t_j^Y and t_j^Z).

Analysis. Informally, we show that an instance of Max 3-DM-3 where all the vertices $X \cup Y \cup Z$ can be covered by a feasible collection of triples T' corresponds to an instance of MEC with a large optimal value, i.e., the graph can be partitioned into colorful components inducing a large transitive closure. On the other hand, we show that an instance of Max 3-DM-3 where no more than $(1 - \varepsilon)$ fraction of the vertices $X \cup Y \cup Z$ can be covered by any feasible set of triples corresponds to an instance of MEC with a much smaller optimal value. We now analyze both cases.

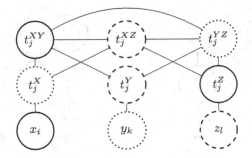

Fig. 1. A subgraph corresponding to a triple $t_j = (x_i, y_k, z_l)$. Colors of the vertices are denoted using the line styles: solid, dotted and dashed lines respectively corresponds to colors 1, 2 and 3.

Lemma 1. *Let (X, Y, Z, T) be an instance of Max 3-DM-3 where $OPT_{3DM} = n$, i.e., where all the vertices of $X \cup Y \cup Z$ can be covered by a feasible collection of triples. Then for the corresponding instance of MEC, we have $OPT_{MEC} \geq 6N + 3n$.*

Proof. The colorful components of the MEC instance are constructed as follows. For each triple $t_j \in T'$ (there are n of them), we add three colorful components, each component consisting of three vertices. Given a triple $t_j = (x_i, y_k, z_l)$, the colorful components are $\{x_i, t_j^X, t_j^{XZ}\}$, $\{y_k, t_j^Y, t_j^{XY}\}$ and $\{z_l, t_j^Z, t_j^{YZ}\}$. For each triple $t_{j'} \in T \setminus T'$ (there are $N - n$ of them), we create two colorful components, each consisting of three vertices: $\{t_{j'}^X, t_{j'}^{XZ}, t_{j'}^Z\}$ and $\{t_{j'}^{XY}, t_{j'}^Y, t_{j'}^{YZ}\}$. See Fig. 2 for an illustration.

As T' is a feasible collection of triples, that is a set of triples such that no two elements agree on any coordinate, we obtain a feasible partition of the graph into colorful components. Clearly, the total number of edges in the transitive closure equals $9n + 6(N - n) = 6N + 3n$, since each of the n triples in T' induces three colorful components of size three and each of the $N - n$ other triples induces two colorful components of size three. $\square$

Lemma 2. *Let (X, Y, Z, T) be an instance of Max 3-DM-3 where $OPT_{3DM} < (1 - \varepsilon)n$, i.e., where every feasible collection of triples covers less than a $(1 - \varepsilon)$ w of vertices $X \cup Y \cup Z$. Then, for the corresponding instance of MEC, we have $OPT_{MEC} < 6N + 3n(1 - \varepsilon/2)$.*

Proof. Let $(G = (V, E), \sigma)$ be the instance of the MEC problem corresponding to an instance of Max 3-DM-3 as defined in the lemma statement. For any triple $t_j = (x_i, y_k, z_l) \in T$, let G_{t_j} be a subgraph of G induced by the following set of vertices $\{x_i, y_k, z_l, t_j^X, t_j^Y, t_j^Z, t_j^{XY}, t_j^{XZ}, t_j^{YZ}\}$, as shown in Fig. 1.

Let us fix an optimal solution $\hat{S}$ for the MEC problem for (G, σ). This solution defines a partition Γ of G into colorful components. First, notice that each colorful component is contained within some subgraph G_t. Indeed, by construction, the only vertices which belong to multiple subgraphs G_{t_j} are the vertices

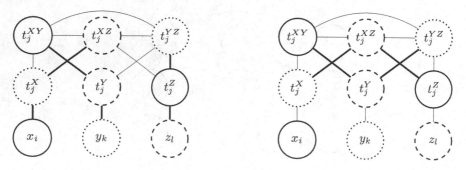

Fig. 2. Colorful components (defined by bold edges) for a triple from T' (left) and a triple from $T \setminus T'$ (right).

$\{x_i, y_i, z_i \mid 1 \le i \le n\}$. Moreover, for $1 \le i \le n$, each vertex x_i (resp. y_i and z_i) has only neighbours of color 2 (resp. 3 and 1) in G, and therefore is a leaf in its colorful component. Finally, there are no edges between vertices (excluding $\{x_i, y_i, z_i \mid 1 \le i \le n\}$) belonging to different subgraphs G_{t_j} and $G_{t_{j'}}$, with $1 \le j, j' \le N$.

We will now partition G into vertex-disjoint subgraphs G'_t for $t \in T$, such that G'_t is a subgraph of G_t (possibly G_t itself), and each colorful component from our fixed partition Γ is contained within a single graph G'_t. We proceed as follows. Each vertex from $\{t_j^X, t_j^Y, t_j^Z, t_j^{XY}, t_j^{XZ}, t_j^{YZ}\}$ belongs to a single subgraph G_{t_j}, therefore it belongs also to the subgraph G'_{t_j}. We assign each vertex x_i (resp. y_k and z_l) to one subgraph G'_{t_j}, in such a way that we do not split any colorful component. That means that if x_i (resp. y_k and z_l) is in the same colorful component as t_j^X (resp. t_j^Y and t_j^Z), we assign it to G'_{t_j}. If the colorful component containing x_i (resp. y_k and z_l) is a singleton, we assign the vertex arbitrarily.

Now each subgraph G'_{t_j}, with $t_j = (x_i, y_k, z_l)$, contains 6, 7, 8 or 9 vertices ($\{t_j^X, t_j^Y, t_j^Z, t_j^{XY}, t_j^{XZ}, t_j^{YZ}\}$ and a possibly empty subset of $\{x_i, y_k, z_l\}$). Let us denote by g_6, g_7, g_8 and g_9 the number of subgraphs G'_t containing 6, 7, 8 and 9 vertices, respectively. We have $g_6 + g_7 + g_8 + g_9 = N$. Moreover, $g_7 + 2 \cdot g_8 + 3 \cdot g_9 = 3n$, since each vertex x_i, y_k and z_l belongs to exactly one subgraph G'_t.

As we observed earlier, all colorful components from the partition Γ are contained within the above-defined subgraphs G'_t. Therefore the solution S to the MEC problem on (G, σ) is a union of solutions S_t for the MEC problem for the graphs $(G'_t, \sigma|_{G'_t})$. Now observe the following.

1. The value of an optimal MEC solution for $(G'_t, \sigma|_{G'_t})$ for a subgraph G'_t on 6 vertices is at most 6 (value 6 is achieved when the solution consists of two colorful components of 3 vertices each).
2. The value of an optimal MEC solution for $(G'_t, \sigma|_{G'_t})$ for a subgraph G'_t on 7 vertices is at most 6 (value 6 is achieved when the solution consists of two colorful components of 3 vertices each, and one singleton).

3. The value of an optimal MEC solution for $(G'_t, \sigma|_{G'_t})$ for a subgraph G'_t on 8 vertices is at most 7 (value 7 is achieved when the solution consists of two colorful components of 3 vertices and one component of 2 vertices).
4. The value of an optimal MEC solution for $(G'_t, \sigma|_{G'_t})$ for a subgraph G'_t on 9 vertices is at most 9 (value 9 is achieved when the solution consists of three colorful components of 3 vertices each).

Now, assume toward a contradiction that $\mathrm{OPT}_{\mathrm{MEC}} \geq 6N + 3n(1 - \varepsilon/2)$, i.e., the value of S is at least $6N + 3n(1 - \varepsilon/2)$. We just mentioned that each subgraph G'_t (there are N of them in total) of size 6 (7,8,9) contributes at most a value of 6 (6,7,9, respectively) towards the value of S. Thus, we must have

$$g_8 + 3 \cdot g_9 \geq 3n(1 - \varepsilon/2). \tag{1}$$

We already know that

$$2 \cdot g_8 + 3 \cdot g_9 \leq 3n. \tag{2}$$

Multiplying inequality (1) by 2 and inequality (2) by -1 and adding them yields the inequality $g_9 \geq n(1 - \varepsilon)$.

Notice that a subgraph G'_t on 9 vertices corresponds to a triple from T. Moreover, any two such triples are disjoint. As $g_9 \geq n(1-\varepsilon)$, in the corresponding Max 3-DM-3 instance, we can cover at least a $(1-\varepsilon)$ fraction of elements $X \cup Y \cup Z$ by disjoint triples, which contradicts the lemma statement. Therefore, we must have $\mathrm{OPT}_{\mathrm{MEC}} < 6N + 3n(1 - \varepsilon/2)$.   □

Let us now derive the APX-hardness of MEC for instances where vertices are colored using 3 colors from the above lemmas.

Theorem 2. *The Maximum Edges in Transitive Closure problem is APX-hard, even for $|C| = 3$.*

Proof. From Theorem 1, we know that it is NP-hard to distinguish between instances of Max 3-DM-3 such that $\mathrm{OPT}_{\mathrm{3DM}} = n$ or $\mathrm{OPT}_{\mathrm{3DM}} < (1 - \varepsilon)n$. From Lemmas 1 and 2, it is NP-hard to distinguish between instances of MEC for which $\mathrm{OPT}_{\mathrm{MEC}} \geq 6N + 3n$ and instances for which $\mathrm{OPT}_{\mathrm{MEC}} < 6N + 3n(1 - \varepsilon/2)$.

Since $N \leq 3n$, we get that it is NP-hard to approximate MEC within some constant factor, i.e., MEC is APX-hard. As all MEC instances considered here use only 3 colors, MEC is APX-hard already for $|C| = 3$.   □

3   Approximation of MEC for an Unbounded Number of Colors

3.1   A Positive Result

In this section we show that the MEC problem for an unbounded number of colors admits approximation within a factor of $\sqrt{2 \cdot \mathrm{OPT}}$. The algorithm is an exact polynomial time algorithm for the Minimum Singleton Vertices (MSV) problem from [1]. Let us first restate the definition of the MSV problem.

MINIMUM SINGLETON VERTICES: Given a simple, undirected graph $G = (V, E)$ and a coloring $\sigma : V \to C$ of the vertices, remove a collection of edges $E' \subseteq E$ from the graph such that each connected component in $G' = (V, E \setminus E')$ is colorful and the number of isolated vertices is minimum.

Theorem 3 ([1]). *The MSV problem can be solved exactly in polynomial time.*

We are now ready to prove our result.

Theorem 4. *The MEC problem admits a polynomial-time $\sqrt{2 \cdot OPT}$ approximation algorithm.*

Proof. We show that the exact MSV algorithm is a $\sqrt{2 \cdot OPT}$-approximation algorithm for MEC. Let $G = (V, E)$ be the input graph and let OPT be the value of an optimal solution (i.e., the number of edges in the transitive closure) of the MEC problem on G.

Let G_{MSV} be the graph obtained by running the exact MSV algorithm on G. Clearly, as each connected component of G_{MSV} is colorful, G_{MSV} is a feasible solution for the MEC problem.

Let I_{MEC} be the number of isolated vertices in an optimal solution of the MEC problem, and let I_{MSV} be the number of isolated vertices in G_{MSV}. We have $I_{\mathrm{MSV}} \le I_{\mathrm{MEC}}$.

We have $\mathrm{OPT} \le \binom{|V| - I_{\mathrm{MEC}}}{2}$, since the largest possible value of OPT is achieved when all the vertices that are not isolated are in the same connected component.

Define $\mathrm{Val}_{\mathrm{MSV}}$ to be the number of edges in the transitive closure of G_{MSV}. We get that $\mathrm{Val}_{\mathrm{MSV}} \ge (|V| - I_{\mathrm{MSV}})/2$. Thus, we have

$$\frac{\mathrm{OPT}}{\mathrm{Val}_{\mathrm{MSV}}} \le \frac{\sqrt{\mathrm{OPT}} \cdot \frac{1}{\sqrt{2}}(|V| - I_{\mathrm{MEC}})}{\frac{1}{2}(|V| - I_{\mathrm{MSV}})} \le \sqrt{2 \cdot \mathrm{OPT}} \ ,$$

as $|V| - I_{\mathrm{MEC}} \le |V| - I_{\mathrm{MSV}}$. □

3.2 A Negative Result

In this section, we show that the MEC problem is NP-hard to approximate within a factor of $|V|^{1/3 - \varepsilon}$ for any constant $\varepsilon > 0$. This result holds even if the input graph is a tree and each color appears at most twice in the graph. We use the same reduction as Rizzi and Sikora for proving hardness of approximation of the Graph Motif problem [7].

Reduction. We make a reduction from the Maximum Independent Set problem (MIS). Let $G = (V, E)$ be a MIS instance, and let $n = |V|$. We create an instance $G' = (V', E')$ of MEC in the following way. See Fig. 3 for an illustration.

The set of vertices V' consists of the following vertices:

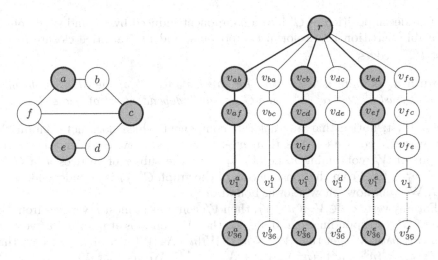

Fig. 3. Reduction from an instance G of MIS (left) to an instance G' of MEC (right). The only pairs of vertices in G' sharing the same color are vertices v_{uw} and v_{wu} for $u, w \in \{a, b, c, d, e, f\}$. An independent set in G corresponds to a colorful component in G' containing the root vertex r (gray vertices).

1. a special vertex r colored with a unique color c_r,
2. for each edge $uw \in E$, vertices v_{uw} and v_{wu} colored with the same color c_{uw},
3. for each vertex $u \in V$, a collection of n^2 vertices $v_1^u, v_2^u, \ldots, v_{n^2}^u$ colored with unique colors $c_1^u, c_2^u, \ldots, c_{n^2}^u$.

The resulting graph G' will be a tree on the set of vertices V', rooted at r. For each vertex $u \in V$, we add to G' a path starting at r which visits all vertices v_{uw} (in an arbitrary order), and then all vertices $v_1^u, v_2^u, \ldots, v_{n^2}^u$.

Analysis

Lemma 3. *If $G = (V, E)$ has an independent set of size α, then there is a solution for the corresponding instance G' of the MEC problem with value at least $\binom{\alpha n^2}{2}$.*

Proof. Let $V_I \subseteq V$ be an independent set in G consisting of α vertices. We will show that there is a colorful component in G' consisting of at least $\alpha \cdot n^2$ vertices.

We construct the set V_C' (see Fig. 3) in the following way. It consists of the root vertex r, together with all vertices lying on the paths corresponding to the vertices $u \in V_I$ (i.e., the vertices v_{uw} where $w \in V$ and $uw \in E$, and the vertices v_i^u for $i = 1, \ldots, n^2$). The subgraph of G' induced by V_C' is connected and consists of at least $\alpha \cdot n^2$ vertices. The subgraph is colorful, as from the construction of G' if two vertices lying on two paths of G' have the same color, then the vertices of G corresponding to these paths are connected by an edge, and therefore they cannot belong to an independent set.

The decomposition of G' into a component induced by V'_C and singletons is a feasible partition into colorful components, and its transitive closure has at least $\binom{\alpha n^2}{2}$ edges. □

Lemma 4. *If there is a solution for the instance G' of the MEC problem of value at least $n^5/2 + \alpha^2 \cdot n^4$, then G has an independent set of size at least α.*

Proof. First, notice that any colorful component which does not contain the root vertex r consists of less than $n^2 + n$ vertices. Now, consider the colorful component V'_C containing r. Let $V_I \subseteq V$ be the subset of vertices u of G for which $v_1^u \in V'_C$. From the construction of the graph G', V_I is an independent set in G. We will, now, show a lower bound on $|V_I|$.

For any vertex $u \in V$, if $u \notin V_I$, then V'_C contains at most n vertices from the path of G' corresponding to u. If $u \in V_I$, then V'_C contains at most $n + n^2$ vertices from this path. We get that $|V'_C| \leq n^2 + |V_I| n^2$. As $|V'| < n^3 + n^2$, we have that $OPT_{MEC} < (n^3 + n^2) \cdot (n^2 + n)/2 + |V_I|^2 \cdot n^4/2$. We get that $|V_I| \geq \alpha$. □

Theorem 5. *It is NP-hard to approximate the MEC problem within a factor of $|V|^{1/3-\varepsilon}$.*

Proof. The MIS problem is NP-hard to approximate within a factor of $n^{1-\varepsilon}$ for any constant $\varepsilon > 0$ [11]. In particular, it is NP-hard to distinguish whether a graph G has a maximum independent set of size at most n^ε, or at least $n^{1-\varepsilon}$.

In the first case, by Lemma 4, $OPT_{MEC} \leq n^5/2 + n^{4+2\varepsilon}$. In the second case, by Lemma 3, $OPT_{MEC} \geq n^{6-2\varepsilon}/2$. As the number of vertices of G' is in $\Theta(n^3)$, we get that approximating MEC within a factor of $|V|^{1/3-\varepsilon}$ is NP-hard. □

4   Conclusions and Future Work

In this paper we show several approximation and hardness results for the Maximum Edges in Transitive Closure Problem. First we prove that the problem is NP-hard to approximate within a factor of $|V|^{1/3-\varepsilon}$, for any constant $\varepsilon > 0$. Additionally, we show that the problem is APX-hard already for the case when the number of vertex colors equals 3. We complement these results by showing the first approximation algorithm for the problem, with approximation factor $\sqrt{2} \cdot OPT$.

There are several directions for future work. First, it would be interesting to close the gap between the approximation upper and lower bounds by showing an approximation algorithm with a better ratio or improving the hardness result. Another way to extend the current set of results would be to also consider the problem of maximizing the number of edges in the connected components and to identify and highlight similarities and differences between the two variants. Maximising the number of edges in the components has the same complexity as problem which asks to delete the minimum number of edges (it is the dual problem which is known to be NP-hard). Nevertheless, from the approximation point of view, these two problems are different.

Another direction would be to consider approximation guarantees that take the number of colors into account. However, we believe that there is not much room for improvement in this direction. The presented approximation algorithm has a ratio of $(C - 1)/4$, where C is the number of colors and, due the hardness results, we cannot hope for a much better approximation.

References

1. Adamaszek, A., Popa, A.: Algorithmic and hardness results for the colorful components problems. In: Pardo, A., Viola, A. (eds.) LATIN 2014. LNCS, vol. 8392, pp. 683–694. Springer, Heidelberg (2014)
2. Avidor, A., Langberg, M.: The multi multiway cut problem. Theoret. Comput. Sci. **377**(1–3), 35–42 (2007)
3. Bruckner, S., Hüffner, F., Komusiewicz, C., Niedermeier, R.: Evaluation of ILP-based approaches for partitioning into colorful components. In: Demetrescu, C., Marchetti-Spaccamela, A., Bonifaci, V. (eds.) SEA 2013. LNCS, vol. 7933, pp. 176–187. Springer, Heidelberg (2013)
4. Bruckner, S., Hüffner, F., Komusiewicz, C., Niedermeier, R., Thiel, S., Uhlmann, J.: Partitioning into colorful components by minimum edge deletions. In: Kärkkäinen, J., Stoye, J. (eds.) CPM 2012. LNCS, vol. 7354, pp. 56–69. Springer, Heidelberg (2012)
5. He, G., Liu, J., Zhao, C.: Approximation algorithms for some graph partitioning problems. J. Graph Algorithms Appl. **4**(2), 1–11 (2000)
6. Petrank, E.: The hardness of approximation: gap location. Comput. Complex. **4**(2), 133–157 (1994)
7. Rizzi, R., Sikora, F.: Some results on more flexible versions of graph motif. In: Hirsch, E.A., Karhumäki, J., Lepistö, A., Prilutskii, M. (eds.) CSR 2012. LNCS, vol. 7353, pp. 278–289. Springer, Heidelberg (2012)
8. Sankoff, D.: OMG! orthologs for multiple genomes - competing formulations. In: Chen, J., Wang, J., Zelikovsky, A. (eds.) ISBRA 2011. LNCS, vol. 6674, pp. 2–3. Springer, Heidelberg (2011)
9. Savard, O.T., Swenson, K.M.: A graph-theoretic approach for inparalog detection. BMC Bioinform. **13**(S–19), S16 (2012)
10. Zheng, C., Swenson, K., Lyons, E., Sankoff, D.: OMG! orthologs in multiple genomes – competing graph-theoretical formulations. In: Przytycka, T.M., Sagot, M.-F. (eds.) WABI 2011. LNCS, vol. 6833, pp. 364–375. Springer, Heidelberg (2011)
11. Zuckerman, D.: Linear degree extractors and the inapproximability of max clique and chromatic number. Theory Comput. **3**(1), 103–128 (2007)

Quantifying Privacy: A Novel Entropy-Based Measure of Disclosure Risk

Mousa Alfalayleh and Ljiljana Brankovic(✉)

School of Electrical Engineering and Computer Science,
The University of Newcastle, Callaghan, NSW 2308, Australia
{mousa.alfalayleh,ljiljana.brankovic}@newcastle.edu.au

Abstract. It is well recognised that data mining and statistical analysis pose a serious treat to privacy. This is true for financial, medical, criminal and marketing research. Numerous techniques have been proposed to protect privacy, including restriction and data modification. Recently proposed privacy models such as differential privacy and k-anonymity received a lot of attention and for the latter there are now several improvements of the original scheme, each removing some security shortcomings of the previous one. However, the challenge lies in evaluating and comparing privacy provided by various techniques. In this paper we propose a novel entropy based security measure that can be applied to any generalisation, restriction or data modification technique. We use our measure to empirically evaluate and compare a few popular methods, namely query restriction, sampling and noise addition.

1 Introduction

Over the last few decades, proliferation of computer, network and communication technology, and in particular social networking and cloud computing, had a great impact on the way personal data is collected, stored and used [44]. Data collected in one location (e.g., hospital) can now be stored remotely in a cloud and accessed from anywhere in the world. These advances have undoubtedly changed the way we think about privacy [3–5,9,23,27,38] and what once could have been regulated by legislative measures alone now requires a sophisticated suite of privacy enhancing technologies. In this study we are concerned with a situation where confidential personal data is made available to a wide range of users who are authorised to perform data mining and statistical analysis, but not to access any individual data. There are various *Statistical Disclosure Control (SDC)* techniques that can be used to alleviate this problem [1,10,13,26,55] but, unfortunately, none of them is able to solve it completely, due to its intrinsic contradictory nature. On one hand, one must keep the risk of individual data disclosure as low as possible. On the other hand, the utility (usefulness) of the data must remain high. However, low risk implies low utility and high utility implies high risk. A good SDC technique aims at finding a right balance between the two. In order to achieve this balance, it is crucial to adequately measure both utility and disclosure risk. While measuring data utility has been well studied in

© Springer International Publishing Switzerland 2015
J. Kratochvíl et al. (Eds.): IWOCA 2014, LNCS 8986, pp. 24–36, 2015.
DOI: 10.1007/978-3-319-19315-1_3

the literature [7,12,16–18,24,33,55], measuring disclosure risk is still considered as a difficult problem and has been only partly solved.

Contribution: (1) In this paper we propose a novel entropy based measure of disclosure risk, which we refer to as Confidential Attribute Equivocation (CAE), and which is independent of the underlying SDC technique and thus can always be used. The main novelty and advantage of our technique over similar ones is that it takes into account the candidate confidential values themselves, rather than just their probabilities, and is thus able to capture the risk of approximate disclosure of confidential values, rather than the exact disclosure alone. (2) We develop an efficient dynamic programming algorithm to evaluate the CAE. (3) We show how our measure can be applied to evaluate a few common SDC techniques, including sampling, query restriction and noise addition.

The paper is organised as follows. In the next section we present related work on disclosure risk measures. In Sect. 3 we present a scenario that is not adequately covered by any of the previous work and we show how our novel entropy-based measure CAE covers it. In Sect. 4 we present a dynamic programming algorithm for calculating the disclosure risk with CAE, in Sect. 5 we use CAE to empirically evaluate a few existing SDC techniques, and in Sect. 6 we discuss the experimental results and give some concluding remarks.

2 Introduction to Statistical Disclosure Control and Related Work on Disclosure Risk Measures

Privacy is an elusive concept, and many privacy models have been proposed, with varying success. We next present a few of the most prominent models.

In a data set, some attributes may be considered public knowledge and used to identify records. They are refereed to as "identifying" attributes or "quasi-identifiers" (QI). A class of records where values of all QI attributes are the same is called an *equivalence class*. To limit disclosure, Samarati and Sweeney [47] proposed a so-called k-anonymity that requires each equivalence class to have no less than k records. The main problem with k-anonymity arises when all the records in an equivalence class share the same confidential value, which allows intruder to disclose the confidential value without actually re-identifying the record. In order to alleviate this problem, Machanavajjhala et al. proposed l-diversity [43], which in its simplest form requires every QI to contain at least l distinct values. While this model is indeed a great improvement over k-anonymity, it does not consider how close these values are from each other, and thus leaves room for approximate disclosure of confidential values. Li et al. [40] introduced t-closeness, which considers the distribution of confidential attribute values in each equivalence class and the distribution of the confidential attribute values in the whole dataset, and requires that the distance between these two distributions does not exceed a given threshold t. While it greatly reduces the disclosure risk, t-closeness is overly restrictive and severely impacts the utility of data. In this context, our measure of disclosure risk can be seen as bridging a gap between the l-diversity and the rigid t-closeness.

A similar, yet different model, known as a "k-compromise" and also "g-group compromise", has been developed in a series of papers by Brankovic et al. [7,8,11, 12,15–18,31], Griggs [29,30] and Ahlswede and Aydinian [2]. This model is based on rigorous mathematical proofs and guarantees that a user cannot deduce any statistics based on k or less records. Unlike in k-anonymity, the user can only query the dataset, instead of having a direct access to it, and thus this model does not suffer from approximate compromise. This model has been designed for any combination of SUM, COUNT and AVERAGE queries and it offers almost twice as many queries as k-anonymity for the same level of security. More work is needed to adapt this model for other types of queries.

Another prominent privacy model is differential privacy [21], which requires that the results to all queries allowed on the database do not change significantly if a single record is added or deleted from the database. While this is certainly an efficient model against table linkage attack, it is not design to prevent attribute and record linkage [26] and in practice may be inferior to k-anonymity [51].

Each one of the above models can be implemented using different SDC techniques, which can be classified as modification techniques and query restriction techniques [10,13,19,55]. *Modification techniques* involve some kind of alternation of the original data set before it is released to statistical users. This includes noise addition, data swapping, aggregation, suppression and sampling [10,13,14,28,32,34,35,42]. The common denominator of all modification techniques is that the modified dataset is released to users who are free to perform any query on it, but the answers they get are only approximate and not exact. On the other hand, *query restriction techniques* do not release database to a user but rather provide a query access. The SDC system decides whether or not to answer the query but if the query is answered, the answer will always be exact and not approximate as with modification techniques [10,19].

In this study we are not concerned with SDC techniques or privacy models as such but rather with measuring disclosure risk. In the literature, disclosure risk measures are classified as measures for record re-identification or confidential value disclosure [10,20,39]. The latter focuses on measuring the risk of compromising a confidential value of a particular individual, and the former on measuring the risk of inferring an individual's identity. In either case the disclosure risk measures may be applied to the database as a whole, or to individual records.

Several methods have been proposed to estimate the disclosure risk in sampling and they fall under the category of record identification. Winkler [56] refers to these methods as Sample-Unique-Population-Unique (SUPU) methods as disclosure risk estimation requires assessing the uniqueness of records in the released sample and in the population. Skinner and Eliot [49] introduced a new disclosure risk measure for microdata which falls under SUPU methods [56]. Their measure is based on the probability Pr(correct match|unique match) that a microdata record and a population unit are correctly matched. Additionally, they introduced a simple variance estimator and claimed that their measure is able to evaluate the different ways of releasing microdata from a sample survey. Truta et al. [54] introduced other SUPU measures and named them minimal, maximal, and weighted disclosure risk measures. The minimal disclosure risk measure is the percentage of records in a population that can be correctly re-identified by an intruder. All these records must be population unique. The maximum disclosure risk measure takes into account records that are not population unique while the weighted disclosure risk measure assigns more weights to unique records over other records. These measures are not linked to a certain individual but are rather used to compute the overall disclosure risk for the database. As a drawback, they can only be applied to limited SDC methods such as sampling and microaggregation, and it is considered hard to choose the disclosure risk weight matrix [54]. However, assigning weights enables a data owner to setup different levels of confidentiality. These measures are useful in deciding the order of applying more than one SDC method on the initial data.

Trottini and Fienberg [53] proposed a simple Bayesian model for capturing user uncertainty after releasing the data by an agency. They distinguish between the legitimate user (researcher) uncertainty and the malicious user (intruder) uncertainty. This distinction is used as the basis of defining appropriate disclosure risk measure. The proposed measure is an arbitrary decreasing function of the user's uncertainty about a confidential attribute value.

Spruill [50] measured confidentiality as a percentage of records in the released data where a link with the original data can not be made. In order to decide if there is such a link, for each released record, we add up either the square or the absolute value of the difference between the released value and the true value for all common numerical attributes. A link is said to be made if a released record was derived from the true record that has the minimum sum of differences. Spruill's early work gave rise to *record linkage*, much studied in recent years [26].

There are some recent proposals that use information-theoretic approach to measure privacy and utility of various SDC techniques [45,48]; however, none of them measures the "approximate" compromise.

3   A Novel Entropy-Based Measure

Out of all disclosure risk measures, the closest to our proposal is a measure introduced by Onganian and Domingo-Ferrer [45] that evaluates the security of releasing tabular data. The measure is equal to the reciprocal of conditional entropy given the knowledge of an intruder:

$$DR(X) = \frac{1}{H(X|Y=y)} = \frac{1}{(-\sum_x p(x|y) \log_2 p(x|y))} \tag{1}$$

where X represents a confidential attribute for a given record and Y represents intruder's knowledge. The disclosure risk is inversely proportional to the uncertainty about the confidential attribute given intruder's knowledge. The measure performs *a posteriori*, that is, after applying one of SDC methods to the tables. It is a complement to *a priori* measures such as some sensitivity rules including (n,k)-dominance and pq-rule, which help a data owner in deciding whether to release the data or not. The main strength of the above a posteriori measure is its generality: it is applicable to various SDC methods such as Cell Suppression, Rounding, and Table Redesign. In order to evaluate this disclosure risk, one has to find a set of the possible confidential attribute values and their probabilities given the condition $Y = y$. A down side of this measure is that it does not capture accurately the knowledge that an intruder has about a confidential attribute, as it does not give careful consideration to the attribute values but only the probabilities with which the values occur. Our proposed method considers the attribute values in addition to their probabilities.

Before we proceed to describe our measure in more detail and compare some of the SDC techniques in the experimental section, we need to introduce the concept of "database compromise". We say that a database is *compromised* if a *sensitive statistic* is disclosed [19]. There are several distinct types of compromise, depending upon what is considered to be sensitive. For example, if only exact individual values are considered sensitive, we have the so-called *exact compromise*. *Approximate compromise* occurs when a user is able to infer that a confidential individual value X lies within a range $[X_0 - \frac{\varepsilon}{2}, X_0 + \frac{\varepsilon}{2}]$ for some predefined value of ε. Approximate compromise will prove crucial for the definition of our new security measure.

We consider a scenario where an intruder is trying to unlawfully disclose confidential information from a database. She uses all the available information she can get from the database, as well as any external knowledge she may have. At the end of her analysis, the intruder is able to reduce the possibilities and limit her suspicion to certain data values. Shannon's entropy can measure the intruder's uncertainty, but does not take into consideration how far or close these values are from each other. The first example in Table 1 shows the queries submitted by an intruder and the database responses to them. Assuming that there are only two female academics, the intruder learns that Layla's salary has one of the two values: It is either 107K or 50K. In the second example we assume that there are only two academics aged 37 and the intruder knows that Qay's salary is either 80K or 77K. If we use Shannon's entropy to evaluate the intruder's uncertainty in examples 1 and 2, we get the same result, 1 bit in each case. However, we argue that the intruder learns more in example 2, as he can pretty accurately estimate the salary to be 78.5K $\pm$ 1.5K. This highlights the need for more accurate measure than Shannon's entropy, which would be able to capture such differences. We introduce a notion of privacy for the so-called *approximate compromise range* (ε). In the approximate compromise an intruder learns that the confidential value X lies within a range $[X_0 - \frac{\varepsilon}{2}, X_0 + \frac{\varepsilon}{2}]$. For the two example above we have $X \in [X_0 - 28.5K, \ X_0 + 28.5K]$ for Layla and $X \in [X_0 - 1.5K, \ X_0 + 1.5K]$ for Qay, where in both cases $X_0 = 78.5K$. Obviously, the intruder knows more about Qay's than Layla's salary, as in the former the approximate compromise range is 3K, while in the latter it is 57K.

To capture the information about the range ε, we use Shannon's entropy H as a function of ε. The graphs in Fig. 1 correspond to the intruder's uncertainty $H(\varepsilon)$ in the above examples. We notice that the entropy $H(0)$ is the same in both cases, that is, the disclosure risks are the same for exact compromise. However,

Table 1. Two examples

	Example 1	Example 2
Q1:	SELECT MAX (Salary) FROM AcademicStaff WHERE Sex = F	SELECT MAX (Salary) FROM AcademicStaff WHERE Age = 37
A1:	Maximum salary = 107,000	Maximum salary = 80,000
Q2:	SELECT AVG (Salary) FROM AcademicStaff WHERE Sex = F	SELECT AVG (Salary) FROM AcademicStaff WHERE Age = 37
A2:	Average salary = 78,500	Average salary = 78,500

the area under $H(\varepsilon)$ is much larger for Layla implying that this case is more resistant against approximate compromise.

In general, we can evaluate intruder's uncertainty for any given ε. In particular, we use $H_0 = H(0)$ to denote intruder's uncertainty in the case of exact compromise, that is, approximate compromise range of "0" and we call it *initial entropy*. Additionally, in what follows we examine the area (A) determined as an integral: $A = \int_0^{\varepsilon_{max}} H(\varepsilon)$, where "$\varepsilon_{max}$" is the value of ε for which entropy $H(\varepsilon)$ drops to zero. Formally, $H(\varepsilon_{max}) = 0$ and $H(\varepsilon) > 0$ for all $\varepsilon < \varepsilon_{max}$.

We next explain how $H(\varepsilon)$ is calculated in general. We introduce a "window" of length ε. When a window "covers" two or more values, then they are replaced with a single value whose probability is equal to the sum of probabilities of all the values covered by the window. In general, there will be more than one way to cover the values with windows of length ε and we need to select the way that minimises the entropy $H(\varepsilon)$. Computing the minimum entropy $H(\varepsilon)$ as a function of ε is not straightforward and in the next section we introduce a dynamic programming algorithm to find it, and hence calculate the area (A) that together with the initial entropy (H_0) represents our disclosure risk.

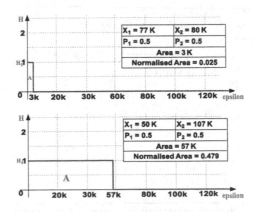

Fig. 1. Our security measure: for (77K, 80K) and (50K, 107K).

4   A Dynamic Programming Algorithm to Compute $H(\varepsilon)$

We are given as input a set of values x_i in increasing order ($x_1 < x_2 < x_3 < \cdots < x_n$) where each x_i has a given probability p_i, $p_i \geq 0$ and $\Sigma p_i = 1$. In order to produce our security measure, we consider a collection C of subsets $(x_1, ... x_{y_1}), (x_{y_1+1}, ..., x_{y_2}), ..., (x_{y_m+1}, ..., x_n)$, such that (1) $x_{y_1} - x_1 \leq \varepsilon$, (2) $x_{y_i} - x_{y_{i-1}+1} \leq \varepsilon$, $2 \leq i \leq m$, (3) $x_n - x_{y_m+1} \leq \varepsilon$ and the corresponding probabilities $q_1 = p_1 + ... + p_{y_1}, ..., q_{m+1} = p_{y_m+1} + ... + p_n$. We need to calculate minimum $H(\varepsilon)$ over probabilities q, such that $H(\varepsilon)$ is maximised over all collections C satisfying conditions above, for each ε. We break the problem into stages (rows) and states (columns). Each row in the table corresponds to a stage or ε. Column "i" in the table corresponds to the subproblem containing values $x_1, x_2, \cdots, x_i$. For a given row (stage) in the table, each cell in this row is viewed as a subproblem $H(\varepsilon, i)$ of the original problem $H(\varepsilon)$. For a given stage and state, $H(\varepsilon, i)$ is computed by the following recurrence:

$$H(\varepsilon, i) = min[(H(\varepsilon, i-1) + a_i), (H(\varepsilon, i-2) + \dot{a}_{i-1}), ..., (H(\varepsilon, j-1) + a_j)]$$

where $a_j = (\sum_{k=j}^i p_k) \cdot log(\frac{1}{(\sum_{k=j}^i p_k)})$, $X_i - X_j \leq \varepsilon$ and $X_i - X_{j-1} > \varepsilon$ for $1 \leq j \leq i \leq n$, $H(0, \varepsilon) = 0$ and $H(1, \varepsilon) = p_1 \cdot log(\frac{1}{p_1})$.

Input: $x[\ ]$: a set of integer values in ascending order;
  $\quad p[\ ]$: a set of probabilities corresponding to the above integer values.
Output: $H(\varepsilon)$
$H_0 \leftarrow \sum_{i=1}^n p(x_i) \cdot log(\frac{1}{p(x_i)})$;
foreach ε **do**
$\quad H(\varepsilon, 0) \leftarrow 0$;
$\quad H(\varepsilon, 1) \leftarrow p_1 \cdot log(\frac{1}{p_1})$;
$\quad$ **for** $i \leftarrow 2$ **to** n **do**
$\quad\quad j \leftarrow i$;
$\quad\quad p_{partial} \leftarrow 0$;
$\quad\quad H(\varepsilon, i) \leftarrow H(\varepsilon - 1, i)$;
$\quad\quad$ **while** $(x_i - x_j \leq \varepsilon)$ *and* $(j \neq 0)$ **do**
$\quad\quad\quad p_{partial} \leftarrow p_{partial} + p_j$;
$\quad\quad\quad H_{temp} \leftarrow p_{partial} \cdot log(\frac{1}{p_{partial}}) + H(\varepsilon, j-1)$;
$\quad\quad\quad$ **if** $H_{temp} < H(\varepsilon, i)$ **then**
$\quad\quad\quad\quad H(\varepsilon, i) \leftarrow H_{temp}$;
$\quad\quad\quad$ **end**
$\quad\quad\quad j \leftarrow j - 1$;
$\quad\quad$ **end**
$\quad$ **end**
$\quad H(\varepsilon) \leftarrow H(\varepsilon, n)$;
$\quad$ **Display:** $H(\varepsilon)$
end

Algorithm 1. A dynamic programming algorithm to compute $H(\varepsilon)$

5   The Experiments: Description and Implementation

In this section we apply our proposed security measure to a few common Statistical Disclosure Control (SDC) techniques. In all instances we assume that the intruder has *supplementary knowledge* (SK) about an individual whose corresponding record is stored in the original dataset, which can be as limited as one attribute or can be as extensive as all attributes except the confidential one. The comparative study is performed on PUMS dataset [46].

Sampling. Instead of the whole dataset, we release a random sample without replacement where each record in the original dataset is equally likely to be included in the produced sample and duplicates are not allowed. The size of the produced sample is specified as a percentage of the total size and referred to as a "sampling size" (or a "sampling factor"). In deciding on the structure of the sampling experiment, we follow work by Truta et al. [54] on disclosure risk measures for sampling, where we compute the overall disclosure risk for the database, rather than for a certain individual.

In order to study the effect of sample size on the security we use four different sampling factors: $5\%, 10\%, 20\%, 50\%$. For each sample size, we generate 30 different sample files. Additionally, we study the effect of the intruder's supplementary knowledge. We start with supplementary knowledge as little as one attribute and extended it to reach all attributes except the confidential one. For each attribute we performed experiments for all possible values. The results in Fig. 2 are the averaged over all 30 samples, all attributes and all values.

Query Restriction. In this experiment, we consider a scenario where an intruder submits a set of range queries to a DBMS. The intruder performs an analysis using the answers to the submitted queries as well as the supplementary knowledge with an aim to infer a confidential attribute value for the given record, e.g., salary in PUMS dataset. We assume that the intruder has built a system of linear equations out of the responses to range queries. We use $Q = 2l$ to denote the query set size for the queries a user (i.e., an intruder) is permitted to ask. For simplicity, we only consider even query set sizes. We use k to denote the number of queries and thus also the number of linear equations: $k = \lfloor \frac{2n}{Q} \rfloor - 1$.

We run the experiment for 5 different query set sizes $\{2, 4, 8, 16, 32\}$ and for each size we shuffle the records in the original dataset to produce randomly 30 different systems of linear equations. The results in Fig. 2 are the average results, over all 30 systems of linear equations, all SK attributes and all values.

Noise Addition. In this scenario the noise is added to all attributes in the dataset, sensitive and non-sensitive, categorical and numerical. We use additive noise studied in [25,36,37,52,58]. The amount of noise is drawn randomly from binomial probability distribution as the nature of attributes in our dataset is discrete. The DBMS then releases the perturbed version of the dataset and an intruder obtains a copy of it. The intruder analyses the released perturbed dataset using their supplementary knowledge in a bid to infer a confidential attribute value, e.g. salary in PUMS dataset, corresponding to the individual of concern. We assume there is only one confidential attribute; the generalisation to more than one confidential attribute is straightforward.

6   Discussion and Conclusion

As expected, for all three SDC techniques our privacy measure, CAE, increases with decrease in utility (Fig. 2). In Sampling, utility is proportional to the sample size, in Query Restriction it is inversely proportional to the query size, and in Noise Addition it is inversely proportional to the amount of noise. CAE declines with additional supplementary knowledge that intruder might have, which is expressed on horizontal axis as the number of known attributes. We note that this decline is sometimes gentle and sometimes sharp, depending on the utility which is in Fig. 2 given as a parameter: for low utility privacy only gently declines with supplementary knowledge, while for higher utility the decline is typically sharp.

Importantly, our experiments demonstrate how we can compare different SDC techniques and select the most suitable one for specific applications and requirements.

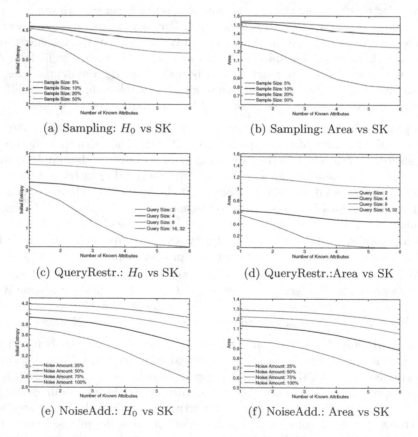

(a) Sampling: H_0 vs SK   (b) Sampling: Area vs SK

(c) QueryRestr.: H_0 vs SK   (d) QueryRestr.:Area vs SK

(e) NoiseAdd.: H_0 vs SK   (f) NoiseAdd.: Area vs SK

Fig. 2. Sampling, query restriction and noise addition (Color figure online)

For example, in the absence of supplementary knowledge sampling with size 50 % and query restriction with query set size 8 provide similar level of privacy (blue line in Fig. 2(a) and red line in Fig. 2(c)); however, the privacy sharply drops as the supplementary knowledge increases in sampling, while it remains flat in query restriction. Moreover, Figs. 2(b) and (d) that indicate approximate compromise show a slight superiority of sampling in the absence of supplementary knowledge, but as supplementary knowledge grows sampling becomes much more vulnerable than query restriction.

In summary, unlike previously proposed privacy measures, our novel information theoretic privacy measure (CAE) has the ability to capture approximate compromise; it can also be applied to any SDC technique, as long as the probabilities of different confidential values can be estimated. In this paper we considered the most common SDC techniques and showed how CAE can be used to evaluate the privacy they offer, and how the privacy relates to both utility and supplementary knowledge.

References

1. Adam, N.R., Worthmann, J.C.: Security-control methods for statistical databases: a comparative study. ACM Comput. Surv. **21**(4), 515–556 (1989)
2. Ahlswede, R., Aydinian, H.: On security of statistical databases. SIAM J. Discrete Math. **25**(4), 1778–1791 (2011)
3. Al-Saggaf, Y., Islam, M.Z.: Privacy in social network sites (SNS) - the threats from data mining. Ethical Space: J. Commun. Ethics **9**(4), 32–40 (2012)
4. Al-Saggaf, F., Islam, M.Z.: A malicious use of a clustering algorithm to threaten the privacy of a social networking site user. World J. Comput. Appl. Technol. **1**(2), 29–34 (2013)
5. Al-Saggaf, Y., Islam, M.Z.: Data mining and privacy of social network sites users: implications of the data mining problem. Sci. Eng. Ethics (2014)
6. Blake, C.L.: Wine Recognition Data (1998)
7. Brankovic, L.: Usability of secure statistical databases. Ph.D. Thesis, Newcastle, Australia (1998)
8. Brankovic, L., Cvetkovic, D.: The eigenspace of the eigenvalue -2 in generalized line graphs and a problem in security of statistical databases. Publikacije ETF, Serija: matematika. **14**, 37–48 (2003)
9. Brankovic, L., Estivill-Castro, V.: Privacy issues in knowledge discovery and data mining. In: Australian Institute of Computer Ethics Conference, pp. 89–99 (1999)
10. Brankovic, L., Giggins, H.: Statistical database security. In: Petković, M., Jonker, W. (eds.) Security, Privacy, and Trust in Modern Data Management, pp. 167–181. Springer, Heidelberg (2007)
11. Brankovic, L., Horak, P., Miller, M.: An optimization problem in statistical databases. SIAM J. Discrete Math. **13**(3), 46–353 (2000)
12. Brankovic, L., Horak, P., Miller, M., Wrightson, G.: Usability of compromise-free statistical databases for range sum queries. In: 9th International Conference on Scientific and Statistical Database Management, pp. 144–154. IEEE Computer Society (1997)
13. Brankovic, L., Islam, M.Z., Giggins, H.: Security, privacy, and trust in modern data management. In: Petković, M., Jonker, W. (eds.) Privacy-Preserving Data Mining, pp. 151–165. Springer, Heidelberg (2007)

14. Brankovic, L., Lopez, N., Miller, M., Sebe, F.: Triangle randomization for social network data anonymization. Ars Math. Contemp. **7**(2), 461–477 (2014)
15. Brankovic, L., Miller, M., Siran, J.: Graphs, 0–1 matrices, and usability of statistical databases. Congressus Numerantium **12**, 169–182 (1996)
16. Brankovic, L., Miller, M., Siran, J.: Usability of k-compromise-free statistical databases. In: Proceedings of the 11th Australasian Workshop on Combinatorial Algorithms (AWOCA 2000), Hunter Valley, pp. 159–166 (2000)
17. Brankovic, L., Miller, M., Siran, J.: Range query usability of statistical databases. Int. J. Comput. Math. **79**(12), 1265–1271 (2002)
18. Brankovic, L., Sirán, J.: 2-compromise usability in 1-dimensional statistical databases. In: Ibarra, O.H., Zhang, L. (eds.) COCOON 2002. LNCS, vol. 2387, pp. 448–455. Springer, Heidelberg (2002)
19. Denning, D.E.: Cryptography and Data Security. Addison-Wesley Longman Publishing Co., Inc., Boston (1982)
20. Duncan, G.T., Lambert, D.: Disclosure-limited data dissemination. J. Am. Stat. Assoc. **81**, 10–28 (1986)
21. Dwork, C.: Differential privacy. In: Bugliesi, M., Preneel, B., Sassone, V., Wegener, I. (eds.) ICALP 2006. LNCS, vol. 4052, pp. 1–12. Springer, Heidelberg (2006)
22. Estivill-Castro, V., Brankovic, L.: Data swapping: balancing privacy against precision in mining for logic rules. In: Mohania, M., Tjoa, A.M. (eds.) DaWaK 1999. LNCS, vol. 1676, pp. 389–398. Springer, Heidelberg (1999)
23. Estivill-Castro, V., Brankovic, L., Dowe, D.L.: Privacy in data mining. Privacy - Law Policy Reporter **9**(3), 33–35 (1999)
24. Fletcher, S., Islam, M.Z.: Measuring information quality for privacy preserving data mining. Int. J. Comput. Theory Eng. **7**(1), 21–28 (2015)
25. Fuller, W.A.: Masking procedures for microdata disclosure limitation. J. Off. Stat. **9**(2), 383–406 (1993)
26. Fung, C.C.M., Wang, K., Chen, R., Yu, P.S.: Privacy-preserving data publishing: a survey of recent developments. ACM Comput. Surv. **42**(4), 14.2–14.53 (2010)
27. Giggins, H.: Security of genetic databases. Ph.D. Thesis, Newcastle, Australia (2009)
28. Giggins, H., Brankovic, L.: VICUS - a noise addition technique for categorical data. In: 10th Australasian Data Mining Conference. CRPIT, vol. 134, pp. 139–148 (2012)
29. Griggs, J.R.: Concentrating subset sums at k points. Bull. Inst. Comb. Appl. **20**, 65–74 (1997)
30. Griggs, J.R.: Database security and the distribution of subset sums in R^m. In: Proceedings of the International Colloquium on Combinatorics and Graph Theory (1998)
31. Horak, P., Brankovic, L., Miller, M.: A combinatorial problem in database security. Discrete Appl. Math. **91**(1–3), 119–126 (1999)
32. Islam, M.Z.: Privacy preservation in data mining through noise addition. Ph.D. Thesis, Newcastle, Australia (2008)
33. Islam, M.Z., Barnaghi, P.M., Brankovic, L.: Measuring data quality: predictive accuracy vs. similarity of decision trees. In: 6th International Conference on Computer and Information Technology, Dhaka, Bangladesh, pp. 457–462 (2003)
34. Islam, M.Z., Brankovic, L.: Noise addition for protecting privacy in data mining. In: 6th Engineering Mathematics and Applications Conference, Sydney, pp. 85–90 (2003)

35. Islam, M.Z., Brankovic, L.: Detective: a decision tree based categorical value clustering and perturbation technique in privacy preserving data mining. In: 3rd International IEEE Conference on Industrial Informatics, Australia, pp. 701–708 (2005)
36. Kim, J.J.: A method for limiting disclosure in microdata based on random noise and transformation. In: Proceedings of the Section on Survey Research Methods, pp. 303–308. American Statistical Association (1986)
37. Kim, J.J., Winkler, W.E.: Masking microdata files. In: Proceedings of the Section on Survey Research Methods, pp. 114–119. American Statistical Association (1995)
38. King, T., Brankovic, L., Gillard, P.: Perspectives of Australian adults about protecting the privacy of their health information in statistical databases. Int. J. Med. Inform. **81**(4), 279–289 (2012)
39. Lambert, D.: Measures of disclosure risk and harm. J. Off. Stat. **9**, 313–331 (1993)
40. Li, N., Li, T., Venkatasubramanian, S.: t-closeness: privacy beyond k-anonymity and l-diversity. In: IEEE International Conference on Data Engineering (2007)
41. Lopez, N., Sebe, F.: Privacy preserving release of blogosphere data in the presence of search engines. Inf. Process. Manage. **49**(4), 833–851 (2013)
42. López, N., Sebé, F.: Degree sequences of pagerank uniform graphs and digraphs with prime outdegrees. In: Lecroq, T., Mouchard, L. (eds.) IWOCA 2013. LNCS, vol. 8288, pp. 303–313. Springer, Heidelberg (2013)
43. Machanavajjhala, A., Kifer, D., Gehrke, J., Venkitasubramaniam, M.: l-diversity: privacy beyond k-anonymity. ACM Trans. Knowl. Discov. Data. **1** (2007)
44. Morris, S., Cooper, J., Bomba, D., Brankovic, L., Miller, M., Pacheco, F.: Australian healthcare: a smart card for a clever country. Int. J. Biomed. Comput. **40**(2), 101–105 (1995)
45. Oganian, A., Domingo-Ferrer, J.: A posteriori disclosure risk measure for tabular data based on conditional entropy. SORT - Stat. Oper. Res. Trans. **27**(2), 175–190 (2003)
46. Public Use Microdata Sample (PUMS) (2006)
47. Samarati, P., Sweeney, L.: Protecting privacy when disclosing information: k-anonymity and its enforcement through generalization and suppression. Technical Report SRI-CSL-98-04. SRI Computer Science Laboratory, Palo Alto, CA (1998)
48. Sankar, L., Rajagopalan, S.R., Poor, H.V.: Utility-privacy tradeoffs for databases: an information-theoretic approach. IEEE Trans. Inf. Forensics Secur. **9**(6), 838–852 (2013). Special Issue on Privacy and Trust Management in the Cloud and Distributed Data Systems
49. Skinner, C.J., Elliot, M.J.: A measure of disclosure risk for microdata. J. Roy. Stat. Soc. B **64**(4), 855–867 (2002)
50. Spruill, N.L.: Measures of Confidentiality, Statistics of Income and Related Administrative Record Research, pp. 131–136 (1982)
51. Sramka, M., Safavi-Naini, R., Denzinger, J., Askari, M.: A Practice-oriented framework for measuring privacy and utility in data sanitization systems. In: EDBT/ICDT2010 Workshops, Lausanne, Switzerland, pp. 315–333 (2010)
52. Tendick, P.: Optimal noise addition for preserving confidentiality in multivariate data. J. Stat. Plan. Inference **27**, 341–353 (1991)
53. Trottini, M., Fienberg, S.E.: Modelling user uncertainty for disclosure risk and data utility. Int. J. Uncertain. Fuzz. Knowl. Based Sys. **10**(5), 511–527 (2002)
54. Truta, T.M., Fotouhi, F., Barth-Jones, D.: Disclosure risk measures for the sampling disclosure control method. In: 2004 ACM symposium on Applied computing (SAC 2004), NY, USA, pp. 301–306 (2004)
55. Willenborg, L., de Waal, T.: Elements of Statistical Disclosure Control. Lecture Notes in Statistics, p. 155. Springer-Verlag, New York (2001)

56. Winkler, W.E.: Masking and re-identification methods for public-use microdata: overview and research problems. In: Domingo-Ferrer, J., Torra, V. (eds.) PSD 2004. LNCS, vol. 3050, pp. 231–246. Springer, Heidelberg (2004)
57. Wolberg, W.H., Street, W.N., Mangasarian, O.L.: Wisc. Diag. Breast Can. (1995)
58. Yancey, W.E., Winkler, W.E., Creecy, R.H.: Disclosure risk assessment in perturbative microdata protection. In: Domingo-Ferrer, J. (ed.) Inference Control in Statistical Databases. LNCS, vol. 2316, p. 135. Springer, Heidelberg (2002)

On the Galois Lattice of Bipartite Distance Hereditary Graphs

Nicola Apollonio[1], Massimiliano Caramia[2], and Paolo Giulio Franciosa[3]([✉])

[1] Istituto per le Applicazioni del Calcolo M. Picone, CNR,
via dei Taurini 19, 00185 Rome, Italy
nicola.apollonio@cnr.it
[2] Dipartimento di Ingegneria dell'Impresa, Università di Roma "Tor Vergata",
via del Politecnico 1, 00133 Rome, Italy
caramia@disp.uniroma2.it
[3] Dipartimento di Scienze Statistiche, Sapienza Università di Roma,
piazzale Aldo Moro 5, 00185 Rome, Italy
paolo.franciosa@uniroma1.it

Abstract. We give a complete characterization of bipartite graphs having tree-like Galois lattices. We prove that the poset obtained by deleting bottom and top elements from the Galois lattice of a bipartite graph is tree-like if and only if the graph is a Bipartite Distance Hereditary graph. We show that the lattice can be realized as the containment relation among directed paths in an arborescence. Moreover, a compact encoding of Bipartite Distance Hereditary graphs is proposed, that allows optimal time computation of neighborhood intersections and maximal bicliques.

Keywords: Galois lattice · Transitive reduction · Distance hereditary graph · Ptolemaic graph

1 Introduction

Galois lattices are a well established topic in applied lattice theory. Their importance is widely recognized [9], and its applications span across theoretical computer science and discrete mathematics as well as artificial intelligence, data mining and data-base theory. There is a growing interest on the interplay between finite Galois lattices and other discrete structures in combinatorics and computer science, and new relationships have been (and are to be) discovered between graphs and Galois lattices. This paper follows this stream: we characterize a class of bipartite graphs by the Galois lattice of their maximal cliques.

Distance Hereditary graphs are graphs with the *isometric property*, i.e., the distance function of a distance hereditary graph is inherited by its connected

The first author was partially supported by Italian MIUR project "La Matematica per la società e l'innovazione tecnologica–MATHTECH". The second author was partially supported by Italian MIUR projects PRIN 2012C4E3KT "AMANDA – Algorithmics for MAssive and Networked DAta" and "Sottografi fault resilient e algoritmi per modelli di calcolo con memory faults".

© Springer International Publishing Switzerland 2015
J. Kratochvíl et al. (Eds.): IWOCA 2014, LNCS 8986, pp. 37–48, 2015.
DOI: 10.1007/978-3-319-19315-1_4

induced subgraphs. This important class of graphs was introduced and thoroughly investigated by Howorka in [10,11]. A *comparability graph* is the graph of the comparability relation among elements of a poset. In [7], Cornelsen and Di Stefano proved that by intersecting the class of Distance Hereditary graphs with the class of comparability graphs one obtains precisely the comparability graphs of tree-like posets, i.e., those posets whose transitive reduction is a tree. Here we investigate another relation between comparability graphs and distance hereditary graphs: inspired on the one hand by the work of Amilhastre, Vilarem and Janssen [1] and on the other hand by the work of Berry and Sigayret [5] and the work of Brucker and Gély [6]. In [1], *Galois lattices* of domino-free bipartite graphs are investigated. In [5] it is shown that the Hasse diagram of the *Galois lattice* of *chordal bipartite* graphs is *dismantlable* [13], while an analogous result is shown in [6] for the *clique lattice* of *strongly chordal graphs*. Both [5,6] use a dismantlability property of these lattices proved in [13]. Recall that a graph G is *strongly chordal* if and only if its *vertex-clique graph*, namely, the incidence bipartite graph of the maximal cliques of G over $V(G)$, is a *bipartite chordal* graph and that a graph is *bipartite chordal* if it does not contain an induced copy of a chordless cycle on more than four vertices—the reader is referred to Sect. 2 for undefined terms and notions.

In this paper we study the transitive reduction of the *Galois lattice* of those bipartite graphs that are chordal (as in [5]) and domino-free (as in [1]). It follows by Theorem 2 in Sect. 2 that these graphs are precisely the Bipartite Distance Hereditary (BDH for shortness) graphs, namely, those distance hereditary graphs which are bipartite. Essentially in the same way as chordal bipartite graphs are related to strongly chordal graphs, BDH graphs are related to the so called *Ptolemaic graphs*. If **CH** denotes the class of chordal graphs, namely, those graphs that do not contain an induced copy of the chordless cycle on more than three vertices, and if **DH** is the class of distance hereditary graphs, then the class **Pt** of Ptolemaic graphs is the intersection between **CH** and **DH**. Actually, by the results of [12], **Pt** is the intersection between **SC** and **DH**, where **SC** is the class of strongly chordal graphs. Let $\mathcal{L}(G)$ denote the *Galois lattice* of a bipartite graph G and let $\mathcal{C}(H)$ denote the *clique lattice* of a graph H. As shown by Wu (as credited in [12]), if G is Ptolemaic, then the vertex-clique graph of G is a BDH graph. Hence there is a map $\lambda : \mathbf{Pt} \to \mathbf{BDH}$ and it is not difficult to see that $\mathcal{L}(\lambda G) \cong \mathcal{C}(G)$, where $\cong$ is lattice isomorphism. In a sense, it can be shown that the converse statement holds as well, namely, there is a mapping μ that takes a BDH graph G into a Ptolemaic graph μG so that $\mathcal{C}(\mu G) \hookrightarrow \mathcal{L}(G)$ in such a way that $\mathcal{L}(G - I) \cong \mathcal{C}(\mu G)$ for a certain set I of *join-irreducible* (or *meet-irreducible*) elements of $\mathcal{L}(G)$, where $\hookrightarrow$ denotes *order embedding*. In other words, the following diagram applies (and commutes):

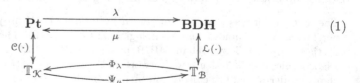

$$\text{(1)}$$

where $\mathbb{T}_{\mathcal{K}}$ and $\mathbb{T}_{\mathcal{B}}$ are the classes of tree-shaped clique lattices and Galois lattices, respectively, Φ_λ is lattice isomorphism induced by λ, and Ψ_μ is an order embedding induced by μ.

Our Result. Let us recall what is the Galois lattice of a bipartite graph G. Let G have color classes X and Y. A *biclique* of G is a set $B \subseteq V(G)$ which induces a complete bipartite graph. Let $\mathcal{B}(G)$ be the set of the (inclusionwise) maximal bicliques of G and for $B \in \mathcal{B}(G)$ let $X(B) = B \cap X$ and $Y(B) = B \cap Y$. $X(B)$ and $Y(B)$ are called the *shores* of B. Throughout the rest of the paper we assume that G does not contain universal vertices, where a *universal vertex* in a bipartite graph is a vertex that is adjacent to all vertices in the opposite color class. This assumption, while it does not cause loss of generality, leads to simpler statements and proofs. Following [1], we endow $\mathcal{B}(G)$ by a partial order $\preceq$ defined by $B \preceq B' \Leftrightarrow X(B) \subseteq X(B')$. Equivalently, the same partial order can be defined as $B \preceq B' \Leftrightarrow Y(B) \supseteq Y(B')$, since $X(B) \subseteq X(B') \Leftrightarrow Y(B) \supseteq Y(B')$. If we extend $\mathcal{B}(G)$ by adding two dummy elements $\perp$ and $\top$ acting as bottom and top element respectively, the poset $\mathcal{L}(G) = (\mathcal{B}(G) \cup \{\perp, \top\}, \preceq)$ is a lattice known as the *Galois lattice* of G. The two dummy elements are respectively defined by: $X(\perp) = \emptyset$, $Y(\perp) = Y$ and $X(\top) = X$, $Y(\top) = \emptyset$.

In this paper we prove that the shape of $\mathcal{L}(G)$ can be used to characterize BDH graphs. More precisely, we show the following.

Theorem 1. *Let G be a connected bipartite graph and let $\mathbf{H}(G)$ be the transitive reduction of $(\mathcal{B}(G), \preceq)$. Then $\mathbf{H}(G)$ is a tree if and only if G is a BDH graph.*

Otherwise stated: after deleting $\perp$ and $\top$, $\mathcal{L}(G)$ is a tree-like poset. This is a very strong property: for instance, it allows efficient enumeration of linear extensions [2]. The question of studying bipartite graphs (binary relations) whose Galois lattice is tree-like (arborescence-like in a sense) was raised first in [4]. Here we completely solve the problem from a graph-theoretical view-point. Theorem 1 implies that $(\mathcal{B}(G), \preceq)$ has linear dimension at most 2, since this can be derived by known properties of planar lattices [17]. Although Theorem 1 can be deduced with some extra work from other known results on graphs and hypergraphs (by taking the longest dipath in Diagram 1) the proof we present here is direct and self-contained. Besides Ptolemaic graphs, BDH are related to series parallel graphs. Indeed it can be proved that BDH graphs are *fundamental graphs* of series parallel graphs. This result—via a theorem due to Shinoda, Chen, Yasuda, Kajitani, and W. Mayeda (as credited by Syslo [16]) and Syslo himself—leads to an implicit representation of the Galois lattice of a BDH graph as a collection of paths in an arborescence. We discuss this representation in Sect. 4, where we exploit it to show how the Galois lattice of a BDH graph, and the BDH graph itself, can be efficiently encoded. The encoding of the BDH graph requires $O(n)$ space in the worst case, n being the order of the graph, still allowing the retrieval of the neighborhood of any vertex in time linear in the size of the neighborhood. Moreover, intersections of neighborhoods can be listed in optimal linear time in the size of the intersection, in the worst case.

For the sake of brevity, most proofs are omitted and will be given in the full paper.

2 Preliminaries

If $\mathcal{H}$ is a family of subsets of a given ground set V, then $\Gamma(\mathcal{H})$ is the *bipartite incidence graph of* $\mathcal{H}$ over V, that is, the bipartite graph with color classes V and $\mathcal{H}$ where there is an edge between $v \in V$ and $F \in \mathcal{H}$ if and only if $v \in F$.

If G is a graph, then $V(G)$ denotes its vertex-set if G is undirected, or its node-set if G is directed. Similarly, $E(G)$ denotes the edge-set of G if G is undirected, or the arc-set of G if G is directed. The distance between two vertices u and v of an undirected graph G, denoted by $d_G(u,v)$, equals the minimum length of a path having u and v as end-vertices, or is ∞ if no such path exists. For a graph G and a vertex $v \in V(G)$, $N_G(v)$ (or simply $N(v)$ when G is understood) is the set of vertices adjacent to v in G. The *degree* of v is the number of vertices in $N_G(v)$. The graph induced by $V(G) - \{v\}$ is denoted by $G - v$. Let G be a directed graph and v be a node of G. We split the neighborhood of v into $N^-(v) = \{u \in V(G) \mid (u,v) \in E(G)\}$ and $N^+(v) = \{w \in V(G) \mid (v,w) \in E(G)\}$. The *outdegree* of v in G, denoted by $\deg_G^+(v)$, is the number $|N^+(v)|$ of arcs leaving v. Analogously, the *indegree* of v in G, $\deg_G^-(v) = |N^-(v)|$, is the number of arcs entering v. A node in G is a *source* if its indegree in G is zero, a *sink* if its outdegree in G is zero, or a *flow-node* if it is neither a source nor a sink. A *dipath* P of G is a path of G with exactly one source in P and exactly one sink in P. A *circuit* C in G is a cycle in G with no source and no sink in C.

The chordless cycle on $n \geq 4$ vertices is denoted by C_n, and a *hole* in a bipartite graph is an induced subgraph isomorphic to C_n for some $n \geq 6$. A *domino* is a subgraph isomorphic to the graph obtained from C_6 by joining two antipodal vertices by a chord (see Fig. 1). A (l, k)-*chordal graph* is a graph such that every cycle of length at least l has at least k chords. Bipartite $(6, 1)$-chordal graphs are simply called *chordal bipartite*. A twin of a vertex v in a graph is a vertex with the same neighbors as v.

Theorem 2 (Bandelt and Mulder [3]**).** *The following statements are equivalent for a bipartite graph G:*

(i) *G is a BDH graph;*
(ii) *G is constructed from a single vertex by a sequence of adding pending vertices and twins of existing vertices;*
(iii) *G contains neither holes nor induced dominoes;*
(iv) *G is a bipartite $(6, 2)$-chordal graph.*

If $G, H_1, H_2 \ldots, H_n$ are graphs, we say that G is $H_1, \ldots, H_n$-*free* if G contains no induced copy of H_i, $i = 1, \ldots, n$. Funny enough, after Theorem 2, one can say that a graph is BDH if and only if it is DH-free: just solve the latter acronym as Domino Hole.

In a poset $(X, \leq)$ an element y covers an element x if $x < y$ and moreover $x \leq z \leq y \Rightarrow z = x$ or $z = y$. If x, y are incomparable we write $x \parallel y$. The *least* or *bottom element* of a poset $(X, \leq)$ is the unique element $x \in X$ such that $x \leq x'$ for every $x' \in X$. This element is usually denoted by $\bot$. The *greatest* or *top element* of $(X, \leq)$, usually denoted by $\top$, is defined dually. The transitive reduction of a poset $(X, \leq)$ is the directed acyclic graph on X where there is

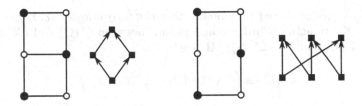

Fig. 1. Domino and C_6 and the corresponding Galois lattices.

an arc leaving x and entering y if and only if y covers x. The meet and the join operators in a lattice are denoted as customary by $\wedge$ and $\vee$, respectively. An element x in a poset $(X, \leq)$ is *meet-irreducible* (resp., *join-irreducible*) if $x = y \wedge z$ (resp., $x = y \vee z$) implies $x = y$ or $x = z$. Let $(X_1, \leq_1)$ and $(X_2, \leq_2)$ be two posets. An order embedding of $(X_1, \leq_1)$ into $(X_2, \leq_2)$ is a map $f : X_1 \to X_2$ satisfying the following condition

$$x \leq_1 y \iff f(x) \leq_2 f(y).$$

An *order isomorphism* is a bijective order embedding.

For a bipartite graph G let $\mathcal{L}^\circ(G) = (\mathcal{B}(G), \preceq)$ and recall that $\mathcal{L}(G)$ denotes $(\mathcal{B}(G) \cup \{\top, \bot\}, \preceq)$. Thus $\mathbf{H}(G)$ is the transitive reduction of $\mathcal{L}^\circ(G)$. Throughout the rest of the paper we represent a biclique B of a bipartite graph G by the ordered pair of its shores, i.e., we write $B = (U, W)$ to mean that $U = X(B)$, $W = Y(B)$ and that $X(B) \cup Y(B)$ induces a complete bipartite subgraph of G. Moreover, with some abuse of notation, if $v \in V(G)$, then we write $v \in B$ to mean that $v \in X(B) \cup Y(B)$ and, analogously, we write $B - v$ for the biclique induced by $(X(B) \cup Y(B)) - \{v\}$. A biclique B *dominates* a biclique B' if $X(B') \subseteq X(B)$ and $Y(B') \subseteq Y(B)$. As an example, let G be either the domino or the C_6 (see Fig. 1). If G is the domino, then $\mathcal{B}(G)$ contains four members: the vertex-sets of the two stars centered at vertices of degree three and the vertex-sets of two squares; $\mathbf{H}(G)$ is thus a directed square with one source and one sink; if G is the C_6 then the members of $\mathcal{B}(G)$ are the vertex-sets of the subpaths of G of length 2; therefore, $\mathbf{H}(G)$ is a directed C_6 with three sources and three sinks.

Remark 1. For $B, B' \in \mathcal{B}(G)$ one has $B \parallel B'$ if and only if $\{X(B), X(B')\}$ and $\{Y(B), Y(B')\}$ both have inclusionwise incomparable members. Indeed, if $X(B) \subseteq X(B')$, say, then $X(B) \cup (Y(B) \cup Y(B'))$ is a biclique of G dominating B.

Remark 2. If $\mathcal{L}(G)$ is the Galois lattice of G then $\mathcal{L}^*(G)$ (the lattice dual of $\mathcal{L}(G)$) is the Galois lattice of G with color classes interchanged. We often use this fact later in the following way: if we prove a property of the lattice for the X-shores of maximal bicliques, then the same property holds by duality for the Y-shores.

If $X_0 \subseteq X$, then there is a biclique $B_0 \in \mathcal{L}^\circ(G)$ such that $X(B_0) = X_0$ if and only if $X_0 = \bigcap_{y \in Y_0} N(y)$ for some $Y_0 \subseteq Y$. Analogously if $Y_0 \subseteq Y$, then there is a biclique $B_0 \in \mathcal{L}^\circ$ such that $Y(B_0) = Y_0$ if and only if $Y_0 = \bigcap_{x \in X_0} N(x)$ for some $X_0 \subseteq X$. Using these facts one has that the projections $(X_0, Y_0) \mapsto X_0$ and $(X_0, Y_0) \mapsto Y_0$ are actually order isomorphisms between $\mathcal{L}^\circ(G)$ and $\{X(B) \mid B \in \mathcal{L}^\circ(G)\}$ and $\{Y(B) \mid B \in \mathcal{L}^\circ(G)\}$. Hence

$$\mathcal{L}^\circ(G) \cong \left(\{X(B) \mid B \in \mathcal{L}^\circ(G)\}, \subseteq \right) \tag{2}$$

and

$$\mathcal{L}^\circ(G) \cong \left(\{Y(B) \mid B \in \mathcal{L}^\circ(G)\}, \supseteq \right) \tag{3}$$

(see also [9]).

To prove the necessity in Theorem 1 we need a sort of "convexity property" for the neighborhood of the vertices of a BDH graph (see Theorem 3). Such a property is interesting on its own and it is equivalent to one of Fagin's results [8]. Let G be a connected BDH graph. For $v, v' \in V(G)$, let $G \star \{v, v'\}$ be the graph defined as follows:

- if v and v' are in different color classes, then $G \star \{v, v'\}$ is G;
- if v and v' are in the same color class, then $G \star \{v, v'\}$ is obtained from G by adding a new vertex $\widehat{vv'}$ to the color class of v and v'. Vertex $\widehat{vv'}$ is adjacent to every vertex in $N(v) \cap N(v')$.

Theorem 3. *Let G be a BDH graph and let $v, v' \in V(G)$. Then $G \star \{v, v'\}$ is a BDH graph, that is the class of BDH graphs is closed under $\star$.*

Let $\mathcal{X}_G = (X(B) \mid B \in \mathcal{B}(G))$ and $\mathcal{Y}_G = (Y(B) \mid B \in \mathcal{B}(G))$.

Corollary 1. *If G is a BDH graph then so are the graphs $\Gamma(\mathcal{X}_G)$ and $\Gamma(\mathcal{Y}_G)$.*

Proof. By duality it suffices to prove the lemma only for $\mathcal{Y}_G$. One has $W \in \mathcal{Y}$ if and only if $(U, W) \in \mathcal{B}(G)$ for some $U \subseteq X$ and $W = \bigcap_{u \in U} N_G(u)$. Therefore, $\mathcal{Y}_G$ is a subfamily of the family $\mathcal{C} = (\bigcap_{u \in U} N_G(u) \mid U \subseteq X)$ and $\Gamma(\mathcal{Y}_G)$ is an induced subgraph of $\Gamma(\mathcal{C})$. Observe that $\Gamma(\mathcal{C}) \cong \Gamma(\{N_{\tilde{G}}(x) \mid x \in \tilde{X}\})$ for a certain graph $\tilde{G}$ with color classes $\tilde{X}$ and Y arising from G by a repeated application of operation $\star$. Such an operation preserves the property of being a BDH graph. Thus $\Gamma(\mathcal{C})$ (and hence $\Gamma(\mathcal{Y}_G)$) is BDH.

3   Characterizing BDH Graphs by Their Galois Lattices

In this section we prove Theorem 1. The proof of the *if part* is given in Sect. 3.1 while the *only if part* is proved in Sect. 3.2.

3.1   Proof of the *if Part*

Let us exploit now the structure of BDH graphs to prove the *if part* of Theorem 1. We remark that the next two results (whose proof is omitted) apply to the more general class of domino-free bipartite graphs.

Lemma 1. *If G is a domino-free bipartite graph then for any $B^1, B^2 \in \mathcal{B}(G)$ such that $B^1 \parallel B^2$ one has*

$$\perp \neq B^1 \wedge B^2 \Rightarrow B^1 \vee B^2 = \top \quad \text{and} \quad B^1 \vee B^2 \neq \top \Rightarrow B^1 \wedge B^2 = \perp.$$

Lemma 2. *Let G be a domino-free bipartite graph and $\mathbf{H}(G)$ be the transitive reduction of $\mathcal{L}^\circ(G)$. Then, any cycle of $\mathbf{H}(G)$ that does not contain $\perp$ or $\top$ has at least six non-flow-nodes.*

We are now ready the prove the *if part* of Theorem 1.

Proof of the *if part* of Theorem 1. We assume $\perp$ and $\top$ have been deleted from $\mathbf{H}(G)$. Since $\mathbf{H}(G)$ is connected we have only to show that it does not contain cycles. Suppose by contradiction that $\mathbf{H}(G)$ contains some cycle, and let $\mathbf{C}$ be a cycle having the least possible number of non-flow-nodes. Let $2t$, $t \in \mathbb{N}$, be such a number. As G is a BDH graph it is domino-free. Therefore, by Lemma 2, $t \geq 3$. Let $B^1, \ldots, B^{2t-1}$ and $B^2, \ldots, B^{2t}$ be the sources and the sinks of $\mathbf{C}$, respectively, as they are met traversing the cycle in a chosen direction. By definition of transitive reduction one has

$$\emptyset \neq X(B^1) \subsetneq X(B^2) \cap X(B^{2t})$$

and

$$\emptyset \neq X(B^{2i+1}) \subset X(B^{2i}) \cap X(B^{2(i+1)}), \ i = 1, \ldots, t-1.$$

Moreover, for $i \in \{0, \ldots, t-1\}$ and $j \in \{1, \ldots, t\}$ such that $|i-j| \notin \{0, 1, t\}$ one has $X(B^{2i+1}) \cap X(B^{2j}) = \emptyset$. Otherwise $X(B^{2i+1}) \wedge X(B^{2j}) \in V(\mathbf{H}(G))$ and one of the two subpaths of $\mathbf{C}$ connecting $X(B^{2i+1})$ and $X(B^{2j})$ along with the two paths of $\mathbf{H}(G)$ connecting $X(B^{2i+1}) \wedge X(B^{2j})$ to $X(B^{2i+1})$ and $X(B^{2j})$ respectively, would define a cycle $\mathbf{C}'$ of $\mathbf{H}(G)$ with fewer non-flow-nodes than $\mathbf{C}$. Now for $i = 0, \ldots, t-1$, pick $x_{2i+1} \in X(B^{2i+1})$ and let $U = \{x_1, x_3 \ldots, x_{2t-1}\}$ and $\mathcal{U} = \{X(B^{2i}) \mid i = 1, \ldots t\}$. Thus $U \cup \mathcal{U}$ induces a hole in $\Gamma(\mathcal{X}_G)$, contradicting Corollary 1.

3.2 Proof of the *only if* Part

To complete the proof of Theorem 1 we need some more properties of $\mathbf{H}(G)$ stated without proof in the following lemmata.

Lemma 3. *Let G be a BDH graph with at least three vertices. Then $x \in X$ is a cut-vertex of G if and only if $(\{x\}, N(x)) \in \mathcal{B}(G)$. Analogously, $y \in Y$ is a cut-vertex of G if and only if $(N(y), \{y\}) \in \mathcal{B}(G)$.*

Recall that in poset that has a bottom element $\perp$, an *atom* is an element of the poset that covers $\perp$. Dually, if the poset has a top element $\top$, a *co-atom* is an element which is covered by $\top$. After this terminology we can say that the cut vertices of G are either atoms or co-atoms.

We now study the behavior of $\mathbf{H}(G - v)$ for $v \in V(G)$. Let us begin with an easy but useful property of $\mathbf{H}(G)$ in the general case. The next lemma proves that if the deletion of a vertex v from a maximal biclique of G does not cause loss of maximality in the biclique, then $\mathbf{H}(G - v)$ inherits from $\mathbf{H}(G)$ as much adjacency as possible.

Lemma 4. *Let G be a bipartite graph, $B_0 \in \mathcal{B}(G)$ and $v \in B_0$. If $B_0 - v \in \mathcal{B}(G - v)$, then*

- *there is an arc $(B - v, B_0 - v)$ in $\mathbf{H}(G - v)$ for every $B \in \mathcal{B}(G)$ such that $B - v \in \mathcal{B}(G - v)$ and (B, K_0) is an arc of $\mathbf{H}(G)$;*
- *there is an arc $(B_0 - v, B - v)$ in $\mathbf{H}(G - v)$ for every $B \in \mathcal{B}(G)$ such that $B - v \in \mathcal{B}(G - v)$ and (B_0, B) is an arc of $\mathbf{H}(G)$;*

In other words,

$$\phi : \mathbf{H}(G - v) \ni B - v \longmapsto B \in \mathbf{H}(G)$$

embeds $\mathbf{H}(G - v)$ in $\mathbf{H}(G)$ as a sub-digraph.

The next lemma shows instead that if the deletion of a vertex v from a maximal biclique B of G causes loss of maximality in the biclique, then the role of B in $\mathcal{L}(G)$ is not really relevant.

Lemma 5. *Let G be a bipartite graph and let $v \in V(G)$ and $B \in \mathcal{B}(G)$ be such that $B - v \notin \mathcal{B}(G - v)$. If $v \in X$ and B is not an atom in $\mathcal{L}(G)$, then $\deg^-_{\mathbf{H}(G)}(B) = 1$. Moreover, if (B', B) is the unique arc entering B in $\mathbf{H}(G)$ then $B' \in \mathcal{B}(G - v)$. Analogously, if $v \in Y$ and B is not a co-atom in $\mathcal{L}(G)$, then $\deg^+_{\mathbf{H}(G)}(B) = 1$. Moreover, if (B, B') is the unique arc leaving B in $\mathbf{H}(G)$ then $B' \in \mathcal{B}(G - v)$.*

Using standard terminology, as in [5], a maximal biclique B that satisfies the hypothesis of Lemma 5 corresponds either to a meet irreducible or to a join irreducible concept in the context associated to the bipartite graph. The results of Lemmas 3, 4, and 5 imply:

Theorem 4. *Let G be a BDH graph and let $v \in V(G)$. Then one of the following conditions holds:*

1. *$\mathbf{H}(G - v)$ has more connected components than $\mathbf{H}(G)$;*
2. *$\mathbf{H}(G - v)$ is an induced subgraph of $\mathbf{H}(G)$;*
3. *$\mathbf{H}(G - v)$ is a contraction of $\mathbf{H}(G)$.*

Proof. Let $\mathcal{B}_0(v) \subseteq \mathcal{B}(G)$ be the set of maximal bicliques B containing v such that $B - v \notin \mathcal{B}(G - v)$. If v is a cut-vertex of G, then condition 1 holds. Otherwise, by Lemmas 4 and 5, $\mathbf{H}(G - v)$ can be derived from $\mathbf{H}(G)$ by the following operations:

- if $v \in X$ and $B \in \mathcal{B}_0(v)$ delete B if it is a sink in $\mathbf{H}(G)$, otherwise contract the unique arc (B', B) with $B' \in \mathcal{B}(G - v)$ to the single node B';
- if $v \in Y$ and $B \in \mathcal{B}_0(v)$ delete B if it is a source in $\mathbf{H}(G)$, otherwise contract the unique arc (B, B') with $B' \in \mathcal{B}(G - v)$ to the single node B'.

In both cases, either conditions 2 or 3 holds.

Proof of the *only if part* of Theorem 1. Let us assume that $\mathbf{H}(G)$ is a tree and let us prove that G is a BDH graph. By Theorem 4, it follows in particular that if G_0 is an induced connected subgraph of G, then $\mathbf{H}(G_0)$ is a contraction of $\mathbf{H}(G_1)$ for some connected induced subgraph G_1 of G such that G_0 is an induced subgraph of G_1. Hence, $\mathbf{H}(G_0)$ is a tree, being the contraction of some subtree of $\mathbf{H}(G)$. Now, to establish the thesis, it suffices to observe that if G_0 is either a domino or a chordless cycle with length greater than four then $\mathbf{H}(G_0)$ is not a tree (see Fig. 1).

4 Encoding $\mathcal{L}(G)$ and Efficiently Computing Maximal Bicliques

In this section, we show how the Galois lattice of a BDH graph can be realized as the containment relation among directed paths in an arborescence. The results are achieved by further exploiting the interplay between BDH graphs and series-parallel graphs. As credited by Syslo [16], Shinoda, Chen, Yasuda, Kajitani, and W. Mayeda, proved that series-parallel graphs can be completely characterized by a property of their spanning trees. They proved that every spanning tree of a series-parallel graph S is a depth-first search tree of a 2-isomorphic copy of S, where 2-isomorphism of graphs (in the sense of Whitney [18]) is isomorphism of binary vector spaces between cycle-spaces of graphs. We can avoid to enter details of such notions and we can content ourselves of restating in our terminology a direct consequence of the result. Recall that an *arborescence* is a directed tree with a single special node distinguished as the *root* such that, for each other vertex, there is a dipath from the root to that vertex.

Theorem 5. *Let G be a connected BDH graph with color classes X and Y. There exists an arborescence ϕT with root r and $X = E(T)$ such that for each $y \in Y$, the set $\{\phi x \mid x \in N_G(y)\}$ is the arc set of a directed path in ϕT. The same holds with the role of X and Y interchanged.*

The main consequence of Theorem 5 is that the Galois lattice of a BDH graph is completely determined by some pairwise intersections of neighborhoods, plus some simple neighborhoods.

Corollary 2. *Let G be a BDH graph with color classes X and Y. Let $\mathcal{F} = \{N(x) \cap N(x') \mid x \neq x',\ x, x' \in X\} \cup \{N(x),\ x \in X\}$. Then $\mathcal{L}^\circ(G) \cong (\mathcal{F}, \subseteq)$.*

Notice that, there are containment orders among paths in an arborescence that are not isomorphic to the Galois lattice of any BDH graph. For example, the Galois lattice of a domino is isomorphic to the containment among sets $\{a, b\}, \{b, c\}, \{a, b, c\}$, and it is immediate to see that these sets are the edge sets of three subpaths of a path with edges a, b, c, which is clearly an arborescence.

We discuss now some algorithmic consequences of the encoding described in Theorem 5 and exploited in Corollary 2. By the results of [15], there exists an algorithm that given a BDH graph computes a supporting arborescence ϕT for

G as in Theorem 5. The algorithm runs in almost linear time in the size of G, that is in time $O(\alpha(|X|, m) \cdot m)$ where m is the number of edges of G and α is an inverse of the Ackermann function, which behaves essentially as a small constant even for very large values of its arguments. We propose a compact encoding of the BDH graph that requires $O(n)$ space in the worst case, where n is the number of vertices in G, that allows to answer the following queries in optimal worst case time, where $x \in X$ and $X' \subseteq X$:

1. list $N(x)$, in time $O(|N(x)|)$;
2. check whether $\bigcap_{x \in X'} N(x) = \emptyset$, in time $O(|X'|)$;
3. list $\bigcap_{x \in X'} N(x)$, in time $O\left(|X'| + |\bigcap_{x \in X'} N(x)|\right)$;
4. check whether $\left(X', \bigcap_{x \in X'} N(x)\right)$ is a maximal biclique, in time $O\left(|X'| + |\bigcap_{x \in X'} N(x)|\right)$.

Note that the size of the encoding is only $O(n)$, while the number of edges in a BDH graph can be $\Theta(n^2)$, and still allows the computation of the maximal biclique containing a given set X' on one side in time linear in the in the number of vertices in the biclique.

The algorithm to solve query 3 (queries 1 and 2 are special cases of query 3) is described in Fig. 2.

Given $X' \subseteq X$, compute $\bigcap_{x \in X'} N(x)$. We assume the arborescence T is given, and a data structure for solving lowest common ancestor queries according to $\leq_T$, as described in [15], has been built.
(a_i, b_i), for $1 \leq i \leq |X'|$, are the end-arcs of the path in T associated to $N(x_i)$

```
1.    let a_max = a_1
2.    for i = 2 to |X'|
3.        if a_i >_T a_max
4.            let a_max = a_i
5.        else if a_i ≰_T a_max
6.            return ∅
7.    let b_min = ⋀_T{b_1, b_2, …, b_k}
8.    if b_min ≥_T a_max
9.        return [a_max, b_min]
10.   else
11.       return ∅
```

Fig. 2. Algorithm NeighborIntersection.

Let $X' = \{x_1, x_2, \ldots, x_k\}$, and let (a_i, b_i), for $1 \leq i \leq k$, be the end-arcs of the path associated to x_i in T. It can be seen that algorithm NeighborIntersection requires $O\left(|X'| + |\bigcap_{x \in X'} N(x)|\right)$ worst case time, since tests in Lines 3 and 5 are performed in constant time starting from the encoding of the 2-dimensional partial order $\leq_T$. The computation of the lowest common ancestor $\bigwedge_T$ at Line

7 is computed in time $O(|X'|)$ using the data structure proposed in [14], which is built in $O(n)$ time. Path retrieval in Line 9 requires $O(|\bigcap_{x \in X'} N(x)|)$ worst case time, starting from $b_{\min}$ and following parent pointers in the arborescence T up to $a_{\max}$.

As a special case, query 3 can be used to list the neighbors of a vertex. In order to solve query 2, we can still use algorithm NeighborIntersection, without listing the path in Line 9, thus requiring $O(|X'|)$ worst case time. Query 4 can be solved using the same algorithm, provided that the same encoding is stored both for side X and for side Y. In fact, $(X', \bigcap_{x \in X'} N(x))$ is a maximal biclique if and only if $X' = \bigcap_{y \in Y_0} N(y)$, where $Y_0 = \bigcap_{x \in X'} N(x)$, that can be checked by computing Y_0 and then computing $\bigcap_{y \in Y_0} N(y)$, i.e., solving two queries of type 3.

References

1. Amilhastre, J., Vilarem, M.C., Janssen, P.: Complexity of minimum biclique cover and minimum biclique decomposition for bipartite domino-free graphs. Discrete Appl. Math. **86**, 125–144 (1998)
2. Atkinson, M.D.: On computing the number of linear extensions of a tree. Order **7**, 23–25 (1990)
3. Bandelt, H.J., Mulder, H.M.: Distance-hereditary graphs. J. Combin. Theory Ser. B **41**, 182–208 (1986)
4. Belohlavek, R., De Baets, B., Outrata, J., Vychodil, V.: Trees in concept lattices. In: Torra, V., Narukawa, Y., Yoshida, Y. (eds.) MDAI 2007. LNCS (LNAI), vol. 4617, pp. 174–184. Springer, Heidelberg (2007)
5. Berry, A., Sigayret, A.: Dismantlable lattices in the mirror. In: Cellier, P., Distel, F., Ganter, B. (eds.) ICFCA 2013. LNCS, vol. 7880, pp. 44–59. Springer, Heidelberg (2013)
6. Brucker, F., Gély, A.: Crown-free lattices and their related graphs. Order **28**(3), 443–454 (2011)
7. Cornelsen, S., Di Stefano, G.: Treelike comparability graphs. Discrete Appl. Math. **157**, 1711–1722 (2009)
8. Fagin, R.: Degrees of acyclicity for hypergraphs and relational database schemes. J. ACM **30**(3), 514–550 (1983)
9. Ganter, B., Wille, R.: Formal Concept Analysis - Mathematical Foundations. Springer, Heidelberg (1999)
10. Howorka, E.: A characterization of distance-hereditary graphs. Q. J. Math. **2**(26), 417–420 (1977)
11. Howorka, E.: A characterization of Ptolemaic graphs, survey of results. In: Proceedings of the 8th SE Conference Combinatorics, Graph Theory and Computing, pp. 355–361 (1977)
12. Peled, U.N., Wu, J.: Restricted unimodular chordal graphs. J. Graph Theory **30**(2), 121–136 (1999)
13. Rival, I.: Lattices with doubly irreducible elements. Can. Math. Bull. **17**(1), 91–95 (1974)
14. Schieber, G., Vishkin, U.: On finding lowest common ancestors: simplification and parallelization. SIAM J. Comput. **17**(6), 1253–1262 (1988)

15. Swaminathan, R.P., Wagner, D.B.: The arborescence-realization problem. Discrete Appl. Math. **59**, 267–283 (1995)
16. Syslo, M.M.: Series-parallel graphs and depth-first search trees. IEEE Trans. Circuits Syst. **31**(12), 1029–1033 (1984)
17. Trotter, W.T.: Combinatorics and Partially Ordered Sets: Dimension Theory. The Johns Hopkins University Press, Baltimore, Maryland (1992)
18. Whitney, H.: 2-isomorphic graphs. Am. Math. J. **55**, 245–254 (1933)

Fast and Simple Computations Using Prefix Tables Under Hamming and Edit Distance

Carl Barton[1], Costas S. Iliopoulos[1,3], Solon P. Pissis[1(✉)],
and William F. Smyth[2]

[1] King's College London, London, UK
{carl.barton,c.iliopoulos,solon.pissis}@kcl.ac.uk
[2] McMaster University, Hamilton, Canada
smyth@mcmaster.ca
[3] University of Western Australia, Crawley, Australia

Abstract. In this article, we introduce a new and simple data structure, the prefix table under Hamming distance, and present two algorithms to compute it efficiently: one asymptotically fast; the other very fast on average and in practice. Because the latter approach avoids the computation of global data structures, such as the suffix array and the longest common prefix array, it yields algorithms much faster in practice than existing methods. We show how this data structure can be used to solve two string problems of interest: (a) approximate string matching under Hamming distance; and (b) longest approximate overlap under Hamming distance. Analogously, we introduce the prefix table under edit distance, and present an efficient algorithm for its computation. In the process, we also define the border array under both distance measures, and provide an algorithm for conversion between prefix tables and border arrays.

1 Introduction

We begin with a few definitions, generally following [19]. We think of a *string* x of *length* n as an array $x[0 . . n-1]$, where every $x[i]$, $0 \leq i < n$, is a *letter* drawn from some finite *alphabet* Σ of size $\sigma = |\Sigma| = \mathcal{O}(1)$. The *empty string* of length 0 is denoted by ε. If $x = uvw$, then u is a *prefix*, v a *substring*, w a *suffix* of x; *proper* in each case if $u \neq x$, $v \neq x$, $w \neq x$, respectively. If x has a proper prefix u that equals a suffix of x, u is said to be a *border* of x.

The *border array* $\beta = \beta[0 . . n-1]$ of x gives the length $\beta[i]$ of the longest border of every prefix $x[0 . . i]$, $0 \leq i < n$, of x. It is computed by an elegant algorithm in time $\Theta(n)$ [4,19], and has the property that for every rth longest border $\beta^r[i] > 0$, $\beta^{r+1}[i]$ is the length of the $(r+1)$th longest border, where β^r denotes r applications $\beta[\beta[\ldots \beta[i] \ldots]]$ of this function. Thus β specifies *all* the borders of every prefix of x. The *prefix table* $\pi = \pi[0 . . n-1]$ of x gives the length $\pi[i]$ of the longest substring beginning at position i, $0 \leq i < n$, of x, that equals a prefix of x. The prefix table was introduced in [16] to compute repetitions; it has since prominently appeared in [4,20].

© Springer International Publishing Switzerland 2015
J. Kratochvíl et al. (Eds.): IWOCA 2014, LNCS 8986, pp. 49–61, 2015.
DOI: 10.1007/978-3-319-19315-1_5

The *Hamming distance* between strings x and y, both of length n, is the number of positions i, $0 \leq i < n$, such that $x[i] \neq y[i]$. Given an integer $k > 0$, we write $x \equiv_k^H y$ if the Hamming distance between x and y is at most k.

Observation 1. *If $x \equiv_k^H y$, then for every $i, j \in 0, \ldots, n-1$, $i \leq j$, $x[i..j] \equiv_k^H y[i..j]$.*

We can now define the *k-prefix table* π_k^H of x: for every i, $0 \leq i < n$, $\pi_k^H[i] = \ell$ is the length of the longest prefix of x such that $x[i..i+\ell-1] \equiv_k^H x[0..\ell-1]$. By Observation 1, if $\pi_k^H[i] = \ell$, it follows that every prefix $x[0..j]$, $i \leq j \leq i+\ell-1$, has a k-border of length $j-i+1$. Thus, as for regular prefix tables, π_k^H determines all the k-borders of x [4]. Similarly we can define the *k-border array* β_k^H, but it should be noted that β_k^H specifies only the length of the *longest* border at each position i, not the lengths of shorter borders. This is a consequence of the nontransitivity under the distance model.

Given a string x of length m and a string y of length $n \geq m$, the *edit distance* is the minimum total cost of operations required to transform one string into the other. For simplicity, we consider the cost of each to be 1 [15]. The allowed edit operations are as follows: *insert* a letter in y, not present in x; *delete* a letter in y, present in x; and *replace* a letter in y with a letter in x. We write $x \equiv_k^E y$ if the edit distance between x and y is at most k. Equivalently, if $x \equiv_k^E y$, we say that x and y have at most k *differences*. We refer to the *standard dynamic programming matrix* of x and y defined by $D[i, 0] = i$, for $0 \leq i \leq m$, $D[0, j] = j$, for $0 \leq j \leq n$, and for $1 \leq i \leq m, 1 \leq j \leq n$:

$$D[i, j] = \min \begin{cases} D[i-1, j-1] + 1 \text{ (if } x[i-1] \neq y[j-1]) \\ D[i-1, j] + 1 \\ D[i, j-1] + 1 \end{cases}$$

Analogously, we can define the *k-prefix table* π_k^E of x and the *k-border array* β_k^E of x under edit distance.

In Sect. 2, we present two algorithms to compute π_k^H: a practical one requiring average-case time $\Theta(kn)$; and another requiring worst-case time $\Theta(kn)$; we then show how to compute β_k^H from π_k^H in time $\Theta(n)$.

In Sects. 3 and 4, we show how the computation of π_k^H can be used to greatly speed up two computations of interest in computational biology and elsewhere. The first of these is approximate string matching with k-mismatches (see [4] for a definition) where given a text t of length n the problem is to search for occurrences of a pattern x of length $m < n$ at Hamming distance at most k from x. The original algorithms proposed for this problem [9,14] require time $\mathcal{O}(kn)$. Shortly thereafter a $\mathcal{O}(\sqrt{m \log m} n)$-time algorithm was proposed [1], with a time requirement independent of k and asymptotically faster than its predecessors for $k \geq \sqrt{m \log m}$. About 13 years ago the asymptotically fastest algorithm was proposed, executing in time $\mathcal{O}(\sqrt{k \log k} n)$, as well as an alternative $\mathcal{O}((n + (nk^3)/m) \log k)$-time algorithm [2]. About 10 years ago an optimal average-case algorithm was proposed, executing in time $\mathcal{O}(n(k + \log_\sigma m)/m)$, only if $k/m < 1/2 - \mathcal{O}(1/\sqrt{\sigma})$ [8]. Section 3 shows how to use π_k^H of xt to solve this problem in average-case time $\mathcal{O}\big(n + k(k+1)n/\sigma^{m/(k+1)}\big)$ — in practice, for moderate k, essentially linear in n. We also consider the well-known problem of computing the longest approximate overlap of two strings x of length m and y of length $n \geq m$ with k-mismatches. This overlap can be found in time $\mathcal{O}(kn)$ [13]. In Sect. 4, we present a very simple algorithm, based on the computation of β_k^H from π_k^H, that in time $\Theta(kn)$ not only solves the overlap problem for two strings, but also for every prefix of those strings.

Finally, in Sect. 5, we present an algorithm based on incremental string comparison techniques to compute π_k^E in worst-case time $\Theta(kn)$.

2 Efficient Computation of π_k^H and β_k^H

We present two algorithms that iteratively overwrite $\pi = \pi_0$ with π_j^H, $j \geq 1$, until $j = k$. The first requires average-case time $\Theta(kn)$ and the second worst-case time $\Theta(kn)$. We then compute β_k^H from π_k^H in time $\Theta(n)$.

2.1 Average-Case Algorithm for Computing π_k^H

The first algorithm is very simple and fast in practice. As we show below, it executes in average-case time $\Theta(n)$ for each of the k iterations.

Fact 2. *The expected number of letter comparisons required for each i in algorithm k-PrefixTable-Simple is less than* 3.

Proof. On an alphabet of size σ, the probability that two random strings of length ℓ are equal is $(1/\sigma)^\ell$. Let $r = 1/\sigma$, there is probability r^ℓ the first ℓ symbols match. Thus the expected number of positions matched before inequality occurs is $S = r + 2r^2 + \cdots + (n-1)r^{n-1}$, for some $n \geq 2$. Hall &Knight [10, p. 44] tell us that $S = r(1 - r^{n-1})/(1-r)^2 - (n-1)r^n/(1-r)$, which as $n \to \infty$ approaches $r/(1-r)^2 < 2$ for all r. Thus S, the expected number of matching positions for each i, is less than 2, and hence the expected number of letter comparisons required for each i in algorithm k-PrefixTable-Simple is less than 3. □

By Fact 2, we obtain the following.

Theorem 3. *Given a string x of length n, the prefix table π of x, and an integer threshold $k < n$, algorithm k-PrefixTable-Simple computes π_k^H in average-case time $\Theta(kn)$ and space $\Theta(n)$.*

2.2 Worst-Case Algorithm for Computing π_k^H

Observation 4. *If $\pi_k^H[i] = \ell$, $0 \leq i < n$, then $x[0 \, . . \, \ell - 1] \equiv_k^H x[i \, . . \, i + \ell - 1]$ and $x[\ell] \neq x[i + \ell]$.*

ALGORITHM. k-PrefixTable-Simple(x, n, π, k)
  for $j \leftarrow 1$ **to** k **do**
    — *Nothing to do for $i = 0$.*
    for $i \leftarrow 1$ **to** $n - 1$ **do**
      $\delta \leftarrow i + \pi[i];$
      — *Nothing to do if $i + \pi[i] > n$.*
      if $\delta \leq n$ **then**
        repeat
          $\delta \leftarrow \delta + 1;$
        until $\delta > n$ **or** $x[\delta - i] \neq x[\delta]$
      end if
      $\pi[i] \leftarrow \min(\delta - i, n - i);$
    end for
  end for
  return $\pi;$

ALGORITHM. k-PrefixTable(x, n, π, k)
  Compute SA, iSA, LCP, and RMQ$_{\text{LCP}}$ of x.
  for $j \leftarrow 1$ **to** k **do**
    — *Nothing to do for $i = 0$.*
    for $i \leftarrow 1$ **to** $n - 1$ **do**
      $\delta \leftarrow \pi[i] + 1 + lce(\pi[i] + 1, i + \pi[i] + 1);$
      $\pi[i] \leftarrow \min(\delta, n - i);$
    end for
  end for
  return $\pi;$

Computing the value of ℓ is equivalent to finding the longest common extension, denoted by lce, of the suffixes starting at $\pi_{j-1}^{H}[i]+1$ and $i+\pi_{j-1}^{H}[i]+1$, $1 \leq j \leq k$. To achieve $\Theta(n)$-time computation of each table we must be able to compute the lce of two suffixes in constant time.

Let SA denote the array of positions of the sorted suffixes of x, i.e. for all $1 \leq r < n$, we have $x[\text{SA}[r - 1]..n - 1] < x[\text{SA}[r]..n - 1]$. The inverse iSA of the array SA is defined by $\text{iSA}[\text{SA}[r]] = r$, for all $0 \leq r < n$. Let $\text{lcp}(r, s)$ denote the length of the longest common prefix of the strings $x[\text{SA}[r]..n - 1]$ and $x[\text{SA}[s]..n - 1]$, for all $0 \leq r, s < n$, and 0 otherwise. Let LCP denote the array defined by $\text{LCP}[r] = \text{lcp}(r - 1, r)$, for all $1 < r < n$, and $\text{LCP}[0] = 0$. We perform the following linear-time and linear-space preprocessing: (i) compute arrays SA and iSA of x [17]; (ii) compute array LCP of x [6]; and (iii) preprocess array LCP for range minimum queries, that we denote by RMQ$_{\text{LCP}}$ [7]. With the preprocessing complete, the lce of two suffixes of x starting at positions p and q can be computed in constant time in the following way (see [12] for details).

$$\text{lce}(p, q) = \text{LCP}[\text{RMQ}_{\text{LCP}}(\text{iSA}[p] + 1, \text{iSA}[q])]$$

Therefore, we obtain the following.

ALGORITHM. k-BorderArray(π_k^H, n)

  $\beta_k^H[0] \leftarrow 0$;
  $\ell \leftarrow 0$;
  for $i \leftarrow 1$ **to** $n - 1$ **do**
    — Nothing to do if $i + \pi_k^H[i] - 1 < \ell$.
    if $i + \pi_k^H[i] - 1 \geq \ell$ **then**
      for $r \leftarrow 0$ **to** $i + \pi_k^H[i] - 1 - \ell$ **do**
        $\beta_k^H[i + \pi_k^H[i] - r - 1] \leftarrow \pi_k^H[i] - r$;
      end for
      $\ell \leftarrow i + \pi_k^H[i]$;
    end if
  end for
  return β_k^H;

Theorem 5. *Given a string x of length n, the prefix table π of x, and an integer threshold $k < n$, algorithm k-PrefixTable computes π_k^H in worst-case time $\Theta(kn)$ and space $\Theta(n)$.*

2.3 Computing β_k^H from π_k^H

Lemma 6. *Let ℓ be the largest index in β_k^H which has been correctly updated, and let i be the smallest index such that $i + \pi_k^H[i] > \ell$ and $i \leq \ell$. Then $\beta_k^H[i + \pi_k^H[i] - r - 1] = \pi_k^H[i] - r$, for all $0 \leq r < i + \pi_k^H[i] - 1 - \ell$.*

Proof. We update $\beta_k^H[i + \pi[i] - r - 1]$, for all $0 \leq r < i + \pi_k^H[i] - 1 - \ell$, when we find some index $i \leq \ell$ such that $i + \pi_k^H[i] > \ell$. As $\ell < i + \pi_k^H[i] - r - 1 < i + \pi_k^H[i]$, no $\beta_k^H[i + \pi_k^H[i] - r - 1]$ has been assigned a value, and for all j, such that $i < j < i + \pi_k^H[i]$ and $j + \pi_k^H[j] > \ell$, the k-borders given by j must be smaller than the k-borders given by i for the same prefix. Therefore, the longest k-border is given by $\pi_k^H[i] - r$ and $\beta_k^H[i + \pi_k^H[i] - r - 1] = \pi_k^H[i] - r$, for all $0 \leq r < i + \pi_k^H[i] - 1 - \ell$. $\square$

By Lemma 6, we have no more than $2n$ operations executed in total by algorithm k-BorderArray. Therefore, we obtain the following.

Theorem 7. *Given array π_k^H of string x of length n, algorithm k-BorderArray computes β_k^H in worst-case time and space $\Theta(n)$.*

A similar approach in a different context was shown in [3]. Moreover, we can compute all k-borders of x directly from π_k^H, thus in time $\Theta(kn)$: For every $0 \leq i < n$, $\pi_k^H[i]$ is the length of a k-border of x if and only if $\pi_k^H[i] + i = n$.

3 Application I: Approximate String Matching with k-Mismatches via Filtering π_k^H

In this section, we present FPT, an algorithm for approximate string matching with k-mismatches. Algorithm FPT is based on Filtering the k-Prefix Table.

Given a pattern x of length m, a text t of length $n > m$, and an integer threshold $k < m$, an outline of algorithm FPT is as follows.

1. Construct $T = xt$, and compute the prefix table π_0 of T.
2. The pattern x is split in $k+1$ fragments of length $\lfloor m/(k+1) \rfloor$ and $\lceil m/(k+1) \rceil$.
3. Match the $k + 1$ fragments against the text t using Aho Corasick automaton [5]. Let $\mathcal{L}$ be a list of tuples of size Occ, where $< id, p > \in \mathcal{L}$ is a tuple such that $0 \leq id \leq k$ is the fragment identifier, and $0 \leq p < n$ is the position that the fragment occurs in t.
4. Using these occurrences we could *invalidate* (filter out) the positions on π_0 that can never give a match if we extend them, i.e. we apply the partitioning technique [24]. Equivalently, for each tuple $< id, p > \in \mathcal{L}$, we *validate* $\pi_0[m + p - id \times \ell_{id}]$, where ℓ_{id} is the length of the respective fragment.
5. Compute only the *valid* positions of π_i^H, for all $1 \leq i \leq k$, using algorithm k-PrefixTable.
6. If $\pi_k^H[i] \geq m$, x occurs at starting position $i - m$ of t, for all $m \leq i \leq n$.

Theorem 8. *Given a pattern x of length m drawn from alphabet Σ, $\sigma = |\Sigma|$, a text t of length $n > m$ drawn from Σ, and an integer threshold $k < m$, algorithm FPT requires average-case time $\mathcal{O}(n + k(k+1)n/\sigma^{m/(k+1)})$ and space $\mathcal{O}(n)$.*

Proof. The computation of the prefix table π_0 of $T = xt$ requires time and space $\mathcal{O}(n + m)$ (Step 1) [4]. Splitting the pattern x takes time $\mathcal{O}(k)$ (Step 2). The Aho-Corasick automaton of the $k + 1$ fragments requires time $\mathcal{O}(m)$ with search time $\mathcal{O}(n + Occ)$ (Step 3) [5]. Validating positions of π_0 takes time $\mathcal{O}(Occ)$ (Step 4). Computing the valid positions of $\pi_1^H, \ldots, \pi_k^H$ requires time $\mathcal{O}(kOcc)$ (Step 5)—see Sect. 2.2. Reporting the output requires time $\mathcal{O}(n)$ (Step 6). Since the expected number Occ of occurrences of the $k + 1$ fragments is $\mathcal{O}((k + 1)n/\sigma^{m/(k+1)})$, algorithm FPT requires average-case time $\mathcal{O}(n + k(k+1)n/\sigma^{m/(k+1)})$. □

Corollary 9. *Given a pattern x of length m drawn from alphabet Σ, $\sigma = |\Sigma|$, a text t of length $n > m$ drawn from Σ, and an integer threshold $k = \mathcal{O}(m/\log m)$, algorithm FPT requires average-case time $\mathcal{O}(n)$.*

Proof. Algorithm FPT achieves average-case time $\mathcal{O}(n)$ iff

$$k(k + 1)n/\sigma^{m/(k+1)} \leq cn$$

for some fixed constant c. Let $r = m/(k+1)$. We have $k(k+1)n/\sigma^r \leq cn$. Since $k < m$, we can (pessimistically) replace k by $m - 1$. Then we have

$$m(m - 1)n/\sigma^r \leq cn.$$

Solving for r, and using $k \leq m/r - 1$, gives the maximum value of k, that is $k = \mathcal{O}(m/\log m)$. □

By FPT-Simple, we denote the same algorithm apart from Step 5, where algorithm k-PrefixTable is replaced by algorithm k-PrefixTable-Simple. By applying Fact 2, it requires average-case time $\mathcal{O}(n)$, but because this approach avoids the computation of global data structures, it can be implemented in space $\mathcal{O}(m)$.

Corollary 10. *Given a pattern* $\boldsymbol{x}$ *of length m drawn from alphabet Σ, $\sigma = |\Sigma|$, a text $\boldsymbol{t}$ of length $n > m$ drawn from Σ, and an integer threshold $k = \mathcal{O}(m/\log m)$, algorithm FPT-Simple requires average-case time $\mathcal{O}(n)$ and space $\mathcal{O}(m)$.*

3.1   Experimental Results

We implemented FPT and FPT-Simple as library functions to perform approximate string matching with k-mismatches. They were implemented in the C programming language and developed under GNU/Linux operating system. Keeping in mind we wish to evaluate the practical efficiency of these two algorithms, we compared their performance to the respective performance of the following:

- Naive, an algorithm that considers all $\Theta(n)$ alignments of the text and the pattern, and counts mismatches at each alignment, stopping if more than k of them are found. This algorithm has worst-case time complexity $\mathcal{O}(mn)$, but average-case time complexity $\mathcal{O}(kn)$.
- Abrahamson, the algorithm presented in [1]. Even though this algorithm has worst-case time complexity $\mathcal{O}(\sqrt{m\log m}n)$, we preferred it to the $\mathcal{O}(\sqrt{k\log k}n)$-time algorithm presented in [2]. Both algorithms make extensive use of the Fast Fourier Transform to find the frequently occurring letters, however, the one proposed in [2] also requires the construction of the generalised suffix tree of $\boldsymbol{x}$ and t which is processed to allow constant-time *lowest common ancestor* queries, making it slower in practice. Due to this we opted to use the algorithm proposed in [1].
- FredNava, the algorithm with average-case optimal search time presented in [8]. The search-time complexity is $\mathcal{O}(n(k+\log_\sigma m)/m)$ and the space complexity is $\mathcal{O}(m^5\sigma^{\mathcal{O}(1)})$.

The experiments were conducted on a Desktop PC using 1 Intel Core Quad CPU Q9650 at 3.00 GHz and 8 GB of RAM and running under GNU/Linux. The implementation of algorithms FPT and FPT-Simple is distributed under the GNU General Public License (GPL) and is available at a website[1], which is set up for maintaining the source code. The implementations of algorithms Naive and Abrahamson were obtained from library StringPedia [21]; the implementation of algorithm FredNava was obtained via a personal communication with its author. Tables 1, 2 and 3 illustrate elapsed-time comparisons for various pattern sizes and moderate values of k, using as text a corpus of English, protein, and DNA data taken from the Pizza&Chili website [18]. Different patterns were randomly picked from the text and the average elapsed time for each implementation with these patterns as input is presented.

---

[1] http://www.inf.kcl.ac.uk/research/projects/asmf/.

Table 1. Elapsed-time and speed-up comparisons of algorithms Naive, Abrahamson, FPT, and FPT-Simple using English data ($\sigma = 128$) for $n = 50$MB. *Algorithm FredNava was terminated by a segmentation fault

		Elapsed Time (s)					Speed-up of FPT-Simple			
m	k	Naive	Abrahamson	FredNava	FPT	FPT-Simple	Naive	Abrahamson	FredNava	FPT
2000	10	1.38	14.92	*	19.73	3.11	0.44	4.79	*	6.34
4000	25	2.54	26.28	*	20.15	3.48	0.72	7.55	*	5.79
8000	50	5.08	38.37	*	20.55	3.79	1.34	10.12	*	5.42
16000	100	9.67	52.32	*	20.86	4.17	2.31	12.54	*	5.00
32000	200	18.99	63.85	*	21.35	4.54	4.18	14.06	*	4.70
2000	25	2.93	14.90	*	20.50	3.73	0.78	3.99	*	5.49
4000	50	4.87	26.21	*	20.74	4.08	1.19	6.42	*	5.08
8000	100	9.70	38.62	*	20.98	4.20	2.30	9.19	*	4.99
16000	200	18.99	52.87	*	21.34	4.38	4.33	12.07	*	4.89
32000	400	37.40	64.64	*	22.12	4.54	8.23	14.23	*	4.87
2000	50	5.15	14.92	*	20.84	4.13	1.24	3.61	*	5.04
4000	100	9.28	26.59	*	20.96	4.18	2.22	6.36	*	5.01
8000	200	18.75	38.57	*	21.42	4.42	4.24	8.72	*	4.84
16000	400	37.13	52.48	*	22.37	4.50	8.25	11.66	*	4.97
32000	800	73.02	64.71	*	25.57	4.55	16.04	14.22	*	5.61

Table 2. Elapsed-time and speed-up comparisons of algorithms Naive, Abrahamson, FredNava, FPT, and FPT-Simple using protein data ($\sigma = 20$) for $n = 50$MB

		Elapsed Time (s)					Speed-up of FPT-Simple			
m	k	Naive	Abrahamson	FredNava	FPT	FPT-Simple	Naive	Abrahamson	FredNava	FPT
2000	10	1.11	19.98	13.34	22.32	2.74	0.41	7.29	4.87	8.15
4000	25	2.50	32.39	14.85	23.24	3.72	0.67	8.71	3.99	6.25
8000	50	4.61	57.20	14.92	24.12	4.16	1.11	13.75	3.59	5.80
16000	100	8.80	70.61	15.16	24.46	4.76	1.85	14.83	3.18	5.14
32000	200	17.20	81.77	15.16	24.73	4.97	3.46	16.45	3.05	4.98
2000	25	2.44	19.84	15.01	25.04	3.54	0.69	5.60	4.24	7.07
4000	50	4.55	32.00	14.97	23.72	4.24	1.07	7.55	3.53	5.59
8000	100	8.66	56.80	15.04	24.25	4.64	1.87	12.24	3.24	5.23
16000	200	17.21	70.71	15.18	24.88	4.86	3.54	14.55	3.12	5.12
32000	400	33.45	81.19	15.12	26.25	4.92	6.80	16.50	3.07	5.34
2000	50	4.67	19.88	14.93	23.88	4.17	1.12	4.77	3.58	5.73
4000	100	8.59	32.47	15.10	24.58	4.72	1.82	6.88	3.20	5.21
8000	200	17.18	56.93	15.00	25.16	4.78	3.59	11.91	3.14	5.26
16000	400	33.33	70.81	15.19	28.44	4.78	6.97	14.81	3.18	5.95
32000	800	66.76	80.90	15.22	36.03	5.08	13.14	15.93	3.00	7.09

Table 3. Elapsed-time and speed-up comparisons of algorithms Naive, Abrahamson, FredNava, FPT, and FPT-Simple using DNA data ($\sigma = 4$) for $n = 50$MB

		Elapsed Time (s)				Speed-up of FPT-Simple			
m	k	Naive	Abrahamson	FredNava	FPT	FPT-Simple	Naive	Abrahamson	FredNava	FPT
2000	10	3.14	14.88	3.36	22.71	3.36	0.93	4.42	1.00	6.75
4000	25	6.73	16.00	4.50	22.81	3.35	2.00	4.77	1.34	6.80
8000	50	12.74	16.69	4.32	22.96	3.48	3.66	4.79	1.24	6.59
16000	100	24.86	19.01	4.40	23.18	3.62	6.86	5.25	1.21	6.40
32000	200	49.28	20.38	4.40	23.19	3.86	12.76	5.27	1.13	6.00
2000	25	6.83	14.82	4.49	22.89	3.36	2.03	4.41	1.33	6.81
4000	50	12.82	15.83	4.28	22.91	3.43	3.73	4.61	1.14	6.67
8000	100	24.78	16.72	4.31	22.94	3.50	7.08	4.77	1.23	6.55
16000	200	49.17	19.01	4.47	23.15	3.64	13.50	5.22	1.22	5.16
32000	400	98.23	20.29	4.40	23.31	3.88	25.31	5.22	1.21	6.00
2000	50	12.89	14.86	4.31	23.25	3.42	3.76	4.34	1.26	6.79
4000	100	25.05	15.65	4.31	24.02	3.50	7.15	4.47	1.23	6.86
8000	200	48.90	18.98	4.31	25.30	3.68	13.28	5.15	1.17	6.87
16000	400	97.55	19.04	4.40	26.06	3.78	25.80	5.03	1.16	6.89
32000	800	195.18	20.26	4.40	27.53	4.10	47.60	4.94	1.07	5.55

As demonstrated by the experimental results, algorithm FPT-Simple is in most cases the fastest. Algorithm Naive is the fastest for small m and k. Algorithm FredNava with English data was terminated by a segmentation fault during preprocessing stage due to lack of memory. Algorithms FredNava and FPT-Simple with DNA data perform very similarly. Another observation, also suggested by Corollaries 9 and 10, is that the FPT-based algorithms are essentially *independent* of m for moderate values of k.

4   Application II: Longest Approximate Overlap of Two Strings with k-Mismatches

Finding approximate overlaps is the first phase of many sequence assembly methods. Given a set of r strings and an error rate ϵ, the goal is to find, for all pairs of strings, their suffix/prefix matches (overlaps) that are within edit or Hamming distance $k = \lceil \epsilon \ell \rceil$, where ℓ is the length of the overlap. Many existing solutions focus on applications where r is large, the average string length is small, and k is small; and therefore make use of techniques such as *backward backtracking* and/or *suffix filters* to save space [23]. However, algorithms are also needed to *merge* overlapping paired-end reads, in the case when $r = 2$, while correcting mismatches and uncalled bases [25]. Here we focus on the case where $r = 2$, although our algorithm can be used to compute the approximate overlap between r strings in time $\Theta(r^2 Nk)$, where N is the average length of the r strings.

Given a string x of length m, a string y of length $n \geq m$, and an integer threshold $k < m$, this overlap under edit or Hamming distance can be found in time $\mathcal{O}(kn)$ by the algorithm of [13]. Here, we propose a simple alternative algorithm, for Hamming distance, that requires time $\Theta(kn)$ and space $\Theta(n)$. Furthermore, notice that the proposed algorithm not only computes the longest approximate overlap of x and y with k-mismatches, but also of all their prefixes.

1. Construct $T = yx$, and compute the prefix table π_0 of T.
2. Compute the arrays $\pi_1^H, \ldots, \pi_k^H$ of T using algorithm k-PrefixTable.
3. Compute the arrays $\beta_0^H, \beta_1^H, \ldots, \beta_k^H$ of T using algorithm k-BorderArray.
4. $\beta_0^H[m+n-1], \beta_1^H[m+n-1], \ldots, \beta_k^H[m+n-1]$ give the longest approximate overlap of x and y with $0, 1, \ldots, k$ mismatches, respectively.

As mentioned above, existing solutions for the overlap problem consider very different sets of parameters, and so are not directly comparable with ours. Similar to Sect. 3, we anticipate that using algorithm k-PrefixTable-Simple to compute the k-prefix table and, then, algorithm k-BorderArray to compute the k-border array would yield a very fast and simple solution.

5   Efficient Computation of π_k^E and β_k^E

In this section, we consider the prefix table under edit distance and present an efficient algorithm for its computation. The computation is heavily based on incremental string comparison techniques so first we give an overview of these techniques. The incremental string comparison problem was introduced by Landau *et al.* in [13]. The authors considered the following problem: given the edit distance between two strings A and B, how can the edit distance between A and bB; or Bb be efficiently derived, where b is an additional letter? Given a threshold on the number of differences k, they solve this problem and allow prepending and appending of letters in time $\mathcal{O}(k)$ per operation. Later in [11] a generalisation of the problem was considered where prefixes can be deleted and prepended to A or B with time complexity of $\mathcal{O}(k)$ per letter.

The idea in both [11,13] is the efficient computation of h-*waves*. In the standard dynamic programming matrix, we say that a cell $D[i, j]$ is on the diagonal d iff $j - i = d$. For each diagonal, we may have a lowest cell with value h; if $D[i, j] = h$ and $D[i + 1, j + 1] = h + 1$ then $D[i, j]$ is this cell for diagonal $j - i$. The h-wave, for all $0 \le h \le k$, is the position of all these cells across all diagonals, that is, a list H_h of length $\mathcal{O}(k)$, where each entry is a pair (i, j) such that $D[i, j] = h$ and $D[i + 1, j + 1] = h + 1$. Note that the i-th wave can only contain entries on diagonal zero and the i diagonals either side of it, so for $0 \le i \le k$ every wave has size $\mathcal{O}(k)$. These h-waves define the entire dynamic programming matrix due to monotonicity properties. For any diagonal d, if we know the position of the lowest cell on d with value h and $h + 1$, then we also know the value of every cell between these two cells: it must be $h + 1$. So given the h-waves of the matrix, for all $0 \le h \le k$, we have all the information from the standard dynamic programming matrix. The key result from our perspective is the following. Let $\mathsf{cat}(u', u)$ denote the string obtained by concatenating string u' and string u. Let $\mathsf{del}(\alpha, u)$ denote the string obtained by deleting the prefix of length α from string u. Further let D' denote the standard dynamic programming matrix of $cat(A', A)$ and $del(t_2, B)$, where $|A'| = t_1$.

Theorem 11 ([11]). *The 0-wave, 1-wave, ... , and k-wave of matrix D' can be computed in time $\mathcal{O}((t_1 + t_2)k)$.*

ALGORITHM. k-PrefixTable-ED(x, n, k)

$\pi_k^E[0] \leftarrow n$;
$D \leftarrow DP(x, x[1..n-1], k)$; $H_{0,...,k} \leftarrow GH(D)$;
for $i \in \{1, n-1\}$ **do**
  $\ell \leftarrow -1$;
  for $(u, v) \in H_k$ **do**
   if $v > u$ **then**
    $w \leftarrow u$;
   else
    $w \leftarrow v$;
   end if
   if $w \geq \ell$ **then**
    $\ell \leftarrow w$; $\delta \leftarrow v - u$;
   end if
  end for
  if $\delta > 0$ **then**
   $\pi_k^E[i] \leftarrow \ell$;
  else
   $\pi_k^E[i] \leftarrow \ell - \delta$;
  end if
  if $i < n-1$ **then**
   $H_{0,...,k} \leftarrow ISC(H_{0,...,k}, x, x[i..n-1], 1)$;
  end if
end for
return π_k^E;

Let $DP(x, y, k)$ denote the dynamic programming algorithm for computing the edit distance (at most k) between strings x and y. This algorithm requires time $\Theta(kn)$ [22]. Let D denote the resulting dynamic programming matrix of size $\Theta(kn)$. Further, let $GH(D)$ denote the function to extract $H_{0,...,k}$ from D, and let $ISC(H_{0,...,k}, x, y, \alpha)$ denote the incremental string comparison function that updates $H_{0,...,k}$ for x and $del(\alpha, y)$. We are now in a position to outline the computation of the prefix table under edit distance. For each position i, for all $1 \leq i < n$, we compute $H_{0,...,k}$ for x and $x[i..n-1]$. We then check the k-wave of the dynamic programming matrix to find the length ℓ of the longest prefix of x such that $x[i..i+j-1] \equiv_k^E x[0..\ell-1]$ for $\ell \geq k$ and $\ell - k \leq j \leq \ell + k$.

The k-wave is stored as a linked list of size $2k$ that specifies for each diagonal the lowest cell with value k. To find this longest prefix, we simply iterate through the linked list of the k-wave and keep track of the diagonal δ with the lowest cell on the k-wave. If a diagonal has no cell with value k then clearly that diagonal has reached the last row of the dynamic programming matrix. This procedure can be seen in algorithm k-PrefixTable-ED. Hence we obtain the following.

Theorem 12. *Given a string x of length n and an integer threshold $k < n$, algorithm k-PrefixTable-ED computes π_k^E in worst-case time and space $\Theta(kn)$.*

The conversion between π_k^E and β_k^E is performed in exactly the same way as for Hamming distance (algorithm k-BorderArray).

References

1. Abrahamson, K.: Generalized string matching. SIAM J. Comput. **16**(6), 1039–1051 (1987)
2. Amir, A., Lewenstein, M., Porat, E.: Faster algorithms for string matching with k mismatches. In: Proceedings of the Eleventh Annual ACM-SIAM Symposium on Discrete Algorithms (SODA 2000), pp. 794–803. Society for Industrial and Applied Mathematics, USA (2000)
3. Bland, W., Kucherov, G., Smyth, W.F.: Prefix table construction and conversion. In: Lecroq, T., Mouchard, L. (eds.) IWOCA 2013. LNCS, vol. 8288, pp. 41–53. Springer, Heidelberg (2013)
4. Crochemore, M., Hancart, C., Lecroq, T.: Algorithms on Strings. Cambridge University Press, New York (2007)
5. Dori, S., Landau, G.M.: Construction of Aho Corasick automaton in linear time for integer alphabets. Inf. Process. Lett. **98**(2), 66–72 (2006)
6. Fischer, J.: Inducing the LCP-array. In: Dehne, F., Iacono, J., Sack, J.-R. (eds.) WADS 2011. LNCS, vol. 6844, pp. 374–385. Springer, Heidelberg (2011)
7. Fischer, J., Heun, V.: Space-efficient preprocessing schemes for range minimum queries on static arrays. SIAM J. Comput. **40**(2), 465–492 (2011)
8. Fredriksson, K., Navarro, G.: Average-optimal single and multiple approximate string matching. J. Exp. Algorithmics **9**, 1–47 (2004). http://doi.acm.org/10.1145/1005813.1041513
9. Galil, Z., Giancarlo, R.: Improved string matching with k mismatches. ACM SIGACT News **17**(4), 52–54 (1986)
10. Hall, H.S., Knight, S.R.: Higher Algebra. MacMillan, London (1950)
11. Hsu, P.-H., Chen, K.-Y., Chao, K.-M.: Finding all approximate gapped palindromes. In: Dong, Y., Du, D.-Z., Ibarra, O. (eds.) ISAAC 2009. LNCS, vol. 5878, pp. 1084–1093. Springer, Heidelberg (2009). http://dx.doi.org/10.1007/3-540-12689-9_129
12. Ilie, L., Navarro, G., Tinta, L.: The longest common extension problem revisited and applications to approximate string searching. J. Discrete Algorithms **8**(4), 418–428 (2010)
13. Landau, G.M., Myers, E.W., Schmidt, J.P.: Incremental string comparison. SIAM J. Comput. **27**–**2**, 557–582 (1998)
14. Landau, G.M., Vishkin, U.: Efficient string matching in the presence of errors. In: IEEE (ed.) Proceedings of the 26th Annual Symposium on Foundations of Computer Science (FOCS 1985), USA, pp. 126–136. IEEE Computer Society (1985)
15. Levenshtein, V.I.: Binary codes capable of correcting deletions, insertions, and reversals. Technical report 8 (1966)
16. Main, M.G., Lorentz, R.J.: An $\mathcal{O}(n \log n)$ algorithm for finding all repetitions in a string. J. Algs **5**, 422–432 (1984)
17. Nong, G., Zhang, S., Chan, W.H.: Linear suffix array construction by almost pure induced-sorting. In: Proceedings of the 2009 Data Compression Conference, DCC 2009, pp. 193–202, IEEE Computer Society, Washington, DC (2009)

18. Pizza & Chili, April 2013. http://pizzachili.dcc.uchile.cl/
19. Smyth, B.: Computing Patterns in Strings. Pearson Addison-Wesley, London (2003)
20. Smyth, W.F., Wang, S.: New perspectives on the prefix array. In: Amir, A., Turpin, A., Moffat, A. (eds.) SPIRE 2008. LNCS, vol. 5280, pp. 133–143. Springer, Heidelberg (2008)
21. StringPedia, April 2013. http://stringpedia.bsmithers.co.uk
22. Ukkonen, E.: On approximate string matching. In: Karpinski, M. (ed.) Foundations of Computation Theory. Lecture Notes in Computer Science, vol. 158, pp. 487–495. Springer, Heidelberg (1983). http://dx.doi.org/10.1007/3-540-12689-9_129
23. Välimäki, N., Ladra, S., Mäkinen, V.: Approximate all-pairs suffix/prefix overlaps. In: Amir, A., Parida, L. (eds.) CPM 2010. LNCS, vol. 6129, pp. 76–87. Springer, Heidelberg (2010)
24. Wu, S., Manber, U.: Fast text searching: allowing errors. Commun. ACM **35**(10), 83–91 (1992)
25. Zhang, J., Kobert, K., Flouri, T., Stamatakis, A.: PEAR: a fast and accurate Illumina paired-end reAd mergeR. Bioinformatics **30**(5), 614–620 (2013)

Border Correlations, Lattices, and the Subgraph Component Polynomial

Francine Blanchet-Sadri[1]([⊠]), Michelle Cordier[2], and Rachel Kirsch[3]

[1] Department of Computer Science, University of North Carolina,
P.O. Box 26170, Greensboro, NC 27402–6170, USA
blanchet@uncg.edu
[2] Department of Mathematical Sciences, Kent State University,
P.O. Box 5190, Kent, OH 44242, USA
mcordie1@kent.edu
[3] Department of Mathematics, University of Nebraska-Lincoln,
203 Avery Hall, P.O. Box 880130, Lincoln, NE 68588–0130, USA
rkirsch2@math.unl.edu

Abstract. We consider the border sets of partial words and study the combinatorics of specific representations of them, called border correlations, which are binary vectors of same length indicating the borders. We characterize precisely which of these vectors are valid border correlations, and establish a one-to-one correspondence between the set of valid border correlations and the set of valid period correlations of a given length, the latter being ternary vectors representing the strong and strictly weak period sets. It turns out that the sets of all border correlations of a given length form distributive lattices under suitably defined partial orderings. We also investigate the population size, i.e., the number of partial words sharing a given border correlation, and obtain formulas to compute it. We do so using the subgraph component polynomial of an undirected graph, introduced recently by Tittmann et al. (European Journal of Combinatorics, 2011), which counts the number of connected components in vertex induced subgraphs.

1 Introduction

Borders and *periods* are two fundamental concepts of combinatorics on words that play an important role in several research areas including text compression, computational biology, string searching and pattern matching algorithms (see, e.g., [7]). It is well-known that these two word notions do not exist independently from each other. The length of the maximal border of a word is its length minus the length of its minimal period. Equivalently, it is unbordered if it has no proper period. Borders and periods are also well-studied concepts in combinatorics on partial words which allow positions to have don't care characters or holes (see,

This material is based upon work supported by the National Science Foundation under Grants DMS–0754154 and DMS–1060775. The Department of Defense is also gratefully acknowledged.

J. Kratochvíl et al. (Eds.): IWOCA 2014, LNCS 8986, pp. 62–73, 2015.
DOI: 10.1007/978-3-319-19315-1_6

e.g., [2]), as well as indeterminate strings which allow positions to have subsets of the alphabet (see, e.g., [12]).

The combinatorics of specific representations of the border sets and period sets of (partial) words of length n over a finite alphabet have been studied. Among them are the *period correlations*, which are n-bit vectors indicating the periods, and the *border correlations*, which are n-bit vectors indicating the presence of borders of a certain length. Guibas and Odlyzko [8] introduced the period correlations, so-called (auto)correlations, provided characterizations of them, asymptotic bounds on their number, and a recurrence for calculating the *population size* of a period correlation, that is, the number of words sharing a given period correlation. Rivals and Rahmann [11] showed that the set of all period correlations of words of a given length is a lattice under set inclusion, proposed the first efficient algorithm for enumerating them, and improved upon Guibas and Odlyzko's asymptotic lower bounds on their number. They also provided a new recurrence to compute the population size. Harju and Nowotka [9] studied *refined border correlations* which specify the lengths of all the words' bordered cyclic shifts' minimal borders. Extensions of these results to partial words appear in [4,5]. In particular, ternary vectors representing the strong and strictly weak periods of partial words were considered.

On the other hand, Tittmann et al. [13] introduced the *subgraph component polynomial* $Q(G; x, y)$ of an undirected graph G with n vertices as the bivariate generating function which counts the number of connected components in vertex induced subgraphs, i.e., $Q(G; x, y) = \sum_{j=0}^{n} \sum_{i=0}^{n} q_{ji}(G) x^j y^i$, where $q_{ji}(G)$ is the number of vertex induced subgraphs of G with exactly j vertices and i connected components. They showed that the power of the subgraph component polynomial for distinguishing graphs is quite different from the power of other graph polynomials that appear in the literature (see, e.g., [10] for more information on graph polynomials). Recently, Blanchet-Sadri et al. [3] showed the use of the subgraph component polynomial to count the number of primitive partial words of a given length over an alphabet of a fixed size, which leads to a method for enumerating such partial words.

In this paper, we establish connections between border correlations, lattices, and the subgraph component polynomial. We use the subgraph component polynomial to count the number of partial words of length n over a k-letter alphabet that have the same border correlation, or its population size. We associate an undirected graph $G_{\mathbf{bc}}$ with any border correlation $\mathbf{bc}$ of length n as follows: the vertices represent the positions $0, \ldots, n-1$ and the edges are the pairs $\{i, j\}$ that are $(n - \ell)$-apart, where ℓ indicates the length of a border. It turns out that the population size of $\mathbf{bc}$ can be expressed in terms of the $Q(G_{\mathbf{bc}}; 1, k)$'s.

The contents of our paper are as follows: In Sect. 2, we review some basic concepts on partial words such as borders, strong periods and weak periods, and discuss relationships between them. In Sect. 3, we first recall the fact that the set of period correlations of partial words over an arbitrary alphabet of cardinality at least two is the same as the set of period correlations of partial words over a binary alphabet. We then introduce border correlations and characterize precisely which vectors are valid border correlations. We establish a one-to-one correspondence between the set of valid border correlations and the set of valid period correlations of a given length giving an algorithm for generating the valid period correlations of a given length. We also prove that all valid border correlations correspond to partial words over the binary alphabet. In Sect. 4, we look at a more optimal way of counting valid colorings of undirected graphs by using the subgraph component polynomial and looking at different graph structures. In Sect. 5, we give formulas to calculate the population size of a given border correlation. One approach is based on the previous section and another approach is based on the fact that the sets of all border correlations of a given length form distributive lattices under suitably defined partial orderings. Finally in Sect. 6, we conclude with some open problems.

2   Preliminaries

Let Σ be a finite and non-empty set of characters, called an *alphabet*. Each element a of Σ is referred to as a *letter*, and a sequence of letters from Σ as a *word*, or *total word*, over Σ. A *partial word* over Σ is a sequence of characters from $\Sigma_\diamond = \Sigma \cup \{\diamond\}$, where $\diamond$, a new character which is not in Σ, is the "don't care" or "hole" character (it represents an undefined position). Note that a total word is a partial word with no holes. The *length* of a partial word w, denoted by $|w|$, is the number of characters in w. For example, if $w = abbac\diamond\diamond c$ then $|w| = 8$. The *empty word* is the word of length zero and we denote it by ε. The set of all words over Σ is denoted by Σ^*. We denote by Σ^n the set of all words of length n over Σ. We can similarly define $\Sigma_\diamond^*$ and $\Sigma_\diamond^n$ for partial words over Σ.

A partial word u is a *factor* of a partial word w if there exist (possibly empty) partial words x, y such that $w = xuy$. We say that u is a *prefix* of w if $x = \varepsilon$. Similarly, u is a *suffix* of w if $y = \varepsilon$. Starting numbering positions from 0, we denote the character in position i of w by $w[i]$ and the factor of w from position i to position j (inclusive) by $w[i..j]$ and from position i to position j (non-inclusive) by $w[i..j)$. We denote w concatenated with itself m times as w^m.

If u and v are partial words of equal length over Σ, then u is *contained* in v, denoted by $u \subset v$, if $u[i] = v[i]$ for all i such that $u[i] \in \Sigma$. Partial words u and v are *compatible*, denoted by $u \uparrow v$, if there exists a partial word w such that $u \subset w$ and $v \subset w$. A non-empty partial word w is *unbordered* if no non-empty partial words x_1, x_2, u, v exist such that $w = x_1 u = v x_2$ and $x_1 \uparrow x_2$. If such non-empty partial words exist, then x exists such that $x_1 \subset x$ and $x_2 \subset x$ and we call w *bordered* and x a *border* of w. It is easy to see that if w is unbordered and $w \subset w'$, then w' is unbordered as well. Note that there are two types of borders: writing $w = x_1 u = v x_2$, where $x_1 \subset x$ and $x_2 \subset x$, we say that x is a *simple* border if $|x| \leq |u|$, and a *non-simple* border otherwise. For example, $a\diamond\diamond ab$ is bordered with the simple border ab and non-simple border aab, the first one being minimal, while $ab\diamond c$ is unbordered.

A *strong period* of a partial word w over Σ is a positive integer p such that $w[i] = w[j]$ whenever $w[i], w[j] \in \Sigma$ and $i \equiv j \pmod{p}$; w is called strongly

p-periodic. A *weak period* of w is a positive integer p such that $w[i] = w[i + p]$ whenever $w[i], w[i + p] \in \Sigma$; w is weakly p-periodic. A *strictly* weak period is a weak period that is not a strong period. The set of all strong periods (respectively, weak periods) of w is denoted by $\mathbf{SP}(w)$ (respectively, $\mathbf{WP}(w)$). The following two lemmas are useful for our purposes. The first one states that a weak period is a strong period if and only if all of its multiples are also weak periods, and the second one establishes relationships between borders, and strong and weak periods. Note that $aaa\diamond aba$ has a border of length 5 but is not strongly 2-periodic, hence the bound on ℓ in Lemma 2(b).

Lemma 1 ([5]). *Let w be a partial word and let $p \in \mathbf{WP}(w)$. Then $p \in \mathbf{SP}(w)$ if and only if for all $0 < i \leq \frac{|w|}{p}$, $ip \in \mathbf{WP}(w)$.*

Lemma 2 ([1]). *Let w be a partial word.*

(a) If $0 < \ell < |w|$, then w has a border of length ℓ if and only if w has weak period $|w| - \ell$.

(b) If $0 < \ell \leq \lfloor \frac{|w|}{2} \rfloor$, then w has a border of length ℓ if and only if w has strong period $|w| - \ell$.

3   Period and Border Correlations

Period correlations are defined according to the following definition.

Definition 1 ([5]). *The period correlation of a partial word w of length n is the ternary vector $\mathbf{pc}_w$ of length n such that $\mathbf{pc}_w[0] = 1$ and for $1 \leq i < n$:*

$$\mathbf{pc}_w[i] = \begin{cases} 1 & \text{if } i \in \mathbf{SP}(w), \\ 2 & \text{if } i \in \mathbf{WP}(w) \setminus \mathbf{SP}(w), \\ 0 & \text{otherwise.} \end{cases}$$

Considering the partial word $abaca\diamond\diamond acaba$ which has strong periods 9 and 11 (and 12) and strictly weak period 5, its period correlation vector is 100002000101. For any partial word w, note that both w and its reversal share the same period correlation. We say that a ternary vector of length n is a *valid* period correlation if it is the period correlation of some partial word of length n. The following theorem implies that the sets of all valid period correlations are independent of the alphabet size.

Theorem 1 ([5]). *If w is a partial word over an alphabet Σ, then there exists a binary partial word w' of length $|w|$ such that $\mathbf{SP}(w') = \mathbf{SP}(w)$ and $\mathbf{WP}(w') = \mathbf{WP}(w)$.*

Border correlations are now defined according to the following definition.

Definition 2. *The border correlation of a partial word w of length n is the binary vector $\mathbf{bc}_w$ of length n such that $\mathbf{bc}_w[n-1] = 1$ and for $0 \le i < n-1$:*

$$\mathbf{bc}_w[i] = \begin{cases} 1 & \text{if } w \text{ has a border of length } i+1, \\ 0 & \text{otherwise.} \end{cases}$$

Considering again the partial word $abaca\diamond\diamond acaba$, its border correlation vector is 101000100001. The following theorem gives a characterization of the possible border length sets of partial words of arbitrary length.

Theorem 2. *Given a binary number $q = q_0 q_1 \cdots q_{n-1}$ with $q_{n-1} = 1$, the binary partial word $w = w[0..n-1]$ defined by $w[n-1] = b$ and for $0 \le i < n-1$,*

$$w[i] = \begin{cases} \diamond \text{ if } q_i = 1, \\ a \text{ otherwise} \end{cases}$$

satisfies the equation $\mathbf{bc}_w = q$. In other words, every binary number that ends in 1 is a valid border correlation.

Given the binary number 10101, the previous theorem builds the partial word $\diamond a \diamond ab$ having 10101 as its border correlation.

Theorem 3. *There is a one-to-one correspondence between the set of valid border correlations and the set of valid period correlations of a given length.*

The correspondence of Theorem 3 leads to an algorithm that can generate a list of all the valid border correlations of a given length, and their corresponding period correlations. For example, suppose we want to find the period correlation $\mathbf{pc}$ corresponding to the border correlation 1111001111. We first assign $\mathbf{pc}[0] = 1$ by definition. The border correlation gives the border set $\{1, 2, 3, 4, 7, 8, 9\}$, which corresponds to the weak period set $\{9, 8, 7, 6, 3, 2, 1\}$. Thus, 4 and 5 are not weak periods, so $\mathbf{pc}[4]$ and $\mathbf{pc}[5]$ are zeros. Then we check whether each weak period's multiples are also weak periods. All of the multiples of 9, 8, 7, 6, and 3 are also in the weak period set, so $\mathbf{pc}[3]$, $\mathbf{pc}[6]$, $\mathbf{pc}[7]$, $\mathbf{pc}[8]$, and $\mathbf{pc}[9]$ are all ones. The weak period set does not contain all of the multiples of 2 or of 1, so $\mathbf{pc}[1]$ and $\mathbf{pc}[2]$ are twos. Now $\mathbf{pc}$ has been determined to be 1221001111.

Theorem 4. *If w is a partial word over an alphabet Σ, then there exists a binary partial word w' such that $\mathbf{bc}_{w'} = \mathbf{bc}_w$.*

4   Valid Colorings of Undirected Graphs Using the Subgraph Component Polynomial

We adopt the notation Σ_y for an arbitrary y-letter alphabet.

Definition 3. *Let $G = (V, E)$ be an undirected graph such that $V = [0..n-1]$:*

– A valid coloring *over* Σ_y *of* G *is one in which the colors of adjacent vertices are the same or one is the hole color. Here, the colors are the* y *letters of the alphabet* Σ_y *as well as the hole color,* $\diamond$. *The number of* valid colorings *over* Σ_y *of* G, *denoted by* $\text{VC}_y(G)$, *is the number of partial words* w *of length* n *over* Σ_y *such that if* $\{i, j\} \in E$, *then* $w[i] \uparrow w[j]$.

– *If* s *is a sequence of pairs of the form* (i, c), *where* $i \in V$ *and* $c \in \Sigma_y \cup \{\diamond\}$, *then the number of* s-*valid colorings over* Σ_y *of* G, *denoted by* $s\text{-}\text{VC}_y(G)$, *is the number of valid colorings* w *over* Σ_y *of* G *subject to the restrictions imposed by* s, *that is, if* (i, c) *is in* s, *then* $w[i] = c$. *If* s *consists of a singleton* (i, c), *then we simply write* $(i, c)\text{-}\text{VC}_y(G)$.

Definition 4. *Let* $G = (V, E)$ *be an undirected graph such that* $V = [0..n - 1]$. *The* connected component vector *of* G, *denoted by* $\mathbf{cc}_G$, *is a vector of length* $n+1$ *such that for* $0 \le i \le n$, $\mathbf{cc}_G[i]$ *is the number of ways to remove a set of vertices from* G *such that the resulting induced subgraph has* i *connected components.*

For example, consider the graph $G = (\{0, 1, 2, 3, 4\}, \{\{0, 2\}, \{1, 3\}, \{2, 4\}\})$. Then $\text{VC}_2(G) = 119$, one of the valid colorings over $\{a, b\}$ of G being $aa\diamond ab$ (we show later how to calculate 119). The connected component vector of G is $19(19)300$. For example, the three ways to remove a set of vertices from G to get induced subgraphs with three connected components are to remove the sets $\{2\}$, $\{1, 2\}$, and $\{2, 3\}$, so $\mathbf{cc}_G[3] = 3$.

Let $P(G; y) = \sum_{i=0}^{n} y^i \mathbf{cc}_G[i]$ be the generating function for the connected component vector $\mathbf{cc}_G$ of an undirected graph G on n vertices. We can obtain this by plugging in $x = 1$ into the *subgraph component polynomial* $Q(G; x, y)$, as defined by Tittmann et al. [13]. The next theorem shows that there are $P(G; y)$ partial words of length n over Σ_y that satisfy the required compatibilities to be valid colorings over Σ_y of an undirected graph G on n vertices.

Theorem 5. *If* $G = (V, E)$ *is an undirected graph with* $V = [0..n - 1]$, *then*
$$\text{VC}_y(G) = P(G; y) = \sum_{i=0}^{n} y^i \mathbf{cc}_G[i].$$

Results on subgraph component polynomials facilitate the computation of connected component vectors.

Theorem 6 ([13]). *If* $G = G_1 \sqcup G_2 \sqcup \cdots \sqcup G_n$ *is the disjoint union of the graphs* $G_1, G_2, \ldots, G_n$, *then* $P(G; y) = \prod_{j=1}^{n} P(G_j; y)$.

We adopt the following notations for graphs on n vertices: E_n is the empty graph, K_n is the complete graph, P_n is a path graph, and C_n is the cycle graph. A wheel graph (respectively, broken wheel graph) of order n is a graph consisting of a cycle graph on n vertices and an additional vertex, called a hub, that is adjacent to all (respectively, at least one) of the vertices in the cycle graph. We denote a wheel graph of order n by W_n and a broken wheel graph of order n by W_s, where s is a sequence of numbers whose sum is n. The first number in s is a number of consecutive hub-adjacent vertices, the second number in s is the number of

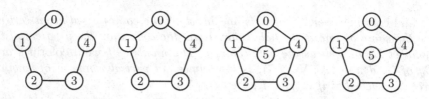

Fig. 1. From left to right: the path graph P_5, the cycle graph C_5, the broken wheel graph $W_{3,2}$, and the broken wheel graph $W_{1,2,1,1}$. They are associated with the border correlations 00011, 10011, 110011, and 010011 respectively (see Sect. 5).

consecutive non-hub-adjacent vertices following the hub-adjacent ones, and so on around the cycle. See Fig. 1 for examples.

Although the subgraph component polynomials of P_n, C_n, and W_n have been studied and formulas have been given, we give new formulas.

Proposition 1 ([13]). *The equalities (a) $P(E_n; y) = (1+y)^n$ and (b) $P(K_n; y) = 1 + (2^n - 1)y$ hold.*

Proposition 2. *The following equality holds:*

$$P(P_n; y) = 1 + \sum_{h=0}^{n-1} \sum_{i=0}^{h} \binom{n-h-1}{i} \binom{h+1}{h-i} y^{i+1}.$$

Moreover, let c be one of y non-hole colors, let $s = \langle (0,c), (n-1,c) \rangle$, and let $s' = \langle (0,c) \rangle$. Then the following hold:

(a) $s\text{-}VC_y(P_n) = 1 + \sum_{h=1}^{n-2} \sum_{i=1}^{h} \binom{n-h-1}{i} \binom{h-1}{h-i} y^{i-1}$,

(b) $s'\text{-}VC_y(P_n) = 1 + \sum_{h=1}^{n-1} \sum_{i=1}^{h} \binom{n-h-1}{i} \binom{h}{h-i} y^{i}$.

Proposition 3. *If c is one of y non-hole colors, then the following hold:*

(a) $P(C_n; y) = 2P(P_{n-1}; y) - P(P_{n-2}; y) + (s\text{-}VC_y(P_n))y$, where $s = \langle (0,c), (n-1,c) \rangle$,

(b) $P(W_n; y) = P(C_n; y) + 2^n y$,

(c) $P(W_{1,n-1}; y) = P(C_n; y) + (s\text{-}VC_y(P_{n+1}) + P(P_{n-1}; y))y$, where $s = \langle (0,c), (n,c) \rangle$,

(d) If $n - r > 1$, then $P(W_{n-r,r}; y)$ is

$$P(C_n; y) + (s\text{-}VC_y(P_{r+2}) + 2(s'\text{-}VC_y(P_{r+1})) + P(P_r; y))y2^{n-r-2},$$

where $s = \langle (0,c), (r+1,c) \rangle$ and $s' = \langle (0,c) \rangle$.

The graph G may not have any of the above forms, but its complement $\overline{G}$ may have.

Theorem 7. *If $G = (V, E)$ is an undirected graph with $V = [0..n-1]$, then $VC_y(G) = P(G; y) = 1 + (2^n - 1)y + \overline{P}(\overline{G}; y)$, where $1 + (2^n - 1)y$ counts the number of valid unary colorings of G, i.e., those with at most one non-hole color, and $\overline{P}(\overline{G}; y)$ counts the number of valid non-unary colorings of G, i.e., those with at least two and at most y non-hole colors.*

We give formulas when $\overline{G} = P_n$ or $\overline{G} = C_n$.

Proposition 4. *Let $G = (V, E)$ be an undirected graph such that $V = [0..n-1]$. If $\overline{G} = P_n$, then $\overline{P}(\overline{G}; y) = y(y - 1)(2n - 3)$.*

Proposition 5. *Let $G = (V, E)$ be an undirected graph such that $V = [0..n-1]$. If $\overline{G} = C_n$, then*

$$\overline{P}(\overline{G}; y) = \begin{cases} y(y-1)(2n+y-2) & \text{if } n = 3 \text{ and } y \geq 3, \\ y(y-1)(2n+1) & \text{if } n = 4, \\ 2y(y-1)n & \text{otherwise.} \end{cases}$$

If a graph G has only one connected component and $\overline{G}$'s connected components are only path graphs, cycle graphs, and vertices of degree zero, we can calculate the number of valid non-unary colorings of G by looking at each connected component of $\overline{G}$, ignoring the vertices of degree zero, and then add the results because the non-unary colorings of these components are mutually exclusive. Considering the graph $G = (\{0, 1, 2, 3, 4\}, \{\{0, 2\}, \{1, 3\}, \{2, 4\}\})$ again, we calculate $\mathrm{VC}_y(G)$ as follows. First, consider P_2 on vertices 1 and 3, and P_3 on vertices 0, 2 and 4. We have $P(P_2; y) = 1 + 3y$ and $P(P_3; y) = 1 + 6y + y^2$ by using Proposition 2. Applying Theorems 5 and 6, we obtain $\mathrm{VC}_y(G) = P(P_2 \sqcup P_3; y) = P(P_2; y) P(P_3; y) = 1 + 9y + 19y^2 + 3y^3$, as is also computed from the connected component vector of G. If $y = 2$, we get $\mathrm{VC}_2(G) = 119$. On the other hand, $\mathrm{VC}_y(\overline{P_2 \sqcup P_3}) = 1 + (2^5 - 1)y + \overline{P}(P_2 \sqcup P_3; y) = 1 + (2^5 - 1)y + \overline{P}(P_2; y) + \overline{P}(P_3; y) = 1 + 27y + 4y^2$ by using Theorem 7 and Proposition 4.

We can modify Definitions 3 and 4 to count valid colorings when restricting the number of holes. For an undirected graph $G = (V, E)$ such that $V = [0..n-1]$ and an integer h in $[0..n]$, a h-*valid coloring* over Σ_y of G is one in which no vertices colored with two different colors from Σ_y are adjacent and in which exactly h vertices are colored with the hole color. The number of h-*valid colorings* over Σ_y of G, denoted by $\mathrm{VC}_{h,y}(G)$, is the number of partial words w of length n with h holes over Σ_y such that if $\{i, j\} \in E$, then $w[i] \uparrow w[j]$. The $(n - h)$-*connected component vector* of G, denoted by $\mathbf{cc}_{n-h,G}$, is a vector of length $n + 1$ such that for $0 \leq i \leq n$, $\mathbf{cc}_{n-h,G}[i]$ is the number of ways to remove a set of exactly h vertices from G such that the resulting induced subgraph has i connected components. Referring to the subgraph component polynomial $Q(G; x, y)$, we use the notation $Q_{n-h}(G; x, y)$ when we restrict to induced subgraphs with exactly $n - h$ vertices, and similarly for $P_{n-h}(G; y)$.

Theorem 8. *If $G = (V, E)$ is an undirected graph with $V = [0..n-1]$ and h is an integer in $[0..n]$, then*

$$\mathrm{VC}_{h,y}(G) = P_{n-h}(G; y) = \sum_{i=0}^{n} y^i \, \mathbf{cc}_{n-h,G}[i].$$

5   Population Size

We derive formulas to compute the *population size* over Σ_k, an arbitrary k-letter alphabet, of a given border correlation $\mathbf{bc}$, i.e., the number of partial words over Σ_k sharing $\mathbf{bc}$ as their border correlation, which we denote by $\mathrm{PS}_k(\mathbf{bc})$. Given m undirected graphs $G_1 = (V, E_1), \ldots, G_m = (V, E_m)$ with the same vertex set V, we let $\cup(G_1, \ldots, G_m) = (V, E_1 \cup \cdots \cup E_m)$.

We first describe a graphical approach based on the subgraph component polynomial to calculate the population size. We define two types of graph collections that are associated with a border correlation.

Definition 5. *Let* $\mathbf{bc} = \mathbf{bc}[0..n)$ *be a border correlation. For each* $\ell \in [1..n)$, *associate an undirected graph whose set of vertices consists of* $[0..n-1]$ *and whose set of edges consists of the pairs* $\{i, j\}$ *such that* $|i - j| = n - \ell$.

– *Let* $\mathbf{G}_C(\mathbf{bc})$ *be the collection of all graphs associated with integers* ℓ *such that* $\mathbf{bc}[\ell - 1] = 1$. *Each graph in this collection is called a* compatibility graph. *If* $\mathbf{G}_C(\mathbf{bc}) = \{G_1, \ldots, G_m\}$, *then let* $G_{\mathbf{bc}}$ *be* $\cup(G_1, \ldots, G_m)$.
– *Let* $\mathbf{G}_I(\mathbf{bc})$ *be the collection of all graphs associated with integers* ℓ *such that* $\mathbf{bc}[\ell - 1] = 0$. *Each graph in this collection is called an* incompatibility graph. *If* $\mathbf{G}_I(\mathbf{bc}) = \{G_1, \ldots, G_m\}$, *then let* $\overline{G}_{\mathbf{bc}}$ *be* $\cup(G_1, \ldots, G_m)$ *(note that* $\overline{G}_{\mathbf{bc}}$ *is the complement graph of* $G_{\mathbf{bc}}$*).*

Note that the graph associated with ℓ records the pairs of positions that are $(n-\ell)$-apart. A compatibility graph associated with ℓ encodes the compatibilities a partial word w must satisfy to have a border of length ℓ, i.e., it must satisfy $w[0..\ell - 1] \uparrow w[n - \ell..n - 1]$, while an incompatibility graph associated with ℓ has a set of edges such that at least one edge must correspond to an incompatibility in order for w not to have a border of length ℓ. Figure 2 gives an example.

The graphs $G_{\mathbf{bc}}$ and $\overline{G}_{\mathbf{bc}}$ may have the forms we discussed in Sect. 5 (see Fig. 1). Letting $\overline{0} = 1$ and $\overline{1} = 0$, note that if $\mathbf{bc} = \mathbf{bc}[0..n)$ is a border correlation and $\mathbf{bc}' = \mathbf{bc}'[0..n)$ is the border correlation such that $\mathbf{bc}'[i] = \overline{\mathbf{bc}[i]}$ for $0 \le i < n - 1$, then $\overline{G}_{\mathbf{bc}} = G_{\mathbf{bc}'}$. Also note that the graph $G_{0001000100010001000010001}$ is the disjoint union of four K_6's. This observation generalizes to the following.

Proposition 6. *Let* $\mathbf{bc} = [0..n)$ *be a border correlation and let* p *be a positive integer. Let* $\mathbf{bc}' = [0..pn)$ *be the border correlation such that* $\mathbf{bc}'[j] = \mathbf{bc}[i]$ *for*

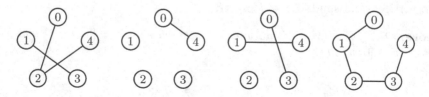

Fig. 2. First graph is the compatibility graph of border correlation 00101 that corresponds to a border of length 3. Three other graphs are the incompatibility graphs of border correlation 00101 that correspond to a border of length 1, 2, and 4, respectively.

$j = ip + p - 1, 0 \leq i < n$ and $\mathbf{bc}'[j] = 0$ otherwise. Then $G_{\mathbf{bc}'}$ is the disjoint union of p graphs isomorphic to $G_{\mathbf{bc}}$ and $P(G_{\mathbf{bc}'}; y) = (P(G_{\mathbf{bc}}; y))^p$.

We now have the necessary definitions to count the number of partial words sharing a given border correlation.

Theorem 9. *Let* $\mathbf{bc}$ *be a border correlation. Let* $\mathbf{G}_I(\mathbf{bc}) = \{G_1, G_2, \ldots, G_m\}$ *and let* $G'_i = \cup(G_{\mathbf{bc}}, G_i)$ *for* $1 \leq i \leq m$. *Then,*

$$PS_k(\mathbf{bc}) = VC_k(G_{\mathbf{bc}}) + \sum_{j=1}^{m}(-1)^j \sum_{\{i_1,\ldots,i_j\}} VC_k(\cup(G'_{i_1}, \ldots, G'_{i_j})),$$

where $\{i_1, \ldots, i_j\}$ *is a subset of j distinct elements of* $\{1, \ldots, m\}$.

To illustrate the theorem, consider the border correlation 00101 of Fig. 2. Using our earlier computations, $VC_k(G_{00101}) = 1 + 9k + 19k^2 + 3k^3$. We obtain

$$\begin{aligned}
VC_k(G'_1) &= VC_k(P_2 \sqcup C_3) = 1 + 10k + 21k^2 \\
VC_k(G'_2) &= VC_k(\overline{C_5}) &= 1 + 21k + 10k^2 \\
VC_k(G'_3) &= VC_k(\overline{P_4}) &= 1 + 26k + 5k^2 \\
VC_k(\cup(G'_1, G'_2)) &= VC_k(\overline{P_5}) &= 1 + 24k + 7k^2 \\
VC_k(\cup(G'_1, G'_3)) &= VC_k(\overline{P_2 \sqcup P_2}) = 1 + 29k + 2k^2 \\
VC_k(\cup(G'_2, G'_3)) &= VC_k(\overline{P_2}) &= 1 + 30k + k^2 \\
VC_k(\cup(G'_1, G'_2, G'_3)) &= VC_k(K_5) &= 1 + 31k.
\end{aligned}$$

Thus, $PS_k(00101) = 4k - 7k^2 + 3k^3$.

We next describe a lattice approach to calculate the population size of a given border correlation and its corresponding period correlation. We denote the set of all partial word border correlations of length n by $\mathbf{BC}_n$. For $\mathbf{bc} \in \mathbf{BC}_n$, define $\mathcal{B}(\mathbf{bc}) = \{i \mid 0 \leq i < n$ and $\mathbf{bc}[i] = 1\}$, and for $\mathbf{bc}, \mathbf{bc}' \in \mathbf{BC}_n$, define $\mathbf{bc} \leq \mathbf{bc}'$ if $\mathcal{B}(\mathbf{bc}) \subseteq \mathcal{B}(\mathbf{bc}')$. We use the symbolism $\mathbf{bc} < \mathbf{bc}'$ to denote $\mathbf{bc} \leq \mathbf{bc}'$ and $\mathbf{bc} \neq \mathbf{bc}'$. Referring to Theorem 2, the pair $(\mathbf{BC}_n, \leq)$ is a distributive lattice.

Let $\mathbf{bc}$ be a border correlation. A partial word w satisfying $\mathbf{bc}_w = \mathbf{bc}$ corresponds to a valid coloring of the vertices of $G_{\mathbf{bc}}$. Each valid coloring of $G_{\mathbf{bc}}$ corresponds to a partial word having a border correlation $\mathbf{bc}'$ such that $\mathbf{bc} \leq \mathbf{bc}'$. The number of valid colorings of $G_{\mathbf{bc}}$ is then the sum of the population sizes of all border correlations $\mathbf{bc}'$ such that $\mathbf{bc} \leq \mathbf{bc}'$. To find the population size of $\mathbf{bc}$, we subtract the population sizes of all border correlations $\mathbf{bc}'$ such that $\mathbf{bc} < \mathbf{bc}'$ from the number of valid colorings of $G_{\mathbf{bc}}$.

Theorem 10. *If* $\mathbf{bc} \in \mathbf{BC}_n$, *then*

$$\begin{aligned}
PS_k(\mathbf{bc}) &= VC_k(G_{\mathbf{bc}}) - \textstyle\sum_{\mathbf{bc}' \in \mathbf{BC}_n, \mathbf{bc} < \mathbf{bc}'} PS_k(\mathbf{bc}'), \\
&= VC_k(\overline{G}_{\mathbf{bc}}) - \textstyle\sum_{\mathbf{bc}' \in \mathbf{BC}_n, \mathbf{bc} < \mathbf{bc}', \mathbf{bc}' \neq 1^n} PS_k(\mathbf{bc}').
\end{aligned}$$

Figure 3 illustrates the lattice $(\mathbf{BC}_5, \leq)$ along with the population size of each of its border correlations, calculated over a binary alphabet. For example, $\overline{G}_{01011}$ contains two connected components, P_2 and C_3. Thus, its number of valid

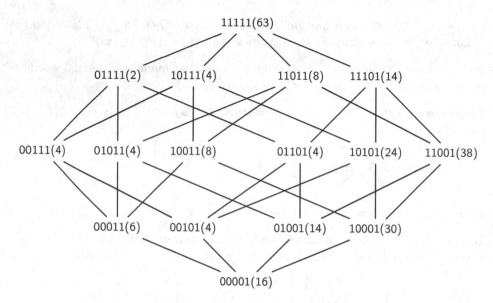

Fig. 3. The distributive lattice $(\mathbf{BC}_5, \leq)$ along with the population size over the binary alphabet of each of its border correlations (population size written in parentheses).

colorings with at most two non-hole colors is $1 + (2^5 - 1)2 + \overline{P}(P_2; 2) + \overline{P}(C_3; 2) = 63 + 2 + 12$. To find the population size of 01011, we must subtract the population sizes of the border correlations $\mathbf{bc}'$ such that $01011 < \mathbf{bc}'$. Thus, we must subtract $\mathrm{PS}_2(01111) + \mathrm{PS}_2(11011) + \mathrm{PS}_2(11111) = 2 + 8 + 63$, so $\mathrm{PS}_2(01011) = 4$.

We can also calculate the population size when restricting the number of holes. Given a border correlation $\mathbf{bc}$ of length n, denote by $\mathrm{PS}_{h,k}(\mathbf{bc})$ the number of partial words of length n with h holes over Σ_k sharing $\mathbf{bc}$ as their border correlation. Referring to Theorems 8 and 9, we obtain the following.

Theorem 11. *Let* $\mathbf{bc} = \mathbf{bc}[0..n)$ *be a border correlation,* h *be an integer in* $[0..n]$, $\mathbf{G}_I(\mathbf{bc}) = \{G_1, G_2, \ldots, G_m\}$, *and* $G_i' = \cup(G_{\mathbf{bc}}, G_i)$ *for* $1 \leq i \leq m$. *Then,*

$$\mathrm{PS}_{h,k}(\mathbf{bc}) = \mathrm{VC}_{h,k}(G_{\mathbf{bc}}) + \sum_{j=1}^{m} (-1)^j \sum_{\{i_1, \ldots, i_j\}} \mathrm{VC}_{h,k}(\cup(G_{i_1}', \ldots, G_{i_j}')),$$

where $\{i_1, \ldots, i_j\}$ *is a subset of* j *distinct elements of* $\{1, \ldots, m\}$.

Note that the population size of the border correlation $0^{n-1}1$, numbering unbordered partial words, can be computed using our results.

6   Conclusion

Our method for computing population sizes of border correlations also applies to computing population sizes of period correlations (see Theorem 3). It also applies to computing population sizes of refined border correlations. It would be worthwhile to find faster methods of computing the population size of a border correlation or to find efficient ways to determine the connected component vector.

Other recent papers have established connections between border/prefix arrays and undirected graphs (see, e.g., [6]). Our method has the advantage that it also applies to indeterminate strings.

There are several open problems involving the correlations that are binary vectors indicating only the strong period sets of partial words w for which $\mathbf{SP}(w) = \mathbf{WP}(w)$: listing the valid correlations, finding the number of valid correlations, and computing their population sizes.

References

1. Allen, E., Blanchet-Sadri, F., Byrum, C., Cucuringu, M., Mercaş, R.: Counting bordered partial words by critical positions. Electron. J. Comb. **18**, P138 (2011)
2. Blanchet-Sadri, F.: Algorithmic Combinatorics on Partial Words. Chapman & Hall/CRC Press, Boca Raton (2008)
3. Blanchet-Sadri, F., Bodnar, M., Fox, N., Hidakatsu, J.: A graph polynomial approach to primitivity. In: Dediu, A.-H., Martín-Vide, C., Truthe, B. (eds.) LATA 2013. LNCS, vol. 7810, pp. 153–164. Springer, Heidelberg (2013)
4. Blanchet-Sadri, F., Clader, E., Simpson, O.: Border correlations of partial words. Theory Comput. Syst. **47**, 179–195 (2010)
5. Blanchet-Sadri, F., Fowler, J., Gafni, J.D., Wilson, K.H.: Combinatorics on partial word correlations. J. Comb. Theory Ser. A **117**, 607–624 (2010)
6. Christodoulakis, M., Ryan, P.J., Smyth, W.F., Wang, S.: Indeterminate strings, prefix arrays and undirected graphs, 12 Jun 2014. arXiv:1406.3289v1 [cs.DM]
7. Crochemore, M., Hancart, C., Lecroq, T.: Algorithms on Strings. Cambridge University Press, Cambridge (2007)
8. Guibas, L.J., Odlyzko, A.M.: Periods in strings. J. Comb. Theory Ser. A **30**, 19–42 (1981)
9. Harju, T., Nowotka, D.: Border correlation of binary words. J. Comb.Theory Ser. A **108**, 331–341 (2004)
10. Makowsky, J.A.: From a zoo to a zoology: towards a general study of graph polynomials. Theory Comput. Syst. **43**, 542–562 (2008)
11. Rivals, E., Rahmann, S.: Combinatorics of periods in strings. J. Comb. Theory Ser. A **104**, 95–113 (2003)
12. Smyth, W.F.: Computing Patterns in Strings. Pearson Addison-Wesley, London (2003)
13. Tittmann, P., Averbouch, I., Makowsky, J.: The enumeration of vertex induced subgraphs with respect to number of components. Eur. J. Comb. **32**, 954–974 (2010)

Computing Minimum Length Representations
of Sets of Words of Uniform Length

Francine Blanchet-Sadri[1]([✉]) and Andrew Lohr[2]

[1] Department of Computer Science, University of North Carolina,
P.O. Box 26170, Greensboro, NC 27402-6170, USA
blanchet@uncg.edu
[2] Department of Mathematics, Rutgers University,
110 Frelinghuysen Rd, Piscataway, NJ 08854-8019, USA

Abstract. Motivated by text compression, the problem of representing sets of words of uniform length by partial words, i.e., sequences that may have some wildcard characters or holes, was recently considered and shown to be in $\mathcal{P}$. Polynomial-time algorithms that construct representations were described using graph theoretical approaches. As more holes are allowed, representations shrink, and if representation is given, the set can be reconstructed. We further study this problem by determining, for a binary alphabet, the largest possible value of the size of a set of partial words that is important in deciding the representability of a given set S of words of uniform length. This largest value, surprisingly, is $\Sigma_{i=0}^{|S|-1} 2^{\chi(i)}$ where $\chi(i)$ is the number of ones in the binary representation of i, a well-studied digital sum, and it is achieved when the cardinality of S is a power of two. We show that circular representability is in $\mathcal{P}$ and that unlike non-circular representability, it is easy to decide. We also consider the problem of computing minimum length representation (circular) total words, those without holes, and reduce it to a cost/flow network problem.

1 Introduction

A sequence over an alphabet Σ *represents* Σ^n, the set of all words of length n over Σ, if each of the elements in Σ^n appears in it. For example, 1101000111 represents all the eight words of length 3 over the binary alphabet $\{0, 1\}$. Such sequences of minimum length are the De Bruijn sequences and have found a number of important applications such as modern public-key cryptographic schemes [9], pseudo-random number generation [10], and non-linear shift registers [5].

In some applications however, such as text compression, it is desirable to consider sequences that represent only a subset S of Σ^n (each of the words in S, and only those words, appear in the sequence). *Partial words* over Σ become useful in such applications. They are sequences from $\Sigma_\diamond = \Sigma \cup \{\diamond\}$, where $\diamond \notin \Sigma$ is the hole character *compatible* with each letter in Σ. *Total words* are sequences without holes. The partial word $0\diamond0\diamond$ with two holes represents the

This material is based upon work supported by the National Science Foundation under Grant No. DMS–1060775.

© Springer International Publishing Switzerland 2015
J. Kratochvíl et al. (Eds.): IWOCA 2014, LNCS 8986, pp. 74–85, 2015.
DOI: 10.1007/978-3-319-19315-1_7

five words $000, 001, 010, 100, 101$ (we fill in the two $\diamond$'s with letters from $\{0, 1\}$). We say that a partial word is a *representation* for its set of length-n *subwords* (as defined in Sect. 2 and a partial word with exactly h holes, where $h \geq 0$, is a *h-representation* for that set. A set S of words of uniform length, i.e., a subset of Σ^n for some Σ and n, is *representable* (resp., *h-representable*) if there exists a representation (resp., h-representation) word for S. We can say that $S = \{000, 001, 010, 100, 101\}$ is 2-represented by $0\diamond0\diamond$ and that $0\diamond0\diamond$ is a *minimum length representation* for S (no other representation has shorter length).

Why do we consider partial words? First, they can be used for the *compression of representations*, e.g., the set $\{000, 001, 010, 100, 101\}$ is representable by the partial words 00010100, $\diamond00101$, and $0\diamond0\diamond$. As more holes are allowed, representations shrink, and if representation is given, the set can be reconstructed. Second, they can be used for the *representation of non-0-representable sets*, e.g., $\{001, 010, 011\}$ has no 0-representation. However, it is 1-representable by $001\diamond$.

Let REP (resp., h-REP) be the problem of deciding whether a given subset S of Σ^n is representable (resp., h-representable). Using a decomposition of the Rauzy graph of S into subgraphs, Blanchet-Sadri and Simmons [3] showed that h-REP is in $\mathcal{P}$ and Blanchet-Sadri and Munteanu [2] showed that REP is in $\mathcal{P}$. They provided polynomial-time algorithms that construct representations (resp., h-representations) when S is representable (resp., h-representable). However, REP and h-REP are not easy to decide (the actual exponent grows quickly with h). Variations of these problems have previously been studied under the name "shortest common superstring", e.g., Gallant, Maier, and Storer [6] proved that given a set S of words and an integer ℓ, whether there exists a word of length at most ℓ that contains as factors the words in S (and maybe some words not in S) is $\mathcal{NP}$-complete.

We further study these concepts of representability and variations on them. In Sect. 2, we recall some basic concepts on partial words, and then discuss the Rauzy graph associated with a set S of words of uniform length n and its relation to representability of S. In Sect. 3, we consider Comp(S), the set of partial words all of whose completions lie in S (a completion is a word obtained by filling in the holes). This construction appears in deciding representability because every representation partial word for S must have its length-n factors in Comp(S) [2,3]. For the binary alphabet we compute the largest possible value of $|\text{Comp}(S)|$, that is, $\Sigma_{i=0}^{|S|-1} 2^{\chi(i)}$, where $\chi(i)$ is the number of ones in i's binary representation, a well-studied digital sum. Though the exact formula is very complicated, it achieves the bound of $|S|^{\log_2 3}$ when $|S|$ is a power of two. In Section (Rauzy Graphs) we show that *circular* representability, CREP, is in $\mathcal{P}$ (as discussed above, *non-circular* representability, REP, can be tested in polynomial time). Here we show that unlike non-circular representability, any set that can be circularly represented by a word with a single hole can be circularly represented by a word with any number of holes and also that every set circularly representable by a partial word is circularly representable by a total word. This leads to CREP being easy to decide. In Sect. 5, we consider the problem of computing minimal representation (circular) total words. We reduce it to a cost/flow network problem that can be done in $O(|S|^2 \log |S|^2)$ time for the circular case. Finally in Sect. 6, we conclude with some open problems for future work.

2   Representable Sets and Rauzy Graphs

Let Σ be a finite alphabet. We denote the set of all *(total) words* of any length formed by concatenating elements of Σ by Σ^*, and similarly we denote the set of words of a finite length n by Σ^n. The *empty word* ε is the unique word of length zero. On the other hand, we denote the set of all *partial words* of any length formed by concatenating elements of $\Sigma_\diamond = \Sigma \cup \{\diamond\}$ by $\Sigma_\diamond^*$ and the set of partial words of length n by $\Sigma_\diamond^n$. Here $\diamond \notin \Sigma$ stands for the "hole character" and represents any undefined position. The character at position i of a partial word w is denoted by $w[i]$, and i is a hole when $w[i] = \diamond$. The *length* of w, denoted by $|w|$, is the number of characters in w (including the hole characters).

If w and w' are partial words of same length, w is *contained* in w', and write $w \subset w'$, if $w[i] = w'[i]$ for all non-hole positions i in w, w is *compatible* with w', and write $w \uparrow w'$, if $w[i] = w'[i]$ for all non-hole positions i in both w and w', and w is *equal* to w', and write $w = w'$, if $w[i] = w'[i]$ for all i. A *completion* of a partial word w is a total word obtained by filling in all the holes of w with letters from the alphabet, while a *strengthening* is a partial word obtained by filling in a (possibly trivial) subset of the holes of w. Taking the binary alphabet $\{0,1\}$, $0\diamond\diamond1 \subset 0\diamond01$, $0\diamond\diamond1 \uparrow 00\diamond\diamond$, 0011 is one of the four completions of $0\diamond\diamond1$, and $0\diamond01$ is one of the nine strengthenings of $0\diamond\diamond1$.

A partial word w of length n or greater has a set of *factors* $\mathrm{fac_n}(w)$ whose elements are sequences of length n that consist of consecutive characters of w. It has also a set of *subwords* $\mathrm{sub}_n(w)$ whose elements are total words compatible with factors of w of length n. For instance, $0\diamond\diamond1$ is a factor of $000\diamond\diamond10\diamond$, while $0001, 0011, 0101, 0111$ are the subwords compatible with that factor. We abbreviate the factor $w[i]w[i+1]\cdots w[j-1]$ by $w[i..j)$.

Let $S \subseteq \Sigma^n$. We say that S is *representable* if there exists a partial word $w \in \Sigma_\diamond^*$ whose set of length-n subwords $\mathrm{sub}_n(w)$ is exactly equal to S, and we call w a *representation partial word* for S. Letting h be a non-negative integer, S is *h-representable* if there exists a partial word $w \in \Sigma_\diamond^*$ with h holes such that $\mathrm{sub}_n(w) = S$, and w is a *h-representation partial word* for S.

The *Rauzy graph of S* is the digraph $\mathrm{RAU}(S) = (V, E)$, where the set of vertices V consists of the length-$(n-1)$ prefixes and suffixes of elements of S and the set of edges E consists of the elements of S, i.e., if $s \in S$ then there is an edge labelled by s from the vertex $s[0..n-1)$ to the vertex $s[1..n)$. For each $v \in V$,

$$\mathrm{pred}(v) = \{u \mid u \in V, u = au', v = u'b, \ a, b \in \Sigma\},$$
$$\mathrm{succ}(v) = \{u \mid u \in V, v = av', u = v'b, \ a, b \in \Sigma\}.$$

A path through a Rauzy graph corresponds to a word with the ith edge corresponding to the length-n subword starting at the ith position. It is obtained by adding on the last letter of each edge traversed. In other words, if $u = u_0, u_1, \ldots, u_\ell = v$ is a path from u to v in $\mathrm{RAU}(S)$, then the word

$w = u_0 u_1[n-2]u_2[n-2]\cdots u_\ell[n-2]$ corresponds to it. Using this correspondence between paths and words, we refer also to w as a path in $\mathrm{RAU}(S)$. Figure 1 gives an example of a Rauzy graph. The word 0011011010 corresponds to the path $001 \to 011 \to 110 \to 101 \to 011 \to 110 \to 101 \to 010$ and so S is 0-representable. However $S' = \{0011, 0101, 0110, 1011, 1101\}$ is not.

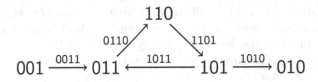

Fig. 1. Rauzy graph of S = {0011,0101,0110,1011,1101}

Lemma 1. *For $S \subseteq \Sigma^n$, there is a bijection between the 0-representation words w for S and the paths in $\mathrm{RAU}(S)$ of length $|w| - n + 1$ that include every edge.*

3   Bound on the Cardinality of Comp(S)

Let $S \subseteq \Sigma^n$ and let $\mathrm{Comp}(S)$ be the set of partial words all of whose completions are in S. For example, if S consists of the six words $v_1 = 0000$, $v_2 = 0001$, $v_3 = 0010$, $v_4 = 0011$, $v_5 = 0100$, and $v_6 = 1010$, then $\mathrm{Comp}(S)$ consists of

$$0000, 0001, 0010, 0011, 0100, 1010, \diamond010, 00\diamond\diamond, 000\diamond, 001\diamond, 00\diamond0, 00\diamond1, 0\diamond00. \quad (1)$$

In this section, we only consider the binary alphabet $\Sigma = \{0, 1\}$. We show that the inequality $|\mathrm{Comp}(S)| \leq |S|^{\log_2 3}$ holds. Set $T(S) = |\mathrm{Comp}(S)|$ and $T(m) = \max\{T(S) \mid |S| = m\}$.

The *hypercube graph of order* $n > 0$ is the digraph $H_n = (V, E)$, where the set of vertices V consists of Σ^n and the set of edges E consists of the pairs (u, v) such that u and v have Hamming distance 1, i.e., u and v differ at only one position. We define H_0 to be the singleton graph. For $S \subseteq \Sigma^n$, let $\mathrm{HAM}(S)$ denote the subgraph of H_n induced by the words in S.

The following lemma establishes a bijection that allows us to refer to sets of partial words of length n (that are closed under strengthening) and the corresponding subgraphs of H_n interchangeably. Thus, $|\mathrm{Comp}(S)|$ is the number of copies of H_h, $0 \leq h \leq n$, in $\mathrm{HAM}(S)$. Returning to our example in (1), there is one copy of H_2 from the partial word with two holes $00\diamond\diamond$, six copies of H_1 from the six partial words with one hole $\diamond010, 000\diamond, 001\diamond, 00\diamond0, 00\diamond1, 0\diamond00$, and six copies of H_0 from the six words $0000, 0001, 0010, 0011, 0100, 1010$.

Lemma 2. *For $S \subseteq \Sigma^n$, there is a bijection mapping a partial word w with h holes in $\mathrm{Comp}(S)$ to a subgraph of $\mathrm{HAM}(S)$ isomorphic with H_h; the completions of w correspond to the vertices in the subgraph.*

We show how to construct Comp(S) starting from the empty set. Let S_j consist of the first j words of S taken lexicographically. At Step 1 of the process of building Comp(S), add the element of S_1 to Comp(S). At Step j, first add the element of $S_j \setminus S_{j-1}$, say v_j. Then add all the partial words in Comp(S_j) that do not already appear in Comp(S). The partial words added at Step j can be constructed from v_j by replacing a (possibly trivial) subset of the 1's of v_j with $\diamond$'s. Note that these partial words were not added to Comp(S) before, because they can be completed by filling all their $\diamond$'s with 1's. Also note that they must be added to Comp(S), because, any completion in which some of their $\diamond$'s are filled with 0's comes earlier lexicographically, hence is in S.

To illustrate the construction, let $S_j = \{v_1, \ldots, v_j\}$ where the v_j's refer to our example set S from (1).

Step j	Added to Comp(S)	Added to HAM(S)
1	0000	H_0
2	0001, 000$\diamond$	H_0, H_1
3	0010, 00$\diamond$0	H_0, H_1
4	0011, 001$\diamond$, 00$\diamond$1, 00$\diamond\diamond$	H_0, H_1, H_1, H_2
5	0100, 0$\diamond$00	H_0, H_1
6	1010, $\diamond$010	H_0, H_1

Neither 10$\diamond$0 nor $\diamond$0$\diamond$0 is added at Step 6 since the completion 1000 is not in S.

Theorem 1. *For $0 \leq h \leq n$, the equality $T(H_h) = \Sigma_{i=0}^{|H_h|-1} 2^{\chi(i)}$ holds, where $\chi(i)$ denotes the number of 1's in the binary representation of i.*

Proof. For $S = H_h$, the partial words added at Step j are constructed by replacing all subsets of the 1's of v_j with $\diamond$'s. Since the number of 1's in the binary representation of v_j is $\chi(j-1)$ in this case, there are $2^{\chi(j-1)}$ new partial words added to Comp(S) or $2^{\chi(j-1)}$ new sub-hypercubes added to HAM(S) at Step j. $\square$

Now we want to establish an upper bound on $T(S)$ for all $S \subseteq \Sigma^n$. For $S \subseteq H_n$, we start with a manipulation on subsets $X \cong H_{n-1} \subset H_n$ and $Y \cong H_{n-1} \subset H_n$, where $X \cap Y = \emptyset$, which, intuitively, pushes the elements of S from X to Y, when the corresponding positions are not already occupied. Let $\varphi : X \rightarrow Y$ be a graph isomorphism. We define push$(X, Y, \varphi, S) \subseteq H_n$ as follows:

- for all $v \in X$, $v \in$ push(X, Y, φ, S) if and only if $v \in S$ and $\varphi(v) \in S$;
- for all $v \in Y$, $v \in$ push(X, Y, φ, S) if and only if $\varphi^{-1}(v) \in S$ or $v \in S$.

Note that $|$push$(X, Y, \varphi, S)| = |S|$.

To illustrate the above manipulation and the following lemma, consider our example set $S = \{0000, 0001, 0010, 0011, 0100, 1010\}$, $X = 1\{0,1\}^3$, and $Y = 0\{0,1\}^3$. Take φ to be the graph isomorphism that relabels 1's with 0's and

0's with 1's. Here push $(X, Y, \varphi, S) = \{0000, 0001, 0010, 0011, 0100, 0101\}$. As noticed earlier $T(S) = 13$, and we can check that $T(\text{push}(X, Y, \varphi, S)) = 15$.

Lemma 3. Let $S \subset H_n$. Let $X \cong H_{n-1} \subset H_n$ and $Y \cong H_{n-1} \subset H_n$ be such that $X \cap Y = \emptyset$, $Y \neq S$, and $0 \neq T(X \cap S) \leq T(Y \cap S)$. Then, there exists a graph isomorphism $\varphi : X \rightarrow Y$ such that $T(S) \leq T(\text{push}(X, Y, \varphi, S))$ and $S \neq \text{push}(X, Y, \varphi, S)$.

By Theorem 1, $T(H_h) = \Sigma_{i=0}^{|H_h|-1} 2^{\chi(i)}$. But, $T(H_h) = 3^h$ because, looking at the partial word with h holes corresponding to H_h, each of the hole positions can be filled with one of $\{\diamond, 0, 1\}$. So, since $|H_h| = 2^h$, we have $\Sigma_{i=0}^{2^h-1} 2^{\chi(i)} = 3^h$. We next show that no other way of selecting a subgraph S with a fixed number of vertices results in a larger value for $T(S)$.

Theorem 2. For all n' and $|S| < 2^{n'}$, where $S \subset H_n$ and $S \subseteq G \cong H_{n'}$, the inequality $T(S) \leq \Sigma_{i=0}^{|S|-1} 2^{\chi(i)}$ holds, where $\chi(i)$ denotes the number of 1's in the binary representation of i.

Proof. We proceed by induction on n'. If $n' = 1$, then $S = \{w\}$ for some word w, in which case, $T(S) = |\text{Comp}(S)| = |\{w\}| = 1 = 2^0$. If $n' > 1$, then we have the following two cases. First, suppose that $|S| < 2^{n'-1}$. Fix a subset $Y \subset H_{n'}$ where $Y \cong H_{n'-1}$. Repeatedly use Lemma 3 until all the elements of S have been moved into Y. Then, apply the inductive hypothesis. Next, suppose that $2^{n'} > |S| \geq 2^{n'-1}$. Consider first the case when our subgraph S on n vertices has a subgraph, say Z, isomorphic to $H_{n'-1}$. Then, $|S \setminus Z| < 2^{n'-1}$, and, $S \setminus Z$ is contained in an adjacent copy of $H_{n'-1}$. Thus,

$$
\begin{aligned}
T(S) &= 3^{n'-1} + 2 \cdot T(S \setminus Z) \\
&\leq \Sigma_{i=0}^{2^{n'-1}-1} 2^{\chi(i)} + 2 \cdot \Sigma_{i=0}^{|S|-1-2^{n'-1}} 2^{\chi(i)} \\
&= \Sigma_{i=0}^{2^{n'-1}-1} 2^{\chi(i)} + \Sigma_{i=2^{n'}-1}^{|S|-1} 2^{\chi(i)} \\
&= \Sigma_{i=0}^{|S|-1} 2^{\chi(i)}.
\end{aligned}
$$

Consider now the case when there is no subgraph $Z \subseteq S$ such that $Z \cong H_{n'-1}$. We can find $S_j \subseteq H_{n'}$ where $|S| = |S_j|$, $T(S) \leq T(S_j)$, and S_j has a subgraph Y isomorphic to $H_{n'-1}$. Indeed, by repeatedly applying Lemma 3 until Y is in S_j, we define $S_0 = S$ and $S_i := \text{push}(X, Y, \varphi, S_{i-1})$ for $i > 0$. Then, we have $T(S) \leq T(S_1) \leq \cdots \leq T(S_j)$. So, we apply the previous argument to S_j to obtain $T(S_j) \leq \Sigma_{i=0}^{|S_j|-1} 2^{\chi(i)}$. The desired bound on $T(S)$ follows easily. □

Corollary 1. For $S \subseteq \Sigma^n$, the inequality $|\text{Comp}(S)| \leq |S|^{\log_2 3}$ holds. This bound is achieved when S consists of the completions of a single partial word.

Proof. We have shown that $T(m) = \Sigma_{i=0}^{m-1} 2^{\chi(i)}$, which is a well-studied digital sum. We have from [4] that

$$
T(m) = m^{\log_2 3} \cdot F(\log_2 m), \tag{2}
$$

where F is a 1-periodic function defined by a Fourier series that we omit here. Combining this with a result from [8], $F(x) \leq 1$ for all x, so $T(m) \leq m^{\log_2 3}$. We get that $|\text{Comp}(S)| \leq T(|S|) \leq |S|^{\log_2 3}$, and the largest possible value for $|\text{Comp}(S)|$ given $|S|$ can be found by leaving in the F in Eq. (2).   □

4 Membership of Circular Representability in $\mathcal{P}$

In this section, we drop the restriction of a binary alphabet, and take $\Sigma = \{0, 1, \ldots, k-1\}$, where $k \geq 2$ is an integer.

For any partial word w and integer $n \geq 0$, denote by $\text{csub}_n(w)$ the set of length-n *circular subwords* of w, i.e., if $u \in \text{csub}_n(w)$ and u occurs at some position j in w such that $j + n \leq |w|$, we have $u \uparrow w[j..j+n)$, otherwise we have $u[0..|w|-j) \uparrow w[j..|w|)$ and $u[|w|-j..n) \uparrow w[0..j+n-|w|)$. Now, let S be a subset of Σ^n. A partial word w such that $\text{csub}_n(w) = S$ is a *circular representation word* for S and a partial word w with h holes such that $\text{csub}_n(w) = S$ is a *h-circular representation word* for S. The set S is *circularly representable* if there exists a circular representation word for S and is *h-circulary representable* if there exists a h-circular representation word for S. For example if we consider $S = \{000, 001, 010, 100, 101\}$, then S can be 0-circularly represented by 000101, 1-circularly represented by $\diamond$0010, and 2-circularly represented by $\diamond$00$\diamond$0.

Let CREP be the problem of deciding whether a given subset is circularly representable and h-CREP be the one of deciding whether it is h-circularly representable. We also denote by h-CREP the class of all the h-circularly representable sets. Using the following lemma, a subset S of Σ^n is 0-circularly representable if and only if $\text{RAU}(S)$ has a cycle that visits every edge at least once implying that 0-CREP is in $\mathcal{P}$.

Lemma 4. *For all $S \subseteq \Sigma^n$, there is a bijection between 0-circular representation words w for S and the cycles in $\text{RAU}(S)$ of length $|w|$ that include every edge.*

We need the following lemmas to prove that for each non-negative integer h, h-CREP, and thus CREP, are in $\mathcal{P}$.

Lemma 5. *For all partial words w and integers $n, i \geq 1$, $\text{csub}_n(w^i) = \text{csub}_n(w)$.*

For $S \subseteq \Sigma^n$, the *De Bruijn graph* of S is the digraph $\text{DEB}(S) = (V, E)$, where V consists of the elements of S and E consists of the pairs (v_1, v_2) such that there exist $u \in \Sigma^{n-1}, a, b \in \Sigma$ such that $v_1 = au$ and $v_2 = ub$. Figure 2 gives an example of a set S that is circularly representable by $w = 001101101$. Note that $\text{DEB}(S)$ is strongly connected, illustrating Lemma 6.

Lemma 6. *A subset S of Σ^n is circularly representable if and only if $\text{DEB}(S)$ is strongly connected.*

Lemma 7. *We have 0-CREP $\supsetneq$ 1-CREP $=$ 2-CREP $=$ 3-CREP $= \cdots$.*

Lemma 8. *For every $S, S \subseteq \Sigma^n$, that is 0-circularly representable, $S \in$ 1-CREP if and only if there exist vertices u, v in $\text{RAU}(S)$ such that there are $|\Sigma|$ distinct paths of length n from u to v.*

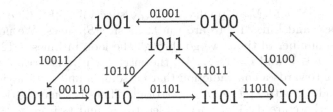

Fig. 2. De Bruijn graph of $S = \{0011, 0100, 0110, 1001, 1010, 1011, 1101\}$

Algorithm 1 determines when the condition of Lemma 8 is satisfied. This is part of our proof for the memberships of h-CREP and CREP in $\mathcal{P}$.

Algorithm 1. Deciding membership in 1-CREP of a given S in 0-CREP

Ensure: returns **true** if $S \in$ 1-CREP, otherwise returns **false**
1: $(V, E) \leftarrow \text{RAU}(S)$
2: assign each vertex v a unique number $\eta(v) \in \{1, \ldots, |V|\}$
3: associate an array arr(v) of size $|V|$ with each vertex v
4: $F \leftarrow$ empty queue
5: **for** $v \in V$ **do**
6:     **for** $u \in \text{pred}(v)$ **do**
7:       arr(u)$[\eta(v)] \leftarrow 1$ and push(F, $(u, \eta(v))$)
8: **while** F is not empty **do**
9:     $(v, j) \leftarrow pop(F)$
10:    **if** $j = 1, \ldots, |V|$ **then**
11:      **for** $u \in \text{pred}(v)$ **do**
12:       arr(u)$[j] \leftarrow$ arr(v)$[j] + 1$ and push(F, (u, j))
13: **for** $u \in V$ **do**
14:    **for** $i = 1, \ldots, |V|$ **do**
15:      *numpaths* $\leftarrow 0$
16:      **for** $u' \in \text{succ}(u)$ **do**
17:       **if** arr(u')$[i] = n - 1$ **then**
18:        *numpaths* $\leftarrow$ *numpaths* $+ 1$
19:      **if** *numpaths* $= |\Sigma|$ **then**
20:       **return true**
21: **return false**

Theorem 3. *For all h, h-CREP is in $\mathcal{P}$. Thus CREP is in $\mathcal{P}$.*

Proof. For $h = 0$, the result follows from Lemma 6 which implies that a subset S of Σ^n is in 0-CREP if and only if DEB(S) is strongly connected. Testing that a digraph is strongly connected can be done in linear time by Tarjan's algorithm [12]. For $h > 0$, the result follows from Lemma 7 which states that deciding h-CREP is equivalent to deciding 1-CREP. We show that decising 1-CREP can be done by Algorithm 1 which determines when the condition of Lemma 8 is satisfied. The size to represent an input set S, $S \subseteq \Sigma^n$, being $n|S|$, we show that Algorithm 1 runs in $O(n|S|^2)$ time.

Since $|E| = |S|$ and $\Sigma_{v \in V}|\mathrm{pred}(v)| = \Sigma_{v \in V}|\mathrm{succ}(v)| = |E|$, Line 7 is run at most $|S|$ times, and Lines 17–18 are run at most $n|S|$ times. We now prove a bound on the number of times we go through the loop in Lines 8–12. We show that for every $u \in V$ and $i = 1, \ldots, |V|$, the entry $\mathrm{arr}(u)[i]$ is written to at most once. Suppose towards a contradiction that for some u and i, we assign distinct integers ℓ_1 and ℓ_2, with $\ell_1, \ell_2 < n$, to $\mathrm{arr}(u)[i]$. There is some $v \in V$ such that $\eta(v) = i$, and there are distinct paths of lengths ℓ_1 and ℓ_2 from u to v. By the correspondence between paths and words, we have two distinct words of lengths $n - 1 + \ell_1$ and $n - 1 + \ell_2$ having the same length-$(n-1)$ prefix and the same length-$(n-1)$ suffix. Since these words are both of length at most $2n - 2$, we obtain a contradiction. Since each time a pair is pushed onto F, there is a write to some $\mathrm{arr}(u)[i]$, we have that there are at most $|V|^2$ times that a pair is pushed onto F. This gives our bound.

By the time the last loop is run, $\mathrm{arr}(u)[\eta(v)] = \ell$ if and only if there is a path of length $\ell < n$ from u to v. So, the property of having $|\Sigma|$ distinct paths of length n from some u to some v is equivalent to there being a path of length $n - 1$ from each of the successors of u to v, which is checked using arr.   □

5   Computing Minimal (Circular) Representation Words

By Lemma 4, to compute a minimum length 0-circular representation word for a given subset S of Σ^n if one exists, we want to find a shortest cycle in $\mathrm{RAU}(S) = (V, E)$ that uses every edge at least once. We reduce this problem to a minimum-cost flow network problem (see [1] for more information on flow networks). A *cost/flow network* is a digraph (V', E') having a distinguished *source vertex* s, a distinguished *sink vertex* t, and such that every $e \in E'$ has a *capacity*, a *flow*, and a *cost*, respectively denoted by $\mathrm{capacity}(e)$, $\mathrm{flow}(e)$, and $\mathrm{cost}(e)$, associated with it. There are polynomial-time algorithms for finding the minimum-cost maximum-flow, i.e., finding a flow that is maximum, but has a cost that is minimum among all the maximum flows. In other words, the min-cost max-flow problem is to minimize the total cost of the flow $\Sigma_{e \in E'}\mathrm{flow}(e) \cdot \mathrm{cost}(e)$ with the constraints $\mathrm{flow}(e) \le \mathrm{capacity}(e)$ for all $e \in E'$, and $\Sigma_{v \in V'}\mathrm{flow}(s, v) = \Sigma_{v \in V'}\mathrm{flow}(v, t) = f$, where f is the amount of flow to be sent from s to t.

We construct the following cost/flow network (V', E') from $\mathrm{RAU}(S) = (V, E)$. For each $v \in V$, let b_v be the out-degree of v minus the in-degree of v, and let $\mathrm{Imb}(S) = \Sigma_{\{v \in V | b_v > 0\}} b_v$. Since we need to use each edge of $\mathrm{RAU}(S)$, think of the vertices with more edges coming in than out as supplying, and those with more going out as consuming. Flows along this network correspond to repeated subwords of length n. We need to keep them to a minimum. So, let $V' = V \cup \{s, t\}$. Put each $e \in E$ in E', with cost 1 and unlimited capacity (or some capacity at least $\mathrm{Imb}(S)$). Then, for each $v \in V$ with $b_v < 0$, add an edge (v, t) of capacity $-b_v$ and cost 0 to E'. Similarly, for each $v \in V$ with $b_v > 0$, add an edge (s, v) of capacity b_v and cost 0 to E'.

Then we run a max-flow min-cost algorithm with (V', E'). We call the flow amount f, the cost c, and the set of unit flows F.

Algorithm 2. Computing a minimal 0-circular representation word for $S \subseteq \Sigma^n$

Ensure: returns a minimum length total word that circularly represents S, or returns
    false if no such word exists
1: $(V, E) \leftarrow \text{RAU}(S)$
2: construct the cost/flow network (V', E') from (V, E)
3: run a max-flow min-cost algorithm with (V', E'), call the flow amount f, the cost
    c, and the set of unit flows F
4: **if** $f < \Sigma_{\{v \in V | b_v > 0\}} b_v$ **then**
5:     **return false**
6: $E'' \leftarrow E$
7: **for all** $p \in F$ **do**
8:     add to E'' an edge from $p[1..n)$ to $p[|p| - n..|p| - 1)$
9: run an Eulerian cycle algorithm on (V, E''), call the path u
10: $w \leftarrow u[0..n)$
11: **for** $i = 1, \ldots, |u| - n$ **do**
12:     **if** $u[i..i + n) \in E$ **then**
13:         $a \leftarrow u[i + n - 1]$
14:     **else**
15:         $p \leftarrow$ the path that made us add $u[i..i + n)$ to E''
16:         $a \leftarrow p[n..|p| - 1)$
17:     $w \leftarrow wa$
18: **return** $w[0..|w| - n + 1)$

Lemma 9. *Let $S \subseteq \Sigma^n$ and let (V', E') be the cost/flow network constructed
from $\text{RAU}(S)$, with capacity $\text{Imb}(S)$ and cost c. Then there exists a word of length
$|S| + c - n + 1$ that 0-circularly represents S.*

Proof. We can view each of the unit flows in F as an edge connecting the vertex
immediately after s, and the vertex immediately before t, with length equal to
the cost of the unit flow. Doing this, we now have a graph with total edge length
$|S| + c$ where every vertex has equal in- and out-degrees. So, there is an Eulerian
cycle. To recover a 0-representation word for S from this cycle, we take the start
vertex, and then, for each edge in the cycle, append the last letter of that edge
until we are back at the start vertex. Call this word w. This implies that $\text{sub}_n(w)$
is equal to the set of edges in the cycle, which, since it is Eulerian, is all of S. Note
that the total cost of the edges in this graph is $|S| + c$, and so, that is $|w|$. Since we
want a 0-circular representation for S, and $w[0..n - 1) = w[|S| + c - n + 1..|S| + c)$,
we can take $w' = w[0..|S| + c - n + 1)$ and, $\text{csub}_n(w') = \text{sub}_n(w)$. $\square$

Lemma 10. *Let $S \subseteq \Sigma^n$ and let (V', E') be the cost/flow network constructed
from $\text{RAU}(S)$. Given an all-edge-visiting cycle of length $|S| + c$ in (V', E'), there
exists a flow of capacity $\text{Imb}(S)$ with cost at most c.*

By Lemmas 9 and 10, any all-edge-visiting path, that is, a 0-representation word,
must correspond to a flow of capacity $\text{Imb}(S)$ with its length a constant off from
the cost of the flow in the network. This means that if we have a 0-representation
word of shorter length than that computed by Algorithm 2, then the min-cost

flow that we find is not actually min-cost. Since $\mathrm{RAU}(S)$ has $|S|$ edges, and even fewer vertices, we can find the min-cost flow in $O(|S|^2 \log |S|^2)$ time using the min-cost max-flow algorithm of Goldberg and Tarjan [7]. For Algorithm 3, the most time-consuming step is computing the min-cost flow, which gets computed for every pair of distinguished start and end vertices, so the algorithm's running time picks up a factor of $|S|^2$.

Theorem 4. *Given as input a set S of words of uniform length, Algorithm 2 computes a minimum 0-circular representation word in $O(|S|^2 \log |S|^2)$ time and Algorithm 3 computes a minimum 0-representation word in $O(|S|^4 \log |S|^2)$ time.*

Algorithm 3. Computing a minimal 0-representation word for $S \subseteq \Sigma^n$

Ensure: returns a minimum length total word that represents S, or returns **false** if no such word exists
1: $(V, E_0) \leftarrow \mathrm{RAU}(S)$
2: $m \leftarrow \infty$
3: **for all** $v_1, v_2 \in V$ **do**
4:     $E \leftarrow E_0 \cup \{(v_2, v_1)\}$ with cost $|S|^2$ and unlimited capacity
5:     run Lines 2 to 9 of Algorithm 2 ; if it returns **false** or has a min-cost $\geq |S|^2$, try a new pair v_1, v_2, otherwise, we have an Eulerian cycle u
6:     since u visits every edge, and (v_2, v_1) is an edge, rotate u so that $u = v_1 \cdots v_2$
7:     run Lines 10 to 17 of Algorithm 2 to get w
8:     **if** $|w| < m$ **then**
9:        $w_{\min} \leftarrow w$
10:       $m \leftarrow |w|$
11: **if** $m = \infty$ **then**
12:    **return false**
13: **else**
14:    **return** $w_{\min}$

For example, the set consisting of the 12 words

$$\begin{array}{cccccc} 00010 & 00011 & 00101 & 00111 & 01011 & 01111 \\ 10110 & 11110 & 01100 & 11100 & 11000 & 10001 \end{array}$$

has minimum total circular representation word 00010110001111 and minimum partial circular representation word 0001◊11.

6   Conclusion and Open Problems for Future Work

For $S \subseteq \Sigma^n$, we gave a bound on the size of Comp(S) when $|\Sigma| = 2$. A larger $|\Sigma|$ should make it harder for all of a partial word's completions to be in S.

Conjecture 1. The inequality $|\mathrm{Comp}(S)| \leq |S|^{\log_{|\Sigma|}(|\Sigma|+1)}$ holds.

Other than the above conjecture, some open problems include: (1) Characterize the sets of words of uniform length that are representable or h-representable. (2) If a subset S of Σ^n is representable, how long a partial word do we need to represent it? (3) Can partial words of minimum length that produce all words in S be constructed efficiently? We gave an efficient construction for (3) using total words but the length could be reduced further by using partial words.

Tan and Shallit [11] focused on sets representable by total words. They considered the following problems: How many subsets of Σ^n are representable by a total word? If a subset is representable, how long a total word do we need to represent it? How many such subsets are represented by words of a fixed length ℓ? For the first problem, they gave upper and lower bounds in the binary case. For the second prob , they gave a weak upper bound and some experimental data. For the third problem, they gave a closed-form formula in the case where $n \leq \ell \leq 2n$. They also left open a number of questions. We suggest extending Tan and Shallit's work to partial words: (4) How many subsets of Σ^n are representable by a partial word? How many such subsets are there if we fix the number of holes or the length of the representation?

References

1. Ahuja, R.K., Magnanti, T.L., Orlin, J.B.: Network Flows: Theory, Algorithms, and Applications. Prentice-Hall, New Jersey (1993)
2. Blanchet-Sadri, F., Munteanu, S.: Deciding representability of sets of words of equal length in polynomial time. In: Lecroq, T., Mouchard, L. (eds.) IWOCA 2013. LNCS, vol. 8288, pp. 28–40. Springer, Heidelberg (2013)
3. Blanchet-Sadri, F., Simmons, S.: Deciding representability of sets of words of equal length. Theoret. Comput. Sci. **475**, 34–46 (2013)
4. Flajolet, P., Grabner, P., Kirschenhofer, P., Prodinger, H., Tichy, F.: Mellin transforms and asymptotics: digital sums. Theoret. Comput. Sci. **123**, 291–314 (1994)
5. Fredericksen, H.: A survey of full length nonlinear shift register cycle algorithms. SIAM Rev. **24**, 195–221 (1982)
6. Gallant, J., Maier, D., Storer, J.A.: On finding minimal length superstrings. J. Comput. Syst. Sci. **20**, 50–58 (1980)
7. Goldberg, A.V., Tarjan, R.E.: Finding minimum-cost circulations by successive approximation. Math. Oper. Res. **15**, 430–466 (1990)
8. Harborth, H.: Number of odd binomial coefficients. Proc. Amer. Math. Soc. **62**, 19–22 (1977)
9. Katz, J., Lindell, Y.: Introduction to Modern Cryptography: Principles and Protocols. Cryptography and Network Security. Chapman & Hall/CRC, Boca Raton (2008)
10. van Lint, J.H., MacWilliams, F.J., Sloane, N.J.A.: On pseudo-random arrays. SIAM J. Appl. Math. **36**, 62–72 (1979)
11. Tan, S., Shallit, J.: Sets represented as the length-n factors of a word. In: Karhumäki, J., Lepistö, A., Zamboni, L. (eds.) WORDS 2013. LNCS, vol. 8079, pp. 250–261. Springer, Heidelberg (2013)
12. Tarjan, R.: Depth-first search and linear graph algorithms. SIAM J. Comput. **1**, 146–160 (1972)

Computing Primitively-Rooted Squares
and Runs in Partial Words

Francine Blanchet-Sadri[1]([⊠]), Jordan Nikkel[2], J.D. Quigley[3],
and Xufan Zhang[4]

[1] Department of Computer Science, University of North Carolina,
P.O. Box 26170, Greensboro, NC 27402–6170, USA
blanchet@uncg.edu
[2] Department of Mathematics, Vanderbilt University, 1326 Stevenson Center,
Nashville, TN 37240, USA
[3] Department of Mathematics, University of Illinois - Urbana-Champaign,
Altgeld Hall, 1409 W. Green Street, Urbana, IL 61801, USA
[4] Department of Mathematics, Princeton University, Fine Hall,
Washington Road, Princeton, NJ 08544, USA

Abstract. This paper deals with two types of repetitions in strings:
squares, which consist of two adjacent occurrences of substrings, and
runs, which are periodic substrings that cannot be extended further to
the left or right. We show how to compute all the primitively-rooted
squares in a given partial word, which is a sequence that may have
undefined positions, called holes or wildcards, that match any letter of
the alphabet over which the sequence is defined. We also describe an
algorithm for computing all primitively-rooted runs in a given partial
word.

1   Introduction

Repetitions in strings, or words, have been extensively studied, both from the
algorithmic point of view and the combinatorial point of view (see, for
example, [9]). Applications can be found in many important areas such as computational biology, data compression, to name a few [8]. Repetitions are characterized by their periods, lengths, and starting positions. There are many equivalent
characterizations of repetitions, and in this paper a repetition in a word w is
a triple (f, l, p), where $w[f..l]$ is p-periodic and the *exponent* of the repetition,
$\frac{l-f+1}{p}$, is at least 2.

Squares are the special case of repetitions when $\frac{l-f+1}{p}$ is 2. In other words, a
square in a word is a factor uu for some word u, called the *root* of the square. It
is *primitively-rooted* or *PR* if u is *primitive*, i.e., u is not a power of another word.
There can be as many as $\Theta(n \log n)$ occurrences of *primitively-rooted* squares in

This material is based upon work supported by the National Science Foundation
under Grant No. DMS–1060775. We thank the referees of preliminary versions of
this paper for their very valuable comments and suggestions.

© Springer International Publishing Switzerland 2015
J. Kratochvíl et al. (Eds.): IWOCA 2014, LNCS 8986, pp. 86–97, 2015.
DOI: 10.1007/978-3-319-19315-1_8

a word of length n, and several $O(n \log n)$ time algorithms have been developed for finding all the repetitions [2,7,16]. A major breakthrough was to compute them in $O(n)$ time; this was achieved in two steps: (1) all repetitions are encoded in *maximal repetitions* or *runs* and (2) there is a linear bound on the number of runs [14].

A repetition (f, l, p) is maximal, or is a run, if it is non-extendible, i.e., neither $(f - 1, l, p)$ nor $(f, l + 1, p)$ are repetitions in the word w. Note that since every run has exponent at least 2, the square at the beginning of the run uniquely defines the run. This is because that square contains the starting position and period of the run, and the property of the run being maximal gives a unique ending position. A *PR-run* in w is a maximal repetition with a primitive root. If $w = 00011010110101101010$, then $w[2..18] = (01101)^{\frac{17}{5}}$ is a PR-run with period 5, root 01101, and exponent $\frac{17}{5}$ (indexing in the word starts at 0). The maximum number of runs in a string of length n is bounded from above by cn, for some constant c. A first proof was given by Kolpakov and Kucherov [14] who provided an $O(n)$ time algorithm for detecting all runs. The number of runs being linear has applications to the analysis of any optimal algorithm for computing all repetitions.

Partial words, also referred to as *strings with don't-cares* which allow for incomplete or corrupted data [1,11], are sequences that may contain undefined positions, called holes and represented by $\diamond$'s, that are *compatible* with, or *match*, any letter in the alphabet. *Total words* are partial words without holes. Here, a *factor* is a consecutive sequence of symbols in a partial word w, while a *subword* is a total word compatible with a factor in w. A factor uv is a square if some *completion*, i.e., a filling in of the holes with letters from the alphabet, turns uv into a total word that is a square; equivalently, u and v are compatible.

Repetitions in partial words have also recently been studied, both from the algorithmic point of view and the combinatorial point of view (see, for example, [5,6,10,12,17]). However, no work has been dedicated to computing all occurrences of PR-squares and PR-runs in partial words. The known algorithms for detecting them in total words do not extend easily to partial words, one of the most important culprits being that the compatibility relation is not transitive, as opposed to the equality relation being transitive, e.g., 0 is compatible with $\diamond$ and $\diamond$ is compatible with 1, but 0 is not compatible with 1.

So we adopt an approach, for our algorithms, based on the *large divisors* of the length n of the input partial word, i.e., divisors of n, distinct from n, whose multiples are either n itself or not divisors of n. Every distinct prime divisor i of n gives rise to exactly one large divisor of n, namely $\frac{n}{i}$, and hence the number of large divisors of n, $\omega(n)$, is the number of distinct prime divisors of n, e.g., 30 has large divisors 6, 10, and 15, giving $\omega(30) = 3$. Recently, a formula for the number of primitive partial words was given in terms of the large divisors [4]. In fact, the maximum number of holes in a primitive partial word of length n over a k-letter alphabet is $n - \omega(n) - 1$, for all $n, k \geq 2$, and this bound is tight.

The contents of our paper are as follows: In Sect. 2, we review a few basic concepts on partial words such as periodicity and primitivity. In Sect. 3, we present efficient algorithms for computing all occurrences of PR-square factors and PR-runs in any partial word over a k-letter alphabet. In Sect. 4, we describe an efficient algorithm for counting the number of occurrences of PR-square subwords. Finally in Sect. 5, we conclude with some open problems.

2   Preliminary Definitions and Results

A *partial* word over an alphabet A is a sequence over the extended alphabet $A_\diamond = A \cup \{\diamond\}$, where the symbol $\diamond$ represents an undefined position or a hole. A *total* word is a partial word with no holes. A *letter* will always refer to an element of A, while a *symbol* will refer to an element of $A_\diamond$. The symbol at position i of a partial word w is denoted $w[i]$, with position numbers starting at 0. We denote by $|w|$ the *length* of w or the number of symbols in w. Two partial words u and v of equal length are *compatible*, denoted $u \uparrow v$, if $u[i] = v[i]$ for every $u[i], v[i] \in A$, and u is *contained* in v, denoted $u \subset v$, if $u[i] = v[i]$ for every $u[i] \in A$. The *least upper bound* of two compatible partial words u and v is the partial word $u \vee v$ where $u \subset (u \vee v)$, $v \subset (u \vee v)$, and if $u \subset w$ and $v \subset w$ then $(u \vee v) \subset w$. A *completion* of a partial word u is any total word v such that $u \subset v$.

For a positive integer p, a length-n partial word w has a *strong period* of p or is *strongly p-periodic* if $i \equiv j \mod p$ implies $w[i] \uparrow w[j]$ for every $0 \leq i < j < n$. It has a *weak period* of p or is *weakly p-periodic* if $w[i] \uparrow w[i+p]$ for every $0 \leq i < n - p$. For total words, since weak periodicity implies strong periodicity, we often do not write "strong(ly)" or "weak(ly)". A partial word w is *primitive* if there exists no total word v such that $w \subset v^m$ with $m \geq 2$, equivalently, if there is no proper divisor p of $|w|$ such that w is strongly p-periodic. Clearly, if w is primitive and $w \subset v$, then v is primitive.

A *factor* of a partial word w is a consecutive sequence of symbols in w, while a *subword* of w is a total word compatible with a factor in w. We denote by $w[i..j]$ the factor of w starting at position i and ending at position $j - 1$. A *square factor* of w is a factor of the form uv with u and v compatible. The *root* of uv is $u \vee v$. Now uv is a *PR-square factor* if its root is primitive, while uu is a *PR-square subword* if it is a total square word, with primitive root u, that is compatible to a square factor (not necessarily a PR-square factor). For example, if $w = 0\diamond\diamond10\diamond\diamond111$ then $0\diamond\diamond10\diamond\diamond1$ is a factor occurring at position 0 of w but it is not a PR-square factor, however, 01110111 is a PR-square subword occurring at position 0 of w. Note that $w[3..7]$ has three PR-square factors, $0\diamond$, $\diamond\diamond$, and $10\diamond\diamond$, with primitive roots 0, $\diamond$, and 10, respectively. It has four PR-square subword occurrences, 0^2, 1^2, and $(10)^2$, with 0^2 occurring twice, at positions 1 and 2.

To extend the definition of repetition, we use *strong* periodicity. The root of a repetition (f, l, p) in w is the length-p partial word u such that for all $0 \leq i < p$, $u[i] = w[f + i] \vee w[f + p + i] \vee \cdots \vee w[f + cp + i]$ with the smallest integer c satisfying $f + (c + 1)p + i > l$. A *PR-run* is then defined as in the case of total words. For example, in $110\diamond0\diamond0\diamond1$, $(2, 5, 2)$ is a square with root $0\diamond$, $(1, 7, 2)$ is a maximal repetition with root 10, and $(2, 7, 1)$ is a PR-run with root 0. The next proposition gives a condition for a maximal repetition to be PR.

Proposition 1. *A maximal repetition* (f, l, p) *in a partial word* w *is not PR if and only if there exists a maximal repetition* (f, l, p') *in* w *such that* p' *is a large divisor of* p.

3   Computing All PR-Square Factor and Run Occurrences

Algorithm 1 finds the positions of every PR-square factor occurrence in a partial word of length $n \geq 2$ over a k-letter alphabet. Let SqQ, $RepQ$ be empty queues which hold integer pairs, and $isPR$ be a boolean array of size $n \times \lfloor \frac{n}{2} \rfloor$, where each entry is initialized to *true*. The idea behind Algorithm 1 is to find all squares by increasing root length, and use the information from the squares of smaller root lengths to determine the primitivity of the roots of squares of larger root lengths. Given a partial word w of length n, for every m from 1 to $\lfloor \frac{n}{2} \rfloor$, do Procedures 1, 2, and 3.

Procedure 1. PSFBlocks(m)

Ensure: blocks and positions of PR-square factors of root length m
1: $start \leftarrow 0$
2: **for** $i = 0, \ldots, n - m - 1$ **do**
3:     **if** $w[i]$ is incompatible with $w[i + m]$ **then**
4:         **if** $i - start \geq m$ **then**
5:             push $(start, i + m)$ onto SqQ
6:             **for** $j = start, \ldots, i - m$ **do**
7:                 **if** $isPR[j][m]$ **then**
8:                     output that a PR-square factor of root length m occurs at j
9:             $start \leftarrow i + 1$
10:     **else if** $i = n - m - 1$ and $i - start + 1 \geq m$ **then**
11:         push $(start, n)$ onto SqQ
12:         **for** $j = start, \ldots, i + 1 - m$ **do**
13:             **if** $isPR[j][m]$ **then**
14:                 output that a PR-square factor of root length m occurs at j

---

Procedure 1: All squares of root length m can be found easily in maximal weakly m-periodic factors of w (i.e., if we extend the factor to the left or right, it is no longer weakly m-periodic), which we refer to as *blocks*. Computing all such blocks can be done by simply iterating once through w and checking positions that are m-apart for compatibility. When an incompatibility is found, the current block is put onto a queue and the next block is started, skipping over the incompatibility. To check if a square of root length m is PR, we look at $isPR[j][m]$ which indicates whether or not the square factor beginning at j with root length m is PR. The array $isPR[j][m]$ is updated in Procedure 3.

Procedure 2: Finding all maximal repetitions of root length m can be done by iterating through every block, maintaining the current period, and keeping track of the last position where a letter was seen for each position in the period.

Procedure 2. PSFReps(m)

Require: *root* (resp., *lastLetter*) is a symbol (resp., an integer) array of length m
Ensure: maximal repetitions of root length m
1: **while** $SqQ.first$ is not empty **do**
2:      $block \leftarrow$ poll(SqQ)
3:      $start \leftarrow block.first$
4:      $rootPos \leftarrow 0$
5:      **for** $i = 0, \ldots, m - 1$ **do**
6:          $root[i] \leftarrow w[start + i]$
7:          **if** $root[i]$ is not a hole **then**
8:              $lastLetter[i] \leftarrow start + i$
9:          **else**
10:             $lastLetter[i] \leftarrow start$
11:     **for** $i = block.first + m, \ldots, block.last - 1$ **do**
12:         **if** ($w[i]$ is not compatible with $root[rootPos]$) and ($lastLetter[rootPos] \geq start$) **then**
13:             push $(start, i - start)$ onto $RepQ$
14:             $start \leftarrow \max\{lastLetter[rootPos], start\} + 1$
15:             $root[rootPos] \leftarrow w[i]$ and $lastLetter[rootPos] \leftarrow i$
16:         **else if** $i - block.last - 1$ **then**
17:             push $(start, block.last - start)$ onto $RepQ$
18:         **else**
19:             **if** $w[i]$ is not a hole **then**
20:                 $root[rootPos] \leftarrow w[i]$ and $lastLetter[rootPos] \leftarrow i$
21:         $rootPos \leftarrow (rootPos + 1) \bmod m$

When an incompatibility is found, the current repetition is put onto a queue, the period position is updated to the most recent letter, and the process continues from that same spot.

Procedure 3: Since any square that is not PR is contained in a repetition of some root length m, using the maximal repetitions to mark off which squares of larger root lengths are not PR provides a good alternative to checking every factor individually for primitivity. For each maximal repetition found from left to right, and for each prime p, mark off the squares of root length pm contained in the current maximal repetition but not in any later maximal repetition.

To illustrate Algorithm 1, consider the partial word 101◊◊◊◊01◊012◊12112000◊. In Procedure 1, $m = 1$ finds blocks (2,7), (8,10), (12,14), (16,17), (19,22); $m = 2$ finds blocks (0,9), (13,16), (19,22); $m = 3$ finds the block (0,18); $m = 4$ finds the block (0,10); $m = 5$ finds no blocks; $m = 6$ finds the block (0,15); and $m \geq 7$ finds no blocks. Once a block is found in Procedures 1, and 2 finds maximal repetitions. For $m = 1$, it finds maximal repetitions (2,6), (3,7), (8,9), (9,10), (12,13), (13,14), (16,17), (19,22); $m = 2$ finds (0,9), (13,16), (19,22); $m = 3$ finds (0,11), (1,15), (11,18); $m = 4$ finds (0,9), (3,10); $m = 5$ finds no repetitions; $m = 6$ finds (0,11), (1,15); $m \geq 7$ finds no repetitions. Once a maximal repetition is found in Procedures 2, and 3 updates $isPR[j][m]$, e.g., when considering the maximal repetition $(2, 6, 1)$, $isPR[2][2]$ and $isPR[3][2]$ are set to *false*. Similarly,

---

Procedure 3. PSFPR(m)

Ensure: mark off, using the maximal repetitions, which squares of larger root length
    are not PR

1: **while** $RepQ$ is not empty **do**
2:    $rep \leftarrow \texttt{poll}(RepQ)$
3:    $start \leftarrow rep.first$ and $runLength \leftarrow rep.length$
4:    $nextLength \leftarrow 0$
5:    **if** $RepQ$ is not empty **then**
6:       $nextRep \leftarrow \texttt{peek}(RepQ)$
7:       $next \leftarrow nextRep.first$ and $nextLength \leftarrow nextRep.length$
8:    **for** each prime p less than or equal to $\frac{runLength}{2m}$ **do**
9:       $maxPos \leftarrow runLength - 2pm$
10:      **if** $p \leq \frac{nextLength}{2m}$ **then**
11:        $maxPos \leftarrow \min\{maxPos, next - start - 1\}$
12:      **for** $i = 0, \ldots, maxPos$ **do**
13:        $isPR[start + i][pm] \leftarrow false$

---

the following are set to $false$ when considering the run $(1, 15, 3)$: $isPR[1][6]$,
$isPR[2][6]$, $isPR[3][6]$, and $isPR[4][6]$.

Theorem 1. *Algorithm 1 outputs exactly the PR-square factors.*

Proof. We claim that Procedure 1 finds every block of root length m. If Lines 3–9
push a block, then every position in the block has been checked for compatibility
with the position m positions away, and thus the block contains squares of root
length m. Moreover, the block is cut off because of an incompatibility on the
right side, and it was started directly after an incompatibility on the left side,
thus the block is maximal. The block is also guaranteed to have at least one
square since $i - start \geq m$ implies that $i + m - start \geq 2m$, and thus the first
position of the block starts a square. If a block is pushed on in Lines 10–14, then
we are on the last iteration, and hence the same argument applies, noting that
$i - start + 1 \geq m$ is the exact condition for when the block contains at least one
square, since getting to this conditional statement means that $w[i]$ is compatible
with $w[i+m]$. Furthermore, if u is a block, then the same basic logic applies and
u is found by the procedure. Thus exactly every block of root length m is found
by Procedure 1.

Next we claim that Procedure 2 finds all maximal repetitions of root length m.
Since every such repetition is weakly m-periodic, every maximal repetition is contained in a block, so it suffices to show that the procedure finds every run within
a given block. By checking compatibilities with a given root, and pushing the
current repetition only when we reach the end of the block or an incompatibility
with the root, we ensure that every repetition is maximal on the right. Moreover,
if an incompatibility between $w[i]$ and $root[rootPos]$ is reached, then the next
run must have $root[rootPos] = w[i]$, and the start of the next run must be at
least one position after the previous run as well as at least one position after
the last non-hole that was compatible with $root[rootPos]$. Thus the repetition

cannot be extended on either side and is maximal. Moreover, since every position starts a square, every repetition found has at least length $2m$, and is thus a valid repetition. This implies that the procedure only outputs valid, maximal repetitions of root length m. Similarly it finds every such repetition.

Finally, to prove that Procedure 1 outputs only PR-square factors, we must prove that Procedure 3 updates $isPR$ completely. Suppose a square uv of root length m that is not PR occurs at some position. Then $uv \subset x^{2q}$ for some x^q such that q is a proper divisor of m. If q is not a large divisor of m, then there exists a prime p such that $\frac{m}{p}$ is a multiple of q. This implies that $uv \subset (x^{\frac{q}{p}})^{2p}$, so uv is contained in a repetition that has a period length equal to a large divisor of m. Thus for every square of root length m that is not PR, there is a repetition of period length $\frac{m}{p}$ that witnesses the non-primitivity of the root of uv such that $\frac{m}{p}$ is a large divisor of m. As a result, consider a given repetition of root length m. Then for every prime p such that a factor uv of length $2pm$ occurs in the repetition, m is a large divisor of pm and is a witness to the non-primitivity of uv. Also, since two repetitions of the same period length may overlap, it suffices to only use a rightmost repetition as the witness for a given position and square root length. Therefore Procedure 3 updates $isPR$ completely and correctly so that Procedure 1 outputs every PR-square factor. □

Theorem 2. *Given a partial word w of length n, Algorithm 1 computes the number of occurrences of PR-square factors of w in $O(n^2 \log \log n)$ time.*

Proof. First consider Procedure 1. The outer loop (Line 2) iterates less than n times, and every command other than the inner loops takes constant time. After any going through the first inner loop (Line 3), $start$ is always set to larger than i, and since i is always increasing, $start$ takes on each value from 1 to $n - m$ at most once. Moreover, when the second inner loop (Line 10) iterates, the outer loop is on its last iteration, and hence it adds no more than $O(n)$ time. Thus Procedure 1 takes $O(n)$ time.

Next consider Procedure 2. The outer loop (Line 1) iterates over each block found by Procedure 1. Note that each block has at least length $2m$ since it must contain a square. The first m positions are iterated over by the first for loop (Lines 5–10), and the remaining positions are iterated over by the second for loop (Lines 11–21), both of these loops taking at most constant time per iteration. Thus it remains to show that the number of positions in all the blocks combined is $O(n)$. To see this, observe that for every block except the last, there must be m positions which cannot start squares because of at least one incompatibility. Moreover, there are two types of positions in every block: those which start a square, and those which do not but are included because they are contained in the last square of the block. The last $2m - 1$ positions fit this latter category, but the first m of them cannot begin square positions. Thus at most the last $m - 1$ positions are double-counted in a second block. Since the starts of every block are thus at least m-apart, there are at most $\frac{n}{m}$ blocks of root length m. Thus the number of positions double-counted is at most $\frac{n}{m}(m - 1)$, which is $O(n)$. Since Algorithm 1 has $\lfloor \frac{n}{2} \rfloor$ iterations, the total time spent on Procedures 1 and 2 is $O(n^2)$.

Consider Procedure 3's total running time over every root length m. First, the primes less than n can easily be computed in $O(n^2)$ time using a sieve [3]. Second, by considering the next values from $RepQ.first$ and $RepQ.length$, we check in constant time for each prime less than or equal to the current value of $\frac{runLength}{2m}$ whether or not the positions which this repetition and the next could update have overlap, and if they do, we save the positions for the next repetition to update. Thus no two distinct repetitions of the same root length m access $isPR[i][m]$ for any position i.

Now consider how many times any given value $isPR[i][m]$ is accessed: once at the beginning when it is set to the default value $true$, once if a square of root length m occurs at position i, and once for each distinct prime factor p of m for which there is a repetition of root length $\frac{m}{p}$ that contains $w[i] \cdots w[i + 2m - 1]$. This gives a total of $2 + \omega(m)$ updates at most. Summing over all root lengths at position i gives at most $\sum_{j=1}^{\lfloor \frac{n-i}{2} \rfloor} \omega(j)$ updates, which by [13] is $\Theta(\frac{n-i}{2} \log\log \frac{n-i}{2})$. Summing over all positions gives $\sum_{i=0}^{n-1} c(n-i) \log\log(n-i)$ as an upper bound, and this is in turn $O(n^2 \log\log n)$. $\square$

Remark 1. Algorithm 1 has a best case running time of $O(n^2)$ because Procedures 1 and 2 are $O(n^2)$. By examining the algorithm, we see that Procedure 3 is $O(n^2)$ when the number of maximal repetitions is constant, so in this case, Algorithm 1 has running time $O(n^2)$.

Algorithm 1 can be modified for computing all PR-runs in a partial word. The only modifications that are necessary are to remove the outputs from Procedure 1, to remove Procedure 3, and to check which maximal repetitions are PR using Proposition 1. Thus, all PR-runs can be found by sorting the repetitions based on their first and last positions using a bucket sort, and then checking which periods are large divisors of which for each starting and ending position. To illustrate this PR-runs algorithm, looking at $101\diamond\diamond\diamond01\diamond012\diamond12112000\diamond$ again, we have the same maximal repetitions from Procedures 1 and 2. For $m = 1$, every maximal repetition will be marked as PR. Since $(0, 9, 1)$ is not a PR-run, $(0, 9, 2)$ is marked as PR. Then $(0, 9, 4)$ is marked as not PR because 2 is a large divisor of 4.

Theorem 3. *Given a partial word w of length n, all PR-run occurrences in w can be computed in $O(n^2 \log\log n)$ time.*

Proof. By the previous theorems, Procedures 1 and 2 correctly identify all maximal repetitions for all root lengths $1 \leq m \leq \lfloor \frac{n}{2} \rfloor$ in $O(n^2)$ time. Maximal repetitions are characterized by their starting position f, ending position l, and root length m. Proposition 1 states that a maximal repetition (f, l, p) in w is not PR if and only if there exists a maximal repetition (f, l, p') in w such that p' is a large divisor of p. We can sort the runs by their start and end positions in $O(n^2)$ time using a bucket sort. Since we process m in increasing order, the runs in each bucket are naturally ordered by increasing root length, and we examine them again in this order. To ensure that only PR-runs are counted, we check each run

(f, l, p) against the already examined runs (f, l, p') with the same starting and ending positions. Since $p > p'$, we have the cases: (1) if there is no (f, l, p') such that p' is a large divisor of p, (f, l, p) is PR (2) if there is (f, l, p') such that p' is a large divisor of p, (f, l, p) is not PR. For each p, we only need to check all its large divisors p', the number of which is $\omega(p)$. So it suffices to bound $\sum_{p=1}^{n} \omega(p)$, this sum is $O(n \log \log n)$. There are $O(n)$ repetitions for any fixed p.   □

4   Counting All PR-Square Subword Occurrences

Algorithm 2 finds the number of PR-square subword occurrences compatible with each square factor of a partial word w of length $n \geq 2$ over a k-letter alphabet. In addition to SqQ, $RepQ$, and $isPR$ from Algorithms 1, and 2 requires an array $wRoot$ of symbols of length n. Finding all PR-square subword occurrences reduces to first finding all square factor occurrences and determining if they are primitively-rooted, which Algorithm 1 does, and then finding all primitive completions of the roots of these square factor occurrences, which Procedure 5 does. Given a partial word w of length n, for every m from 1 to $\lfloor \frac{n}{2} \rfloor$, do Procedures 2, 3, and 4. Procedure 4 calls Procedure 5. We first outline Procedure 5, which takes as input a partial word u of length n with h holes over a k-letter alphabet and outputs the number of primitive completions of u.

If u is primitive, then Line 2 returns the correct value and if u is non-primitive, then Line 19 does so. To see the latter, first recall that the number of primitive total words of length n over a k-letter alphabet is $\sum_{d \mid n} k^d \mu(\frac{n}{d})$, where μ is the Möbius function [15]. Since Procedure 5 is called by Algorithm 2, which takes at least $\Omega(n^2)$ time, we can take $O(n^2)$ time to pre-compute all divisors of the integers 1 to n as well as their values for the Möbius function.

Suppose that u is non-primitive. Then u is strongly d-periodic for some proper divisor d of n. Thus we can classify every primitive completion of u by the partial word v with the maximum number of holes such that $u \subset v^{\frac{n}{d}}$. Lines 5–18 consider periods v whose lengths are divisors of n and Line 18 counts the primitive completions of u corresponding to each period found. Note that it is possible to double-count primitive completions of u (e.g., $11\diamond\diamond1\diamond$). So, it is a matter of using inclusion-exclusion to ensure no double-counting. We can apply the number-theoretic inclusion-exclusion (involving the Möbius function). To find the number of primitive completions of u it suffices to find, for each divisor d of n, the number of total words of length n over a k-letter alphabet that are an $\frac{n}{d}$-th power, not necessarily primitively-rooted, compatible with u, and then sum up these results with the appropriate $\mu(\frac{n}{d})$ coefficients. Assuming that all the divisors of n have been pre-computed as well as their values for the Möbius function, this requires stepping through u once for each divisor which takes $O(nd(n))$ time, where $d(n)$ is the number of divisors of n. Procedure 5 is run $O(n)$ times for any fixed $|u|$.

Now, consider Procedure 5 when applied in Line 12 of Procedure 4, for two consecutive values of j. Then the body of the loop of Procedure 5 actually does almost the same thing for both calls. This redundancy can be easily replaced

---

Procedure 4. PSSBlocks(m)

Ensure: blocks and number of PR-square subword occurrences of root length m

1: $start \leftarrow 0$
2: **for** $i = 0, \ldots, n - 1$ **do**
3:   **if** $i + m < n$ and $w[i] \uparrow w[i + m]$ **then**
4:     $wRoot[i] \leftarrow w[i] \vee w[i + m]$
5:   **else**
6:     $wRoot[i] \leftarrow w[i]$
7: **for** $i = 0, \ldots, n - m - 1$ **do**
8:   **if** $w[i]$ is incompatible with $w[i + m]$ **then**
9:     **if** $i - start \geq m$ **then**
10:       push $(start, i + m)$ onto SqQ
11:       **for** $j - start, \ldots, i - m$ **do**
12:         output $\text{PC}(wRoot[j] \cdots wRoot[j + m - 1], m, h, k, isPR[j][m])$
13:         **if** $wRoot[j]$ is a hole and $wRoot[j + m]$ is not a hole **then**
14:           $h \leftarrow h - 1$
15:         **else if** $wRoot[j]$ is not a hole and $wRoot[j + m]$ is a hole **then**
16:           $h \leftarrow h + 1$
17:     $start \leftarrow i + 1$
18:     $h \leftarrow 0$
19:   **else if** $i = n - m - 1$ and $i - start + 1 \geq m$ **then**
20:     push $(start, n)$ onto SqQ
21:     **for** $j = start, \ldots, i + 1 - m$ **do**
22:       output $\text{PC}(wRoot[j] \cdots wRoot[j + m - 1], m, h, k, isPR[j][m])$
23:       **if** $wRoot[j]$ is a hole and $wRoot[j + m]$ is not a hole **then**
24:         $h \leftarrow h - 1$
25:       **else if** $wRoot[j]$ is not a hole and $wRoot[j + m]$ is a hole **then**
26:         $h \leftarrow h + 1$
27:     **if** $i - start < m$ and $wRoot[i]$ is a hole **then**
28:       $h \leftarrow h + 1$

---

with a sliding window approach, which already proved useful in Procedure 1. For any factor of length m and any divisor d of m, we count the number of total words which are an $\frac{m}{d}$-th power, not necessarily primitively-rooted, and are compatible with the factor. Since we are ultimately interested in PR-squares, we can restrict to even values of m and divisors d of $\frac{m}{2}$. These values appear in the number-theoretic inclusion-exclusion, with coefficients $\mu(\frac{m}{2d})$, counting the PR-square subwords compatible with a given factor.

For fixed m and d, we can compute these values for all length-m factors in $O(n)$ time. It suffices to have a sliding window of fixed length m, which for any remainder modulo d tracks the situation at positions inside the sliding window with that remainder modulo d. There can be either a conflict (at least two different letters), just one letter, or just holes. To maintain the situation for a fixed remainder, it suffices to store the last two letters encountered along with the positions where they were encountered for the last time. Then a conflict in some remainder is equivalent to $isCompatible = false$ in Procedure 5, and if there is

---

Procedure 5. PC($u, n, h, k, isPrimitive$)

Require: u is a partial word of length n with h holes over a k-letter alphabet, $isPrimitive$ is true if u is primitive

Ensure: the number of primitive completions of u

1: **if** $isPrimitive$ **then**
2:     **return** k^h
3: **else**
4:     $result \leftarrow 0$
5:     **for** each divisor d of n in increasing order **do**
6:         $isCompatible \leftarrow true$
7:         $i \leftarrow 0$
8:         let $period$ be a symbol array of length d
9:         **while** $i < n$ and $isCompatible$ **do**
10:            **if** $i < d$ **then**
11:              $period[i] \leftarrow u[i]$
12:            **else**
13:              **if** $period[i \bmod d]$ is a hole and $u[i]$ is not a hole **then**
14:                $period[i \bmod d] \leftarrow u[i]$
15:              $isCompatible \leftarrow period[i \bmod d] \uparrow u[i]$
16:            $i \leftarrow i + 1$
17:         **if** $isCompatible$ **then**
18:            $result \leftarrow result + k^{\text{\# holes in period}} \mu(\frac{n}{d})$
19:     **return** $result$

---

no conflict, the answer is $k^{\text{\# of remainders with just holes}}$. The sliding window needs to be applied $\sum_{m=1}^{n} d(m)$ times.

This sum is $O(n \log n)$; there are $O(\frac{n}{i})$ multiplicities of i not exceeding n, this leads to the harmonic series and $H_n = \sum_{i=1}^{n} \frac{1}{i} = O(\log n)$.

Theorem 4. *Given a partial word w of length n, Algorithm 2 computes the number of occurrences of PR-square subwords of w in $O(n^2 \log n)$ time.*

Reexamining 101◇◇◇◇01◇012◇12112000◇, the square occurrence 1◇◇◇ at position 2 has a PR-square completion 1010, but was ruled out before because it is not a PR-square factor. Similarly, ◇◇◇◇ at position 3 has PR-square completions 0101 and 1010, but was ruled out before because it is not a PR-square factor. Additionally, 01◇◇◇◇01◇012 at position 1 is not a PR-square factor, but it has PR-square completions 011012011012 and 010012010012.

Remark 2. One might try to apply dynamic programming to find the runs in all the algorithms above, but since it must be at least quadratic, it will not improve the time complexity of the current algorithms. Also, to keep the transition in dynamic programming efficient, extra work has to be done, which makes it essentially the same as the current algorithms.

Remark 3. If h is large, the result does not fit a RAM word, but instead we can represent it as $\sum_{i=0}^{h} a_i k^i$ for some (small) integer coefficients a_i.

5 Conclusion

We suggest the following problems for future work: How many occurrences of primitively-rooted squares are there in a given partial word of length n? What is the time complexity of computing all the squares and runs (not necessarily primitively-rooted) in a given partial word of length n?

Our algorithms can be useful in the combinatorial analysis of repetitions in partial words where computer checks are often applied.

References

1. Abrahamson, K.: Generalized string matching. SIAM J. Comput. **16**(6), 1039–1051 (1987)
2. Apostolico, A., Preparata, F.P.: Optimal off-line detection of repetitions in a string. Theor. Comput. Sci. **22**(3), 297–315 (1983)
3. Bach, E., Shallit, J.: Algorithmic Number Theory. Efficient Algorithms, vol. 1. MIT Press, Cambridge (1996)
4. Blanchet-Sadri, F., Bodnar, M., Fox, N., Hidakatsu, J.: A graph polynomial approach to primitivity. In: Dediu, A.-H., Martín-Vide, C., Truthe, B. (eds.) LATA 2013. LNCS, vol. 7810, pp. 153–164. Springer, Heidelberg (2013)
5. Blanchet-Sadri, F., Jiao, Y., Machacek, J.M., Quigley, J.D., Zhang, X.: Squares in partial words. Theor. Comput. Sci. **530**, 42–57 (2014)
6. Blanchet-Sadri, F., Merçaş, R., Rashin, A., Willett, E.: Periodicity algorithms and a conjecture on overlaps in partial words. Theor. Comput. Sci. **443**, 35–45 (2012)
7. Crochemore, M.: An optimal algorithm for computing the repetitions in a string. Inf. Process. Lett. **12**(5), 244–250 (1981)
8. Crochemore, M., Hancart, C., Lecroq, T.: Algorithms on Strings. Cambridge University Press, New York (2007)
9. Crochemore, M., Ilie, L., Rytter, W.: Repetitions in strings: Algorithms and combinatorics. Theor. Comput. Sci. **410**, 5227–5235 (2009)
10. Diaconu, A., Manea, F., Tiseanu, C.: Combinatorial queries and updates on partial words. In: Kutyłowski, M., Charatonik, W., Gębala, M. (eds.) FCT 2009. LNCS, vol. 5699, pp. 96–108. Springer, Heidelberg (2009)
11. Fischer, M., Paterson, M.: String matching and other products. In: Karp, R. (ed.) 7th SIAM-AMS Complexity of Computation, pp. 113–125 (1974)
12. Halava, V., Harju, T., Kärki, T.: On the number of squares in partial words. RAIRO-Theor. Inf. Appl. **44**(1), 125–138 (2010)
13. Hardy, G.H., Wright, E.M.: An Introduction to the Theory of Numbers. Oxford University Press, London (2008)
14. Kolpakov, R., Kucherov, G.: Finding maximal repetitions in a string in linear time. In: FOCS 1999, pp. 596–604. IEEE Computer Society Press, Los Alamitos (1999)
15. Lothaire, M.: Combinatorics on Words. Cambridge University Press, Cambridge (1997)
16. Main, M.G., Lorentz, R.J.: An O(nlog n) algorithm for finding all repetitions in a string. J. Algorithms **5**(3), 422–432 (1984)
17. Manea, F., Merçaş, R., Tiseanu, C.: Periodicity algorithms for partial words. In: Murlak, F., Sankowski, P. (eds.) MFCS 2011. LNCS, vol. 6907, pp. 472–484. Springer, Heidelberg (2011)

3-Coloring Triangle-Free Planar Graphs with a Precolored 9-Cycle

Ilkyoo Choi[1], Jan Ekstein[2], Přemysl Holub[2], and Bernard Lidický[3(✉)]

[1] Korea Advanced Institute of Science and Technology, Daejeon, South Korea
ilkyoo@kaist.ac.kr
[2] University of West Bohemia, Pilsen, Czech Republic
{ekstein,holubpre}@kma.zcu.cz
[3] Iowa State University, Ames, USA
lidicky@iastate.edu

Abstract. Given a triangle-free planar graph G and a cycle C of length 9 in G, we characterize all situations where a 3-coloring of C does not extend to a proper 3-coloring of G. This extends previous results for the length of C up to 8.

1 Introduction

Let $[n] = \{1, 2, \ldots, n\}$. Graphs in this paper are finite and may have loops or parallel edges. Given a graph G, let $V(G)$ and $E(G)$ denote the vertex set and the edge set of G, respectively. We will also use $|G|$ for the size of $E(G)$. A *proper k-coloring* of a graph G is a function $\varphi : V(G) \to [k]$ such that $\varphi(u) \neq \varphi(v)$ for each edge $uv \in E(G)$. A graph is *k-colorable* if there exists a proper k-coloring of the graph, and the minimum k where a graph is k-colorable is the *chromatic number* of the graph.

Garey and Johnson [15] proved that deciding if a graph is k-colorable is NP-complete even when $k = 3$. Moreover, deciding if a graph is 3-colorable is still NP-complete when restricted to planar graphs [9]. Therefore, even though planar graphs are 4-colorable by the celebrated Four Color Theorem [4,5,19], finding sufficient conditions for a planar graph to be 3-colorable has been an active area of research. A landmark result in this area is Grötzsch's Theorem [17], which is the following:

Theorem 1 ([17]). *Every triangle-free planar graph is 3-colorable.*

Ilkyoo Choi—Supported by Basic Science Research Program through the National Research Foundation of Korea (NRF) funded by the Ministry of Science, ICT & Future Planning (2011-0011653).
Jan Ekstein—Supported by P202/12/G061 of the Czech Science Foundation and by the European Regional Development Fund (ERDF), project NTIS - New Technologies for the Information Society, European Centre of Excellence, CZ.1.05/1.1.00/02.0090.
Přemysl Holub—Supported by NSF grants DMS-1266016.

© Springer International Publishing Switzerland 2015
J. Kratochvíl et al. (Eds.): IWOCA 2014, LNCS 8986, pp. 98–109, 2015.
DOI: 10.1007/978-3-319-19315-1_9

We direct readers to a nice survey by Borodin [7] for more results and conjectures regarding 3-coloring planar graphs.

A graph G is *k-critical* if it is not $(k-1)$-colorable but every proper subgraph of G is $(k-1)$-colorable. Critical graphs are important since they are (in a certain sense) the minimal obstacles in reducing the chromatic number of a graph. Numerous coloring algorithms are based on detecting critical subgraphs. Despite its importance, there is no known characterization of k-critical graphs when $k \geq 4$. On the other hand, there has been some success regarding 4-critical planar graphs. Extending Theorem 1, the Grünbaum–Aksenov Theorem [1,6,18] states that a planar graph with at most three triangles is 3-colorable, and we know that there are infinitely many 4-critical planar graphs with four triangles. Borodin, Dvořák, Kostochka, Lidický, and Yancey [8] were able to characterize all 4-critical planar graphs with four triangles.

Given a graph G and a proper subgraph C of G, we say G is *C-critical for k-coloring* if for every proper subgraph H of G where $C \subseteq H$, there exists a proper k-coloring of C that extends to a proper k-coloring of H, but does not extend to a proper k-coloring of G. Roughly speaking, a C-critical graph for k-coloring is a minimal obstacle when trying to extend a proper k-coloring of C to a proper k-coloring of the entire graph. Note that $(k+1)$-critical graphs are exactly the C-critical graphs for k-coloring with C being the empty graph.

In the proof of Theorem 1, Grötzsch actually proved that any proper coloring of a 4-cycle or a 5-cycle extends to a proper 3-coloring of a triangle-free planar graph. This implies that there are no triangle-free planar graphs that are C-critical for 3-coloring when C is a face of length 4 or 5. This sparked the interest of characterizing triangle-free planar graphs that are C-critical for 3-coloring when C is a face of longer length. Since we deal with 3-coloring triangle-free planar graphs in this paper, from now on, we will write "C-critical" instead of "C-critical for 3-coloring" for the sake of simplicity.

The investigation was first done on planar graphs with girth 5. Walls [22] and Thomassen [20] independently characterized C-critical planar graphs with girth 5 when C is a face of length at most 11. The case when C is a 12-face was initiated in [20], but a complete characterization was given by Dvořák and Kawarabayashi in [11]. Moreover, a recursive approach to identify all C-critical planar graphs with girth 5 when C is a face of any given length is given in [11]. Dvořák and Lidický [10] implemented an algorithm and used a computer to generate all C-critical graphs with girth 5 when C is a face of length at most 16. The graphs generated were then used to reveal some structure of 4-critical graphs on surfaces without short contractible cycles.

The situation for planar graphs with girth 4, which are triangle-free planar graphs, is more complicated since the list of C-critical graphs is not finite when C has size at least 6. We already mentioned that there are no C-critical triangle-free planar graphs when C is a face of length 4 or 5. An alternative proof of the case when C is a 5-face was given by Aksionov [1]. Gimbel and Thomassen [16] not only showed that there exists a C-critical triangle-free planar graph when C is a 6-face, but also characterized all of them. Aksenov, Borodin, and Glebov [2] independently proved the case when C is a 6-face using the discharging method, and also characterized all C-critical triangle-free planar graphs when C is a 7-face in [3]. Dvořák and Lidický [14] used properties of nowhere-zero flows to give simpler proofs of the case when C is either a 6-face or a 7-face, and also characterized C-critical triangle-free planar graphs when C is an 8-face. The case where C is a 7-face was used in [8].

In this paper, we push the project further and characterize all C-critical triangle-free planar graphs when C is a face of length 9. For a plane graph G, let $S(G)$ denote the set of multisets of possible lengths of internal faces of G with length at least 5.

Theorem 2. *Let G be a connected plane triangle-free graph with outer face bounded by a cycle C of length 9. The graph G is C-critical for 3-coloring if and only if G contains no separating cycles of length at most five, the interior of every non-facial 6-cycle contains only faces of length four and one of the following propositions is satisfied (see Fig. 1 for an illustration):*

(a) $S(G) = \{5\}$ *and the 5-face of G intersects C in a path of length at least two.*
(b) $S(G) = \{7\}$ *and the 7-face of G intersects C in a path of length at least three.*
(c) $S(G) = \{5,6\}$ *and the 5-face, 6-face, of G intersects C in a path of length at least two, and four, respectively.*
(d) $S(G) = \{5,6\}$ *and G is depicted as (d1) or (d2) in Fig. 1.*
(e) $S(G) = \{5,5,5\}$ *and G is depicted as (Bij) in Fig. 1 for all i,j.*

2 Preliminaries

Our proof of Theorem 2 uses the same method as Dvořák and Lidický [14]. The main idea is to use the correspondence between coloring of a plane graph G and flows in the dual of G. In this paper, we give only a brief description of the correspondence and the lemmas useful in our case. A more detailed and general description can be found in [14].

Let $G^\star$ denote the dual of a plane graph G. Let φ be a proper 3-coloring of the vertices of G by colors $\{1, 2, 3\}$. For every edge uv of G, we orient the corresponding edge e in $G^\star$ in the following way. Let e have endpoints f, h in $G^\star$, where f,v,h is in the clockwise order from vertex u in the drawing of G. The edge e will be oriented from f to h if $(\varphi(u), \varphi(v)) \in \{(1,2),(2,3),(3,1)\}$, and from h to f otherwise.

Since φ is a proper coloring, every edge of $G^\star$ has an orientation. Tutte [21] showed that this orientation of $G^\star$ defines a nowhere-zero $\mathbb{Z}_3$-flow, which means that the in-degree and the out-degree of every vertex in $G^\star$ differ by a multiple of three. Conversely, every nowhere-zero $\mathbb{Z}_3$-flow in $G^\star$ defines a proper 3-coloring of G up to the rotation of colors.

Let h be the vertex in $G^\star$ corresponding to the outer face of G. Edges oriented away from h are called *source edges* and the edges oriented towards h are called *sink edges*. The orientations of edges incident to h depend only on the coloring of C, where C is the cycle bounding the outer face of G.

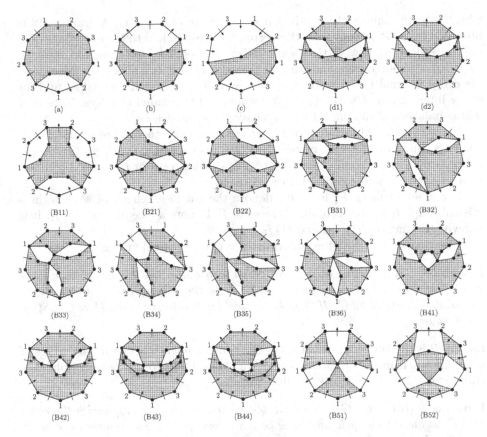

Fig. 1. All C-critical triangle-free plane graphs where C is an outer 9-face. Note that each figure actually represents infinitely many graphs, including ones that can be obtained by identifying some of the depicted vertices. The arrows correspond to source edges and sink edges that are defined in Preliminaries.

For a vertex f of $G^\star$, let $\delta(f)$ denote the difference of the out-degree and in-degree of f. Possible values of $\delta(f)$ depend on the size of the face corresponding to f, denoted by $|f|$. Clearly $|\delta(f)| \leq |f|$ and $\delta(f)$ has the same parity as $|f|$. Hence if $|f| = 4$, then $\delta(f) = 0$. Similarly, if $|f| \in \{5, 7\}$, then $\delta(f) \in \{-3, 3\}$ and if $|f| = 6$ then $\delta(f) \in \{-6, 0, 6\}$.

Next we convert the problem of extending a proper 3-coloring of C to the existence of a $\mathbb{Z}$-flow in an auxiliary graph obtained from $G^\star$. We call a function q assigning an integer to every internal face f of G a *layout* if $q(f) \leq |f|$, $q(f)$ is divisible by 3, and $q(f)$ has the same parity as $|f|$. Notice that $q(f)$ satisfies the same conditions as $\delta(f)$. Therefore it is sufficient to specify the q-values for faces of size at least 5, since $q(f) = 0$ if f is a 4-face.

Let ψ be a proper 3-coloring C. The coloring ψ gives an orientation of the edges corresponding to the edges of C in $G^\star$. Denote by n^s the number of source edges and by n^t the number of sink edges. A layout q is ψ-*balanced* if $n^s + m = n^t$, where m is the sum of the q-values over all internal faces of G. A graph $G^{q,\psi}$ is obtained from G^* by removing the vertex h corresponding to the outer face of G and by adding two new vertices s and t. For every edge hf in G^* we add one edge sf if hf is a source edge and we add one edge tf if it is a sink edge. Moreover, for every internal face f with $q(f) > 0$, we add $q(f)$ parallel sf edges and for every internal face f with $q(f) < 0$, we add $-q(f)$ parallel tf edges. Note that q is ψ-balanced if and only if s and t have the same degree.

For a ψ-balanced layout q of G, let $c(q, \psi)$ denote the degree of the source vertex s (and also the sink vertex t) of $G^{q,\psi}$. For an edge cut K in $G^{q,\psi}$ separating s from t, the component of $G^{q,\psi} \setminus K$ containing s, or t, is called a *source component*, or a *sink component*, respectively.

For a set of faces F, let $\ell(F)$ denote the smallest length of a cycle in a critical graph that may contain all faces of F. Denote a face of size i by f_i. It is known [13] that $\ell(\{f_i\}) = i$ and $\ell(\{f_5, f_6\}) = 9$.

The next lemma describes interiors of cycles in critical graphs.

Lemma 1 ([12]). *Let G be a plane graph with outer face K. Let C be a cycle in G that does not bound a face, and let H be the subgraph of G drawn in the closed disk bounded by C. If G is K-critical for k-coloring, then H is C-critical for k-coloring.*

Lemma 2 is the key lemma that gives the correspondence between 3-colorings of C and flows. It implies that if a 3-coloring of C extends to the entire graph, then there is a $\mathbb{Z}$-flow from s to t of value $c(q, \psi)$.

Lemma 2 ([14]). *Let G be a connected plane triangle-free graph with the outer face C bounded by a cycle and let ψ be a 3-coloring of C. The coloring ψ extends to a 3-coloring of G if and only if there exists a ψ-balanced layout q such that the terminals of $G^{q,\psi}$ are not separated by an edge cut smaller than $c(q, \psi)$.*

The cuts showing that a 3-coloring of C does not extend are described by the following lemma.

Lemma 3 ([14]). *Let G be a connected plane triangle-free graph with the outer face C bounded by a cycle and let ψ be a 3-coloring of C that does not extend to a 3-coloring of G. If q is a ψ-balanced layout in G, then there exists a subgraph $K_0 \subset G$ such that either*

(i) *K_0 is a path with both ends in C and no internal vertex in C, and if P is a path in C joining the end vertices of K_0, n_s is the number of source edges of P, n_t is the number of the sink edges of P and m is the sum of the values of q over all faces of G drawn in the open disk bounded by the cycle $P + K_0$, then $|n_s + m - n_t| > |K_0|$. In particular, $|P| + |m| > |K_0|$. Or,*

(ii) *K_0 is a cycle with at most one vertex in C, and if m is the sum of the values of q over all faces of G drawn in the open disk bounded by K_0, then $|m| > |K_0|$.*

3   Proof of Theorem 2

Let $\mathcal{S}_k$ be the set of possible multisets of sizes of faces of length at least five in a graph of girth at least 4 where the length of the precolored face is k. The result of Dvořák, Král', and Thomas [13] implies among others that $\mathcal{S}_6 = \{\emptyset\}$, $\mathcal{S}_7 = \{\{5\}\}$, $\mathcal{S}_8 = \{\emptyset, \{6\}, \{5, 5\}\}$, and $\mathcal{S}_9 = \{\{7\}, \{5\}, \{6, 5\}, \{5, 5, 5\}\}$.

From now on, G is always a C-critical triangle-free plane graph and C is always the outer face of length 9. By the previous paragraph, we have four cases to consider when C has length 9. The case of one 7-face was already resolved by Dvořák and Lidický [14], and it is described in Theorem 2(b). We resolve the remaining three cases in Lemmas 4, 5, and 6. The proof of Lemma 6 is omitted due to the page limit. In order to simplify the cases, we first solve the case when C has a chord.

If G is C-critical and C has a chord, then Lemma 1 implies that G can be obtained by identifying two edges of the outer faces of two different smaller critical graphs. It is not difficult to show that the converse is also true.

Therefore, we can enumerate C-critical graphs G where C has a chord and has length 9 by identifying edges from two smaller critical graphs with outer faces of length either 4 and 7 or 5 and 6. The resulting graphs are depicted in Fig. 1 (a) and (b), where some of the vertices must be identified.

In the following we assume that C has no chords. In the rest of the paper, ψ will always be a 3-coloring of C. Also, for a subset Z of the edges of C, we will use n_Z^s and n_Z^t to denote the number of source edges and sink edges of Z, respectively.

Lemma 4. *If G contains one 5-face f_5 and one 6-face f_6, and all other faces are 4-faces, then G is described by Theorem 2(c),(d) and depicted in Fig. 1(c),(d1), and (d2).*

Proof. Let G be a C-critical graph containing one 5-face f_5 and one 6-face f_6.

Let $e \in E(G) \setminus E(C)$. We want to find a 3-coloring ψ of C that does not extend to a proper 3-coloring of G but extends to a proper 3-coloring of $G - e$. Note that $G - e$ has either one 5-face and one 8-face, or one 6-face and one 7-face, or one 9-face, or two 6-faces and and one 5-face. We know that the smallest k such that $\mathcal{S}_k$ contains any of $\{5, 8\}, \{6, 7\}, \{9\},$ or $\{5, 6, 6\}$ is at least 11. Hence every precoloring of C extends to $G - e$. In particular, ψ extends to $G - e$. Therefore, we only need to characterize ψ that does not extend to G.

Let ψ be a proper 3-coloring of C that does not extend to a proper 3-coloring of G. By symmetry, we assume that C has more source edges than sink edges. Hence C has either 9 or 6 source edges. Let q be a ψ-balanced layout of G. By Lemma 2, there exists an edge-cut K in $G^{q,\psi}$ separating s from t such that $|K|$ is smaller than $c(q, \psi)$. Let $K_0 \subset G$ be obtained by Lemma 3 and let $k_0 = |K_0|$.

First suppose that K_0 is a cycle. Let m denote the sum of the q-values of the faces in the interior of K_0. By Lemma 3, $|m| > k_0$. If both f_5, f_6 are in the interior of K_0, then $|m| \le 9$, contradicting the fact that $|m| > k_0$ since $k_0 \ge \ell(\{f_5, f_6\}) = 9$. If f_5 is in the interior of K_0, but f_6 is not, then $|m| = 3$, while $\ell(\{f_5\}) = 5$, a contradiction again. Similarly, we obtain a contradiction when f_6 is in the interior of K_0 but f_5 is not, since $\ell(\{f_6\}) = 6$ and $|m| \le 6$. Therefore K_0 is always a path joining two distinct vertices of C.

The graph G bounded by C is divided by K_0 into two closed disks X and Y intersecting at K_0, where faces in X correspond to the vertices in the component containing s in $G^{q,\psi} - K$. For $Z \in \{X, Y\}$, denote by P_Z the subpath of C such that Z is bounded by $P_Z + K_0$. Recall that n_Z^s and n_Z^t denote the number of source edges and sink edges in P_Z, respectively. The described structure is shown in Fig. 2.

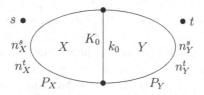

Fig. 2. Structure of a cut in G.

Claim 1. *There are 6 source edges in C.*

Proof. Suppose for a contradiction that C contains 9 source edges. Hence there is just one ψ-balanced layout q with $q(f_5) = -3$, $q(f_6) = -6$, and $c(q, \psi) = 9$. Note that $n_X^s + n_Y^s = 9$ and $n_X^t + n_Y^t = 0$. If both f_5, f_6 belong to X then $|K| = k_0 + n_Y^s + 9 < 9$, a contradiction. If both f_5, f_6 belong to Y then $|K| = k_0 + n_Y^s < 9$, while the length of the boundary cycle of Y is $k_0 + n_Y^s \ge \ell(\{f_5, f_6\}) = 9$, which is a contradiction again. Now suppose that exactly one of f_5, f_6 belongs to X and let f_X denote such a face and f_Y the other one. Then $|K| = k_0 + n_Y^s + |q(f_X)| < 9$ and $k_0 + n_Y^s \ge |f_Y|$. If $f_X = f_5$ then $k_0 + n_Y^s + 3 < 9$ and $k_0 + n_Y^s \ge 6$, which is a contradiction. If $f_X = f_6$ then $k_0 + n_Y^s + 6 < 9$ and $k_0 + n_Y^s \ge 5$, a contradiction. $\square$

Claim 2. *If q is a ψ-balanced layout with $q(f_5) = -3$ and $q(f_6) = 0$, then f_5 belongs to Y and f_6 belongs to X.*

Proof. Assume that $q(f_5) = -3$ and $q(f_6) = 0$. Hence the six source edges are the only edges incident to s, thus $c(q, \psi) = 6$. Note that $n_X^s + n_Y^s = 6$ and $n_X^t + n_Y^t = 3$. First suppose that both f_5, f_6 belong to X. Then $n_X^s + n_X^t + k_0 \ge \ell(\{f_5, f_6\}) = 9$, and the size of the cut K is $3 + k_0 + n_X^t + n_Y^s < c(q, \psi) = 6$. By subtracting the two previous inequalities we get $n_X^s - n_Y^s > 6$, contradicting the fact that $n_X^s + n_Y^s = 6$. Now suppose that both f_5, f_6 belong to Y. Then $n_Y^s + n_Y^t + k_0 \ge \ell(\{f_5, f_6\}) = 9$ and $|K| = k_0 + n_X^t + n_Y^s < 6$. By subtracting them we get $n_Y^t - n_X^t > 3$, a contradiction with $n_Y^t + n_X^t = 3$. Finally we suppose that f_5 belongs to X and f_6 belongs to Y. Then $n_Y^s + n_Y^t + k_0 \ge \ell(\{f_6\}) = 6$ and $|K| = 3 + k_0 + n_X^t + n_Y^s < 6$. But then $n_Y^t - n_X^t > 3$, a contradiction again. Therefore f_5 is in Y and f_6 is in X. $\square$

Claim 3. *If q is a ψ-balanced layout with $q(f_5) = 3$ and $q(f_6) = -6$, then f_5 belongs to X and f_6 belongs to Y.*

Proof. Assume that $q(f_5) = 3$ and $q(f_6) = -6$. Since there are six source edges on C and three edges from s to f_5 in $G^{q,\psi}$, $c(q,\psi) = 9$. Note that $n_X^s + n_Y^s = 6$ and $n_X^t + n_Y^t = 3$. First suppose that both f_5 and f_6 belong to X. Then $k_0 + n_X^s + n_X^t \geq \ell(\{f_5, f_6\}) = 9$ and $|K| = 6 + k_0 + n_Y^s + n_X^t < c(q,\psi) = 9$. But then we obtain $n_X^s - n_Y^s > 6$, contradicting the fact that $n_X^s + n_Y^s = 6$. Now suppose that both f_5 and f_6 belong to Y. Then $k_0 + n_Y^s + n_Y^t \geq \ell(\{f_5, f_6\}) = 9$ and the size of K is $3 + k_0 + n_Y^s + n_X^t < 9$. But then we get $n_Y^t - n_X^t > 3$, contradicting $n_X^t + n_Y^t = 3$. Finally we suppose that f_6 belongs to X and f_5 belongs to Y. Then $|K| = 9 + k_0 + n_X^s + n_X^t < 9$, a contradiction.   □

Since C has 6 source edges, we have two different ψ-balanced layouts. Let q_1 and q_2 be the layouts where $q_1(f_5) = -3$, $q_1(f_6) = 0$, and $q_2(f_5) = 3$, $q_2(f_6) = -6$, respectively. Let K and L be the subgraphs of G obtained by Lemma 3 applied to q_1 and q_2, respectively, and let $k = |K|$ and $l = |L|$. Note that we already showed that each of K and L is a path joining pairs of distinct vertices of C. Denote these vertices by v_1, v_2 for K and by w_1, w_2 for L. The prescribed structure is depicted in Figs. 3 and 4.

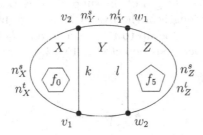

Fig. 3. A structure for two non-crossing cuts.

If we can choose the labels of the endpoints of K and L so that the clockwise order along C is v_1, v_2, w_1, w_2, then we call K and L *non-crossing*, and we call K and L *crossing* otherwise. Notice that K and L are always non-crossing if they have a vertex of C in common.

We treat the cases of K and L being crossing and non-crossing separately.

Claim 4. *If K and L are non-crossing, then G is depicted in Fig. 1(c).*

Proof. Assume that K and L are non-crossing. See Fig. 3. Note that K, L are not necessarily disjoint. The cuts K and L partition G into three parts. Denote by X the region of G containing f_6, by Z the region of G containing f_5, and by Y the rest of G. For an edge cut K' of $G^{q_1, \psi}$ corresponding to K, f_6 belongs to the source subdisk of G while f_5 belongs to the sink subdisk of G by Claim 2.

Analogously, for an edge cut L' of $G^{q_2,\psi}$ corresponding to L, f_5 belongs to the source subdisk of G while f_6 belongs to the sink subdisk of G by Claim 3. For the edge cut K', $|K'| = k + n_X^t + n_Y^s + n_Z^s < c(q_1, \psi) = 6$. For the edge cut L', $|L'| = l + n_X^s + n_Y^s + n_Z^t < c(q_2, \psi) = 9$. By the assumptions that C has no chord, $k \geq 2$ and $l \geq 2$. Since X contains f_6, $k + n_X^s + n_X^t \geq \ell(\{f_6\}) = 6$ and even, and since Z contains f_5, $l + n_Z^s + n_Z^t \geq \ell(\{f_5\}) = 5$ and odd. Clearly $n_X^s + n_Y^s + n_Z^s = 6$ and $n_X^t + n_Y^t + n_Z^t = 3$. Integer solutions to these constraints are in the following table:

n_X^s	n_X^t	n_Y^s	n_Y^t	n_Z^s	n_Z^t	k	l
4	0	0	2	2	1	2	2
4	0	0	3	2	0	2	3

From these solutions we obtain the graphs depicted in Fig. 1(c).   □

Claim 5. *If K and L are crossing, then G is depicted in Fig. 1(d1) or (d2).*

Proof. Assume that K and L cross, hence G is divided by K and L into four regions. Let X be the region of G containing f_6, Z be the region containing f_5, and let W, Y be the two remaining regions. Since K and L cross, they have a common internal vertex v. Note that $K \cap L$ is a path and v can be any vertex on the path. Denote by k_1 the length of the subpath of K between X and Y up to v, and denote by k_2 the length of the rest of K. Denote by l_1 the length of the subpath of L between Y and Z up to v, and denote by l_2 the length of the rest of L. The prescribed structure is depicted in Fig. 4.

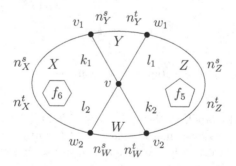

Fig. 4. A structure for two crossed cuts.

Note that $\min\{k_1, k_2, l_1, l_2\} \geq 1$ since v is an internal vertex. For an edge cut K' of $G^{q_1,\psi}$ corresponding to K, f_6 belongs to the source component while f_5 belongs to the sink component by Claim 2. Analogously, for an edge cut L' of $G^{q_2,\psi}$ corresponding to L, f_5 belongs to the source component while f_6 belongs to the sink component by Claim 3.

We obtain the following set of constraints that must be satisfied in this subcase.

$$|K'| = k_1 + k_2 + n_X^t + n_Y^s + n_Z^s + n_W^t < c(q_1, \psi) = 6 \tag{1}$$

$$|L'| = l_1 + l_2 + n_X^s + n_Y^s + n_Z^t + n_W^t < c(q_2, \psi) = 9 \tag{2}$$

$$k_1 + l_2 + n_X^s + n_X^t \geq \ell(\{f_6\}) = 6 \text{ and even} \tag{3}$$

$$l_1 + k_2 + n_Z^s + n_Z^t \geq \ell(\{f_5\}) = 5 \text{ and odd} \tag{4}$$

$$l_2 + k_2 + n_X^s + n_X^t + n_Y^s + n_Y^t + n_Z^s + n_Z^t \geq \ell(\{f_5, f_6\}) = 9 \text{ and odd} \tag{5}$$

$$\min\{k_1, l_1\} + n_Y^s + n_Y^t > \max\{k_1, l_1\} \tag{6}$$

$$\min\{k_2, l_2\} + n_W^s + n_W^t > \max\{k_2, l_2\} \tag{7}$$

$$n_X^s + n_Y^s + n_Z^s = 6 \tag{8}$$

$$n_X^t + n_Y^t + n_Z^t = 3 \tag{9}$$

Inequalities (1) and (2) come from the size of the cut being smaller than $c(q_1, \psi)$ and $c(q_2, \psi)$, respectively. Inequalities (3)–(5) come from the fact that interior of cycles are also critical graphs. Finally, if any of the inequalities (6)–(7) are violated then the cuts K and L can be taken as non-crossing.

We solve the system of constraints by computer programs. From these solutions we get graphs depicted in Fig. 1(d1) and (d2).   □

This finishes the proof of Lemma 4.   □

Lemma 5. *If G contains one 5-face f_5 and all other faces are 4-faces, then G is described by Theorem 2(a) and depicted in Fig. 1(a).*

Proof. Let G be a C-critical graph containing one 5-face f_5. Let $e \in E(G) \backslash E(C)$. We want to find a 3-coloring ψ of C that does not extend to a proper 3-coloring of G but extends to a proper 3-coloring of $G - e$. Note that either $G - e$ has a 5-face and a 6-face or $G - e$ has a 7-face. This gives us two cases to consider.

Case 1: $G - e$ contains a 5-face and a 6-face.
        Let ψ be a 3-coloring of C containing 9 source edges (i.e. the colors around C are $1, 2, 3, 1, 2, 3, 1, 2, 3$). Then ψ extends to a 3-coloring of $G - e$ by Claim 1. However, ψ does not extend to a 3-coloring of G since it is not possible to create a ψ-balanced layout for G.

Case 2: $G - e$ contains a 7-face f_7.
        By Theorem 9 from [14], if ψ is a 3-coloring of C containing 9 source edges, then ψ does not extend to a proper 3-coloring of $G - e$ and if ψ is a 3-coloring of C containing 6 source edges and 3 sink edges, then ψ always extends to a proper 3-coloring of $G - e$. Since ψ must extend to $G - e$, we know that ψ contains 6 source edges and 3 sink edges. Now we need to construct such a proper 3-coloring ψ that does not extend to a proper 3-coloring of G.
        Let q be a ψ-balanced layout of G. The only possibility is $q(f_5) = -3$ and $c(q, \psi) = 6$. By Lemma 2, there exists an edge-cut K in $G^{q, \psi}$ separating

s from t such that $|K|$ is smaller than 6. By a proof of Lemma 3 (for details see [14]), there is a subgraph K_0 of G containing edges of G, which are crossed by edges of K that are not adjacent to any of the terminals in $G^{q,\psi}$. Denote $|K_0|$ by k_0. First suppose that K_0 is a cycle. Let m denote the sum of the q-values of the faces in the interior of K_0. By Lemma 3 $|m| > k_0$. If f_5 is in the interior of K_0, then $|m| = 3$, while $\ell(\{f_5\}) = 5$, a contradiction. Therefore K_0 is a path joining two distinct vertices of C.

From a ψ-balanced layout q we obtain that $n_X^s + n_Y^s = 6$ and $n_X^t + n_Y^t = 3$. This structure is the same as in the proof of Lemma 4 (see Fig. 2). The following two possibilities can occur:

Let f_5 belong to X. For the edge cut K, $|K| = k_0 + n_Y^s + n_X^t + 3 < c(q, \psi) = 6$. Hence $k_0 = 2$, $n_Y^s = 0$, $n_X^t = 0$, $n_X^s = 6$, and $n_Y^t = 3$. The cycle bounding X has length 8. However, it contains only one face of odd size, which is a contradiction.

Let f_5 belong to Y. For the edge-cut K, $|K| = k_0 + n_Y^s + n_X^t < c(q, \psi) = 6$. For X we have $k_0 + n_X^s + n_X^t \geq \ell(\{f_4\}) = 4$ and even. Since Y contains f_5, $k_0 + n_X^s + n_X^t \geq \ell(\{f_5\}) = 5$ and odd. We solve the system of these constraints by computer programs.

From these solutions we obtain that either Y is a 5-face f_5 sharing at least two sink edges with C (the first three solutions) or Y is bounded by a 7-cycle sharing at least three sink edges with C (the last three solutions). The situation is depicted in Fig. 1(a)(b).   □

Lemma 6. *If G contains three 5-faces and all other faces are 4-faces, then G is described by Theorem 2(e) and depicted in Fig. 1(Bij) for all i and j.*

The proof of Lemma 6 is omitted due to the page limit. The proof goes along similar lines as the proof of Lemma 4.

References

1. Aksenov, V.A.: The extension of a 3-coloring on planar graphs. Diskret. Analiz (Vyp. 26 Grafy i Testy), 3–19, 84 (1974)
2. Aksenov, V.A., Borodin, O.V., Glebov, A.N.: Continuation of a 3-coloring from a 6-face to a plane graph without 3-cycles. Diskretn. Anal. Issled. Oper. Ser. 1 **10**(3), 3–11 (2003)
3. Aksenov, V.A., Borodin, O.V., Glebov, A.N.: Continuation of a 3-coloring from a 7-face onto a plane graph without 3-cycles. Sib. Èlektron. Mat. Izv. **1**, 117–128 (2004)
4. Appel, K., Haken, W.: Every planar map is four colorable. I. Discharging. Illinois J. Math. **21**(3), 429–490 (1977)
5. Appel, K., Haken, W., Koch, J.: Every planar map is four colorable. II. Reducibility. Illinois J. Math. **21**(3), 491–567 (1977)
6. Borodin, O.V.: A new proof of Grünbaum's 3 color theorem. Discrete Math. **169**(1–3), 177–183 (1997). http://dx.doi.org/10.1016/0012-365X(95)00984-5

7. Borodin, O.V.: Colorings of plane graphs: A survey. Discrete Math. **33**(4), 517–539 (2013). http://dx.doi.org/10.1016/j.disc.2012.11.011
8. Borodin, O.V., Dvořák, Z., Kostochka, A.V., Lidický, B., Yancey, M.: Planar 4-critical graphs with four triangles. Eur. J. Combin. **41**, 138–151 (2014). http://dx.doi.org/10.1016/j.ejc.2014.03.009
9. Dailey, D.P.: Uniqueness of colorability and colorability of planar 4-regular graphs are NP-complete. Discrete Math. **30**(3), 289–293 (1980). http://dx.doi.org/10.1016/0012-365X(80)90236-8
10. Dvořák, Z., Lidický, B.: 4-critical graphs on surfaces without contractible ($\leq$ 4)-cycles. SIAM J. Discrete Math. **28**(1), 521–552 (2014). http://dx.doi.org/10.1137/130920952
11. Dvořák, Z., Kawarabayashi, K.i.: Choosability of planar graphs of girth 5. ArXiv e-prints, September 2011
12. Dvořák, Z., Král, D., Thomas, R.: Three-coloring triangle-free graphs on surfaces I. Extending a coloring to a disk with one triangle (2013) (Submitted)
13. Dvořák, Z., Král, D., Thomas, R.: Three-coloring triangle-free graphs on surfaces IV. 4-faces in critical graphs (2014) (Manuscript)
14. Dvořák, Z., Lidický, B.: 3-coloring triangle-free planar graphs with a precolored 8-cycle (2014). http://dx.doi.org/10.1002/jgt.21842 (Accepted to J. Graph Theory)
15. Garey, M.R., Johnson, D.S.: Computers and Intractability: A Guide to the Theory of NP-Completeness. W. H. Freeman and Co., San Francisco (1979)
16. Gimbel, J., Thomassen, C.: Coloring graphs with fixed genus and girth. Trans. Amer. Math. Soc. **349**(11), 4555–4564 (1997). http://dx.doi.org/10.1090/S0002-9947-97-01926-0
17. Grötzsch, H.: Ein Dreifarbenzatz für Dreikreisfreie Netze auf der Kugel. Math. Natur. Reihe **8**, 109–120 (1959)
18. Grünbaum, B.: Grötzsch's theorem on 3-colorings. Michigan Math. J. **10**, 303–310 (1963)
19. Robertson, N., Sanders, D., Seymour, P., Thomas, R.: The four-colour theorem. J. Combin. Theory Ser. B **70**(1), 2–44 (1997). http://dx.doi.org/10.1006/jctb.1997.1750
20. Thomassen, C.: The chromatic number of a graph of girth 5 on a fixed surface. J. Combin. Theory Ser. B **87**(1), 38–71 (2003). http://dx.doi.org/10.1016/S0095-8956(02)00027-8 (dedicated to Crispin St. J. A. Nash-Williams)
21. Tutte, W.T.: A contribution to the theory of chromatic polynomials. Canadian J. Math. **6**, 80–91 (1954)
22. Walls, B.H.: Coloring girth restricted graphs on surfaces. ProQuest LLC, Ann Arbor (1999). http://gateway.proquest.com/openurl?url_ver=Z39.88-2004&rft_val_fmt=info:ofi/fmt:kev:mtx:dissertation&res_dat=xri:pqdiss&rft_dat=xri:pqdiss:9953838 (thesis (Ph.D.)–Georgia Institute of Technology)

Computing Heat Kernel Pagerank and a Local Clustering Algorithm

Fan Chung and Olivia Simpson[✉]

University of California, San Diego,
La Jolla, CA 92093, USA
{fan,osimpson}@ucsd.edu

Abstract. Heat kernel pagerank is a variation of Personalized PageRank given in an exponential formulation. In this work, we present a sublinear time algorithm for approximating the heat kernel pagerank of a graph. The algorithm works by simulating random walks of bounded length and runs in time $O\left(\frac{\log(\epsilon^{-1})\log n}{\epsilon^3 \log\log(\epsilon^{-1})}\right)$, assuming performing a random walk step and sampling from a distribution with bounded support take constant time.

The quantitative ranking of vertices obtained with heat kernel pagerank can be used for local clustering algorithms. We present an efficient local clustering algorithm that finds cuts by performing a sweep over a heat kernel pagerank vector, using the heat kernel pagerank approximation algorithm as a subroutine. Specifically, we show that for a subset S of Cheeger ratio ϕ, many vertices in S may serve as seeds for a heat kernel pagerank vector which will find a cut of conductance $O(\sqrt{\phi})$.

Keywords: Heat kernel pagerank · Heat kernel · Local algorithms

1 Introduction

In large networks, many similar elements can be identified to a single, larger entity by the process of clustering. Increasing granularity in massive networks through clustering eases operations on the network. There is a large literature on the problem of identifying clusters in a graph [10,13,14,17], and the problem has found many applications. However, in a variation of the graph clustering problem we may only be interested in a single cluster near one element in the graph. For this, local clustering algorithms are of greater use.

As an example, local clustering is a common tool for identifying communities in a network. A community is loosely defined as a subset of vertices in a graph which are more strongly connected internally than to vertices outside the subset. Properties of community structure in large, real world networks have been studied in [12], for example, where local clustering algorithms are employed for identifying communities of varying quality.

The goal of a local clustering algorithm is to identify a cluster in a graph near a specified vertex. Using only local structure avoids unnecessary computation

© Springer International Publishing Switzerland 2015
J. Kratochvíl et al. (Eds.): IWOCA 2014, LNCS 8986, pp. 110–121, 2015.
DOI: 10.1007/978-3-319-19315-1_10

over the entire graph. An important consequence of this are running times which are often in terms of the size of the small side of the partition, rather than of the entire graph. The best performing local clustering algorithms use probability diffusion processes over the graph to determine clusters (see Sect. 1.1). In this paper we present a new algorithm which identifies a cut near a specified vertex with simple computations over a heat kernel pagerank vector.

The theory behind using heat kernel pagerank for computing local clusters has been considered in previous work. Here we give an efficient approximation algorithm for computing heat kernel pagerank. Note that we use a "relaxed" notion of approximation which allows us to derive a sublinear probabilistic approximation algorithm for heat kernel pagerank, while computing an exact or sharp approximation would require computational complexity of order similar to matrix multiplication. We use this sublinear approximation algorithm for efficient local clustering.

1.1 Previous Work

Heat kernel and approximation of matrix exponentials. Heat kernel pagerank was first introduced in [6] as a variant of personalized PageRank [5]. While PageRank can be viewed as a geometric sum of random walks, the heat kernel pagerank is an exponential sum of random walks. An alternative interpretation of the heat kernel pagerank is related to the heat kernel of a graph as the fundamental solution to the heat equation. As such, it has connections with diffusion and mixing properties of graphs and has been incorporated into a number of graph algorithmic primitives.

Orecchia et al. use a variant of heat kernel random walks in their randomized algorithm for computing a cut in a graph with prescribed balance constraints [18]. A key subroutine in the algorithm is a procedure for computing $e^{-A}v$ for a positive semidefinite matrix A and a unit vector v in time $\tilde{O}(m)$ for graphs on n vertices and m edges. They show how this can be done with a small number of computations of the form $A^{-1}v$ and applying the Spielman-Teng linear solver [20]. Their main result is a randomized algorithm that outputs a balanced cut in time $O(m \text{polylog } n)$. In a follow up paper, Sachdeva and Vishnoi [19] reduce inversion of positive semidefinite matrices to matrix exponentiation, thus proving that matrix exponentiation and matrix inversion are equivalent to polylog factors. In particular, the nearly-linear running time of the balanced separator algorithm depends upon the nearly-linear time Spielman-Teng solver.

Another method for approximating matrix exponentials is given by Kloster and Gleich in [11]. They use a Gauss-Southwell iteration to approximate the Taylor series expansion of the column vector $e^P e_c$ for transition probability matrix P and e_c a standard basis vector. The algorithm runs in sublinear time assuming the maximum degree of the network is $O(\log \log n)$.

Local clustering. Local clustering algorithms were introduced in [20], wherein Spielman and Teng present a nearly-linear time algorithm for finding local partitions with certain balance constraints. Let $\Phi(S)$ denote the cut ratio of a subset

S that we will later define as the Cheeger ratio. Then, given a graph and a subset of vertices S such that $\Phi(S) < \phi$ and $\text{vol}(S) \leq \text{vol}(G)/2$, their algorithm finds a set of vertices T such that $\text{vol}(T) \geq \text{vol}(S)/2$ and $\Phi(T) \leq O(\phi^{1/3} \log^{O(1)} n)$ in time $O(m(\log n/\phi)^{O(1)})$. This seminal work incorporates the ideas of Lovász and Simonovitz [15,16] on isoperimetric properties of random walks, and their algorithm works by simulating truncated random walks on the graph. Spielman and Teng later improve their approximation guarantee to $O(\phi^{1/2} \log^{3/2} n)$ in a revised version of the paper [21].

Andersen et al. [2] give an improved local clustering algorithm using approximate PageRank vectors. For a vertex subset S with Cheeger ratio ϕ and volume k, they show that a PageRank vector can be used to find a set with Cheeger ratio $O(\phi^{1/2} \log^{1/2} k)$. Their local clustering algorithm runs in time $O(\phi^{-1} m \log^4 m)$. The analysis of the above process was strengthened in [1] and emphasized that vertices with higher PageRank values will be on the same side of the cut as the starting vertex.

Andersen and Peres [3] later simulate a volume-biased evolving set process to find sparse cuts. Although their approximation guarantee is the same as that of [2], their process yields a better ratio between the computational complexity of the algorithm on a given run and the volume of the output set. They call this value the *work/volume ratio*, and their evolving set algorithm achieves an expected ratio of $O(\phi^{-1/2} \log^{3/2} n)$. This result is improved by Gharan and Trevisan in [9] with an algorithm that finds a set of conductance at most $O(\epsilon^{-1/2} \phi^{1/2})$ and achieves a work/volume ratio of $O(\varsigma^\epsilon \phi^{-1/2} \log^2 n)$ for target volume ς and target conductance ϕ. The complexity of their algorithm is achieved by running copies of an evolving set process in parallel.

1.2 Our Contributions

In this paper, we give a probabilistic approximation algorithm for computing a vector that yields a ranking of vertices close to the heat kernel pagerank vector. The approximation algorithm, `ApproxHKPRseed`, works by simulating heat kernel random walks of length k, where k is taken according to the Poisson distribution, and then computing contributions of these walks for each vertex in the graph. Assuming access to a constant-time query which returns the destination of a heat kernel random walk starting from a specified vertex, `ApproxHKPRseed` runs in time $O\left(\frac{\log(\epsilon^{-1}) \log n}{\epsilon^3 \log \log(\epsilon^{-1})}\right)$.

Using `ApproxHKPRseed` as a subroutine, we then present a local clustering algorithm that uses a ranking according to an approximate heat kernel pagerank. Let G be a graph and S a proper vertex subset with volume $\varsigma \leq \text{vol}(G)/4$ and Cheeger ratio $\Phi(S) \leq \phi$. Then, with probability at least $1 - \epsilon$, our algorithm outputs either a cutset T with $\text{vol}(T) \geq \text{vol}(S)/2$ and ς-local Cheeger ratio at most $O(\sqrt{\phi})$ or a certificate that no such set exists. The algorithm has work/volume ratio of $O(\varsigma^{-1} \epsilon^{-3} \log n \log(\epsilon^{-1}) \log \log(\epsilon^{-1}))$ (Table 1).

Table 1. Summary of local clustering algorithms

Algorithm	Conductance of output set	Work/volume ratio
[21]	$O(\phi^{1/2}\log^{3/2} n)$	$O(\phi^{-2}\mathrm{polylog}\, n)$
[2]	$O(\phi^{1/2}\log^{1/2} n)$	$O(\phi^{-1}\mathrm{polylog}\, n)$
[3]	$O(\phi^{1/2}\log^{1/2} n)$	$O(\phi^{-1/2}\mathrm{polylog}\, n)$
[9]	$O(\epsilon^{-1/2}\phi^{1/2})$	$O(\varsigma^{\epsilon}\phi^{-1/2}\mathrm{polylog}\, n)$
This work	$O(\phi^{1/2})$	$O(\varsigma^{-1}\epsilon^{-3}\log n \log(\epsilon^{-1})\log\log(\epsilon^{-1}))$

The theory behind finding local cuts with heat kernel pagerank vectors was first presented in [6,7]. Using some of this analysis as a starting point, we provide the algorithm for computing local clusters, called ClusterHKPR.

2 Preliminaries

Let $G = (V, E)$ be an undirected graph on n vertices and m edges. We use $u \sim v$ to denote $u, v \in E$. The *degree*, d_v, of a vertex v is the number of vertices u such that $u \sim v$. The *volume* of a set of vertices $S \subseteq V$ is the total degree of its vertices, $\mathrm{vol}(S) = \sum_{v \in S} d_v$, and the *edge boundary* S is the set of edges with one vertex in S and the other outside of S, $\partial(S) = \{u \sim v \;:\; u \in S, v \notin S\}$. When discussing the full vertex set, V, we write $S \subseteq G$ and $\mathrm{vol}(G) = \mathrm{vol}(V)$.

Let $f \in \mathbb{R}^n$ be a row vector over the vertices of G. Then the support of f is the set of vertices with nonzero values in f, $\mathrm{supp}(f) = \{u \in V \;:\; f(u) \neq 0\}$. For a subset of vertices S, we define $f(S) = \sum_{u \in S} f(u)$.

2.1 A Local Cheeger Inequality

The quality of a cut can be measured by the ratio of the number of edges between the two parts of the cut and the volume of the smaller side of the cut. This is called the *Cheeger ratio* of a set, defined by

$$\Phi(S) = \frac{|\partial(S)|}{\min(\mathrm{vol}(S), \mathrm{vol}(V \setminus S))}.$$

The *Cheeger constant* of a graph is the minimal Cheeger ratio, $\Phi(G) = \min_{S \subseteq G} \Phi(S)$. Finally, for a given subset S of a graph G, the *local Cheeger ratio* is defined

$$\Phi^*(S) = \min_{T \subseteq S} \Phi(T).$$

Our local clustering algorithm relies on a local version of the usual Cheeger inequalities which relate the Cheeger constant of a graph to an eigenvalue of the graph. Namely, let the normalized Laplacian of a graph be the matrix $\mathcal{L} = D^{-1/2}(D - A)D^{-1/2}$, where D is the diagonal matrix of vertex degrees and A is the unweighted, symmetric adjacency matrix. Also, let $\mathcal{L}_S$ be determined by a subset S of size $|S| = s$ and defined as the restricted matrix of $\mathcal{L}$ with rows and columns indexed by vertices in S. Then the eigenvalues $\lambda_S := \lambda_{S,1} \leq \lambda_{S,2} \leq \cdots \leq \lambda_{S,s}$ of $\mathcal{L}_S$ are also known as the *Dirichlet eigenvalues of* S, and are related to $\Phi^*(S)$ by the following local Cheeger inequality [7]:

$$\frac{1}{2}(\Phi^*(S))^2 \leq \lambda_S \leq \Phi^*(S). \tag{1}$$

The inequality (1) will be used to derive a relationship between a ranking according to heat kernel pagerank and sets with good Cheeger ratios. Details will be given in Sect. 4.

2.2   Heat Kernel and Heat Kernel Pagerank

The *heat kernel pagerank* vector has entries indexed by the vertices of the graph and involves two parameters; t, a non-negative real value representing the temperature, and a preference row vector $f : V \to \mathbb{R}$, by the following equation:

$$\rho_{t,f} = e^{-t} \sum_{k=0}^{\infty} \frac{t^k}{k!} f P^k \tag{2}$$

where P is the transition probability matrix

$$(P)_{uv} = \begin{cases} 1/d_u & \text{if } u \sim v \\ 0 & \text{otherwise.} \end{cases}$$

When f is a probability distribution, the heat kernel pagerank can be regarded as the expected distribution of a random walk according to the transition probability matrix P. A starting distribution we will be particularly concerned with is that with all probability initially on a single vertex u, i.e. $f = \chi_u$ where χ_u is the indicator vector for vertex u. We will denote the heat kernel pagerank vector over this distribution by $\rho_{t,u} := \rho_{t,\chi_u}$.

The *heat kernel* of a graph is defined $H_t = e^{-t\Delta}$ where Δ is the Laplace operator $\Delta = I - P$. Then an alternative definition for heat kernel pagerank is $\rho_{t,f} = f H_t$.

We can compare the heat kernel pagerank to the personalized PageRank vector, given by

$$\mathrm{pr}_{\alpha,f} = \alpha \sum_{k=0}^{\infty} (1 - \alpha)^k f P^k.$$

In this definition, α is often called the *jumping* or *reset* constant, meaning that at any step the random walk may jump to a vertex taken from f with probability α. When $f = \chi_u$ for some u, the random walk is "reset" to the first vertex of the walk, u, with probability α. We note that, compared to the personalized PageRank vector, which can be viewed as a geometric sum, we can expect better convergence rates from the heat kernel pagerank, defined as an exponential sum.

3   Heat Kernel Pagerank Approximation

We begin our discussion of heat kernel pagerank approximation with an observation. Each term in the infinite series defining heat kernel pagerank in (2) is of the form $e^{-t}\frac{t^k}{k!}fP^k$ for $k \in [0, \infty]$. The vector fP^k is the distribution after k random walk steps with starting distribution f. Then, if we perform k steps of a random walk given by transition probability matrix P from starting distribution f with probability $p_k = e^{-t}\frac{t^k}{k!}$, the heat kernel pagerank vector can be viewed as the expected distribution of this process.

This suggests a natural way to approximate the heat kernel pagerank. That is, we can obtain a close approximation to the expected distribution with sufficiently many samples. Our algorithm operates as follows. We perform r random walks to approximate the infinite sum, choosing r large enough to bound the error. We also use the fact that very long walks are performed with small probability. As such, we limit the lengths of our random walks by a finite number K. Both r, K depend on a predetermined error bound ϵ.

ApproxHKPRseed(G, t, u, ϵ)

---

input: a graph G, $t \in \mathbb{R}^+$, seed vertex $u \in V$, error parameter $0 < \epsilon < 1$.
output: ρ, an ϵ-approximation of $\rho_{t,u}$.
   initialize a 0 vector ρ of dimension n, where $n = |V|$
   $r \leftarrow \frac{16}{\epsilon^3}\log n$
   $K \leftarrow \frac{\log(\epsilon^{-1})}{\log\log(\epsilon^{-1})}$
   for r iterations **do**
      Start
         simulate a P random walk from vertex u where k steps are taken with probability $e^{-t}\frac{t^k}{k!}$ and $k \leq K$
         let v be the last vertex visited in the walk
         $\rho[v] \leftarrow \rho[v] + 1$
      End
   end for
   return $1/r \cdot \rho$

In our analysis we will use the following definition of an ϵ-approximate vector.

Definition 1. *Let G be a graph on n vertices, and let $f : V \to \mathbb{R}$ be a vector over the vertices of G. Let $\rho_{t,f}$ be the heat kernel pagerank vector according to f and t. Then we say that $\nu \in \mathbb{R}^n$ is an $\epsilon - approximate vector$ of $\rho_{t,f}$ if*

1. for every vertex $v \in V$ in the support of ν,
   $(1 - \epsilon)\rho_{t,f}(v) - \epsilon \leq \nu(v) \leq (1 + \epsilon)\rho_{t,f}(v)$,
2. for every vertex with $\nu(v) = 0$, it must be that $\rho_{t,f}(v) \leq \epsilon$.

We note that this is a rather coarse requirement for an approximation, but satisfies our needs for local clustering. In the following algorithm, we approximate $\rho_{t,u}$ by an ϵ-approximate vector which we denote by $\hat{\rho}_{t,u}$. The running time of the algorithm is $O\left(\frac{\log(\epsilon^{-1})\log n}{\epsilon^3 \log\log(\epsilon^{-1})}\right)$.

Theorem 1. *Let G be a graph and let u be a vertex of G. Then, the algorithm* ApproxHKPRseed*(G, t, u, ϵ) outputs an ϵ-approximate vector $\hat{\rho}_{t,u}$ of the heat kernel pagerank $\rho_{t,u}$ for $0 < \epsilon < 1$ with probability at least $1 - \epsilon$. The running time of* ApproxHKPRseed *is $O\left(\frac{\log(\epsilon^{-1})\log n}{\epsilon^3 \log\log(\epsilon^{-1})}\right)$.*

Our analysis relies on the usual Chernoff bounds as stated below.

Lemma 1. ([4]) *Let X_i be independent Bernoulli random variables with $X = \sum\limits_{i=1}^{r} X_i$. Then,*

1. *for $0 < \epsilon < 1$, $\mathbb{P}(X < (1-\epsilon)r\mathbb{E}(X)) < \exp(-\frac{\epsilon^2}{2}r\mathbb{E}(X))$*
2. *for $0 < \epsilon < 1$, $\mathbb{P}(X > (1+\epsilon)r\mathbb{E}(X)) < \exp(-\frac{\epsilon^2}{4}r\mathbb{E}(X))$*
3. *for $c \geq 1$, $\mathbb{P}(X > (1+c)r\mathbb{E}(X)) < \exp(-\frac{c}{2}r\mathbb{E}(X))$.*

Proof (Theorem 1). Consider the random variable which takes on value fP^k with probability $p_k = e^{-t}\frac{t^k}{k!}$ for $k \in [0, \infty)$. The expectation of this random variable is exactly $\rho_{t,f}$. Heat kernel pagerank can be understood as a series of distributions of weighted random walks over the vertices, and the weights are related to the number of steps taken in the walk. The series can be computed by simulating this process, i.e., draw k according to p_k and compute fP^k with sufficiently many random walks of length k.

We approximate the infinite sum by limiting the walks to at most K steps. We will take K to be $K = \frac{\log(\epsilon^{-1})}{\log\log(\epsilon^{-1})}$. These interrupts risk the loss of contribution to the expected value, but can be upper bounded by $\frac{e^{-t}t^K}{K!} \leq \frac{\epsilon}{2}$ provided that $t > K/\log K$. This is within the error bound for an approximate heat kernel pagerank. If $t \leq K/\log K$, the expected length of the random walk is

$$\sum_{k=0}^{\infty} \frac{e^{-t}t^k}{k!} \cdot k = t < K/\log K.$$

Thus we can ignore walks of length more than K while maintaining $\rho_{t,u}(v) - \epsilon \leq \hat{\rho}_{t,u}(v) \leq \rho_{t,u}(v)$ for every vertex v.

Next we show how many samples are necessary for our approximation vectors. For $k \leq K$, our algorithm simulates k random walk steps with probability $e^{-t}\frac{t^k}{k!}$. To be specific, for a fixed u, let X_k^v be the indicator random variable defined by $X_k^v = 1$ if a random walk beginning from vertex u ends at vertex v in k steps. Let X^v be the random variable that considers the random walk process ending at vertex v in *at most* k steps. That is, X^v assumes the vector X_k^v with probability $e^{-t}\frac{t^k}{k!}$. Namely, we consider the combined random walk $X^v = \sum_{k \leq K} e^{-t}\frac{t^k}{k!}X_k^v$.

Now, let $\rho(k)_{t,u}$ be the contribution to the heat kernel pagerank vector $\rho_{t,u}$ of walks of length at most k. The expectation of each X^v is $\rho(k)_{t,u}(v)$. Then, by Lemma 1,

$$\mathbb{P}(X^v < (1-\epsilon)\rho(k)_{t,u}(v) \cdot r) < \exp(-\rho(k)_{t,u}(v)r\epsilon^2/2)$$
$$= \exp(-(8/\epsilon)\rho(k)_{t,u}(v)\log n)$$
$$< n^{-4}$$

for every component with $\rho_{t,u}(v) > \epsilon$, since then $\rho(k)_{t,u}(v) > \epsilon/2$. Similarly,

$$\mathbb{P}(X^v > (1+\epsilon)\rho(k)_{t,u}(v) \cdot r) < \exp(-\rho(k)_{t,u}(v)r\epsilon^2/4)$$
$$= \exp(-(4/\epsilon)\rho(k)_{t,u}(v)\log n)$$
$$< n^{-2}.$$

We conclude the analysis for the support of $\rho_{t,u}$ by noting that $\hat{\rho}_{t,u} = \frac{1}{r}X^v$, and we achieve an ϵ-multiplicative error bound for every vertex v with $\rho_{t,u}(v) > \epsilon$ with probability at least $1 - O(n^{-2})$.

On the other hand, if $\rho_{t,u}(v) \leq \epsilon$, by the third part of Lemma 1, $\mathbb{P}(\hat{\rho}_{t,u}(v) > 2\epsilon) \leq n^{-8/\epsilon^2}$. We may conclude that, with high probability, $\hat{\rho}_{t,u}(v) \leq 2\epsilon$.

For the running time, we use the assumptions that performing a random walk step and drawing from a distribution with bounded support require constant time. These are incorporated in the random walk simulation, which dominates the computation. Therefore, for each of the r rounds, at most K steps of the random walk are simulated, giving a total of $rK = O\left(\frac{16}{\epsilon^3}\log n \cdot \frac{\log(\epsilon^{-1})}{\log\log(\epsilon^{-1})}\right) = \tilde{O}(1)$ queries. $\qquad\qquad\qquad\qquad\qquad\qquad\qquad\qquad\square$

We note that the algorithm works for any t, but a good choice of t will be related to the size of the local cluster S and a desirable convergence rate. In particular, the constraints put on t are necessary for our local clustering results, presented in the next section.

The algorithm for efficient heat kernel pagerank computation has promise for a variety of applications. It has been shown in [8] how to apply heat kernel pagerank in solving symmetric diagonally dominant linear systems with a boundary condition, for example.

4   Finding Good Local Cuts

The premise of the algorithm is to find a good cut near a specified vertex by performing a *sweep* over a vector associated to that vertex, which we will specify. Let $p : V \to \mathbb{R}$ be a probability distribution vector over the vertices of the graph of support size $N_p = \text{supp}(p)$. Then, consider a *probability-per-degree* ordering of the vertices where $p(v_1)/d_{v_1} \geq p(v_2)/d_{v_2} \geq \cdots \geq p(v_{N_p})/d_{v_{N_p}}$. Let S_i be the set of the first i vertices per the ordering. We call each S_i a *segment*. Then the process of investigating the cuts induced by the segments to find an optimal cut is called performing a sweep over p.

In this section we will show how a sweep over a single heat kernel pagerank vector finds local cuts. Specifically, we show that for a subset S with $\text{vol}(S) \leq \text{vol}(G)/4$ and $\Phi(S) \leq \phi$, and for a large number of vertices $u \in S$, performing a sweep over the vector $\hat{\rho}_{t,u}$, where $\hat{\rho}_{t,u}$ is an ϵ-approximation of $\rho_{t,u}$, will find a set with Cheeger ratio at most $O(\sqrt{\phi})$.

Remark 1. Though all the vertices in the support of the vector are sorted to build segments, in practice the sweep will be aborted after the volume of the current segment is larger than the target size. This is the *locality* of the algorithm, and ensures that the amount of work performed is proportional to the volume of the output set.

The ς-local Cheeger ratio of a sweep over a vector ν is the minimum Cheeger ratio over segments S_i with volume $0 \leq \text{vol}(S_i) \leq 2\varsigma$. Let $\Phi_\varsigma(\nu)$ the ς-local Cheeger ratio of cuts over a sweep of ν that separates sets of volume between 0 and 2ς.

We will make use of the following bounds for heat kernel pagerank in terms of local Cheeger ratios and sweep cuts. The proof is given in the full version of this paper.

Lemma 2. *Let G be a graph and S a subset of vertices of volume $\varsigma \leq \text{vol}(G)/4$. Then the set of $u \in S$ satisfying*

$$\frac{1}{2}e^{-t\Phi^*(S)} \leq \rho_{t,u}(S) \leq \sqrt{\varsigma}e^{-t\Phi_\varsigma(\rho_{t,u})^2/4}$$

has volume at least $\varsigma/2$.

We use Lemma 2 to reason that many vertices u satisfy the above inequalities, and thus can serve as good seeds for performing a sweep.

4.1 A Local Graph Clustering Algorithm

It follows from Lemma 2 that the ranking induced by a heat kernel pagerank vector with appropriate seed vertex can be used to find a cut with approximation guarantee $O(\sqrt{\phi})$ by choosing the appropriate t. To obtain a sublinear time local clustering algorithm for massive graphs, we use `ApproxHKPRseed` to efficiently compute an ϵ-approximate heat kernel pagerank vector, $\hat{\rho}_{t,u}$, to rank vertices.

The ranking induced by $\hat{\rho}_{t,u}$ is not very different from that of a true vector $\rho_{t,u}$ in the support of $\hat{\rho}_{t,u}$. Namely, using the bounds of [7], we have $\hat{\rho}_{t,u}(S) \geq (1-\epsilon)\rho_{t,u}(S) - \epsilon s$. In particular,

$$\frac{1}{2}(1-\epsilon)e^{-t\Phi^*(S)} - \epsilon s \leq \hat{\rho}_{t,u}(S) \leq \sqrt{\varsigma}e^{-t\Phi_\varsigma(\hat{\rho}_{t,u})^2/4} \tag{3}$$

for a set of vertices u of volume at least $\varsigma/2$.

Theorem 2. *Let G be a graph and $S \subset G$ a subset with $\text{vol}(S) = \varsigma \leq \text{vol}(G)/4$, $|S| = s$, and Cheeger ratio $\Phi(S) \leq \phi$. Let $\hat{\rho}_{t,u}$ be an ϵ-approximate of $\rho_{t,u}$ for some vertex $u \in S$. Then there is a subset $S_t \subset S$ with $\text{vol}(S_t) \geq \varsigma/2$ for which a sweep over $\hat{\rho}_{t,u}$ for any vertex $u \in S_t$ with*

1. $t = \phi^{-1}\log(\frac{2\sqrt{\varsigma}}{1-\epsilon} + 2\epsilon s)$ and
2. $\Phi_\varsigma(\hat{\rho}_{t,u})^2 \leq 4/t\log(2)$

finds a set with ς-local Cheeger ratio at most $\sqrt{8\phi}$.

Proof. Let u be a vertex in S_t as described in the theorem statement. Using the inequalities (3), we can bound the local Cheeger ratio by a sweep over $\hat{\rho}_{t,u}$:

$$e^{-t\Phi^*(S)} \leq \frac{2}{1-\epsilon}\left(\sqrt{\varsigma}e^{-t\Phi_\varsigma(\hat{\rho}_{t,u})^2/4} + \epsilon s\right)$$

which implies

$$e^{-t\Phi^*(S)} \leq e^{-t\Phi_\varsigma(\hat{\rho}_{t,u})^2/4}\left(\frac{2\sqrt{\varsigma}}{1-\epsilon} + \epsilon s e^{t\Phi_\varsigma(\hat{\rho}_{t,u})^2/4}\right),$$

and by the assumption 2, we have

$$e^{-t\Phi^*(S)} \leq e^{-t\Phi_\varsigma(\hat{\rho}_{t,u})^2/4}\left(\frac{2\sqrt{\varsigma}}{1-\epsilon} + 2\epsilon s\right)$$

$$\Phi^*(S) \geq \frac{\Phi_\varsigma(\hat{\rho}_{t,u})^2}{4} - \frac{\log(\frac{2\sqrt{\varsigma}}{1-\epsilon} + 2\epsilon s)}{t}.$$

Let $x = \log(\frac{2\sqrt{\varsigma}}{1-\epsilon} + 2\epsilon s)$. Then,

$$\Phi_\varsigma(\hat{\rho}_{t,u})^2 \leq 4\Phi^*(S) + 4x/t.$$

Since $\Phi^*(S) \leq \Phi(S) \leq \phi$ and $t = \phi^{-1}x$, it follows that $\Phi_\varsigma(\hat{\rho}_{t,u}) \leq \sqrt{8\phi}$. In particular, a sweep over $\hat{\rho}_{t,u}$ finds a cut with Cheeger ratio $O(\sqrt{\phi})$ as long as u is contained in S_t. □

We are now prepared to give our algorithm for finding cuts locally with heat kernel pagerank. The algorithm takes as input a starting vertex u, a desired volume ς for the cut set, and a target Cheeger ratio ϕ for the cut set. Then, to find a set achieving a minimum ς-local Cheeger ratio, we perform a sweep over an approximate heat kernel pagerank vector with the starting vertex as a seed.

ClusterHKPR$(G, u, s, \varsigma, \phi, \epsilon)$

---

input: a graph G, a vertex u, target cluster size s, target cluster volume $\varsigma \leq \text{vol}(G)/4$, target Cheeger ratio ϕ, error parameter ϵ.
output: a set T with $\varsigma/2 \leq \text{vol}(T) \leq 2\varsigma$, $\Phi(T) \leq \sqrt{8\phi}$.
1: $t \leftarrow \phi^{-1}\log(\frac{2\sqrt{\varsigma}}{1-\epsilon} + 2\epsilon s)$
2: $\hat{\rho} \leftarrow \texttt{ApproxHKPRseed}(G, t, u, \epsilon)$
3: sort the vertices of G in the support of $\hat{\rho}$ according to the ranking $\hat{\rho}[v]/d_v$
4: **for** $j \in [1, n]$ **do**
5:     $S_j = \bigcup_{i \leq j} v_i$
6:     **if** $\text{vol}(S_j) > 2\varsigma$ **then**
7:         output NO CUT FOUND, break
8:     **else if** $\varsigma/2 \leq \text{vol}(S_j) \leq 2\varsigma$ and $\Phi(S_j) \leq \sqrt{8\phi}$ **then**
9:         output S_j
10:    **end if**
11: **end for**
12: **if** no set was output **then**
13:     output NO CUT FOUND
14: **end if**

Theorem 3. *Let G be a graph which contains a subset S of volume at most $\mathrm{vol}(G)/4$ and Cheeger ratio bounded by ϕ. Further, assume that u is contained in the set $S_t \subseteq S$ as defined in Theorem 2. Then* ClusterHKPR$(G, u, s, \varsigma, \phi, \epsilon)$ *outputs a cutset T with ς-local Cheeger ratio at most $\sqrt{8\phi}$. The running time is the same as that of* ApproxHKPRseed.

Proof. Since it is given that $u \in S_t$ for $t = \phi^{-1}\log(\frac{2\sqrt{\varsigma}}{1-\epsilon} + 2\epsilon s)$, and by the assumptions on G and S, Theorem 2 states that a sweep over the approximate heat kernel pagerank vector $\hat{\rho}$ will find a set with ς-local Cheeger ratio at most $\sqrt{8\phi}$. The checks performed in line 8 of the algorithm discover such a set.

The computational work reduces to the main procedures of computing the heat kernel pagerank vector in line 2 and performing a sweep over the vector in line 4. Performing a sweep involves sorting the support of the vector (line 3) and calculating the conductance of segments. From the guarantees of an ϵ-approximate heat kernel pagerank vector, any vertex with average probability less than ϵ will be excluded from the support. Then the volume of a vector $\hat{\rho}$ output in line 2 is $O(\epsilon^{-1})$, and performing a sweep over $\hat{\rho}$ can be done in $O(\epsilon^{-1}\log(\epsilon^{-1}))$ time. The algorithm is therefore dominated by the time to compute a heat kernel pagerank vector, and the total running time is $O\left(\frac{\log(\epsilon^{-1})\log n}{\epsilon^3 \log\log(\epsilon^{-1})}\right)$. □

References

1. Andersen, R., Chung, F.: Detecting sharp drops in pagerank and a simplified local partitioning algorithm. In: Cai, J.-Y., Cooper, S.B., Zhu, H. (eds.) TAMC 2007. LNCS, vol. 4484, pp. 1–12. Springer, Heidelberg (2007)
2. Andersen, R., Chung, F., Lang, K.: Local graph partitioning using pagerank vectors. In: IEEE 47th Annual Symposium on Foundations of Computer Science, pp. 475–486. IEEE (2006)
3. Andersen, R., Peres, Y.: Finding sparse cuts locally using evolving sets. In: Proceedings of the 41st Annual Symposium on Theory of Computing, pp. 235–244. ACM (2009)
4. Borgs, C., Brautbar, M., Chayes, J., Teng, S.-H.: A sublinear time algorithm for pagerank computations. In: Bonato, A., Janssen, J. (eds.) WAW 2012. LNCS, vol. 7323, pp. 41–53. Springer, Heidelberg (2012)
5. Brin, S., Page, L.: The anatomy of a large-scale hypertextual web search engine. Comput. Netw. ISDN Syst. **30**(1), 107–117 (1998)
6. Chung, F.: The heat kernel as the pagerank of a graph. Proc. Natl. Acad. Sci. **104**(50), 19735–19740 (2007)
7. Chung, F.: A local graph partitioning algorithm using heat kernel pagerank. Internet Math. **6**(3), 315–330 (2009)
8. Chung, F., Simpson, O.: Solving linear systems with boundary conditions using heat kernel pagerank. In: Bonato, A., Mitzenmacher, M., Prałat, P. (eds.) WAW 2013. LNCS, vol. 8305, pp. 203–219. Springer, Heidelberg (2013)
9. Gharan, S.O., Trevisan, L.: Approximating the expansion profile and almost optimal local graph clustering. In: IEEE 53rd Annual Symposium on Foundations of Computer Science, pp. 187–196. IEEE (2012)

10. Kannan, R., Vempala, S., Vetta, A.: On clusterings: good, bad and spectral. J. ACM (JACM) **51**(3), 497–515 (2004)
11. Kloster, K., Gleich, D.F.: A nearly-sublinear method for approximating a column of the matrix exponential for matrices from large, sparse networks. In: Bonato, A., Mitzenmacher, M., Prałat, P. (eds.) WAW 2013. LNCS, vol. 8305, pp. 68–79. Springer, Heidelberg (2013)
12. Leskovec, J., Lang, K.J., Dasgupta, A., Mahoney, M.W.: Statistical properties of community structure in large social and information networks. In: Proceedings of the 17th International Conference on World Wide Web, pp. 695–704. ACM (2008)
13. Lin, F., Cohen, W.W.: Power iteration clustering. In: Proceedings of the 27th International Conference on Machine Learning (ICML 2010), pp. 655–662 (2010)
14. Lin, F., Cohen, W.W.: A very fast method for clustering big text datasets. In: Proceedings of the 19th European Conference on Artificial Intelligence, pp. 303–308 (2010)
15. Lovász, L., Simonovits, M.: The mixing rate of markov chains, an isoperimetric inequality, and computing the volume. In: Proceedings of the 31st Annual Symposium on Foundations of Computer Science, pp. 346–354. IEEE (1990)
16. Lovász, L., Simonovits, M.: Random walks in a convex body and an improved volume algorithm. Random Struct. Algorithms **4**(4), 359–412 (1993)
17. Ng, A.Y., Jordan, M.I., Weiss, Y., et al.: On spectral clustering: analysis and an algorithm. Adv. Neural Inf. Proc. Syst. **2**, 849–856 (2002)
18. Orecchia, L., Sachdeva, S., Vishnoi, N.K.: Approximating the exponential, the lanczos method and an $\tilde{O}(m)$-time spectral algorithm for balanced separator. In: Proceedings of the 44th Symposium on Theory of Computing, pp. 1141–1160. ACM (2012)
19. Sachdeva, S., Vishnoi, N.K.: Matrix inversion is as easy as exponentiation (2013). arXiv preprint arXiv:1305.0526
20. Spielman, D.A., Teng, S.H.: Nearly-linear time algorithms for graph partitioning, graph sparsification, and solving linear systems. In: Proceedings of the thirty-sixth annual ACM symposium on Theory of Computing, pp. 81–90. ACM (2004)
21. Spielman, D.A., Teng, S.H.: A local clustering algorithm for massive graphs and its application to nearly-linear time graph partitioning. CoRR abs/0809.3232 (2008)

A Γ-magic Rectangle Set and Group Distance Magic Labeling

Sylwia Cichacz$^{(\boxtimes)}$

AGH University of Science and Technology,
Al. Mickiewicza 30, 30-059 Kraków, Poland
cichacz@agh.edu.pl

Abstract. A Γ-distance magic labeling of a graph $G = (V, E)$ with $|V| = n$ is a bijection ℓ from V to an Abelian group Γ of order n such that the weight $w(x) = \sum_{y \in N_G(x)} \ell(y)$ of every vertex $x \in V$ is equal to the same element $\mu \in \Gamma$ called the magic constant. A graph G is called a group distance magic graph if there exists a Γ-distance magic labeling for every Abelian group Γ of order $|V(G)|$.

A Γ-magic rectangle set $MRS_\Gamma(a, b; c)$ of order abc is a collection of c arrays $(a \times b)$ whose entries are elements of group Γ, each appearing once, with all row sums in every rectangle equal to a constant $\omega \in \Gamma$ and all column sums in every rectangle equal to a constant $\delta \in \Gamma$.

In the paper we show that if a and b are both even then $MRS_\Gamma(a, b; c)$ exists for any Abelian group Γ of order abc. Furthermore we use this result to construct group distance magic labeling for some families of graphs.

Keywords: Distance magic labeling · Magic constant · Sigma labeling · Graph labeling · Cartesian product · Γ-magic rectangle set

1 Definitions

All graphs $G = (V, E)$ are finite undirected simple graphs. For standard graph theoretic notation and definitions we refer to Diestel [6].

A *distance magic labeling* of a graph G of order n is a bijection $\ell : V \to \{1, 2, \ldots, n\}$ so that there exists a positive integer μ such that the *weight* $w(v) = \sum_{u \in N(v)} \ell(u) = \mu$ for all $v \in V$, where $N(v)$ is the open neighborhood of v. The constant μ is called the *magic constant* of the labeling f. Any graph which admits a distance magic labeling is called a *distance magic graph*.

A *magic rectangle* $\mathrm{MR}(a, b)$ is an $a \times b$ array with entries from the set $\{1, 2, \ldots, ab\}$, each appearing once, with all its row sums equal to a constant δ and with all its column sums equal to a constant η. It was proved in [11,12]:

Theorem 1 ([11,12]). *A magic rectangle $MR(a, b)$ exists if and only if $a, b > 1$, $ab > 4$, and $a \equiv b \pmod 2$.*

The author was supported by National Science Centre Grant Nr 2011/01/D/ST1/04104.

J. Kratochvíl et al. (Eds.): IWOCA 2014, LNCS 8986, pp. 122–127, 2015.
DOI: 10.1007/978-3-319-19315-1_11

The following generalization of magic rectangles was introduced by Froncek in [7].

Definition 2. *A magic rectangle set M= MRS(a, b; c) is a collection of c arrays (a × b) whose entries are elements of $\{1, 2, \ldots, abc\}$, each appearing once, with all row sums in every rectangle equal to a constant δ and all column sums in every rectangle equal to a constant η.*

Moreover it was shown:

Theorem 3 ([7]). *If $a \equiv b \equiv 0$ (mod 2), $a \geq 2$ and $b \geq 4$, then a magic rectangle set MRS(a, b; c) exists for every c.*

Observation 4 ([7]). *If a magic rectangle set MRS(a, b; c) exists, then both MR(a, bc) and MR(ac, b) exist.*

The concept of distance magic labeling has been motivated by the construction of magic rectangles, since we can construct a distance magic complete r partite graph with each part size equal to n by labeling the vertices of each part by the columns of the magic rectangle. Although there does not exist 2×2 magic rectangle, observe that the partite sets of $K_{2,2}$ can be labeled $\{1, 4\}$ and $\{2, 3\}$, respectively, to obtain a distance magic labeling.

A *Cartesian product* $G \square H$ is a graph with the vertex set $V(G) \times V(H)$. Two vertices (g, h) and (g', h') are adjacent in $G \square H$ if and only if $g = g'$ and h is adjacent with h' in H, or $h = h'$ and g is adjacent with g' in G.

The necessary and sufficient condition for Cartesian product of cycles to be distance magic is given.

Theorem 5 ([13]) *The Cartesian product $C_n \square C_m$ is distance magic if and only if $n = m \equiv 2$ (mod 4).*

Let $K(n; r)$ denote the complete r-partite graph $K(n, n, \ldots, n)$. It has been recently proved:

Theorem 6 ([1]). *The Cartesian product $K(n; r) \square C_4$ is distance magic if and only if $n > 2$, $r > 1$ and n is even.*

Theorem 7 ([1]). *The Cartesian product $K(n; r) \square C_4$ is distance magic if and only if there exists a magic rectangle set MRS(2, n; 2r).*

Froncek in [8] defined the notion of group distance magic graphs, i.e. the graphs allowing the bijective labeling of vertices with elements of an Abelian group resulting in constant sums of neighbor labels.

A Γ-*distance magic labeling* of a graph $G = (V, E)$ with $|V| = n$ is a bijection ℓ from V to an Abelian group Γ of order n such that the weight $w(x) = \sum_{y \in N_G(x)} \ell(y)$ of every vertex $x \in V$ is equal to the same element $\mu \in \Gamma$, called the *magic constant*. A graph G is called a *group distance magic graph* if there exists a Γ-distance magic labeling for every Abelian group Γ of order $|V(G)|$.

The connection between distance magic graphs and Γ-distance magic graphs is as follows. Let G be a distance magic graph of order n with the magic constant μ'. If we replace the label n in a distance magic labeling for the graph G by the label 0, then we obtain a $\mathbb{Z}_n$-distance magic labeling for the graph G with the magic constant $\mu = \mu'$ (mod n). Hence every distance magic graph with n vertices admits a $\mathbb{Z}_n$-distance magic labeling. Although a $\mathbb{Z}_n$-distance magic graph on n vertices is not necessarily a distance magic graph. For instance compare Theorem 5 and the one below:

Theorem 8 ([7]). *The Cartesian product $C_m \square C_k$, $m, k \geq 3$, is a $\mathbb{Z}_{mk}$-distance magic graph if and only if km is even.*

There are some graphs not being distance magic but at the same time they are group distance magic (see [3]).

Recall that any group element $\iota \in \Gamma$ of order 2 (i.e., $\iota \neq 0$ such that $2\iota = 0$) is called an *involution*, and that a non-trivial finite group has elements of order 2 if and only if the order of the group is even. Moreover every cyclic group of even order has exactly one involution. The fundamental theorem of finite Abelian groups states that the finite Abelian group Γ can be expressed as the direct sum of cyclic subgroups of prime-power order. This product is unique up to the order of the direct product. When t is the number of these cyclic components whose order is a power of 2, then Γ has $2t - 1$ involutions. Moreover the sum of all the group elements is equal to the sum of the involutions and the neutral element.

Cichacz and Froncek proved:

Theorem 9 ([4]). *Let G be an r-regular distance magic graph on n vertices, where r is odd. There does not exists an Abelian group Γ having exactly one involution ι, $|\Gamma| = n$ such that G is Γ-distance magic.*

The concept of magic squares on an Abelian group was presented in [9]. A Γ-magic square $MS_\Gamma(n)$ is an $n \times n$ array with entries from the an Abalian group Γ of order n^2, each appearing once, with all its row, column and diagonal sums equal to a constant δ. It was proved in [9]:

Theorem 10 ([9]) *Γ-magic squares $MS_\Gamma(n)$ exist for all groups Γ of order n^2 for any $n > 2$.*

To prove our main result, we will need the following generalization of Γ-magic squares $MS_\Gamma(n)$.

Definition 11. *A Γ-magic rectangle set $MRS_\Gamma(a, b; c)$ on group Γ of order abc is a collection of c arrays ($a \times b$) whose entries are elements of group Γ, each appearing once, with all row sums in every rectangle equal to a constant $\omega \in \Gamma$ and all column sums in every rectangle equal to a constant $\delta \in \Gamma$.*

In the paper we show that if a and b are both even then $MRS_\Gamma(a, b; c)$ exists for any Abelian group Γ of order abc. Furthermore we use this result to construct group distance magic labeling for some families of graphs.

2   A Γ-magic Rectangle Set $MRS_\Gamma(a, b; c)$

We start with simple observations.

Observation 12. *If a Γ-magic rectangle set $MRS_\Gamma(a, b; c)$ on group Γ exists, then both $MRS_\Gamma(a, bc; 1)$ and $MRS_\Gamma(ac, b; 1)$ exist.*

To construct $MRS_\Gamma(a, bc; 1)$ (or $MRS_\Gamma(ac, b; 1)$), we simply take all $a \times b$ rectangles and "glue" them together into one $a \times bc$ (or $ac \times b$) rectangle.

Observation 13. *If a is even, b is odd then for any c and an Abelian group Γ having exactly one involution, $|\Gamma| = abc$ there does not exist a Γ-magic rectangle set $MRS_\Gamma(a, b; c)$.*

Proof. Assume that there exists a $MRS_\Gamma(ac, b; 1)$ on Γ. Then we can construct a Γ-distance magic complete ac partite graph with each part size equal to b (i.e. $K(b; ac)$) by labeling the vertices of each part by the columns of the magic rectangle, a contradiction with Theorem 9. Therefore there does not exist a $MRS_\Gamma(ac, b; 1)$ on Γ what implies not existence a magic rectangle set by Observation 12. □

Observation 13 implies immediately the following observations:

Observation 14. *If exactly one of numbers the a, b, c is even, then there does not exist a magic rectangle set $MRS_\Gamma(a, b; c)$ on any Abelian group Γ having exactly one involution.*

Observation 15. *If exactly one of numbers the a, b, c is even and moreover congruent to 2 modulo 4, then there does not exist a magic rectangle set $MRS_\Gamma(a, b; c)$ on any Abelian group Γ.*

Observation 16. *If a is even, b is odd and $\Gamma \cong \mathbb{Z}_{abc}$ there does not exist a magic rectangle set $MRS_\Gamma(a, b; c)$ on the group Γ.*

Now we show that for a and b both even, a magic rectangle set $MRS_\Gamma(a, b; c)$ can be constructed for any c and any Abelian group Γ. We start with an elementary step.

Lemma 17. *A magic rectangle set $MRS_\Gamma(2, 2; c)$ exists for every c for any Abelian group Γ.*

Proof. Denote by $x_{i,j}^s$ the entry in the i-th row and j-th column of the s-th rectangle.

Suppose first that $\Gamma \cong \mathbb{Z}_2 \times \mathbb{Z}_2 \times \Psi$ for some Abelian group Ψ of order c. Using the isomorphism $\varphi : \Gamma \rightarrow \mathbb{Z}_2 \times \mathbb{Z}_2 \times \Psi$, we identify every $\gamma \in \Gamma$ with its image $\varphi(\gamma) = (j_1, j_2, g_s)$, where $j_1, j_2 \in \mathbb{Z}_2$ and $g_s \in \Psi$, $s = 0, 1, \ldots, c - 1$. Set $x_{1,1}^s = (0, 0, g_s)$, $x_{1,2}^s = (1, 1, -g_s)$, $x_{2,1}^s = (0, 1, -g_s)$ and $x_{2,2}^s = (1, 0, g_s)$ for $s = 0, 1, \ldots, c - 1$. Apparently, every column adds up to $(1, 0, 0)$ and every row in each rectangle has the sum equal to $(1, 1, 0)$.

Suppose now that $\Gamma \cong \mathbb{Z}_{2^\alpha} \times \Psi$ for some Abelian group Ψ of order $c/2^{\alpha-2}$ and $\alpha > 1$. Using the isomorphism $\varphi : \Gamma \to \mathbb{Z}_{2^\alpha} \times \Psi$, we identify every $\gamma \in \Gamma$ with its image $\varphi(\gamma) = (j, g_s)$, where $j \in \mathbb{Z}_{2^\alpha}$ and $g_s \in \Psi$, $s = 0, 1, \ldots, c/2^{\alpha-2}$. Set

$$a_{i,j}^s = \begin{cases} (s \bmod 2^{\alpha-2} + 1 + 2^{\alpha+1-i}, g_{\lfloor s2^{-\alpha+2} \rfloor}), & \text{for } i = j, \\ (2^\alpha + 1, 0) - a_{i \bmod 2 + 1, j}^s, & \text{for } i \neq j, \end{cases}$$

for $i, j = 1, 2$, $s = 0, 1, \ldots, 2^{\alpha-2}c - 1$. Notice that every column adds up to $(1,0)$ and every row in each rectangle has the sum equal to $(2^{\alpha-1} + 1, 0)$. $\qquad\square$

Theorem 18. *If a, b are both even, then a magic rectangle set $MRS_\Gamma(a, b; c)$ exists for every c for any Γ.*

Proof. There exist $MRS_\Gamma(2, 2; abc/4)$ by Lemma 17. To construct one of $MRS_\Gamma(a, b; c)$, we simply take $ab/2$ of $MRS_\Gamma(2, 2; abc/4)$ rectangles and "glue" them into a rectangle $a \times b$. $\qquad\square$

3   Group Distance Magic Graphs

Theorem 19. *A graph $K(n; r)$ is group distance magic if $n = r$ or n and r are both even or $n \equiv 0 \pmod 4$.*

Proof. Let Γ be an Abelian group of order rn. If $n = r$ then there exists a $MRS_\Gamma(n, n; 1)$ by Theorem 10. If now $n, r \equiv 0 \pmod 2$ or $n \equiv 0 \pmod 4$ there exists a $MRS_\Gamma(n, r; 1)$ or $MRS_\Gamma(n/2, 2r; 1)$, respectively by Theorem 18. For $n = r$ or n and r both even we can construct a distance magic labeling for $K(n; r)$ by labeling the vertices of each part by the columns of the magic rectangle.

For $n \equiv 0 \pmod 4$ let $V(K(n; r)) = \{v_i^j : i = 1, \ldots, n; j = 1, \ldots, r\}$ and denote by $x_{i,j}$ i-th row and j-th column of the Γ-magic rectangle. Let

$$\ell(v_i^j) \begin{cases} x_{i,j}, & \text{for } i \leq n/2 \\ x_{i,j+r}, & \text{for } i > n/2. \end{cases}$$

Obviously ℓ is a Γ-distance magic labeling. $\qquad\square$

Theorem 20. *The Cartesian product $K(n; r) \square C_4$ is group distance magic if n is even, $r > 1$.*

Proof. Let $V(K(n; r)) = \{v_i^j : i = 1, \ldots, n; j = 1, \ldots, r\}$, $C_4 = xuywx$, and $H = K(n; r) \times C_4$. Let Γ be an Abelian group of order $4rn$.

If n is even, then a Γ-magic rectangle set $MRS(2, n; 2r)$ with all row sums in every rectangle equal to a constant $\omega \in \Gamma$ and all column sums in every rectangle equal to a constant $\delta \in \Gamma$ exists by Theorem 18. Denote by $a_{i,h}^j$ the entry in the i-th row and h-th column of the j-th rectangle from the set $MRS(2, n; 2r)$, let:

$$\ell(v_i^j, x) = a_{i,1}^j, \ \ell(v_i^j, y) = a_{i,2}^j,$$

$$\ell(v_i^j, u) = a_{i,1}^{j+r}, \ \ell(v_i^j, v) = a_{i,2}^{j+r}$$

for $i = 1, \ldots, n$ and $j = 1, \ldots, r$. Obviously the labeling ℓ is bijection. Moreover it is distance magic since $\sum_{y \in N(x)} \ell(y) = (r-1)\omega + \delta$ for any $x \in H$. $\qquad\square$

At the end of this section we state a conjecture analogously to Theorem 7:

Conjecture 21. The Cartesian product $K(n; r)\Box C_4$ is Γ-distance magic if and only if there exists a magic rectangle set $\text{MRS}_\Gamma(2, n; 2r)$ for some Abelian group Γ of order $4nr$.

References

1. Barrientos, C., Cichacz, S., Froncek, D., Krop, E., Raridan, C.: Distance Magic Cartesian Product of Two Graphs (preprint)
2. Cichacz, S.: Group distance magic graphs $G \times C_n$. Discrete Appl. Math. **177**(20), 80–87 (2014)
3. Cichacz, S.: Note on group distance magic complete bipartite graphs. Cent. Eur. J. Math. **12**(3), 529–533 (2014)
4. Cichacz, S., Froncek, D.: Distance magic circulant graphs. Preprint Nr MD 071 (2013). http://www.ii.uj.edu.pl/documents/12980385/26042491/MD_71.pdf
5. Combe, D., Nelson, A.M., Palmer, W.D.: Magic labellings of graphs over finite abelian groups. Australas. J. Comb. **29**, 259–271 (2004)
6. Diestel, R.: Graph Theory, Graduate Texts in Mathematics, vol. 173. Springer, Heidelberg (2005)
7. Froncek, D.: Handicap distance antimagic graphs and incomplete tournaments. AKCE Int. J. Graphs Comb. **10**(2), 119–127 (2013)
8. Froncek, D.: Group distance magic labeling of Cartesian product of cycles. Australas. J. Combin. **55**, 167–174 (2013)
9. Sun, H., Yihui, W.: Note on magic squares and magic cubes on Abelian groups. J. Math. Res. Exposition **17**(2), 176–178 (1997)
10. Gallian, J.A.: A dynamic survey of graph labeling. Electron. J. Comb. **17**, 17–20 (2013). #D30
11. Harmuth, T.: Ueber magische Quadrate undÉihnliche Zahlenfiguren. Arch. Math. Phys. **66**, 286–313 (1881)
12. Harmuth, T.: Ueber magische Rechtecke mit ungeraden Seitenzahlen. Arch. Math. Phys. **66**, 413–447 (1881)
13. Rao, S.B., Singh, T., Parameswaran, V.: Some sigma labelled graphs I. In: Arumugam, S., Acharya, B.D., Raoeds, S.B. (eds.) Graphs, Combinatorics, Algorithms and Applications, pp. 125–133. Narosa Publishing House, New Delhi (2004)

Solving Matching Problems Efficiently
in Bipartite Graphs

Selma Djelloul$^{(\boxtimes)}$

LRI, UMR 8623, Bât 650 Université de Paris-Sud,
91405 Orsay Cedex, France
djelloul@lri.fr

Abstract. We investigate the problem maxDMM of computing a largest set of pairwise disjoint maximum matchings in undirected graphs. In this paper, n, m denote, respectively, the number of vertices and the number of edges. We solve maxDMM for bipartite graphs, by providing an $O(n^{1.5}\sqrt{m/\log n} + mn \log n)$-time algorithm. We design better algorithms for complete bipartite graphs, and *bisplit* graphs. (Bisplit graphs are bipartite graphs with the nested neighborhood property.) Specifically, we prove that the problem maxDMM is solvable in complete bipartite graphs in time $O(m)$. A sequence $S = (s_1, \cdots, s_t)$ of positive integers is said to be *color-feasible* for a graph G, if there exists a proper edge-coloring of G with colors $1, \cdots, t$, such that precisely s_i edges have color i, for every $i = 1, \cdots, t$. Actually, for complete bipartite graphs, we prove that, for any sequence S of integers which is color-feasible for a complete bipartite graph G, an edge-coloring of G corresponding to S can be obtained in time $O(m)$. For bisplit graphs, (1) we solve maxDMM in time $O(mn \log n)$, and (2) we design an $O(n^2 \log n)$-time algorithm to count all maximum matchings. This latter time is the same time in which runs the best known algorithm computing the number of maximum matchings in bisplit graphs [17], but our algorithm is much simpler than the one given in [17]. The key idea underlying both results is that bisplit graphs have an $O(n)$-time enumeration of their minimal vertex covers.

1 Introduction

We consider undirected simple graphs. In a graph, a matching is a subset of pairwise independent edges. A maximum matching is a matching of maximum cardinality. For a graph G, we denote by $\nu(G)$ the cardinality of a maximum matching of G, which is called the matching number of G. In all the paper, n and m denote, respectively, the number of vertices and the number of edges. If x is a vertex of a graph G, $d_G(x)$ denotes the degree of x in G. The notation $\Delta(G)$ and $\delta(G)$ is used, respectively, for the maximum degree and the minimum degree of G. If no ambiguity exists, we simply write Δ and δ.

Determining a maximum matching in a graph can be done in time $O(n^{2.5})$ [7] in general graphs, and it can be done in time $O(n^{1.5}\sqrt{m/\log n})$ in bipartite graphs [1].

© Springer International Publishing Switzerland 2015
J. Kratochvíl et al. (Eds.): IWOCA 2014, LNCS 8986, pp. 128–139, 2015.
DOI: 10.1007/978-3-319-19315-1_12

We are interested in the following optimization problem:

maxDMM: Maximum set of Disjoint Maximum Matchings.

Instance: A graph G.

Solution: A set $\mathcal{M}$ of pairwise disjoint maximum matchings.

Measure: $|\mathcal{M}|$.

Let $\mu(G)$ denotes this maximum for G. Note that $\mu(G) \le \lfloor m/\nu(G) \rfloor$, and that any graph G with a perfect matching satisfies $\mu(G) \le \delta$.

A related more general problem is the following one posed in [8]. Given a graph G, and a sequence $(s_1, \cdots, s_t)$ of positive integers, does there exist a proper edge-coloring of G, with colors $1, \cdots, t$, (edges with the same color are pairwise independent) such that precisely s_i edges are colored with color i, for each $i = 1, \cdots, t$? If the answer is "yes", the sequence $(s_1, \cdots, s_t)$ is said to be color-feasible for G. This problem is NP-complete even if G is bipartite, $\Delta = 3$, and sequences are of the form (s_1, s_2, s_3) with $s_2 \le s_1 - 2$, and $s_3 \le s_2 - 2$ (see [2] for references). Stated in terms of color-feasible sequences, maxDMM is the problem of determining for a graph G, the greatest integer t such that a sequence of the form $(s_1, \cdots, s_t)$, or of the form $(s_1, \cdots, s_t, 1, \cdots, 1)$, with $s_i = \nu(G)$, for every $i = 1, \cdots, t$, is color-feasible for G. In [2,8], the following partial order is considered on non-increasing sequences of positive integers which sum to m. The sequence $T = (t_1, \cdots, t_p)$ is said to majorize the sequence $S = (s_1, \cdots, s_q)$, if $p \le q$, and $\sum_{i=1}^{r} t_i \ge \sum_{i=1}^{r} s_i$, for every $r = 1, \cdots, p - 1$. It is proved in [8] that, if P is a color-feasible sequence for G, and S is a sequence majorized by P, then S is color-feasible for G, as well. Moreover, the proof in [8] shows that, given an edge-coloring of G corresponding to P, an edge-coloring of G corresponding to S, can be obtained in polynomial time by a sequence of transfers of edges, one by one, from one matching to another one. Before each transfer, the connected components of some subgraph are determined. This gives an $O(m^2)$-time algorithm for constructing an edge-coloring corresponding to a sequence S, provided we were given an edge-coloring corresponding to a color-feasible sequence for G that majorizes S. As mentioned in [8], characterizing all maximal sequences (in the sense of majorization) that are color-feasible for G, allows us to characterize all sequences that are color-feasible for G. We cite from [8]: "However, even for the case of bipartite graphs, we don't know how to construct one such (maximal) sequence, let alone all of them".

Before outlining the main results in this paper, let us give some additional definitions and notation. For a vertex x, let us denote by $N(x)$ the open neighborhood of x. An ordering $x_1, \ldots, x_k$, of a subset $X = \{x_i, \ 1 \le i \le k\}$, of vertices of a graph G is a *nested neighborhood ordering* if it satisfies $(N(x_1) \setminus X) \subseteq (N(x_2) \setminus X) \subseteq \ldots \subseteq (N(x_k) \setminus X)$ (non-decreasing order), or $(N(x_1) \setminus X) \supseteq (N(x_2) \setminus X) \supseteq \ldots \supseteq (N(x_k) \setminus X)$ (non-increasing order). In a bipartite graph, one side of the bipartition satisfies the nested neighborhood property if and only if both sides have the property [24].

A *bisplit graph* [9] (also called difference graph [11], chain graph [24], and nonseparable bipartite graph [6,10]) is a bipartite graph such that one side of the bipartition satisfies the nested neighborhood property. Bisplit graphs can be recognized in time $O(n + m)$, and the corresponding nested neighborhood orderings computed in the same time [12].

We denote by $\chi'(G)$ the number of colors in an edge-coloring of G with the minimum number of colors. According to Vizing's theorem, $\chi'(G)$ is Δ or $\Delta + 1$, and according to König's *line coloring* theorem, bipartite graphs have an edge-coloring with Δ colors. Bipartite graphs with maximum degree Δ can be Δ-edge-coloured in time $O(m \log \Delta)$ [4].

Our Results

Biregular graphs are bipartite graphs where the vertices of the same partition class have the same degree. We say that a bipartite graph of maximum degree Δ is *semi-biregular* if one partition class satisfies that all the vertices have degree Δ. Clearly, edge-coloring semi-biregular graphs yields a solution to maxDMM in time $O(m \log \Delta)$. Moreover, let $G = (X, Y, E)$ be a semi-biregular bipartite graph, with $|X| \leq |Y|$. Then the sequence $(s_1, \cdots, s_\Delta)$, with $s_i = |X|$, for every $i = 1, \cdots, \Delta$, is the unique maximal (in the sense of majorization introduced above) color-feasible sequence for G. Hence, sequences that are color-feasible for a semi-biregular bipartite graph are all well characterized. They are all non-increasing sequences $(s_1, \ldots, s_t)$ of positive integers which sum to m, and such that $s_1 \leq \min\{|X|, |Y|\}$. For each of them, a corresponding edge-coloring of G can be obtained in time $O(m^2)$.

The complete bipartite graphs form a subclass of semi-biregular graphs for which we prove a more efficient algorithm for constructing edge-colorings corresponding to color-feasible sequences. We do that by constructing, in time $O(m)$, an ordering of the edges of complete bipartite graphs. Once this ordering $e_1, \ldots, e_m$, is done for the edges of complete bipartite graph $G = (X, Y, E)$, then, for any sequence $(s_1, \ldots, s_k)$, of non-increasing positive integers with sum at most m, and such that $s_1 \leq \min\{|X|, |Y|\}$, the subsets $M_1 = \{e_1, \ldots, e_{s_1}\}$, $M_2 = \{e_{s_1+1}, \ldots, e_{s_1+s_2}\}$, ..., $M_k = \{e_{s_1+\ldots+s_{k-1}+1}, \ldots, e_{s_1+\ldots+s_k}\}$, form a set of pairwise disjoint matchings. This gives an edge-coloring in complete bipartite graphs corresponding to any color-feasible sequence in time $O(m)$, without performing any transfer from one matching to another one. The coloring is obtained directly just by picking edges, consecutively, starting from the beginning of the ordering. As a consequence, we obtain that maxDMM is solvable in complete bipartite graphs in time $O(m)$.

Our first main result is the design of an algorithm solving maxDMM in general bipartite graphs and running in time $O(n^{1.5}\sqrt{m/\log n} + mn\log n)$. The term $n^{1.5}\sqrt{m/\log n}$ represents the cost of computing a minimum vertex cover in a bipartite graph. For subclasses of bipartite graphs where the computation of a minimum vertex cover requires less time, the overall running time will be improved. This is what we do next for bisplit graphs.

The graphs whose vertices can be linearly ordered in such a way that all the minimal vertex covers are subsets of consecutive vertices were characterized in [6] in various ways. Among these characterizations, we mention the following one.

The set of vertices of a graph G can be linearly ordered such that all the minimal vertex covers are subsets of consecutive vertices if and only if G is a bisplit graph. We give an explicit $O(n)$-time enumeration of the minimal vertex covers of bisplit graphs using a concise encoding of the graph. This result allows us to derive: (1) an algorithm for maxDMM in bisplit graphs running in time $O(mn \log n)$, and (2) $O(n^2 \log n)$-time algorithms for computing the number of maximum matchings of a bisplit graph, and for computing the number of perfect matchings of a *threshold graph*. This is the same time as for the best known algorithms computing these two numbers ([17]), but we claim that our algorithms are simpler than those given in [17] which need to handle a recursive subdivision structure associated with the graph.

2   Preliminaries

The problem of deciding whether a 2-connected cubic graph has two disjoint maximum matchings is NP-complete. This can be proved using a reduction from the problem of 3-edge-coloring 2-connected cubic graphs, which is NP-complete [14]. (Note that, by Petersen's theorem [3,18], every 2-connected cubic graph has a perfect matching.) It follows that maxDMM is not approximable within $2/3 + \epsilon$, for any $\epsilon > 0$.

A graph H is a minor of a graph G if H is a subgraph of a graph that can be obtained from G by contracting a set of edges. If H is fixed, checking whether H is a minor of G can be done in time $O(n^2)$ [15].

Every 2-connected cubic graph G which has not the Petersen graph as a minor is 3-edge-colorable [20,23]. Hence, the algorithm for 3-edge-coloring a 2-connected cubic graph G with no Petersen minor produces a solution of maxDMM for G. We point out the particular case of 2-connected planar cubic graphs (they do not have the Petersen graph as a minor.) Tait proved that 3-edge-coloring a 2-connected planar cubic graph is equivalent to 4-coloring a plane graph. Hence, the quadratic time algorithm used in [19] to produce a 4-coloring for a plane graph can be used to produce a solution of maxDMM for 2-connected planar cubic graphs.

We end this section by mentioning that there exists an algorithm running in time $O(1.344^n)$ and polynomial space that checks 3-edge-colorability for graphs with $\Delta = 3$, and computes such a coloring if it exists [16]. For this problem, there also exists an $O(1.201^n)$-time and exponential space algorithm [5]. We apply one of these algorithms, say the latter one, on a 2-connected cubic graph G. If the algorithm says: "No 3-edge-coloring", then $\mu(G) = 1$, and a solution to maxDMM is any maximum matching (which is perfect by Petersen's theorem.) If the algorithm outputs a 3-edge-coloring, then it outputs 3 disjoint perfect matchings. Hence, maxDMM is solvable in 2-connected cubic graphs in time $O(1.201^n)$.

3   maxDMM in Arbitrary Bipartite Graphs

Let G be a graph, and let g, and f, be two integer functions on the vertex set of G, such that $0 \leq g(v) \leq f(v) \leq d_G(v)$, for each vertex v of G. A (g, f)-factor of G is a spanning subgraph F of G such that $g(v) \leq d_F(v) \leq f(v)$, for every vertex $v \in V(G)$. The following result has been proved in [13]:

Theorem 1. *If $g(v) < f(v)$ for every vertex v, or if the graph G is bipartite, then, deciding whether there exists a (g, f)-factor in G can be done in time $O(\sqrt{g(V) \cdot m})$, where $g(V)$ denotes $\sum_{v \in V} g(v)$. Such a factor can be constructed in the same time if it exists.*

In [22], Slater gave a characterization of trees with (at least) k disjoint maximum matchings, without mentioning any algorithmic complexity aspect.

We say that a bipartite graph $G = (X, Y, E)$ is X- (Y-) matchable if $\nu(G) = |X|$ ($\nu(G) = |Y|$.) If U is a minimum vertex cover of a bipartite graph G, we let G_U be the spanning subgraph obtained from G by removing the set E_U of edges having both ends in U, and $\delta_U = \min\{d_{G_U}(z), z \in U\}$. We call G_U the *pruning* of G with respect to minimum vertex cover U.

By König's theorem, in bipartite graphs G, the size of a minimum vertex cover is precisely $\nu(G)$. Hence, for any minimum vertex cover U of G, and any maximum matching F of G, each vertex of U is matched by F, and each edge of F has a single end in U. Thus, we have the following:

Fact 1. Let G be a bipartite graph and U be any minimum vertex cover of G. The pruning $G_U = (U, V - U, E - E_U)$ of G with respect to U is U-matchable and the set of maximum matchings of G is exactly the set of maximum matchings of G_U. Moreover, the set of solutions to maxDMM in G_U is exactly the set of solutions to maxDMM in G, and $\mu(G) \leq \delta_U$.

Theorem 2. *maxDMM is solvable in bipartite graphs in time $O(n^{1.5}\sqrt{m/\log n} + mn \log n)$.*

Proof. First, we consider the case where $G = (X, Y, E)$ is X-matchable. There cannot be more than δ_X disjoint maximum matchings. There are k disjoint maximum matchings in G if and only if there does exist a k-edge-colorable (g, f)-factor $F(k)$ in G, where f is k at each vertex, and g is k on X, and 0 elsewhere. The condition on the k-edge-colorability can be omitted since the required (g, f)-factor is a bipartite graph with maximum degree k. It is then k-edge-colorable. Note that the number of edges in $F(k)$ is $k|X|$. We test the values of k in $\{1, \cdots, \delta_X\}$ by binary search, starting with $k = \delta_X$. We obtain the value of optimum k, say OPT, after an amount of running time which is at worst of the form $c \cdot m \sqrt{\delta_X |X|} \cdot \log \delta_X$, for some constant c (Theorem 1). By Fact 1, OPT $\leq \delta_U$, for every minimum vertex cover U. We finish by OPT-edge-coloring the obtained $F(\text{OPT})$ factor, which can be performed in time $c' \cdot \text{OPT} \cdot |X| \cdot \log \text{OPT}$, for some constant c'. Each color appears at each vertex in X. Hence, each color class has size $|X|$. Thus, each color class is a maximum matching of G. Now, we consider an arbitrary bipartite graph G. Given a maximum matching of bipartite graph G, we can compute a minimum vertex cover U for G in time $O(m)$ (see [21], for example). In time $O(m)$, we obtain the pruning G_U of G with respect to U. We apply on G_U the algorithm of X-matchable graphs, taking δ_U as the first tested value for the maximum degree of the searched (g, f)-factor. By Fact 1, this solves maxDMM in G. □

4   Complete Bipartite Graphs

Theorem 3. *Let $G = (X, Y, E)$ be a complete bipartite graph. There exists an $O(m)$-time ordering $e_1, \ldots, e_m$, of the edges of G guaranteeing the following. For any non-increasing sequence $(s_1, \ldots, s_k)$ of positive integers, with sum at most m, and $s_1 \leq \min\{|X|, |Y|\}$, the subsets $M_1 = \{e_1, \ldots, e_{s_1}\}$, $M_2 = \{e_{s_1+1}, \ldots, e_{s_1+s_2}\}$, $\ldots$, $M_k = \{e_{s_1+\ldots+s_{k-1}+1}, \ldots, e_{s_1+\ldots+s_k}\}$, form a set of pairwise disjoint matchings.*

Before proceeding to the proof, let us give some definitions and notation. If a, b, are positive integers, and R is a subset of $\{0, \ldots, a-1\} \times \{0, \ldots, b-1\}$, we denote by $\Pi_1(R), \Pi_2(R)$ the projections of R on $\{0, \ldots, a-1\}$ and $\{0, \ldots, b-1\}$, respectively. That is: $\Pi_1(R) = \{i \mid (i, j) \in R\}$, and $\Pi_2(R) = \{j \mid (i, j) \in R\}$.

For any positive integers a, b, $a \leq b$, an injective mapping $\varphi : \{0, \ldots, ab - 1\} \to \{0, \ldots, a-1\} \times \{0, \ldots, b-1\}$ is called an (a, b)-sequencing of $\{0, \ldots, ab-1\}$ if one of the two holds: (1) $a < b$, and every sequence S of at most a consecutive integers from $\{0, \ldots, ab - 1\}$ satisfies $|\Pi_1(\varphi(S))| = |\Pi_2(\varphi(S))| = |S|$; (2) $a = b$, and every sequence S of at most $a - 1$ consecutive integers from $\{0, \ldots, ab - 1\}$ satisfies $|\Pi_1(\varphi(S))| = |\Pi_2(\varphi(S))| = |S|$.

Lemma 1. *For any positive integers a, b, $a \leq b$, there exists an (a, b)-sequencing of $\{0, \ldots, ab-1\}$. Moreover such a sequencing can be constructed in time $O(ab)$.*

Proof. Let $d = \gcd(a, b)$, and a', b' such that $a = da'$, $b = db'$. Let $I_\lambda = \{\lambda a'b, \lambda a'b + 1, \cdots, (\lambda + 1)a'b - 1\}$, $0 \leq \lambda \leq d - 1$. The sets I_λ are pairwise disjoint, each of them has cardinality $a'b$, and their union is $\{0, \cdots, da'b - 1\}$. Hence $\{I_0, \cdots, I_{d-1}\}$ is a partition of $\{0, \cdots, ab - 1\}$.

Let $\varphi : \{0, \ldots, ab - 1\} \to \{0, \ldots, a-1\} \times \{0, \ldots, b-1\}$ defined by $\varphi(i) = (i \mod a, (i + \lambda) \mod b)$, where λ is such that $i \in I_\lambda$. We want to prove that φ is injective. Let i, i' such that $\varphi(i) = \varphi(i')$. We have $a|(i-i')$ and $b|((i-i')+(\lambda-\lambda'))$ (1), with $i \in I_\lambda$, and $i' \in I_{\lambda'}$. Hence d divides both $(i - i')$ and $(i - i') + (\lambda - \lambda')$. It then divides $(\lambda - \lambda')$. Since $|\lambda - \lambda'| < d$, we have $\lambda = \lambda'$, and therefore $|i - i'| < a'b$, and (1) becomes $b|(i - i')$. Thus $(i - i')$ is a multiple of both a and b, and thus it is a multiple of their lowest common multiple, which is $ab/d = a'b = ab'$. We have $a'b\,|(i - i')$ and $|i - i'| < a'b$. This means that $i = i'$.

For any sequence S of consecutive integers from $\{0, \cdots, ab - 1\}$, $\Pi_1(\varphi(S))$ is a sequence of remainders modulo a of consecutive integers. Hence, if $|S|$ is at most a, all those remainders are pairwise distinct and we have $|\Pi_1(\varphi(S))| = |S|$. Now, we consider $\Pi_2(\varphi(S))$, where S is any sequence of at most a consecutive integers from $\{0, \cdots, ab - 1\}$. If $S \subseteq I_\lambda$ for some λ, $0 \leq \lambda \leq d - 1$, then $\Pi_2(\varphi(S))$ is a sequence of remainders modulo b of at most b consecutive integers, and we have $|\Pi_2(\varphi(S))| = |S|$. Now, assume that $\min S$ and $\max S$ are in distinct $I'_\lambda s$. Since the length of each I_λ is at least b, $\min S \in I_{\lambda-1}$ and $\max S \in I_\lambda$, for some λ, $1 \leq \lambda \leq d - 1$. Therefore, $\Pi_2(\varphi(S))$ is a sequence of remainders modulo b of $|S|$ integers that are all consecutive except the pair $\{(\lambda - 1 + \lambda a'b - 1) \mod b, (\lambda + \lambda a'b) \mod b\}$ corresponding to the second components of $\varphi(\max I_{\lambda-1})$, $\varphi(\min I_\lambda)$, respectively. The latter pair is a pair of remainders modulo b of two integers with a gap of two between them. Hence, if we restrict S to have no more than a elements or no more than $a - 1$ elements according to whether $a < b$, or $a = b$, then, still $\Pi_2(\varphi(S))$ is a sequence of pairwise distinct integers modulo b. $\qquad\square$

Proof (of Theorem 3). Let $G = (X, Y, E)$ be a complete bipartite graph. We assume, w.l.o.g; that $|X| \leq |Y|$. Let $a = |X|$, $b = |Y|$. Let φ be the (a, b)-sequencing of $\{0, \cdots, ab - 1\}$ defined in Lemma 1. Consider X as the set of integers modulo a, Y as the set of integers modulo b, and E as the set $\varphi(\{0, \cdots, ab - 1\})$. Let k, and $s_1 \geq \ldots \geq s_k$ be integers satisfying: $s_1 \leq a$, and $\Sigma_{i=1}^{k} s_i \leq |E|$. We consider the subset $S_1 = \{0, \cdots, s_1 - 1\}$, and if $k > 1$, the subsets $S_l = \{\Sigma_{r=1}^{l-1} s_r, \cdots, \Sigma_{r=1}^{l} s_r - 1\}$, $2 \leq l \leq k$. For every l, $1 \leq l \leq k$, let $M_l = \varphi(S_l)$. If a, b, s_1 satisfy one of the two cases: (1) $a < b$, or (2) $a = b$, and $s_1 < a$, then we are done. The case $a = b = s_1$ is settled by noting that, after picking all the matchings of size a, any integer s among the next integers (if any) in the sequence $(s_1, \cdots, s_k)$ is such that $s < b$. $\qquad\square$

It follows that maxDMM is solvable in complete bipartite graphs $G = (X, Y, E)$ in time $O(m)$, and we have $\mu(G) = \max\{|X|, |Y|\}$. (Take $k = \max\{|X|, |Y|\}$, and $s_1 = \ldots = s_k = \min\{|X|, |Y|\}$.)

5   Bisplit Graphs

Let X be a subset of vertices of a graph G. A vertex v of G is said to be X-*universal* if it is adjacent to every vertex from X.

In the rest of the paper, if $G = (X, Y, E)$ denotes a bisplit graph, then it is assumed that the ordering of X is non-increasing and that the ordering of Y is non-decreasing. We let $|X| = n_x$, and $|Y| = n_y$.

Proposition 1. *Let $G = (X, Y, E)$ be a bisplit graph. The following statements are equivalent: (1) G has no isolated vertices; (2) x_1 is Y-universal and y_{n_y} is X-universal; (3) G is connected.*

Since removing isolated vertices still gives a bisplit graph, from now on, all bisplit graphs are assumed to be connected.

Theorem 4. *maxDMM is solvable in bisplit graphs in time $O(mn \log n)$.*

The key idea is the following result that we prove firstly.

Theorem 5. *The set of all minimal vertex covers of a bisplit graph can be computed in time $O(n)$.*

If n is a positive integer, $[n]$ denotes the set $\{1, \ldots, n\}$. Let $X = \{x_1, \cdots, x_n\}$, and $I \subseteq [n]$. Then $X(I)$ denotes the subset $\{x_i, i \in I\}$ of X.

5.1   The Break Points of a Bisplit Graph

Let $G = (X, Y, E)$ be a bisplit graph. We define two mappings $\alpha : [n_x] \rightarrow [n_y]$, and $\beta : [n_y] \rightarrow [n_x]$ as follows. For every $i \in [n_x]$, $\alpha(i) = \min\{j \mid y_j \in N(x_i)\}$. For every $j \in [n_y]$, $\beta(j) = \max\{i \mid x_i \in N(y_j)\}$. We have $\alpha(1) = 1$, and $\beta(n_y) = n_x$. The two mappings α, β, encode the entire graph, since if $[x_i, y_j]$ is an edge then the pairs of the form $(x_{i'}, y_{j'})$, $i' \leq i$, $j' \geq j$, are all edges as well.

We consider bisplit gaphs that are not complete. Let $i_1 = \beta \circ \alpha(1)$, and for $k \geq 1$, if $i_k \neq n_x$, $i_{k+1} = \beta \circ \alpha(i_k + 1)$. We obtain a sequence of integers from $[n_x]: 1 \leq i_1 < \ldots < i_l = n_x$. Similarly, let $j_1 = \alpha \circ \beta(n_y)$, and for $k \geq 1$, if $j_k \neq 1$, $j_{k+1} = \alpha \circ \beta(j_k - 1)$. We obtain a sequence of integers from $[n_y]: 1 = j_p < \ldots < j_1 \leq n_y$, that we rename in reverse order to get: $1 = j_1 < \ldots < j_p \leq n_y$. We prove that $p = l$, and that for every r, $1 \leq r \leq l$, $j_r = \alpha(i_r)$, and $i_r = \beta(j_r)$. The details are omitted. In fact, the $i'_r s$ are the points where the invariability of α is broken and so are the $j'_r s$ for the invariability of β. We say that the pair $(\{i_1, \cdots, i_l = n_x\}, \{1 = j_1, \ldots, j_l\})$ is the *break points map* of G, and that l is its *breaking multiplicity*. They are determined in time $O(n)$. Hence, bisplit graphs can be given by means of their break points map.

5.2   Vertex Covers

If $(\{i_1, \cdots, i_l = n_x\}, \{1 = j_1, \cdots, j_l\})$ is the break points map of a bisplit graph, we let $U_r = X(\{1, \cdots, i_r\}) \cup Y(\{j_{r+1}, \cdots, n_y\})$, $1 \leq r \leq l - 1$.

Proof (of Theorem 5). We prove that the minimal vertex covers of G are X, Y, and U_r, $1 \leq r \leq l - 1$. Since there are no edges between $X([n_x] - \{1, \cdots, i_r\})$ and $Y(\{1, \cdots, j_{r+1} - 1\})$, U_r is a vertex cover. Let U be a vertex cover of G that contains neither X nor Y. The set U must contain x_1 (which is Y-universal). Let i be the greatest i' such that $\{1, \cdots, i'\} \subseteq U$. We have $i < n_x$ because U does not contain X. There are edges incident with x_{i+1} because G is connected. The neighborhood of x_{i+1} is $Y(\{\alpha(i + 1), \cdots, n_y\})$, and it is then included in U. If $\alpha(i) \neq \alpha(i + 1)$ then i is a break point for α, say i_{r_0}. We have: $\alpha(i + 1) = \alpha(i_{r_0+1}) = j_{r_0+1}$. Hence U contains U_{r_0}. Now, we assume that $\alpha(i) = \alpha(i+1)$. Since U does not contain Y, we have $\alpha(i) \neq 1$. Hence, there are edges between $X(\{1, \cdots, i\})$ and $Y(\{1, \cdots, \alpha(i) - 1\})$. Let i_1 be the greatest i' between 1 and $i - 1$ such that $\alpha(i) \neq \alpha(i')$. From i_1 to $i_1 + 1$, the invariability of α breaks. Then, i_1 is of the form i_r for some r, $1 \leq r \leq l - 1$. We have: $j_{r+1} = \alpha(i_{r+1}) = \alpha(i_r + 1) = \alpha(i) = \alpha(i+1)$. Hence U contains U_r. $\square$

In any graph, the complement of a vertex cover is an independent set. Then, it follows from Theorem 5, that determining the set of all maximal independent sets in bisplit graphs can be done in time $O(n)$. By König's theorem, in bipartite graphs, the matching number equals the size of a minimum vertex cover. Then, it follows from Theorem 5, that the matching number of a bisplit graph can be computed in time $O(n)$.

Proof (of Theorem 4). Pick any minimum vertex cover U. As in the proof of Theorem 2, we solve maxDMM in G in a time which is at worst of the form $c \cdot m\sqrt{\delta_U \cdot |U|} \cdot \log \delta_U$, for some constant c, by taking δ_U, as the first tested value for the maximum degree of the searched (g, f)-factor (Theorem 1). By Fact 1, $\mu(G) \leq \tilde{\delta}$, where $\tilde{\delta} = \min \{\delta_{U'}, U'$ is a minimum vertex cover$\}$. Thus, we can run the algorithm on G_U, starting the series of tests with the better degree value $\tilde{\delta}$. Furthermore, instead of running the algorithm on G_U, we can run it on the U-matchable spanning subgraph $(U, V - U, E')$, obtained from G by removing, for every minimum vertex cover U', all the edges that have both ends in U'.   □

5.3   Matching and Counting All Maximum Matchings

In this section, we prove that:

Theorem 6. *For bisplit graphs, constructing a maximum matching can be done in time $O(n)$.*

We will also give another proof of the following result proved in [17] by designing a simpler algorithm:

Theorem 7. *For bisplit graphs, counting all maximum matchings can be done in time $O(n^2 \log n)$.*

We let $U_0 = Y$, $U_l = X$, $I_1 = \{1, \cdots, i_1\}$, $I_2 = \{i_1 + 1, \cdots, i_2\}, \cdots, I_l = \{i_{l-1} + 1, \cdots, n_x\}$, and $J_1 = \{1, \cdots, j_2 - 1\}$, $J_2 = \{j_2, \cdots, j_3 - 1\}, \cdots, J_l = \{j_l, \cdots, n_y\}$, and $a_r = |I_r|, b_r = |J_r|$, $1 \leq r \leq l$. If U_r is a minimum vertex cover, G_r denotes the pruning of G with respect to U_r.

Lemma 2. *G denotes a bisplit graph, U_r denotes a minimum vertex cover of G, and F denotes a maximum matching of G. If $1 \leq r \leq l - 1$, then F matches all the vertices of $X(\{1, \cdots, i_r\})$ in $Y(\{1, \cdots, j_{r+1} - 1\})$, and all the vertices of $Y(\{j_{r+1}, \cdots, n_y\})$ in $X(\{i_r + 1, \cdots, n_x\})$.*

Proof. By Fact 1, G_r is U_r-matchable, and F is a maximum matching of G_r. □

Lemma 3. *G denotes a bisplit graph, U_r denotes a minimum vertex cover of G.*

(1) If $r \neq 0$, then every s, $0 \leq s < r$, satisfies $\sum_{k=s+1}^{r} a_k \leq \sum_{k=s+1}^{r} b_k$.
(2) If $r \neq l$, then every s, $l \geq s > r$, satisfies $\sum_{k=r+1}^{s} b_k \leq \sum_{k=r+1}^{s} a_k$.

Proof

(1) $U_s = U_r \cup J_{s+1} \cup \ldots \cup J_r - I_{s+1} \cup \ldots \cup I_r$. Hence $|U_s| = |U_r| - \sum_{k=s+1}^{r} a_k + \sum_{k=s+1}^{r} b_k$. Since U_r is of minimum size, the inequality follows.
(2) We proceed similarily by observing that $U_s = U_r \cup I_{r+1} \cup \ldots \cup I_s - J_{r+1} \cup \ldots \cup J_s$.   □

Proof (of Theorems 6 and 7). Seeking readability, we allow denoting the vertices directly by means of their corresponding subscripts in the orderings of X and Y. Pick any minimum vertex cover, say U_r. We have $U_r = X_r \cup Y_r$, where, if not empty, X_r is $I_1 \cup \ldots \cup I_r$, and if not empty, Y_r is $J_{r+1} \cup \ldots \cup J_l$. The pruning G_r of G with respect to U_r is the disjoint union of two bisplit subgraphs: H_0, H_1. By Lemma 2, if not empty, the subgraph H_0 induced by $X_r \cup [n_y] - Y_r$ is X_r-matchable, and if not empty, the subgraph H_1 induced by $Y_r \cup [n_x] - X_r$ is Y_r-matchable. Consider two variables N_0, and N_1 that will hold the numbers of maximum matchings of H_0 and H_1 respectively. Let $N_0 \leftarrow 1$, and $N_1 \leftarrow 1$. We recall that a_k (resp. b_k) denotes $|I_k|$ (resp. $|J_k|$.) If H_0 is not empty, we proceed by induction on t from r down to 1. Variable N_0 is updated when going along the induction. Let $d \leftarrow 0$. By Lemma 3, $a_r \leq b_r$. Hence, in the complete bipartite subgraph induced by $I_r \cup J_r$, we can match all the vertices of I_r with a subset S_r of J_r. Update N_0 by $N_0 \leftarrow N_0 \cdot (d+b_r)!/((d+b_r-a_r)!)$. Now, update d by $d \leftarrow d + b_r - a_r$. If we are just constructing a maximum matching, we don't need to handle the numbers N_0, N_1. We just have to pick, at step k, of the induction, where the vertices of I_k are about to be matched, a set of vertices of size $|a_k|$ in the other side of the current complete bipartite subgraph. Assume we have matched all the vertices of $I_r \cup \ldots \cup I_t$, with a subset S_t of $J_r \cup \ldots \cup J_t$, for $1 \leq t \leq r$ (this is done in time of the form $c \cdot \sum_{k=t}^{r} a_k$, for some constant c, including the time needed to update the other side of the current complete bipartite subgraph.) If $X_r - I_r \cup \ldots \cup I_t$, is not empty, then $t > 1$. The subgraph induced by $I_{t-1} \cup J_r \cup \ldots \cup J_{t-1} - S_t$, is complete bipartite. Since $a_{t-1} \leq b_{t-1} + \sum_{k=t}^{r} b_k - \sum_{k=t}^{r} a_k$ (Lemma 3), we can match all the vertices of I_{t-1} with a subset S_{t-1} of $J_r \cup \ldots \cup J_{t-1} - S_t$ (in time $O(a_{t-1})$). Update N_0: $N_0 \leftarrow N_0 \cdot (d+b_{t-1})!/((d+b_{t-1}-a_{t-1})!)$. Now, update d: $d \leftarrow d + b_{t-1} - a_{t-1}$. Similarly, if H_1 is not empty, we proceed by induction on t, $r+1 \leq t \leq l$, as follows. Reset $d \leftarrow 0$. As done for N_0, we will first have: $N_1 \leftarrow N_1 \cdot (d+a_{r+1})!/((d+a_{r+1}-b_{r+1})!)$, then $d \leftarrow d + a_{r+1} - b_{r+1}$. If we must go further than $r+1$ in the induction, then the update operations from t to $t+1$ are first N_1: $N_1 \leftarrow N_1 \cdot (d+a_{t+1})!/((d+a_{t+1}-b_{t+1})!)$, and then d: $d \leftarrow d + a_{t+1} - b_{t+1}$. Let $N = N_0 \cdot N_1$. Output N as the number of maximum matchings of G.

All factorials $i!$, $1 \leq i \leq n$, are computed beforehand in time $O(n^2 \log n)$ and stored in space $O(n^2 \log n)$. Note that N is bounded up by the number of arrangements of $\sum_{k=1}^{r} a_k + \sum_{k=r+1}^{l} b_k$ elements from a set of $\sum_{k=1}^{r} b_k + \sum_{k=r+1}^{l} a_k \leq n$, elements. Hence, N is at most $n!$. Therefore, each update operation is done in time $O(n \log n)$. There are at most l such update operations, where l is the breaking multiplicity. All update operations are done in the same memory location which is of space $O(n \log n)$. □

5.4 Counting All Perfect Matchings in Threshold Graphs

Cochain and threshold graphs have a structure similar to bisplit graphs. A cochain graph is the complement of a bisplit (chain) graph. Namely, a cochain graph is a graph whose vertex set can be partitioned into two cliques joined by a set of edges that induces a bisplit graph. Split graphs are those graphs whose vertex set can be partitioned into a clique and an independent set. Threshold graphs are split graphs where the independent set part has the nested neighborhood property. All those graphs can be recognized in time $O(n + m)$ ([12]). In [17], the authors used their algorithm for bisplit graphs to derive $O(n^2 \log n)$-time algorithms for counting all perfect matchings in cochain and threshold graphs. Our algorithm for bisplit graphs can be used to derive a simpler $O(n^2 \log n)$-time algorithm for computing the number of perfect matchings in threshold graphs. Let $G = (X, Y, E)$ be a connected threshold graph, where X is the independent set part. Let G' be the underlying bisplit graph. We denote by $N(G)$ (resp. $N(G')$) the number of maximum matchings of G (resp. G'), and we denote by $k(n) = n!/\lceil n/2 \rceil! \cdot 2^{\lceil n/2 \rceil}$ the number of maximum matchings of K_n: the clique on n vertices. We prove that $N(G) \geq N(G') \cdot k(|Y| - \nu(G'))$, and that if G has a perfect matching, every maximum matching of G induces a maximum matching in G'. Hence, if G has a perfect matching, $N(G) = N(G') \cdot k(|Y| - \nu(G'))$.

6   Further Work

A natural question that arises from this work is whether there exist other types of graph classes in which maxDMM is solvable in polynomial time. Most likely, there should be few such graph classes, since even by restricting to graphs with $\Delta = 3$, and having a perfect matching, the problem remains NP-hard. One approach to deal with the problem may be looking for graph classes where we could efficiently find structures that prevent from having k disjoint maximum matchings. This leads to upper bounds on maxDMM. Another approach is to consider fixed parameterized algorithms for maxDMM. For instance, maxDMM is solvable in time $O(f(w) \cdot n \log \delta)$, for graphs of minimum degree at most δ, having a perfect matching. Here, $f(w)$ denotes a function of treewidth w, only. Indeed, for every integer k, the property $\varphi_k(G)$ saying: "G has k disjoint perfect matchings" is expressible in the monadic second order logic that allows quantification on subsets of edges. Hence, for each k, $\varphi_k(G)$ is decidable in time $O(f(w) \cdot n)$. We get the optimum from $\{1, \ldots, \delta\}$ after testing at most $\log \delta$ such formulae.

Acknowledgements. I thank Odile Favaron for her helpful idea to handle the proof of Theorem 3 by arithmetic. I thank Pierre Fraigniaud for his careful reading and advices to improve the paper writing.

References

1. Alt, H., Blum, N., Mehlhorn, K., Paul, M.: Computing a maximum cardinality matching in a bipartite graph in time $O(n^{1.5}\sqrt{m/\log n})$. Inf. Process. Lett. **37**(4), 237–240 (1991)
2. Asratian, A.S.: Some results on an edge-coloring problem of Folkman and Fulkerson. Discret. Math. **223**, 13–25 (2000)

3. Chartrand, G., Zhang, P.: A First Course in Graph Theory. Dover Publications, New York (2012)
4. Cole, R., Ost, K., Schirra, S.: Edge-coloring bipartite multigraphs in $O(E \log D)$ time. Combinatorica **21**(1), 5–12 (2001)
5. Couturier, J.F., Golovach, P.A., Kratsch, D., Liedloff, M., Pyatkin, A.: Colorings with few colors: counting, enumeration and combinatorial bounds. Theory Comput. Syst. **52**, 645–667 (2013). Springer-Verlag New York. Secaucus, NJ, USA
6. Ding, G.: Covering the edges with consecutive sets. J. Graph Theory **15**(5), 559–562 (1991)
7. Even, S., Kariv, O.: An $O(n^{2.5})$ algorithm for maximum matching in general graphs. In: IEEE 16th annual Symposium on Foundations of Computer Science (FOCS), pp. 100–112 (1975)
8. Folkman, J., Fulkerson, D.R.: Edge-colorings in bipartite graphs. In: Bose, R., Dowling, T. (eds.) Combinatorial Mathematics and its Applications, pp. 561–577. University of North Carolina Press, Chapel Hill (1969)
9. Frost, H., Jacobson, M., Kabell, J., Morris, F.R.: Bipartite analogues of split graphs and related topics. Ars Combinatoria **29**, 283–288 (1990)
10. Golumbic, M.R., Goss, C.F.: Perfect elimination and chordal bipartite graphs. J. Graph Theory **2**(2), 155–163 (1978)
11. Hammer, P.L., Peled, U.N., Sun, X.: Difference graphs. Discret. Appl. Math. **28**, 35–44 (1990)
12. Heggernes, P., Kratsch, D.: Linear-time certifying recognition algorithms and forbidden induced subgraphs. Nordic J. Comput. **14**, 87–108 (2007)
13. Heinrich, K., Hell, P., Kirkpatrick, D.G., Liu, G.: A simple existence criterion for $(g < f)$-factors. Discrete Mathematics **85**, 313–317 (1990)
14. Holyer, I.: The NP-completeness of edge-coloring. SIAM J. Comput. **10**(4), 718–720 (1981)
15. Kawarabayashi, K., Kobayashi, Y., Reed, B.: The disjoint paths problem in quadratic time. J. Comb. Theory Ser. B **102**, 424–435 (2012)
16. Kowalik, L.: Improved edge-coloring with three colors. Theoret. Comput. Sci. **410**, 3733–3742 (2009)
17. Okamoto, Y., Uehara, R., Uno, T.: Counting the number of matchings in chordal and chordal bipartite graph classes. In: Paul, C., Habib, M. (eds.) WG 2009. LNCS, vol. 5911, pp. 296–307. Springer, Heidelberg (2010)
18. Petersen, J.: Die theorie der regulären graphen. Acta Mathematica **15**, 193–220 (1891)
19. Robertson, N., Sanders, D., Seymour, P., Thomas, R.: Efficiently four-coloring planar graphs. In: Proceedings of the 28th annual ACM Symposium on Theory of Computing, (STOC), pp. 571–575 (1996)
20. Robertson, N., Seymour, P., Thomas, R.: Tutte's edge-coloring conjecture. J. Comb. Theory Ser. B **70**, 166–183 (1997)
21. Schrijver, A.: Combinatorial Optimization, vol. 1. Springer-Verlag, Berlin (2003)
22. Slater, P.: A constructive characterization of trees with at least k disjoint maximum matchings. J. Comb. Theory Ser. B **25**, 326–338 (1978)
23. Thomas, R.: Recent excluded minor theorem for graphs. In: Surveys in Combinatorics, vol. 267, pp. 201–222 (1999). The electronic journal of combinatorics 8 (2001)
24. Yannakakis, M.: Node deletion problems on bipartite graphs. SIAM J. Comput. **10**(2), 310–327 (1981)

A 3-Approximation Algorithm for Guarding Orthogonal Art Galleries with Sliding Cameras

Stephane Durocher and Saeed Mehrabi[(⊠)]

Department of Computer Science, University of Manitoba,
Winnipeg, Canada
{durocher,mehrabi}@cs.umanitoba.ca

Abstract. A *sliding camera* travelling along a line segment s in a polygon P can see a point p in P if and only if p lies on a line segment contained in P that intersects s at a right angle. The objective of the *minimum sliding cameras (MSC)* problem is to guard P with the fewest sliding cameras possible, each of which is a horizontal or vertical line segment. In this paper, we give an $O(n^3)$-time 3-approximation algorithm for the MSC problem on any simple orthogonal polygon with n vertices. Our algorithm involves establishing a connection between the MSC problem and the problem of guarding simple grids with *periscope guards*.

1 Introduction

Given a polygon P with n vertices in the plane, the art gallery problem is to find a minimum-cardinality set of guards such that every point in P is visible to at least one guard, where each guard g is a point in the plane that sees a point p if the line segment from g to p is contained in P. In the *orthogonal art gallery problem*, the input polygon P is orthogonal; that is, every edge of P is vertical or horizontal. The art gallery problem is NP-hard for both arbitrary [13] and orthogonal polygons [16]. Eidenbenz [4] proved that the art gallery problem is APX-hard on simple polygons, and that no polynomial-time algorithm can guarantee to find a solution with $o(\log n)$ times the minimum number of guards on polygons with holes, unless $P = NP$ [5]. Ghosh [7] gave an $O(\log n)$-approximation algorithm for the art gallery problem that runs in $O(n^4)$ time on simple polygons and $O(n^5)$ time on polygons with holes. Krohn and Nilsson [12] gave a polynomial-time $O(OPT^2)$-approximation algorithm for the orthogonal art gallery problem, where OPT is the cardinality of the optimal solution. Many variants of the art gallery problem have been studied based on different types of visibility [14,19], different polygonal domains (e.g., orthogonal polygons [8], or polyominoes [1]) and different types of guards (e.g., points or line segments). See the surveys by O'Rourke [15] or Urrutia [18] for a history of the art gallery problem.

Recently, Katz and Morgenstern [9] introduced a variant of the art gallery problem in which *sliding cameras* are used to guard an orthogonal polygon. Given an orthogonal polygon P with n vertices, a sliding camera is a point guard that travels back and forth along a horizontal or vertical line segment s inside P. The camera can see a point $p \in P$ if there is a point $q \in s$ such that the

© Springer International Publishing Switzerland 2015
J. Kratochvíl et al. (Eds.): IWOCA 2014, LNCS 8986, pp. 140–152, 2015.
DOI: 10.1007/978-3-319-19315-1_13

line segment $\overline{pq}$ is horizontal or vertical, and is contained in P. In the *minimum sliding cameras (MSC)* problem, the objective is to guard P using the minimum number of sliding cameras.

A grid D is a connected union of vertical and horizontal line segments; each maximal line segment in the grid is called a *grid segment*. We denote the set of grid segments of D by T_D. Moreover, a simple grid is defined as follows:

Definition 1 (Kosowski et al. [10]). *A grid D is simple if there exists $\delta > 0$ such that for every $\epsilon \in (0, \delta)$ and every grid segment d in D, both endpoints of d_ϵ lie in the outer face, where d_ϵ is the extension of d by ϵ units in both directions.*

A *periscope guard* x located on a grid segment s in a grid D is a point on s that sees a point y in D if some path from x to y in D has at most one bend. In other words, points x and y are mutually visible if and only if they lie on respective segments s_x and s_y in D such that $s_x \cap s_y \neq \emptyset$ (it could be that $s_x = s_y$). Periscope guards were introduced by Gewali and Ntafos [6] in their examination of the complexity of the orthogonal art gallery problem (the orthogonal art gallery problem was shown to be NP-hard three years later by Schuchardt and Hecker [16]). In the *minimum periscope guards (MPG)* problem on a grid, the objective is to guard the grid with the minimum number of periscope guards. The MPG problem can be defined on an orthogonal polygon P similarly: the goal is to locate the minimum number of periscope guards in P such that every point in P is guarded by at least one periscope guard.

Related Work. Katz and Morgenstern [9] first considered a restricted version of the MSC problem in which only vertical cameras are allowed; by reducing the problem to the minimum clique cover problem on chordal graphs, they solved the problem exactly in polynomial time. For the generalized case, where both vertical and horizontal cameras are allowed, they gave a 2-approximation algorithm for the MSC problem under the assumption that the polygon P is x-monotone. Durocher and Mehrabi [3] showed that the MSC problem is NP-hard when the polygon P is allowed to have holes (i.e., polygon P is not simple). They also gave an exact polynomial-time algorithm that solves a variant of the MSC problem, called the minimum-length sliding cameras (MLSC) problem, in which the objective is to minimize the sum of the lengths of line segments along which cameras travel. Seddighin [17] considered the MLSC problem under k-visibility, where a camera's line of sight can pass through k edges of the polygon, and proved that the MLSC problem is NP-hard under k-visibility for any fixed $k \geq 2$. Durocher et al. [2] gave an $O(n^{2.5})$-time (3.5)-approximation algorithm for the MSC problem on a simple orthogonal polygon with n vertices. Their algorithm uses different techniques from those used in this paper; specifically, it applies solutions to the minimum edge cover problem in graphs and the guarded mobile guard problem on grids (where each guard must be seen by at least one other guard). The complexity of the MSC problem on simple orthogonal polygons remains unknown.

Gewali and Ntafos [6] showed that the MPG problem is NP-hard on general three-dimensional grids and that it is polynomial-time tractable on simple two-dimensional grids (see Theorem 1). Moreover, Kosowski et al. [11] showed that the problem of guarding a two-dimensional grid with the minimum number of k-periscope guards is NP-hard (a point p on the grid is visible to a k-periscope guard g if there exists a path of at most k bends in the grid from p to g). Our results refer to the following theorem by Gewali and Ntafos [6].

Theorem 1 (Gewali and Ntafos [6]). *Given a simple two-dimensional grid G with n segments, the MPG problem can be solved exactly on G in $O(n^3)$ time.*

Our Result. In this paper, we give an $O(n^3)$-time 3-approximation algorithm for the MSC problem on any simple orthogonal polygon P. To this end, we describe a connection between the MSC problem on simple orthogonal polygons and the MPG problem on simple grids. We first construct a simple grid G_P associated with polygon P and then show that a reduction from the MSC problem on P to the MPG problem on grid G_P gives a set of sliding cameras whose cardinality is at most twice the cardinality of the solution to MPG problem on G_P. However, some new potentially unguarded regions are introduced. We show that the number of such unguarded regions is bounded from above by the cardinality of the optimal solution to the MPG problem, each of which can be guarded with a single sliding camera. Finally, we show that the cardinality of the optimal solution to the MPG problem is a lower bound for any feasible solution for the MSC problem. This results in an approximation factor of 3 (2 for each periscope guard in the solution of the MPG problem and 1 for guarding each unguarded region), improving the previous best approximation factor of 3.5 [2].

2   Preliminaries

Throughout the paper, let P denote a simple orthogonal closed polygonal with n vertices (including the polygon's interior). Observe that every simple orthogonal polygon with at most six vertices can be guarded by a single sliding camera; therefore, we assume throughout the paper that $n > 6$. Let OPT_P and OPT_{PG} denote optimal solutions for the MSC problem on P and the MPG problem on a simple grid, respectively. Let $V(P)$ denote the set of reflex vertices of P and let H_u and V_u be the maximum-length horizontal and vertical line segments, respectively, inside P through a vertex $u \in V(P)$. Let $L(P) = \{H_u \mid u \in V(P)\} \cup \{V_u \mid u \in V(P)\}$. Let L and L' be two orthogonal line segments (with respect to P) inside P; the *visibility region* of L is the union of points in P that are seen by a sliding camera that travels along L. We say L dominates L' if the visibility region of L' is a subset of that of L.

Let r be a reflex vertex of P. The lines through H_r and V_r partition the plane into four quadrants, exactly one of which contains the exterior of P in an ϵ-neighbourhood around r, for some $\epsilon > 0$; we call the quadrant that is opposite to this quadrant the *essential quadrant* of r, denoted by $Q(r)$. See Fig. 1(a) for an example. Let u be a convex vertex of P such that both the vertices v and w of P that are adjacent to u are also convex. Let p and q denote the next vertices of P that are adjacent to v and w, respectively. We call vertex u a *pocket vertex*

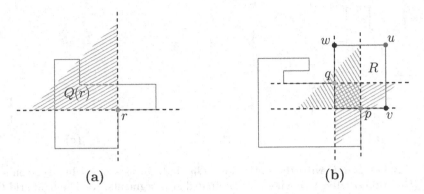

Fig. 1. (a) An example of a reflex vertex r with the essential quadrant $Q(r)$ (i.e., the open hatched quadrant) shown in pink. (b) An example of a pocket vertex u; both vertices p and q are reflex and $Q(p) \cap Q(q) \neq \emptyset$. The edges uv and uw are the pocket edges of the convex pocket R(red rectangle) (Colour figure online).

of P if and only if (i) both the vertices p and q are reflex, and (ii) $Q(p) \cap Q(q) \neq \emptyset$. Moreover, we refer to the edges of P that are incident to a pocket vertex as the *pocket edges* of P and to the rectangular subregion of P whose sides are two of the pocket edges of P as a *convex pocket* of P. See Fig. 1(b).

3   A 3-Approximation Algorithm

In this section, we describe the 3-approximation algorithm for the minimum sliding cameras (MSC) problem on simple orthogonal polygons. Given any simple orthogonal polygon P, we first construct a grid G_P associated with P as follows. Initially, let G_P be the set of all line segments in $L(P)$. Now, for any pair of reflex vertices u and v where H_u dominates H_v (resp., V_u dominates V_v) in P, we remove H_v (resp., V_v) from G_P; if two segments mutually dominate each other, remove one of the two arbitrarily. Next, for each convex pocket R of P, we add a segment into G_P for every pocket edge of R. We call a grid segment in G_P corresponding to a pocket edge of P a *pocket segment* of G_P. Let G_P denote the resulting grid. Each of the pocket segments remains in G_P even if it is dominated by another segment in G_P. See Fig. 2 for an example.

Observe that the number of grid segments in G_P (i.e., $|T_G|$) is at most n, where n is the number of vertices of P. Moreover, G_P is simple because the construction preserves the property that the endpoints of each grid segment in T_G lie on the boundary of the polygon and, therefore, the endpoints of every grid segment in T_G lie on the outer face of G_P. To see that G_P is connected, it suffices to note that (i) the grid induced by the line segments in $L(P)$ is connected, and (ii) for each grid segment $s \in L(P)$ that is removed (due to domination), the set of grid segments that are intersected by s are also intersected by $s' \in T_G$, where s' is the grid segment that dominates s. Therefore, G_P is also connected and we have the following result.

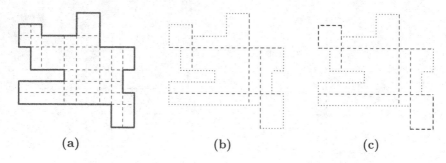

(a)   (b)   (c)

Fig. 2. (a) Polygon P with the initial grid G_P that consists of all line segments in $L(P)$. (b) Grid G_P after removing the dominated grid segments. (c) The final grid G_P after adding the segments corresponding to the pocket edges of the convex pockets of P; these grid segments are shown in green (Colour figure online).

Lemma 1. *Grid G_P is a simple and connected grid.*

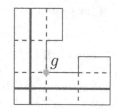

As described in Sect. 1, we reduce the MSC problem on P to the MPG problem on G_P. In general the visibility region of a periscope guard g cannot be guarded entirely by a single sliding camera; see Fig. 3 for an example. Two sliding cameras suffice to guard the visibility region of a periscope guard.

Observation 1. *The visibility region of any periscope guard g in a polygon P can be guarded by the maximal vertical and horizontal line segments through g in P.*

Fig. 3. Although the grid G_P induced by P can be guarded by a single periscope guard g, two sliding cameras (shown in purple) are needed to guard P (Colour figure online).

By Theorem 1 we can obtain a set of periscope guards by solving the MPG problem on G_P in $O(n^3)$ time. Let M denote the set of sliding cameras obtained by placing a pair of sliding cameras on each periscope guard. Since the algorithm of Gewali and Ntafos [6] positions periscope guards only at the intersections of grid segments, this ensures that the sliding cameras located in P are all aligned with line segments in $L(P)$.

The procedure described above can result in a set M of sliding cameras whose cardinality exceeds three times that of an optimal solution. See Fig. 4: the vertical segment in G_P that corresponds to the vertical pocket edge of convex pocket R cannot be guarded by periscope guard g_1, forcing the algorithm to add a second periscope guard, while a single sliding camera suffices to guard polygon P entirely. We now describe how to modify the grid G_P to bound $|M|$.

3.1 Pocket Segments and Desert Regions

As illustrated in Fig. 4, the cardinality of solution M may not be bounded by three times the cardinality of an optimal solution for the MSC problem. To resolve this problem, we add into G_P exactly *one* of the pocket segments

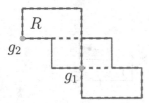

Fig. 4. By adding the two pocket segments into grid G_P corresponding to the pocket edges of every convex pocket of P, an optimal solution to the MPG problem uses two guards (g_1 and g_2). The algorithm for the MSC problem uses four sliding cameras (red), which is four times the size of the optimal solution (purple) (Colour figure online).

corresponding to the pocket edges of every convex pocket of P as follows: let R be a convex pocket of P and let s_1 and s_2 be, respectively, the vertical and horizontal grid segments in G_P whose corresponding maximal line segments in P enter R. Observe that the vertical pocket edge (resp., horizontal pocket edge) of R intersects s_2 (resp., s_1). If the number of grid segments intersected by s_1 is greater than the number of grid segments intersected by s_2, then we remove from G_P the pocket segment that corresponds to the vertical pocket edge of R; otherwise, we remove the pocket segment that corresponds to the horizontal pocket edge of R. Note that we now have exactly one pocket segment in G_P for both the pocket edges of every convex pocket of P. We show later that by this modification the cardinality of M obtained by solving the MPG problem on G_P is at most three times that of OPT_P.

The set M might not still be a feasible solution for the MSC problem. See Fig. 5 for an example. In the following, we characterize such unguarded regions, called the *desert regions*, and show that the number of desert regions is bounded from above by $|M|$. To characterize desert regions, take any unguarded point p in P and let R_p be a maximal axis-aligned rectangle contained in P that covers p and is also not guarded by the line segments in M. Observe that (i) rectangle R_p is visible to some line segments in G_P, and that (ii) no such line segments are in M because R_p is unguarded. Consider the maximal regions in P that lie immediately above, below, left, and right of R_p; we denote the union of these regions by X. See the hatched region in Fig. 6(a) for an example. Any sliding camera that sees any part of R_p must intersect some region of X. Since R_p

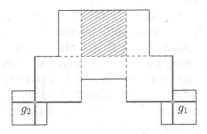

Fig. 5. A simple orthogonal polygon P and its corresponding grid G_P(dashed red). The set $\{g_1, g_2\}$ of periscope guards guard G_P. However, the sliding cameras located by the algorithm (solid purple) do not guard P entirely. The hatched pink region, called a *desert*, is not guarded by any sliding camera (Colour figure online).

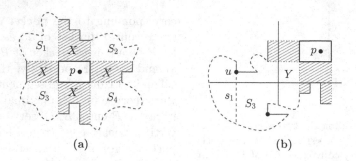

Fig. 6. (a) An unguarded point p inside a polygon P with maximal unguarded rectangle R_p. The hatched pink region of P indicates the region X; the four regions S_1, S_2, S_3 and S_4 are labelled accordingly. (b) An illustration in support of the proof of Lemma 2 (Colour figure online).

is unguarded, region X cannot contain any sliding camera in M; therefore, no periscope guard lies in X. Moreover, region X partitions the polygon into five subregions (see Fig. 6(a)): the union of X and rectangle R_p, the subregion on the upper-left side of X (denoted by S_1), the subregion on the upper-right side of X (denoted by S_2), the subregion on the lower-left side of X (denoted by S_3) and the subregion on the lower-right side of X (denoted by S_4). Note that the periscope guards in S can only lie in regions S_1, S_2, S_3 and S_4. We first show the following results.

Lemma 2. *If for some $1 \leq i \leq 4$, the subregion S_i contains no periscope guards of S, then all reflex vertices of P in S_i face the unguarded rectangle R_p.*

Proof. Without loss of generality, assume that there is no periscope guard in S_3. Suppose, to the contrary of the lemma statement, that there exists a reflex vertex u of P in S_3 that is not faced towards rectangle R_p. Observe that there are only two possibilities for such reflex vertex as shown in Fig. 6(b). We now continue the proof for the upper vertex u shown in Fig. 6(b); the proof for the other vertex is similar. Consider the maximal vertical line segment s_1 that passes through u and let Y be the set of line segments in G_P that enter region S_3 from the other regions. First, note that either $s_1 \in G_P$ or otherwise $s_j \in G_P$, for some s_j that dominates s_1. Without loss of generality, assume that $s_1 \in G_P$ (otherwise, the proof will be similar by replacing s_1 with s_j). Since R_p is unguarded, there is no sliding camera and, therefore, no periscope guard located on L, for all $L \in Y$. Since there is no periscope guard in S_3, we conclude that s_1 is not guarded by any periscope guard, which is a contradiction to the fact that S is a feasible solution to the MPG problem on G_P. Therefore, all the reflex vertices of P inside S_3 must face towards rectangle R_p.   □

By Lemma 2, we conclude that if there is no periscope guard in S_i, for some $1 \leq i \leq 4$, then the region S_i must be bounded by at most two staircases with their reflex vertices all facing towards the unguarded rectangle R_p. There are two possibilities for the staircases to lie in S_i depending on the orientation of the staircases: they can be either both horizontal or both vertical; see Fig. 8 for an illustration in which $S_i = S_1$. Moreover, Lemma 2 implies that region S_i is *orthogonally convex*, because otherwise there must be a reflex vertex in S_i that is not faced toward rectangle R_p and, therefore, there will be a grid segment in S_i that is not guarded by any periscope guard.

Lemma 3. *If the subregion S_i, for some $1 \le i \le 4$, contains no periscope guards of S, then the subregion S_i has no convex pockets.*

Proof. Without loss of generality, assume that there is no periscope guard in S_3. Suppose, to the contrary of the lemma statement, that there exists a convex pocket R inside S_3 and let Y be the set of grid segments in G_P that intersect at least one of the pocket edges s_1 and s_2 of R; see Fig. 7. Without loss of generality, assume that $s_1 \in G_P$. We first show that there is no periscope guard on L, for all $L \in Y$. Take any grid segment L in Y. Note that if L is entirely contained in S_3, then there is no periscope guard on L by the assumption. If L enters S_3 from another region, then it must intersect region X and, therefore, rectangle R_p is visible to L. Since R_p is unguarded, there is no sliding camera (and therefore no periscope guard) on L. This means that s_1 is not guarded by any periscope guard, which is a contradiction to the fact that S is a feasible solution to the MPG problem. This completes the proof of the lemma.

By Lemma 3, we conclude that the staircases of S_i are joined with each other in such a ways that they do not create any convex pocket in S_i.

3.2 Characterizing Desert Regions

Recall that the periscope guards in S can only lie in $S_1 \cup S_2 \cup S_3 \cup S_4$. The structure of a desert region depends on how many of the four regions S_1, S_2, S_3 and S_4 contain at least one periscope guard. In the following, we consider all the four cases and show that the desert region in each case can be guarded entirely by a single sliding camera. Let $Z \subseteq \{S_1, S_2, S_3, S_4\}$ such that $S_i \in Z$, for all $1 \le i \le 4$, if and only if S_i contains at least one periscope guard.

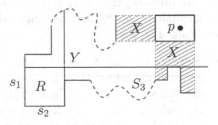

Fig. 7. An illustration in support of the proof of Lemma 3.

Case 1: $|\mathbf{Z}| = 4$. In this case, there is at least one periscope guard in S_i, for all $1 \le i \le 4$. Since (i) the grid segments in G_P are all guarded by at least one periscope guard, and (ii) each part of region X (i.e., the parts that are immediately above, below, to the left and to the right of rectangle R_p) is intersected by at least one grid segment in G_P, we conclude that the region S_i is guarded by sliding cameras in M, for all $1 \le i \le 4$. Therefore, the desert region in this

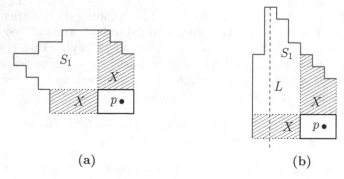

(a)   (b)

Fig. 8. An example of a region S_1 such that it contains no periscope guards of S. The staircases in S_1 must both be either (a) horizontal, or (b) vertical. Note that S_1 is an orthogonally convex region.

case is just the rectangle R_p and can be guarded by a single sliding camera. See Fig. 5 for an example.

Case 2: $|\mathbf{Z}| = 3$. Without loss of generality, assume that there is no periscope guard in region S_1. Note that the desert region in this case is the union of S_1 and rectangle R_p. Recall that the staircases in S_1 must both be horizontal or both vertical. Assume without loss of generality that the staircases are lied vertically in S_1 (i.e., Fig. 8(b)). Since S_1 is an orthogonally convex region the maximal vertical line segment L that crosses the top most horizontal edge of S_1 guards the union of S_1 and rectangle R_p; we call L the *neighbour camera* associated with S_1. Therefore, one sliding camera located on L can guard the desert region.

Case 3: $|\mathbf{Z}| = 2$. Let S_i and S_j, for some $1 \leq i, j \leq 4$ and $i \neq j$, be the regions that contain no periscope guard. We observe that in this case, the desert region is the union of S_i, S_j and rectangle R_p. There are two cases depending on the positions of S_i and S_j:

(a) Suppose regions S_i and S_j are neighbours to each other. Without loss of generality, assume that $S_i = S_1$ and $S_j = S_2$ and that the staircases in S_1 are lied vertically; see Fig. 9(a). Let L_1 and L_2 be the neighbour cameras of S_1 and S_2, respectively. If the staircases in S_2 are also lied vertically, then it is straightforward to see that there exists a maximal horizontal line segment inside P that guards the union S_1, S_2 and rectangle R_p (see Fig. 9(a)). If the staircases in S_2 are lied horizontally, we show that L_1 guards the union of S_1, S_2 and rectangle R_p. First, note that L_1 guards the union of S_1 and rectangle R_p. Now, suppose to the contrary, that there exists a point $q \in S_2$ that is not visible to L_1; see Fig. 9(b). Since S_2 consists of only two staircases and such staircases in S_2 are lied horizontally, they must be joined with each other in S_2 such that they form a convex pocket, which is a contradiction to Lemma 3. Therefore, S_2 is entirely visible to L_1.

(b) Suppose regions S_i and S_j are opposite to each other. Note that (i) each of the regions S_i and S_j consists of at most two staircases, and (ii) the staircases

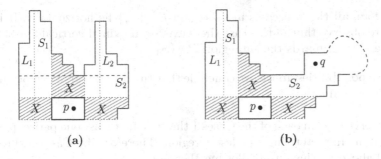

Fig. 9. An illustration in support of Case 3.

in S_i (or in S_j) are either both vertical or both horizontal. By an argument analogous to that given in Case (a), we can conclude that the union of regions S_i, S_j and rectangle R_p can be guarded by one sliding camera.

By the two cases described above, we conclude that the desert region can be guarded by one sliding camera.

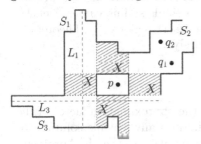

Fig. 10. An illustration in support of Case 4.

Case 4: $|Z| = 1$. Without loss of generality, assume that all the periscope guards lie in S_4. We show that the subregion $P \setminus \{S_4\}$, which forms the desert region, can be guarded by a single sliding camera. Consider the neighbour camera L_3 associated with region S_3 and assume without loss of generality that the staircases in S_3 lie horizontally; see Fig. 10. It is straightforward to see that L_3 guards the union of S_3 and rectangle R_p. We now check to see if L_3 can also guard the union of S_1 and S_2. If L_3 guards the union of S_1 and S_2, then the subregion $P \setminus \{S_4\}$ can be guarded by one sliding camera located on L_3. Otherwise, there are two possibilities:

(a) Suppose exactly one of the regions S_1 or S_2 is guarded by L_3. Without loss of generality, assume that S_2 is not guarded by L_3 entirely. Thus, there is a point $q_1 \in S_2$ that is not visible to L_3; see Fig. 10. Since L_3 guards S_1 the staircases in S_1 must be lied vertically. Therefore, the neighbour camera L_1 (associated with region S_1) is vertical and guards the union of S_1, S_3 and rectangle R_p. Note that L_1 also guards S_2 because otherwise there must be a point $q_2 \in S_2$ that is not visible to L_1 (see Fig. 10). But, the existence of points q_1 and q_2 in S_2 implies that the staircases in S_2 must be joined with each other in such a way that they form a convex pocket in S_2, which is a contradiction to Lemma 3. Therefore, in this case, L_1 guards the desert region entirely.

(b) Suppose neither S_1 nor S_2 is guarded entirely by L_3. Since L_3 is horizontal, the staircases in regions S_1 and S_2 must all have lain horizontally and, therefore, all the staircases in subregion $P \setminus \{S_4\}$ lie horizontally. It is now easy to observe that in this case there exists a maximal vertical line segment inside P that guards the subregion $P \setminus \{S_4\}$.

By the two possibilities above, we conclude that the desert region can be guarded by one sliding camera.

We observe that in each of the Cases 1 through 4, at least one periscope guard is required in characterizing the desert region. Therefore, by Cases 1 through 4 described above, we have the following theorem.

Theorem 2. *Every point in P that is not inside a desert region is guarded by at least one sliding camera in M. Each desert region of P consists of a set of staircases and it can be guarded entirely by a single sliding camera. Moreover, the number of desert regions is at most the number of periscope guards in S.*

To summarize the algorithm, we first solve the MPG problem on G_P and compute the set S of optimal periscope guards. By Observation 1, we locate two sliding cameras inside P for each periscope guard to obtain the set M. By Theorem 2, we then guard each desert region by a single sliding camera; let M' denote the set of sliding cameras that guard the desert regions. By Theorem 2, the set $M \cup M'$ of sliding cameras guards P entirely.

3.3   Analyzing the Algorithm

In this section, we analyze the running time and the approximation factor of the algorithm. To this end, we first give a lower bound on any feasible solution for the MSC problem on P. Recall OPT_P, an optimal solution to the MSC problem, and recall OPT_{PG}, an optimal solution for the MPG problem on G_P. We show the following result whose proof is omitted due to space constraints.

Lemma 4. $|OPT_P| \geq |OPT_{PG}|.$

We know that $|M| \leq 2 \cdot |S|$. By Theorem 2, we have that $|M'| \leq |S|$ and so $|M \cup M'| \leq 3 \cdot |S|$. Therefore, by Lemma 4 we have that $|M \cup M'| \leq 3 \cdot |OPT_P|$. To analyze the running time of the algorithm, we note that the construction of grid G_P can be completed in $O(n^2)$ time [2]. Since $|T_G| = O(n)$, where n is the number of vertices of P, the MPG problem can be solved on G_P in $O(n^3)$ time. Next, the desert regions of P can be detected in $O(n^2)$ time by detecting the visibility region of cameras in M and comparing their union with P. Finally, the desert regions can be guarded in $O(n)$ time by locating a sliding camera inside P, for each desert region. Therefore, we have the main result of this paper.

Theorem 3. *There exists an $O(n^3)$-time 3-approximation algorithm for the MSC problem on any simple orthogonal polygon with n vertices.*

4   Conclusion

In this paper, we gave an $O(n^3)$-time 3-approximation algorithm for the problem of guarding a simple orthogonal polygon P with n vertices using the minimum number of sliding cameras (i.e., the MSC problem). The complexity of the MSC problem is still unknown and remains the main direction for future work. Also, giving algorithms with better approximation factor or showing a hardness of approximation remains open as another direction for future work.

References

1. Biedl, T.C., Irfan, M.T., Iwerks, J., Kim, J., Mitchell, J.S.B.: The art gallery theorem for polyominoes. Disc. Comp. Geom. **48**(3), 711–720 (2012)
2. Durocher, S., Filtser, O., Fraser, R., Mehrabi, A.D., Mehrabi, S.: A (7/2)-approximation algorithm for guarding orthogonal art galleries with sliding cameras. In: Pardo, A., Viola, A. (eds.) LATIN 2014. LNCS, vol. 8392, pp. 294–305. Springer, Heidelberg (2014)
3. Durocher, S., Mehrabi, S.: Guarding orthogonal art galleries using sliding cameras: algorithmic and hardness results. In: Chatterjee, K., Sgall, J. (eds.) MFCS 2013. LNCS, vol. 8087, pp. 314–324. Springer, Heidelberg (2013)
4. Eidenbenz, S.: Inapproximability results for guarding polygons without holes. In: Chwa, K.-Y., Ibarra, O.H. (eds.) ISAAC 1998. LNCS, vol. 1533, p. 427. Springer, Heidelberg (1998)
5. Eidenbenz, S.: Inapproximability of visibility problems on polygons and terrains. Ph.D. thesis, ETH Zurich (2000)
6. Gewali, L., Ntafos, S.C.: Covering grids and orthogonal polygons with periscope guards. Comput. Geom. **2**, 309–334 (1992)
7. Ghosh, S.K.: Approximation algorithms for art gallery problems in polygons. Disc. App. Math. **158**(6), 718–722 (2010)
8. Hoffmann, F.: On the rectilinear art gallery problem. In: Paterson, M.S. (ed.) Automata, Languages and Programming. LNCS, pp. 717–728. Springer, Heidelberg (1990)
9. Katz, M.J., Morgenstern, G.: Guarding orthogonal art galleries with sliding cameras. Int. J. Comp. Geom. App. **21**(2), 241–250 (2011)
10. Kosowski, A., Małafiejski, M., Żyliński, P.: An efficient algorithm for mobile guarded guards in simple grids. In: Gavrilova, M.L., Gervasi, O., Kumar, V., Tan, C.J.K., Taniar, D., Laganá, A., Mun, Y., Choo, H. (eds.) ICCSA 2006. LNCS, vol. 3980, pp. 141–150. Springer, Heidelberg (2006)
11. Kosowski, A., Małafiejski, M., Zylinski, P.: Cooperative mobile guards in grids. Comp. Geom. **37**(2), 59–71 (2007)
12. Krohn, E., Nilsson, B.J.: Approximate guarding of monotone and rectilinear polygons. Algorithmica **66**(3), 564–594 (2013)
13. Lee, D.T., Lin, A.K.: Computational complexity of art gallery problems. IEEE Trans. Info. Theory **32**(2), 276–282 (1986)
14. Motwani, R., Raghunathan, A., Saran, H.: Covering orthogonal polygons with star polygons: the perfect graph approach. In: Proceedings of ACM SoCG, pp. 211–223 (1988)
15. O'Rourke, J.: Art Gallery Theorems and Algorithms. Oxford University Press Inc, New York (1987)

16. Schuchardt, D., Hecker, H.: Two NP-hard art-gallery problems for ortho-polygons. Math. Log. Q. **41**(2), 261–267 (1995)
17. Seddighin, S.: Guarding polygons with sliding cameras. Master's thesis, Sharif University of Technology (2014)
18. Urrutia, J.: Art gallery and illumination problems. Handb. Comp. Geom. **1**(1), 973–1027 (2000). North-Holland
19. Worman, C., Keil, J.M.: Polygon decomposition and the orthogonal art gallery problem. Int. J. Comp. Geom. App. **17**(2), 105–138 (2007)

On Decomposing the Complete Graph into the Union of Two Disjoint Cycles

Saad I. El-Zanati[1]([✉]), Uthoomporn Jongthawonwuth[2], Heather Jordon[1],
and Charles Vanden Eynden[1]

[1] Illinois State University, Normal 61790-4520, USA
{saad,hjordon,cve}@ilstu.edu
[2] Chulalongkorn University, Bangkok 10330, Thailand
aor_utoo@hotmail.com

Abstract. Let G of order n be the vertex-disjoint union of an even and an odd cycle. It is known that there exists a G-decomposition of K_v for all $v \equiv 1 \pmod{2n}$. We use an extension of the Bose construction for Steiner triple systems and a recent result on the Oberwolfach Problem for 2-regular graphs with two components to show that there exists a G-decomposition of K_v for all $v \equiv n \pmod{2n}$, unless $G = C_4 \cup C_5$ and $v = 9$.

Keywords: Graph decomposition · Bose construction · Disjoint cycles

1 Introduction

Let $\mathbb{Z}_n$ be the group of integers modulo n. For integers a and b with $a \leq b$, we denote the set $\{a, a+1, \ldots, b\}$ by $[a, b]$ (if $a > b$, then $[a, b] = \varnothing$). For a graph G, let $V(G)$ and $E(G)$ denote the vertex set of G and the edge set of G, respectively. The *order* and the *size* of a graph G are $|V(G)|$ and $|E(G)|$, respectively. We will denote the complete multipartite graph with n partite sets of order m by $K_{n \times m}$. The vertex-disjoint union of t copies of a graph G will be denoted by tG.

A *decomposition* of a graph K is a set $T = \{G_1, G_2, \ldots, G_t\}$ of subgraphs of K such that the edge sets of the graphs G_i form a partition of the edge set of K. If each G_i is isomorphic to a subgraph G of K, such a decomposition is called a *G-decomposition of K* or a *(K, G)-design*. A (K_v, G)-design is also known as a *G-design of order v*. The study of graph decompositions is known as the study of graph designs or simply as the study of G-designs. For recent surveys on G-designs, we direct to the reader to [2,8].

One of the better-studied problems in G-designs is the case when G is a cycle. Necessary and sufficient conditions for the existence of C_n-designs of order v were found about a decade ago by Alspach and Gavlas [4] and by Šajna [17]. Necessary and sufficient conditions for the existence of a G-design of order v are found in [3] when G is a 2-regular graph of order at most 10. For an arbitrary 2-regular graph G of order n, the problem of finding necessary and sufficient conditions for the existence of a G-design of order v is far from settled. It is expected however that

© Springer International Publishing Switzerland 2015
J. Kratochvíl et al. (Eds.): IWOCA 2014, LNCS 8986, pp. 153–163, 2015.
DOI: 10.1007/978-3-319-19315-1_14

for such a G, there will exist a G-design of order v for all $v \equiv 1 \pmod{2n}$. This has been confirmed when G is bipartite (see [5,11]), when G is rC_m where m is odd (see [12]), and when G has two components (see [1,3,6,9]). If in addition n is odd and $(G, v) \notin \{(C_4 \cup C_5, 9), (C_3 \cup C_3 \cup C_5, 11)\}$, then a G-design of order v for all $v \equiv n \pmod{2n}$ is likely to exist. We confirm this assertion here in the case where G has exactly two components.

Let G be a 2-regular graph of odd order n. The problem of determining whether there exists a G-decomposition of K_n is known as the *Oberwolfach problem*. Although the general problem is far from settled, Traetta [18] recently settled the case when G has two components.

Theorem 1. *Let $a \geq 2$ and $b \geq 1$ be integers and let $n = 2a + 2b + 1$. There exists a $(C_{2a} \cup C_{2b+1})$-decomposition of K_n if and only if $(a, b) \neq (2, 2)$.*

In this article, we use an extension of the Bose construction for Steiner triple systems (see [15]) to show that if G of odd order n is the vertex-disjoint union of two cycles, then there exists a G-decomposition of $K_{(2k+1) \times n}$ for every positive integer k. We also show that there exists a G-decomposition of $K_{k \times 2n}$ for every integer $k \geq 3$. We combine the decomposition of $K_{(2k+1) \times n}$ result with Traetta's result on the Oberwolfach Problem for 2-regular graphs with two components to show that there exists a G-decomposition of K_v for all $v \equiv n \pmod{2n}$, unless $G = C_4 \cup C_5$ and $v = 9$.

2   Quasigroups and the Bose Construction

Let $G = C_m \cup C_\ell$ where $m \geq 4$ is even and $\ell \geq 3$ is odd and let $n = m + \ell$. We use an extension of the Bose construction for Steiner triple systems of order $6k + 3$ to show that there exists a G-decomposition of $K_{(2k+1) \times n}$ for every positive integer k. We also show that there exists a G-decomposition of $K_{k \times 2n}$ for every integer $k \geq 3$. These constructions make use of idempotent commutative quasigroups and of commutative quasigroups with holes.

A *quasigroup* of *order* q is a pair $(Q, \circ)$ where Q is a set of size q, say $Q = [1, q]$, and $\circ$ is a binary operation on Q such that for every pair of elements $a, b \in Q$, the equations $a \circ x = b$ and $y \circ a = b$ have unique solutions. The quasigroup is *idempotent* if $i \circ i = i$ for every $i \in Q$ and it is *commutative* if $i \circ j = j \circ i$ for all $i, j \in Q$. Note that in any idempotent quasigroup, if $a \neq b$, then a, b, and $a \circ b$ are distinct. It is known that an idempotent commutative quasigroup of order q exists if and only if q is odd (see [15]).

Let $Q = [1, 2k]$ and let $H = \{\{1, 2\}, \{3, 4\}, \ldots, \{2k-1, 2k\}\}$. In what follows, the two-element subsets $\{2i-1, 2i\} \in H$ are called *holes*. A *quasigroup with holes* H is a quasigroup $(Q, \circ)$ of order $2k$ in which for each $h \in H$, we have $(h, \circ)$ is a subquasigroup of $(Q, \circ)$. It is known that for every $k \geq 3$, there exists a commutative quasigroup $(Q, \circ)$ of order $2k$ with holes H (see [15]). Commutative quasigroups of order $2k$ with holes H are used to construct C_3-decompositions of $K_{k \times 6}$ for every integer $k \geq 3$.

We give a brief description of Bose's construction for Steiner triple systems of order $6k + 3$. We also describe how to obtain a C_3-decomposition of $K_{k \times 6}$ for $k \geq 3$. We direct the reader to the book by Lindner and Rodger [15] for detailed information on quasigroups and triple systems.

We will define a *Steiner triple system* of *order* v to be a C_3-decomposition of K_v. It has long been known that a Steiner triple system of order v exists if and only if $v \equiv 1$ or $3 \pmod 6$. In 1939, Bose [7] used the existence of an idempotent commutative quasigroup of order $q = 2k + 1$ to construct a C_3-decomposition of K_{6k+3} for every positive integer k. One can view K_{6k+3} as $(2k + 1)K_3 \bigcup K_{(2k+1) \times 3}$. Thus to construct a C_3-decomposition of K_{6k+3}, it suffices to construct a C_3-decomposition of $K_{(2k+1) \times 3}$. Let (a, b, c) denote the C_3 (also called a *triple*) with vertex set $\{a, b, c\}$.

Let $(Q, \circ)$ be an idempotent commutative quasigroup of order $2k + 1$ where $Q = [1, 2k+1]$ and let $V(K_{(2k+1) \times 3}) = \mathbb{Z}_3 \times Q$ with the obvious vertex partition. Let $T = \{((0, i), (0, j), (1, i \circ j)), ((1, i), (1, j), (2, i \circ j)), ((2, i), (2, j), (0, i \circ j)) : 1 \leq i < j \leq 2k + 1\}$. Then the C_3's in T form a C_3-decomposition of $K_{(2k+1) \times 3}$. Figure 1 shows an idempotent commutative quasigroup of order 5 and one triple from the Bose construction of a Steiner triple system of order 15.

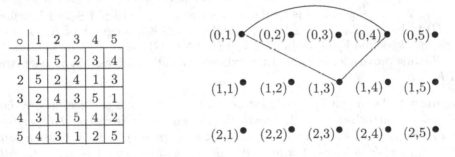

$\circ$	1	2	3	4	5
1	1	5	2	3	4
2	5	2	4	1	3
3	2	4	3	5	1
4	3	1	5	4	2
5	4	3	1	2	5

Fig. 1. An idempotent commutative quasigroup of order 5 and one triple from the Bose construction of a Steiner triple system of order 15.

Alternatively, let $k \geq 3$ be an integer and for $i \in [1, k]$, let $h_i = \{2i - 1, 2i\}$ and $g_i = \mathbb{Z}_3 \times h_i$. Let $Q = [1, 2k]$ and $H = \{h_1, h_2, \ldots, h_k\}$. Let $(Q, \circ)$ be a commutative quasigroup of order $2k$ with holes H. Let $V(K_{k \times 6}) = \mathbb{Z}_3 \times Q$ with the vertex-set partition $\{g_1, g_2, \ldots, g_k\}$. Let $T = \{((0, i), (0, j), (1, i \circ j)), ((1, i), (1, j), (2, i \circ j)), ((2, i), (2, j), (0, i \circ j)) : 1 \leq i < j \leq 2k, \{i, j\} \notin H\}$. Then the C_3's in T form a C_3-decomposition of $K_{k \times 6}$. This process is part of what is known as the *quasigroups with holes construction* for triple systems (see [15]). Figure 2 shows a commutative quasigroup of order 6 with holes and one triple from the corresponding C_3-decomposition of $K_{3 \times 6}$.

3   *G*-decompositions of $K_{(2k+1)} \times n$ and of $K_{k \times 2n}$

Let $n \geq 3$ be an odd integer and let k be a positive integer. Let $K_{(2k+1) \times n}$ have vertex set $\mathbb{Z}_n \times [1, 2k + 1]$ with the obvious vertex partition. For $i \in [1, k]$,

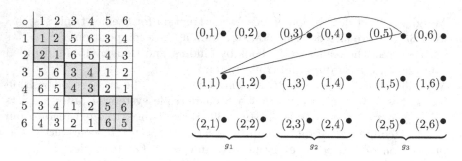

∘	1	2	3	4	5	6
1	1	2	5	6	3	4
2	2	1	6	5	4	3
3	5	6	3	4	1	2
4	6	5	4	3	2	1
5	3	4	1	2	5	6
6	4	3	2	1	6	5

Fig. 2. A commutative quasigroup of order 6 with holes and one triple from the corresponding C_3-decomposition of $K_{3\times 6}$.

let $h_i = \{2i - 1, 2i\}$ and $g_i = \mathbb{Z}_n \times h_i$. Let $V(K_{k\times 2n}) = \mathbb{Z}_n \times [1, 2k]$ with the vertex-set partition $\{g_1, g_2, \ldots, g_k\}$. If $e = \{(i, r), (j, s)\}$, where $r < s$, is an edge in $K_{(2k+1)\times n}$ or in $K_{k\times 2n}$, define the *length* of e to be $j - i$ if $j \geq i$; otherwise the length of e is $n + i - j$. Thus, between any two parts, there are edges of lengths $0, 1, \ldots, n-1$. We will often write $-j$ for edge length $n - j$ when n is understood. Therefore, between any two parts, there are edges of lengths $0, \pm 1, \pm 2, \ldots, \pm \frac{(n-1)}{2}$. If G is a subgraph of $K_{(2k+1)\times n}$ and $1 \leq j \leq 2k+1$, define $V_j(G)$ to be $\{i \in \mathbb{Z}_n : (i, j) \in V(G)\}$. Similarly, if G is a subgraph of $K_{k\times 2n}$ and $1 \leq j \leq 2k$, define $V_j(G)$ to be $\{i \in \mathbb{Z}_n : (i, j) \in V(G)\}$.

We first prove a lemma about the existence of paths with certain edge lengths in $K_{n,n}$.

Lemma 1. *Let $n \geq 3$ be an odd integer and let $x \leq n$ be a positive integer. Let $K_{n,n}$ have vertex set $\mathbb{Z}_n \times [1, 2]$ with the obvious vertex partition. For positive integers $d_1, d_2, \ldots, d_x$ with $d_1 < d_2 < \cdots < d_x \leq (n-1)/2$, there exists a path P in $K_{n,n}$ of length $2x + 1$ whose edges have lengths $0, \pm d_1, \pm d_2, \ldots, \pm d_x$ with endpoints $(0, 1)$ and $(0, 2)$. Furthermore, $V(P) \subseteq [0, d_x] \times [1, 2]$.*

Proof. For $k \in [1, x]$, define $e_k = \sum_{i=0}^{k-1} (-1)^i d_{x-i}$. Note that since $d_1 < d_2 < \cdots < d_x$, we have that $e_1 > e_3 > \cdots$ and $e_2 < e_4 < \cdots$. Consider the path of length x given by P': $(0, 1), (e_1, 2), (e_2, 1), (e_3, 2), \ldots$ where P' ends with $(e_x, 2)$ if x is odd or $(e_x, 1)$ if x is even. Observe that the lengths of the edges on P', in the order encountered, are $d_x, d_{x-1}, \ldots, d_1$. Next consider the path P'': $(0, 2), (e_1, 1), (e_2, 2), (e_3, 1), \ldots$ where P'' ends with $(e_x, 1)$ if x is odd or $(e_x, 2)$ if x is even, and observe that the edges on P'', in the order encountered, are $-d_x, -d_{x-1}, \ldots, -d_1$. Construct the path P from the paths P' and P'' by adding the edge from $(e_x, 1)$ to $(e_x, 2)$. Note that P has length $2x + 1$, the edges of P have lengths $0, \pm d_1, \pm d_2, \ldots, \pm d_x$, and $V(P) \subseteq [0, d_x] \times [1, 2]$. □

Let H be a subgraph of a graph with vertex set $\mathbb{Z}_n \times [1, q]$. For a positive integer ℓ, the graph $H + \ell$ has vertex set $\{(i + \ell, z) : (i, z) \in V(H)\}$ and edge set $\{\{(i + \ell, r), (j + \ell, s)\} : \{(i, r), (j, s)\} \in E(H)\}$.

Given a sequence $W_1, W_2, \ldots, W_t$ of directed edge-disjoint walks (each with 1 vertex or more) in a graph H, by the *connection* of the sequence we mean the directed walk, the edges of which are the edges of W_1, then the edge from the last vertex of W_1 to the first vertex of W_2, then the edges of W_2, etc., ending with the edges of W_t.

Theorem 2. *Let G be a 2-regular graph of odd order n consisting of exactly two cycles. For every positive integer k, there exists a G-decomposition of $K_{(2k+1) \times n}$.*

Proof. Let $G = C_m \cup C_\ell$, where $m \geq 4$ is even and $\ell \geq 3$ is odd. Let k be a positive integer and let $Q = [1, 2k+1]$. Label the vertex set of $K_{(2k+1) \times n}$ with the elements of the set $\mathbb{Z}_n \times [1, 2k+1]$ with the obvious vertex partition. Let $(Q, \circ)$ be an idempotent commutative quasigroup of order $2k+1$.

For integers $j > 0$, a, and b and for $r, s \in Q$, with $r < s$, let $W(a, j, b)$ be the directed path with consecutive vertices

$$(a, r), (b+j-1, s), (a+1, r), (b+j-2, s), (a+2, r), \ldots, (b-1, s), (a+j-1, r), (b, s).$$

Notice that this sequence contains $2j$ vertices, starting with (a, r) and ending with (b, s). We have $V_r(W(a, j, b)) = [a, a+j-1]$, and $V_s(W(a, j, b)) = [b, b+j-1]$. The set of lengths of the edges of this path is $b - a + [-(j-1), j-1]$.

Fix r and s with $1 \leq r < s \leq 2k+1$. In what follows, we will construct a graph $G_{r,s}$, consisting of a cycle $C_{r,s}$ of length m and a cycle $C'_{r,s}$ of length ℓ such that $C_{r,s}$ and $C'_{r,s}$ are vertex disjoint. We proceed by cases depending on the congruence class of m modulo 8.

Case 1: $m \equiv 2 \pmod 8$. Let $m = 8t+2$ for some positive integer t. Let $C_{r,s}$ be the walk with m edges which is the connection of

$$W(0, t+1, 3t+2), W(t+2, t-1, 2t+3), (2t+1, r), (2t-1, s),$$
$$W(2t+2, t-1, t), W(3t+2, t, 0), (4t+2, r), (2t+1, s), (0, r).$$

Notice that we have listed $2(t+1) + 2(t-1) + 2 + 2(t-1) + 2t + 3 = 8t+3 = m+1$ vertices, but $(0, r)$ is listed twice. Here $V_r(C_{r,s})$ is

$$[0, t] \cup [t+2, 2t] \cup \{2t+1\} \cup [2t+2, 3t] \cup [3t+2, 4t+1] \cup \{4t+2\}$$
$$= [0, t] \cup [t+2, 3t] \cup [3t+2, 4t+2],$$

and these are distinct in $\mathbb{Z}_n$. Likewise $V_s(C_{r,s})$ is

$$[3t+2, 4t+2] \cup [2t+3, 3t+1] \cup \{2t-1\} \cup [t, 2t-2] \cup [0, t-1] \cup \{2t+1\}$$
$$= [0, 2t-1] \cup \{2t+1\} \cup [2t+3, 4t+2],$$

and these are also distinct in $\mathbb{Z}_n$. Thus $C_{r,s}$ is a cycle of length m.
The set of lengths of edges in $C_{r,s}$ is

$$[2t+2, 4t+2] \cup \{2t\} \cup [3, 2t-1] \cup \{2, -2, -3\} \cup [-2t, -4] \cup \{-2t-2\} \cup$$
$$[-4t-1, -2t-3] \cup \{-4t-2, -2t-1, 2t+1\} = [-(4t+2), -2] \cup [2, 4t+2].$$

Notice that the difference between the largest and smallest length is $8t + 4$, but $n = m + \ell \geq 8t + 2 + 3$. Thus these lengths are distinct in $\mathbb{Z}_n$.

We next give the construction of $C'_{r,s}$. If $\ell = 3$, then let $C'_{r,s}$ be the 3-cycle $((4t + 3, r), (4t + 3, s), (4t + 4, r \circ s))$.

Suppose that $\ell \geq 5$. By Lemma 2, there exists a path $P_{r,s}$ of length $\ell - 2$ using the edge lengths in $\{-1, 0, 1\} \cup \pm[4t + 3, (n - 3)/2]$ with endpoints $(0, r)$ and $(0, s)$. In the lemma we would use $d_1 = 1, d_2 = 4t + 3, \ldots, d_x = (n - 3)/2$, so $V(P_{r,s}) \subseteq [0, (n - 3)/2] \times \{r, s\}$. Let $P^*_{r,s} = P_{r,s} + 4t + 3$, with endpoints $(4t + 3, r)$ and $(4t + 3, s)$. Then $V(P^*_{r,s}) \subseteq [4t + 3, (n - 3)/2 + 4t + 3] \times \{r, s\}$. Since $(n - 3)/2 + 4t + 3 = (n + m + 1)/2 = (2n - \ell + 1)/2 < n$, we see that $P^*_{r,s}$ is vertex disjoint from $C_{r,s}$.

Finally we construct the cycle $C'_{r,s}$ from $P^*_{r,s}$ by adding the edges $\{(4t + 3, r), (4t + 3 + (n - 1)/2, r \circ s)\}$ and $\{(4t + 3, s), (4t + 3 + (n - 1)/2, r \circ s)\}$. Note that in the subgraph of $K_{(2k+1) \times n}$ with vertex set $\mathbb{Z}_n \times \{r, s\}$, $G_{r,s}$ contains one edge of each length $i \in [-(n - 1)/2, (n - 1)/2] \setminus \{\pm z\}$, where $z = 1$ if $\ell = 3$ and $z = (n - 1)/2$, otherwise. Figure 3 shows an example of $C_{r,s}$ and $C'_{r,s}$ where $m = 10$ and $\ell = 9$.

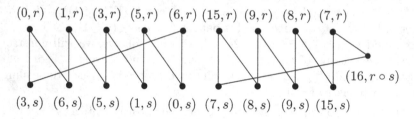

$(0, r)$ $(1, r)$ $(3, r)$ $(5, r)$ $(6, r)$ $(15, r)$ $(9, r)$ $(8, r)$ $(7, r)$

$(16, r \circ s)$

$(3, s)$ $(6, s)$ $(5, s)$ $(1, s)$ $(0, s)$ $(7, s)$ $(8, s)$ $(9, s)$ $(15, s)$

Fig. 3. $C_{r,s}$ and $C'_{r,s}$ where $m = 10$ and $\ell = 9$.

Case 2: $m \equiv 6 \pmod 8$. Let $m = 8t + 6$ for some nonnegative integer t. We will consider two cases.

Case 2.1: $t = 0$. Let $C_{r,s}$ be the C_6 defined by the connection of $(0, r), (3, s)$, $(2, r), (0, s), (3, r), (2, s)$, and $(0, r)$. Note that the set of edge lengths of $C_{r,s}$ is $\pm[1, 3]$.

We next give the construction of $C'_{r,s}$. If $\ell = 3$, then let $C'_{r,s}$ be the 3-cycle $((4, r), (4, s), (8, r \circ s))$.

Suppose that $\ell \geq 5$. By Lemma 2, there exists a path $P_{r,s}$ of length $\ell - 2$ using the edge lengths in $\{0\} \cup \pm[4, (n - 3)/2]$ with endpoints $(0, r)$ and $(0, s)$. In the lemma we would use $d_1 = 4, d_2 = 5, \ldots, d_x = (n - 3)/2$, so $V(P_{r,s}) \subseteq [0, (n - 3)/2] \times \{r, s\}$. Let $P^*_{r,s} = P_{r,s} + 4$, with endpoints $(4, r)$ and $(4, s)$. Then $V(P^*_{r,s}) \subseteq [4, (n - 3)/2 + 4] \times \{r, s\}$. Since $(n - 3)/2 + 4 = (n + 5)/2 = (2n - \ell - 1)/2 < n$, we see that $P^*_{r,s}$ is vertex disjoint from $C_{r,s}$.

Finally we construct the cycle $C'_{r,s}$ from $P^*_{r,s}$ by adding the edges $\{(4, r), (4 + (n - 1)/2, r \circ s)\}$ and $\{(4, s), (4 + (n - 1)/2, r \circ s)\}$. Note that in the subgraph of $K_{(2k+1) \times n}$ with vertex set $\mathbb{Z}_n \times \{r, s\}$, $G_{r,s}$ contains one edge of each length $i \in [-(n - 1)/2, (n - 1)/2] \setminus \{\pm z\}$.

Case 2.2: $t \geq 1$. Let $C_{r,s}$ be the walk with m edges which is the connection of

$$W(0, t+1, 3t+4), W(t+1, t, 2t+3), (2t+2, r), (2t+1, s),$$
$$W(2t+4, t-1, t+2), W(3t+3, t+1, 0), (4t+4, r), (2t+2, s), (0, r).$$

Notice that we have listed $2(t+1)+2t+2+2(t-1)+2(t+1)+3 = 8t+7 = m+1$ vertices, but $(0, r)$ is listed twice. Here $V_r(C_{r,s})$ is

$$[0, t] \cup [t+1, 2t] \cup \{2t+2\} \cup [2t+4, 3t+2] \cup [3t+3, 4t+3] \cup \{4t+4\}$$
$$= [0, 2t] \cup \{2t+2\} \cup [2t+4, 4t+4],$$

and these are distinct in $\mathbb{Z}_n$. Likewise $V_s(C_{r,s})$ is

$$[3t+4, 4t+4] \cup [2t+3, 3t+2] \cup \{2t+1\} \cup [t+2, 2t] \cup [0, t] \cup \{2t+2\}$$
$$= [0, t] \cup [t+2, 3t+2] \cup [3t+4, 4t+4],$$

and these are also distinct in $\mathbb{Z}_n$. Thus $C_{r,s}$ is a cycle of length m.

The set of lengths of edges in $C_{r,s}$ is

$$[2t+4, 4t+4] \cup \{2t+3\} \cup [3, 2t+1] \cup \{1, -1, -3\} \cup [-2t, -4] \cup \{-2t-1\} \cup$$
$$[-4t-3, -2t-3] \cup \{-4t-4, -2t-2, 2t+2\} = [-(4t+4), -3] \cup [3, 4t+4] \cup \{\pm 1\}.$$

Notice that the difference between the largest and smallest length is $8t+8$, but $n = m + \ell \geq 8t + 6 + 3$. Thus these lengths are distinct in $\mathbb{Z}_n$.

We next give the construction of $C'_{r,s}$. If $\ell = 3$, then let $C'_{r,s}$ be the 3-cycle $((4t+5, r), (4t+5, s), (4t+7, r \circ s))$.

Suppose that $\ell \geq 5$. By Lemma 2, there exists a path $P_{r,s}$ of length $\ell - 2$ using the edge lengths in $\{-2, 0, 2\} \cup \pm[4t+5, (n-3)/2]$ with endpoints $(0, r)$ and $(0, s)$. In the lemma we would use $d_1 = 2, d_2 = 4t+5, \ldots, d_x = (n-3)/2$, so $V(P_{r,s}) \subseteq [0, (n-3)/2] \times \{r, s\}$. Let $P^*_{r,s} = P_{r,s} + 4t + 5$, with endpoints $(4t+5, r)$ and $(4t+5, s)$. Then $V(P^*_{r,s}) \subseteq [4t+5, (n-3)/2 + 4t+5] \times \{r, s\}$. Since $(n-3)/2 + 4t+5 = (n+m+1)/2 = (2n - \ell + 1)/2 < n$, we see that $P^*_{r,s}$ is vertex disjoint from $C_{r,s}$.

Finally we construct the cycle $C'_{r,s}$ from $P^*_{r,s}$ by adding the edges $\{(4t+5, r), (4t+5+(n-1)/2, r \circ s)\}$ and $\{(4t+5, s), (4t+5+(n-1)/2, r \circ s)\}$. Note that in the subgraph of $K_{(2k+1) \times n}$ with vertex set $\mathbb{Z}_n \times \{r, s\}$, $G_{r,s}$ contains one edge of each length $i \in [-(n-1)/2, (n-1)/2] \setminus \{\pm z\}$, where $z = 2$ if $\ell = 3$ and $z = (n-1)/2$, otherwise. Figure 4 shows an example of $C_{r,s}$ and $C'_{r,s}$ where $m = 14$ and $\ell = 7$.

Case 3: $m \equiv 0 \pmod 4$. Let $m = 4t$ for some positive integer t. Let $C_{r,s}$ be the walk with m edges which is the connection of

$$W(0, t, t+1), W(t+2, t-1, 2), (2t+1, r), (1, s), (0, r).$$

Notice that we have listed $2t + 2(t-1) + 3 = 4t+1 = m+1$ vertices, but $(0, r)$ is listed twice. Here $V_r(C_{r,s})$ is

$$[0, t-1] \cup [t+2, 2t] \cup \{2t+1\} = [0, t-1] \cup [t+2, 2t+1],$$

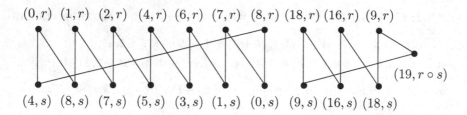

$(0,r)$ $(1,r)$ $(2,r)$ $(4,r)$ $(6,r)$ $(7,r)$ $(8,r)$ $(18,r)$ $(16,r)$ $(9,r)$

$(19, r \circ s)$

$(4,s)$ $(8,s)$ $(7,s)$ $(5,s)$ $(3,s)$ $(1,s)$ $(0,s)$ $(9,s)$ $(16,s)$ $(18,s)$

Fig. 4. $C_{r,s}$ and $C'_{r,s}$ where $m = 14$ and $\ell = 7$.

and these are distinct in $\mathbb{Z}_n$. Likewise $V_s(C_{r,s})$ is

$$[t+1, 2t] \cup [2, t] \cup \{1\} = [1, 2t],$$

and these are also distinct in $\mathbb{Z}_n$. Thus $C_{r,s}$ is a cycle of length m.

The set of lengths of edges in $C_{r,s}$ is

$$[2, 2t] \cup \{-1\} \cup [-2t+2, -2] \cup \{-2t+1, -2t, 1\} = [-2t, -1] \cup [1, 2t].$$

Notice that the difference between the largest and smallest length is $4t$, but $n = m + \ell \geq 4t + 3$. Thus these lengths are distinct in $\mathbb{Z}_n$.

We next give the construction of $C'_{r,s}$. If $\ell = 3$, then let $C'_{r,s}$ be the 3-cycle $((2t+2,r), (2t+2,s), (0, r \circ s))$.

Suppose that $\ell \geq 5$. By Lemma 2, there exists a path $P_{r,s}$ of length $\ell - 2$ using the edge lengths in $\{0\} \cup \pm[2t+1, (n-3)/2]$ with endpoints $(0,r)$ and $(0,s)$. In the lemma we would use $d_1 = 2t+1, d_2 = 2t+2, \ldots, d_x = (n-3)/2$, so $V(P_{r,s}) \subseteq [0, (n-3)/2] \times \{r, s\}$. Let $P^*_{r,s} = P_{r,s} + 2t + 2$, with endpoints $(2t+2, r)$ and $(2t+2, s)$. Then $V(P^*_{r,s}) \subseteq [2t+2, (n-3)/2 + 2t + 2] \times \{r, s\}$. Since $(n-3)/2 + 2t + 2 = (n+m+1)/2 = (2n - \ell + 1)/2 < n$, we see that $P^*_{r,s}$ is vertex disjoint from $C_{r,s}$.

Finally we construct the cycle $C'_{r,s}$ from $P^*_{r,s}$ by adding the edges $\{(2t+2, r), (2t+2+(n-1)/2, r \circ s)\}$ and $\{(2t+2, s), (2t+2+(n-1)/2, r \circ s)\}$. Note that in the subgraph of $K_{(2k+1) \times n}$ with vertex set $\mathbb{Z}_n \times \{r, s\}$, $G_{r,s}$ contains one edge of each length $i \in [-(n-3)/2, (n-3)/2]$. Figure 5 shows an example of $C_{r,s}$ and $C'_{r,s}$ where $m = 8$ and $\ell = 11$.

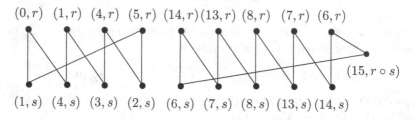

$(0,r)$ $(1,r)$ $(4,r)$ $(5,r)$ $(14,r)$ $(13,r)$ $(8,r)$ $(7,r)$ $(6,r)$

$(15, r \circ s)$

$(1,s)$ $(4,s)$ $(3,s)$ $(2,s)$ $(6,s)$ $(7,s)$ $(8,s)$ $(13,s)$ $(14,s)$

Fig. 5. $C_{r,s}$ and $C'_{r,s}$ where $m = 8$ and $\ell = 11$.

For fixed r and s with $1 \le r < s \le 2k+1$, let $G^*_{r,s} = \{G_{r,s}+x \colon 0 \le x \le n-1\}$. Note that $G^*_{r,s}$ contains n distinct copies of G. Moreover, in the subgraph of $K_{(2k+1)\times n}$ with vertex set $\mathbb{Z}_n \times \{r,s\}$, $G^*_{r,s}$ contains all edges of length i for all $i \in [-(n-1)/2, (n-1)/2] \setminus \{\pm z\}$ for some $z \in [1,(n-1)/2]$. We note that $z = 1$ if $\ell = 3$ and $m \equiv 2 \pmod 8$, and $z = 2$ if $\ell = 3$ and $m \equiv 6 \pmod 8, m \neq 6$. Also, $z = 4$ if $\ell = 3$ and $m = 6$. In all other cases, $z = (n-1)/2$.

Let $\mathcal{C} = \{G_{r,s} + i \colon 1 \le r < s \le 2k+1, 0 \le i \le n-1\}$ and note that $\mathcal{C}$ contains $\binom{2k+1}{2}n$ distinct copies of G. We wish to show that every edge of $K_{(2k+1)\times n}$ appears in some copy of G in $\mathcal{C}$. Let $e = \{(i,r),(j,s)\}$ with $r < s$ be an arbitrary edge of $K_{(2k+1)\times n}$. Let t' be the unique solution to $r \circ t' = s$ and let $\alpha' = \min\{r,t'\}$ and $\beta' = \max\{r,t'\}$. Let t'' be the unique solution to $s \circ t'' = r$ and let $\alpha'' = \min\{r,t''\}$ and $\beta'' = \max\{r,t''\}$. If $j-i \in [-(n-1)/2, (n-1)/2] \setminus \{\pm z\}$, then e belongs to $G_{r,s} + x$ for some x with $0 \le x \le n-1$. If $j - i = z$, then e belongs to $G_{\alpha',\beta'} + x$ where $0 \le x \le n-1$. If $j - i = -z$, then e belongs to $G_{\alpha'',\beta''} + x$ where $0 \le x \le n-1$. Since every edge of $K_{(2k+1)\times n}$ appears in some copy of G in $\mathcal{C}$ and since $\mathcal{C}$ contains $\binom{2k+1}{2}n$ distinct copies of G, it follows that $\mathcal{C}$ is a decomposition of $K_{(2k+1)\times n}$ into copies of G.   □

By using symmetric quasigroups with holes in place of idempotent symmetric quasigroups, we can modify the proof of Theorem 2 to obtain a G-decomposition of $K_{k \times 2n}$ for every integer $k \ge 3$.

Theorem 3. *Let G be a 2-regular graph of odd order n consisting of exactly two cycles. For every integer $k \ge 3$, there exists a G-decomposition of $K_{k \times 2n}$.*

Proof. Let $G = C_m \cup C_\ell$, where $m \ge 4$ is even and $\ell \ge 3$ is odd. Let $k \ge 3$ be an integer and let $Q - [1,2k]$. For $i \in [1,k]$, let $h_i = \{2i-1,2i\}$ and $g_i = \mathbb{Z}_n \times h_i$. Let $V(K_{k \times 2n}) = \mathbb{Z}_n \times [1,2k]$ with the vertex-set partition $\{g_1, g_2, \ldots, g_k\}$. Let $(Q, \circ)$ be a commutative quasigroup of order $2k$ with holes H.

For integers $j > 0$, a, and b and for $r, s \in Q$, with $r < s$ and such that $\{r,s\} \notin H$, let $W(a,j,b)$ be the directed path with consecutive vertices

$$(a,r), (b+j-1,s), (a+1,r), (b+j-2,s), (a+2,r), \ldots, (b-1,s), (a+j-1,r), (b,s).$$

Notice that this sequence contains $2j$ vertices, starting with (a,r) and ending with (b,s). We have $V_r(W(a,j,b)) = [a, a+j-1]$, and $V_s(W(a,j,b)) = [b, b+j-1]$. The set of lengths of the edges of this path is $b - a + [-(j-1), j-1]$.

Fix r and s with $1 \le r < s \le 2k$ and $\{r,s\} \notin H$. We proceed in the same fashion as in the proof of Theorem 2 producing the graph $G_{r,s}$ consisting of a cycle $C_{r,s}$ of length m and a cycle $C'_{r,s}$ of length ℓ such that $C_{r,s}$ and $C'_{r,s}$ are vertex-disjoint.

For fixed r and s with $1 \le r < s \le 2k$, and with $\{r,s\} \notin H$, let $G^*_{r,s} = \{G_{r,s} + x \colon 0 \le x \le n-1\}$. Note that $G^*_{r,s}$ contains n distinct copies of G. Moreover, in the subgraph of $K_{k \times 2n}$ with vertex set $\mathbb{Z}_n \times \{r,s\}$, $G^*_{r,s}$ contains all edges of length i for all $i \in [-(n-1)/2, (n-1)/2] \setminus \{\pm z\}$ for some $z \in [1, (n-1)/2]$. We note that $z = 1$ if $\ell = 3$ and $m \equiv 2 \pmod 8$, and $z = 2$ if $\ell = 3$ and $m \equiv 6 \pmod 8, m \neq 6$. Also, $z = 4$ if $\ell = 3$ and $m = 6$. In all other cases, $z = (n-1)/2$.

Let $\mathcal{C} = \{G_{r,s} + i : 1 \le r < s \le 2k, \{r,s\} \notin H, 0 \le i \le n-1\}$ and note that $\mathcal{C}$ contains $\binom{2k}{2} n$ distinct copies of G. We wish to show that every edge of $K_{k \times 2n}$ appears in some copy of G in $\mathcal{C}$. Let $e = \{(i,r),(j,s)\}$ with $r < s$ be an arbitrary edge of $K_{k \times 2n}$. Let t' be the unique solution to $r \circ t' = s$ and let $\alpha' = \min\{r, t'\}$ and $\beta' = \max\{r, t'\}$. Let t'' be the unique solution to $s \circ t'' = r$ and let $\alpha'' = \min\{r, t''\}$ and $\beta'' = \max\{r, t''\}$. If $j - i \in [-(n-1)/2, (n-1)/2] \setminus \{\pm z\}$, then e belongs to $G_{r,s} + x$ for some x with $0 \le x \le n-1$. If $j - i = z$, then e belongs to $G_{\alpha',\beta'} + x$ where $0 \le x \le n-1$. If $j - i = -z$, then e belongs to $G_{\alpha'',\beta''} + x$ where $0 \le x \le n-1$. Since every edge of $K_{k \times 2n}$ appears in some copy of G in $\mathcal{C}$ and since $\mathcal{C}$ contains $\binom{2k}{2} n$ distinct copies of G, it follows that $\mathcal{C}$ is a decomposition of $K_{k \times 2n}$ into copies of G. □

4 Main Result

By combining the results from Theorems 1 and 2, we obtain the following theorem.

Theorem 4. *Let G be a 2-regular graph of odd order n consisting of exactly two cycles. There exists a G-decomposition of K_v for all $v \equiv n \pmod{2n}$ unless $G = C_4 \cup C_5$ and $v = 9$.*

Proof. In [3], it is shown that there exists a $(C_4 \cup C_5)$-decomposition of K_v if and only if $v \equiv 1$ or $9 \pmod{18}$ and $v \ne 9$. For all other G, let $v = 2kn + n$. Observe that $K_v = (2k + 1)K_n \cup K_{(2k+1) \times n}$. By Theorem 1, there exists a G-decomposition of K_n and hence of $(2k + 1)K_n$ and by Theorem 2, there exists a G-decomposition of $K_{(2k+1) \times n}$. The result follows. □

If n in Theorem 4 is a power of a prime, then we have the following corollary.

Corollary 1. *Let G be a 2-regular graph of odd order n consisting of exactly two cycles. If n is a prime power, then there exists a G-decomposition of K_v if and only if $v \equiv 1$ or $n \pmod{2n}$, unless $G = C_4 \cup C_5$ and $v = 9$.*

Proof. The necessary conditions for the existence of a G-decomposition of K_v include $n | v(v-1)/2$ and $v \ge n$ is odd. If $n = p^k$, where p is prime, then we have $2p^k | v(v-1)$ and $v \ge p^k$ is odd. Since v and $v - 1$ are relatively prime, either $p^k | v$ or $p^k | v - 1$. Thus, $v \equiv 1$ or $p^k \pmod{2p^k}$.

It is known that there exists a G-decomposition of K_v for all $v \equiv 1 \pmod{2n}$ (see [3,6]). By Theorem 4, there exists a G-decomposition of K_v for all $v \equiv n \pmod{2n}$ unless $G = C_4 \cup C_5$ and $v = 9$. The result follows. □

5 Final Comments

This manuscript highlights ways of extending the classical Bose construction for Steiner triple systems to constructions of G-decompositions of complete and complete multipartite graphs where G is the vertex-disjoint union of one even and one odd cycle. In [14], this approach is used to construct G-decompositions of $K_{(2k+1)\times n}$ and of $K_{k'\times 2n}$ as well as of $K_{2nk'+1}$ when G of size n is the union of three vertex-disjoint odd length cycles and k and $k' \neq 2$ are positive integers. Other authors have used similar approaches in the past where G is a single cycle of odd order. In particular, we note the use of this approach in [13,16]. We also note that there are other popular approaches to finding G-decompositions of $K_{(2k+1)\times n}$ for a graph G with n edges. In particular, we note the work on this topic by Buratti and Gionfriddo [10].

References

1. Abrham, J., Kotzig, A.: Graceful valuations of 2-regular graphs with two components. Discrete Math. **150**, 3–15 (1996)
2. Adams, P., Bryant, D., Buchanan, M.: A survey on the existence of G-designs. J. Combin. Des. **16**, 373–410 (2008)
3. Adams, P., Bryant, D., Gavlas, H.: Decompositions of the complete graph into small 2-regular graphs. J. Combin. Math. Combin. Comput. **43**, 135–146 (2002)
4. Alspach, B., Gavlas, H.: Cycle decompositions of K_n and of $K_n - I$. J. Combin. Theory Ser. B **81**, 77–99 (2001)
5. Blinco, A., El-Zanati, I.: A note on the cyclic decomposition of complete graphs into bipartite graphs. Bull. Inst. Combin. Appl. **40**, 77–82 (2004)
6. Blinco, A., El-Zanati, S., Vanden Eynden, C.: On the decomposition of complete graphs into almost-bipartite graphs. Discrete Math. **284**, 71–81 (2004)
7. Bose, R.C.: On the construction of balanced incomplete block designs. Ann. Eugenics **9**, 353–399 (1939)
8. Bryant, D., El-Zanati, S.: Graph decompositions. In: Colbourn, C.J., Dinitz, J.H.(eds.), Handbook of Combinatorial Designs. 2nd edn, pp. 477–485, Chapman & Hall/CRC, Boca Raton (2007)
9. Bunge, R.C., Chantasartrassmee, A., El-Zanati, S., Vanden Eynden, C.: On cyclic decompositions of complete graphs into tripartite graphs. J. Graph Theory **72**, 90–111 (2013)
10. Buratti, M., Gionfriddo, L.: Strong difference families over arbitrary graphs. J. Combin. Des. **16**, 443–461 (2008)
11. El-Zanati, S., Vanden Eynden, C., Punnim, N.: On the cyclic decomposition of complete graphs into bipartite graphs. Australas. J. Combin. **24**, 209–219 (2001)
12. Gannon, D.I., El-Zanati, S.: All 2-regular graphs with uniform odd components admit ρ-labelings. Australas. J. Combin. **53**, 207–219 (2012)
13. Hoffman, D.G., Lindner, C.C., Rodger, C.A.: On the construction of odd cycle systems. J. Graph Theory **13**, 417–426 (1989)
14. Jongthawonwuth, U., El-Zanati, S., Uiyyasathian, C.: On extending the Bose construction for triple systems to decompositions of complete multipartite graphs into 2-regular graphs of odd order. Australas. J. Combin. **59**, 378–390 (2014)
15. Lindner, C.C., Rodger, C.A.: Design Theory: Discrete Mathematics and its Applications, 2nd edn. CRC Press, Boca Raton, FL (2009)
16. Lindner, C.C., Rodger, C.A., Stinson, C.R.: Nesting of cycle systems of odd length. Discrete Math. **77**, 191–203 (1989)
17. Šajna, M.: Cycle decompositions III: complete graphs and fixed length cycles. J. Combin. Des. **10**, 27–78 (2002)
18. Traetta, A.: Complete solution to the two-table Oberwolfach problems. J. Combin. Theory Ser. A **120**, 984–997 (2013)

Reconfiguration of Vertex Covers in a Graph

Takehiro Ito[✉], Hiroyuki Nooka, and Xiao Zhou

Graduate School of Information Sciences, Tohoku University,
Aoba-yama 6-6-05, Sendai 980-8579, Japan
{takehiro,zhou}@ecei.tohoku.ac.jp,
nooka.hiroyuki@ec.ecei.tohoku.ac.jp

Abstract. Suppose that we are given two vertex covers C_0 and C_t of a graph G, together with an integer threshold $k \geq \max\{|C_0|, |C_t|\}$. Then, the VERTEX COVER RECONFIGURATION problem is to determine whether there exists a sequence of vertex covers of G which transforms C_0 into C_t such that each vertex cover in the sequence is of cardinality at most k and is obtained from the previous one by either adding or deleting exactly one vertex. This problem is PSPACE-complete even for planar graphs. In this paper, we first give a linear-time algorithm to solve the problem for even-hole-free graphs, which include several well-known graphs, such as trees, interval graphs and chordal graphs. We then give an upper bound on k for which any pair of vertex covers in a graph G has a desired sequence. Our upper bound is best possible in some sense.

1 Introduction

A *vertex cover* C of a graph G is a vertex subset of G which contains at least one of the two endpoints of every edge in G. (See Fig. 1 which depicts six different vertex covers of the same graph.) Then, the VERTEX COVER problem is a well-known NP-complete problem [5], defined as follows: Given a graph G and an integer k, it determines whether G has a vertex cover of cardinality at most k.

The VERTEX COVER problem has several applications, such as in the SNP assembly problem on the computational biochemistry and in a computer network security problem [12]. In the computer network security problem, each vertex corresponds to a router and each edge corresponds to a link in a computer network, and we wish to pick a subset of routers for monitoring packets flowed on the links; such a subset forms a vertex cover of the corresponding graph.

However, a practical issue in computer network security requires that the formulation should be considered in more dynamic situations: in order to maintain routers, we sometimes need to change the current subset of routers to another subset. Of course, we wish to keep monitoring all links even during the transformation. This situation can be formulated by the concept of reconfiguration problems that have been extensively studied in recent literature [1,2,6–9,11].

1.1 Our Problems

Suppose that we are given two vertex covers C_0 and C_t of a graph $G = (V, E)$, together with an integer threshold $k \geq \max\{|C_0|, |C_t|\}$. Then, the VERTEX

© Springer International Publishing Switzerland 2015
J. Kratochvíl et al. (Eds.): IWOCA 2014, LNCS 8986, pp. 164–175, 2015.
DOI: 10.1007/978-3-319-19315-1_15

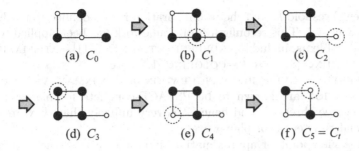

Fig. 1. A sequence $\langle C_0, C_1, \ldots, C_5 \rangle$ of vertex covers of the same graph, where the vertices in vertex covers are depicted by large black circles.

COVER RECONFIGURATION problem is to determine whether there exists a sequence $\langle C_0, C_1, \ldots, C_\ell \rangle$ of vertex covers of G such that

(a) $C_\ell = C_t$, and $|C_i| \leq k$ for all i, $0 \leq i \leq \ell$; and
(b) for each index i, $1 \leq i \leq \ell$, the vertex cover C_i of G is obtained from the previous one C_{i-1} by either deleting or adding a single vertex $u \in V$, that is, $C_{i-1} \triangle C_i = (C_{i-1} \setminus C_i) \cup (C_i \setminus C_{i-1}) = \{u\}$.

Figure 1 illustrates a sequence $\langle C_0, C_1, \ldots, C_5 \rangle$ of vertex covers of the same graph which transforms C_0 into $C_t = C_5$, where the vertex which is deleted from (or added to) the previous vertex cover is surrounded by a dotted circle.

The existence of such a transformation clearly depends on the value of a given threshold k. For example, if $k \geq 4$, then the instance of the two vertex covers C_0 and C_t in Fig. 1 is a yes-instance because all vertex covers in Fig. 1 have the cardinalities at most four. On the other hand, if $k \leq 3$, then the instance in Fig. 1 is a no-instance because there is no transformation between C_0 and C_t that consists only of vertex covers of cardinalities at most three.

Therefore, we can get a natural minimization problem, called the MINMAX VERTEX COVER RECONFIGURATION problem, in which we wish to minimize the maximum cardinality of any vertex cover in a transformation for two given vertex covers C_0 and C_t of a graph G; we denote by $f_G^*(C_0, C_t)$ the optimal value. Then, the answer to VERTEX COVER RECONFIGURATION is "yes" if $k \geq f_G^*(C_0, C_t)$; otherwise "no." (A formal definition will be given in Sect. 2.)

1.2   Related and Known Results

Recently, this type of problems has been studied extensively in the framework of *reconfiguration problems* [8], which arise when we wish to find a step-by-step transformation between two feasible solutions of a problem such that all intermediate solutions are also feasible and each step abides by a prescribed reconfiguration rule (i.e., an adjacency relation defined on feasible solutions of the original problem). For example, in VERTEX COVER RECONFIGURATION, feasible solutions are defined to be all vertex covers of a graph with cardinalities at most a given threshold k; and the reconfiguration rule is defined to be the condition (b) in Sect. 1.1. This reconfiguration framework has been applied to several well-known problems, including INDEPENDENT SET [7–9,11], SATISFIABILITY [6], CLIQUE, MATCHING [8], VERTEX-COLORING [1,2], etc.

Both VERTEX COVER RECONFIGURATION and MINMAX VERTEX COVER RECONFIGURATION are known to be PSPACE-complete for planar graphs of maximum degree three [8], and hence it is very unlikely that they are solvable in polynomial time even for planar graphs.

From the viewpoint of approximation, it is known that the optimal value $f_G^*(C_0, C_t)$ can be approximated within a factor 2 in linear time; indeed, this approximation result can be obtained from a linear-time 2-approximation algorithm for the reconfiguration problem on SET COVER [8, Theorem 6]. However, as far as we know, only this 2-approximation is known for the reconfiguration problem on SET COVER, and hence it is desired to investigate further algorithmic (positive) results for the reconfiguration problems on VERTEX COVER.

One may think that some known results for the reconfiguration problem on INDEPENDENT SET [7–9,11] can be converted into our VERTEX COVER RECONFIGURATION; because, if a vertex subset C is a vertex cover of a graph $G = (V, E)$, then $V \setminus C$ forms an independent set of G, and *vice versa*. There are three types of reconfiguration problems for INDEPENDENT SET which employ different reconfiguration rules. Although one of the three rules corresponds to the one for our VERTEX COVER RECONFIGURATION, almost all (positive) results are given to the other reconfiguration rules. Indeed, as far as we know, there is only one positive result which can be converted into our VERTEX COVER RECONFIGURATION; this will be discussed in Sect. 3.

1.3 Our Contribution

In this paper, we investigate algorithmic results for the VERTEX COVER RECONFIGURATION and MINMAX VERTEX COVER RECONFIGURATION problems.

We first show that both reconfiguration problems can be solved in linear time for even-hole-free graphs. We will define the class of even-hole-free graphs later, but we here note that this graph class contains several well-known graph classes, such as those of trees, interval graphs and chordal graphs.

We then give an upper bound on the optimal value $f_G^*(C_0, C_t)$ for two vertex covers C_0 and C_t of a graph G. Our upper bound holds for any graph; as a corollary, we have $f_G^*(C_0, C_t) \leq \max\{|C_0|, |C_t|\} + 1$ if G is a tree; and $f_G^*(C_0, C_t) \leq \max\{|C_0|, |C_t|\} + 2$ if G is a cactus. We note that our upper bound is best possible in the following sense: there are instances of cacti such that $f_G^*(C_0, C_t) = \max\{|C_0|, |C_t|\} + 2$. (See Sect. 4.2 for details.)

We finally note that our second result gives an approximation for $f_G^*(C_0, C_t)$ with absolute performance guarantee. For an instance of MINMAX VERTEX COVER RECONFIGURATION, let $\mathsf{ap}_G(C_0, C_t)$ be an objective value computed by an algorithm. Then, for two integers $\rho \geq 1$ and $c \geq 0$, we say that the algorithm is (ρ, c)-*approximation* if $\mathsf{ap}_G(C_0, C_t) \leq \rho \cdot f_G^*(C_0, C_t) + c$ holds for any instance.

As we have mentioned above, there is a linear-time $(2,0)$-approximation algorithm for $f_G^*(C_0, C_t)$, which follows from the 2-approximation for the reconfiguration problem on SET COVER [8]. On the other hand, our second result gives a $(1, \alpha)$-approximation for $f_G^*(C_0, C_t)$, where α is some integer defined later. Although this integer α depends on an input graph G, it is remarkable that α can be obtained by only a *local computation*: we just focus on a transformation restricted on 2-connected subgraphs of G, and extend it to a transformation on the whole graph G.

Due to the page limitation, we omit some proofs from this extended abstract.

2   Preliminaries

In this paper, we may assume without loss of generality that graphs are simple, undirected and unweighted. For a graph G, we sometimes denote by $V(G)$ and $E(G)$ the vertex set and the edge set of G, respectively. For a vertex subset V' of a graph G, we denote by $G[V']$ the subgraph of G induced by V'.

A vertex u in a connected graph $G = (V, E)$ is called a *cut vertex* of G if the induced subgraph $G[V \setminus \{u\}]$ is disconnected. A connected graph G is said to be *2-connected* if G has no cut vertex.

A vertex subset C of a graph G is called a *vertex cover* of G if at least one of $v \in C$ and $w \in C$ holds for every edge $vw \in E(G)$. We say that an edge $vw \in E(G)$ is *covered* by v if $v \in C$.

Definitions for VERTEX COVER RECONFIGURATION. Let C_i and C_j be two vertex covers of a graph $G = (V, E)$. We say that C_i and C_j are *adjacent* if their symmetric difference $C_i \triangle C_j$ consists of exactly one vertex. A *reconfiguration sequence* between two vertex covers C and C' of G is a sequence $\mathcal{C} = \langle C_1, C_2, \ldots, C_\ell \rangle$ of vertex covers of G such that $C_1 = C$, $C_\ell = C'$, and C_{i-1} and C_i are adjacent for $i = 2, 3, \ldots, \ell$. For notational convenience, we write $C_i \in \mathcal{C}$ if a vertex cover C_i appears in a reconfiguration sequence $\mathcal{C}$.

For a reconfiguration sequence $\mathcal{C}$, let $f(\mathcal{C}) = \max\{|C_i| : C_i \in \mathcal{C}\}$, that is, the maximum cardinality of a vertex cover that appears in $\mathcal{C}$. For two vertex covers C and C' of a graph G, we define the *reconfiguration index* $f_G^*(C, C')$, as follows:

$$f_G^*(C, C') = \min\{f(\mathcal{C}) : \mathcal{C} \text{ is a reconfiguration sequence between } C \text{ and } C'\}.$$

It should be noted that the reconfiguration index $f_G^*(C, C')$ is well defined, because any pair of vertex covers C and C' of G has a trivial reconfiguration sequence $\mathcal{C}$ such that $f(\mathcal{C}) = |C \cup C'|$, as follows: we add to C the vertices in $C' \setminus C$ one by one, and obtain the vertex cover $C \cup C'$ of G; and we delete from $C \cup C'$ the vertices in $C \setminus C'$ one by one. We note in passing that this trivial reconfiguration sequence $\mathcal{C}$ gives a 2-approximation for $f_G^*(C, C')$ as shown in [8, Theorem 6], because we clearly have

$$f_G^*(C, C') \geq \max\{|C|, |C'|\}. \tag{1}$$

Given two vertex covers C_0 and C_t of a graph G and a positive integer k, the VERTEX COVER RECONFIGURATION problem is to determine whether

$f_G^*(C_0, C_t) \leq k$; while the MINMAX VERTEX COVER RECONFIGURATION problem is to compute the reconfiguration index $f_G^*(C_0, C_t)$. Note that both problems do not ask for an actual reconfiguration sequence between C_0 and C_t. We always denote by C_0 and C_t the *initial* and *target* vertex covers of G, respectively.

3   Even-Hole-Free Graphs

A graph G is *even-hole free* if any induced subgraph of G is not a cycle consisting of an even number of vertices [4]. This graph class includes several well-known classes, such as trees, interval graphs and chordal graphs.

In this section, we give the following theorem.

Theorem 1. *Both* VERTEX COVER RECONFIGURATION *and* MINMAX VERTEX COVER RECONFIGURATION *can be solved in linear time for even-hole-free graphs.*

As a proof of Theorem 1, it suffices to give a linear-time algorithm which computes $f_G^*(C_0, C_t)$ for any pair of vertex covers C_0 and C_t of an even-hole-free graph G. Our algorithm employs the nice property on a reconfiguration sequence for independent sets, given by Kamiński et al. [9]. We first explain this property in Sect. 3.1, and then give our algorithm in Sect. 3.2.

3.1   Reconfiguration of Independent Sets

Kamiński et al. [9] deeply studied three types of reconfiguration problems for independent sets in a graph. In this subsection, we define and explain only one type used for our algorithm, which is called "Token Addition and Removal (TAR) model" in their paper [9].

In the TAR-model, two independent sets I_i and I_j of a graph G are *adjacent* if their symmetric difference $I_i \triangle I_j$ consists of a single vertex u, that is, I_j can be obtained from I_i by either *removing* or *adding* the vertex u. Similarly as for vertex covers, a *reconfiguration sequence* between two independent sets I and I' of G is a sequence $\langle I_1, I_2, \ldots, I_\ell \rangle$ of independent sets of G such that $I_1 = I$, $I_\ell = I'$, and I_{i-1} and I_i are adjacent for $i = 2, 3, \ldots, \ell$. Kamiński et al. [9] gave the following lemma for even-hole-free graphs.

Lemma 1 ([9]). *Let I_0 and I_t be any pair of independent sets of an even-hole-free graph G such that $|I_0| = |I_t|$. Then, there exists a reconfiguration sequence $\mathcal{I}$ between I_0 and I_t such that $|I_i| \geq |I_0| - 1$ for all independent sets $I_i \in \mathcal{I}$.*

Based on Lemma 1, we give the following lemma. Note that two vertex covers do not necessarily have the same cardinality.

Lemma 2. *Let C_0 and C_t be any pair of two vertex covers of an even-hole-free graph G. Then, $f_G^*(C_0, C_t) \leq \max\{|C_0|, |C_t|\} + 1$.*

3.2   Linear-Time Algorithm

We now give our linear-time algorithm. Let C_0 and C_t be two given vertex covers of an even-hole-free graph G. We may assume without loss of generality that $|C_0| \geq |C_t|$. Then, $\max\{|C_0|, |C_t|\} = |C_0|$. Note that Lemma 2 and Eq. (1) imply that $f_G^*(C_0, C_t) \in \{|C_0|, |C_0| + 1\}$.

A vertex cover C of a graph G is said to be *minimal* if there is no vertex $u \in C$ such that $C \setminus \{u\}$ is a vertex cover of G. We can easily check whether a vertex cover of G is minimal or not in linear time. Then, our algorithm is described as follows:

1. **if** C_0 is minimal, then return $f_G^*(C_0, C_t) = |C_0| + 1$;
2. **else if** C_t is minimal, then return $f_G^*(C_0, C_t) = \max\{|C_0|, |C_t| + 1\}$;
3. **otherwise** return $f_G^*(C_0, C_t) = |C_0|$.

The algorithm above clearly runs in linear time, because it just checks the minimality of vertex covers C_0 and C_t. We omit the correctness proof of our algorithm, due to the page limitation.

4   Upper Bound on Reconfiguration Index

In this section, we give an upper bound on the reconfiguration index $f_G^*(C_0, C_t)$.

4.1   Definitions

For a vertex cover C_i of a graph G and a subgraph G' of G, we denote by $C_{i,G'} = C_i \cap V(G')$ the *restriction* of C_i to G'. Observe that $C_{i,G'}$ is a vertex cover of G', because C_i is a vertex cover of G and G' is a subgraph of G.

Let C_i and C_j be two vertex covers of a graph G, and let D be any vertex subset of $C_i \cap C_j$. Then, we introduce the *reconfiguration index* $f_G^*(C_i, C_j; D)$ *under the constraint of D*, as follows:

$$f_G^*(C_i, C_j; D) = \min\{f(\mathcal{C}_D) : \mathcal{C}_D \text{ is a reconfiguration sequence between } C_i$$
$$\text{and } C_j \text{ such that } D \subseteq C_k \text{ for all } C_k \in \mathcal{C}_D\}. \quad (2)$$

Note that $f_G^*(C_i, C_j; D)$ is well defined for any vertex subset $D \subseteq C_i \cap C_j$; recall the trivial reconfiguration sequence $\mathcal{C}_D$ between C_i and C_j via the vertex cover $C_i \cup C_j$ (in Sect. 2), then $D \subseteq C_k$ holds for every $C_k \in \mathcal{C}_D$. We clearly have

$$f_G^*(C_i, C_j) = f_G^*(C_i, C_j; \emptyset). \quad (3)$$

Furthermore, we have the following lemma.

Lemma 3. *Let C_i and C_j be any pair of vertex covers of a graph G, and let D and D' be any vertex subsets such that $D \subseteq D' \subseteq C_i \cap C_j$. Then, $f_G^*(C_i, C_j; D) \leq f_G^*(C_i, C_j; D')$.*

Lemma 3 and Eq. (3) imply that, for any vertex subset $D \subseteq C_0 \cap C_t$,

$$f_G^*(C_0, C_t) \leq f_G^*(C_0, C_t; D).$$

4.2   Our Upper Bound

We now give our upper bound, whose proof will be given in Sect. 4.3.

Theorem 2. *Let α be a fixed integer, and let G be any graph. Suppose that*

$$f_{G''}^*(C_{0,G''}, C_{t,G''}; C_{0,G''} \cap C_{t,G''}) \leq \max\{|C_{0,G''}|, |C_{t,G''}|\} + \alpha$$

for every 2-connected subgraph G'' of G. Then,

$$f_G^*(C_0, C_t; C_0 \cap C_t) \leq \max\{|C_0|, |C_t|\} + \alpha.$$

A graph G is a *cactus* if every edge is part of at most one cycle in G [3]. Using Theorem 2, we give the following corollary.

Corollary 1. *Let C_0 and C_t be any pair of vertex covers of a graph G. Then,*

(a) $f_G^*(C_0, C_t) \leq \max\{|C_0|, |C_t|\} + 1$ *if G is a tree; and*
(b) $f_G^*(C_0, C_t) \leq \max\{|C_0|, |C_t|\} + 2$ *if G is a cactus.*

We note that Corollary 1(a) gives another proof of Lemma 2 for trees. Conversely, Corollary 1(b) cannot be obtained from Lemma 2, because a cactus is not always even-hole free.

Furthermore, we note that our upper bound on $f_G^*(C_0, C_t)$ is best possible in some sense. For example, consider an even-length cycle G and its two vertex covers C_0 and C_t, each of which forms an independent set of G. (See Fig. 2.) Since G is a cycle (and hence a cactus), by Corollary 1(b) we have $f_G^*(C_0, C_t) \leq \max\{|C_0|, |C_t|\} + 2$. Indeed, we have to add at least two vertices to C_0 in order to delete any vertex in C_0. Therefore, $f_G^*(C_0, C_t) = \max\{|C_0|, |C_t|\} + 2$.

(a) C_0 (b) C_t

Fig. 2. Instance for a cycle G such that $f_G^*(C_0, C_t) = \max\{|C_0|, |C_t|\} + 2$.

4.3   Proof of Theorem 2

In this subsection, as a proof of Theorem 2, we construct a reconfiguration sequence $\mathcal{C}$ between C_0 and C_t such that $f(\mathcal{C}) \leq \max\{|C_0|, |C_t|\} + \alpha$ and $C_0 \cap C_t \subseteq C_i$ for all vertex covers $C_i \in \mathcal{C}$. Then, the theorem follows, because $f_G^*(C_0, C_t; C_0 \cap C_t) \leq f(\mathcal{C})$ holds for such a reconfiguration sequence $\mathcal{C}$.

Roughly speaking, our idea is as follows. We first decompose a graph G into its 2-connected subgraphs, and then separately construct a reconfiguration sequence for each 2-connected subgraph G'' which transforms the vertex cover

$C_0 \cap V(G'')$ into the target one $C_t \cap V(G'')$. Of course, we need to extend the reconfiguration sequence for G'' to the one for the whole graph G. Furthermore, we need to find a clever ordering of 2-connected subgraphs of G to be transformed so that all intermediate vertex covers C_i of G satisfy $|C_i| \leq \max\{|C_0|, |C_t|\} + \alpha$.

Therefore, we first introduce some notions and properties of vertex covers in subgraphs, and then give our reconfiguration sequence.

Notion and Properties. Let C_i and C_j be two vertex covers of a graph G. Then, for a subgraph G' of G, we define the *difference* $\delta(G', C_i, C_j)$ from C_i to C_j, as follows:

$$\delta(G', C_i, C_j) = |C_{j,G'}| - |C_{i,G'}|,$$

that is, the cardinality of the vertex cover of G' is increased by $\delta(G', C_i, C_j)$ if we transform the vertex cover $C_i \cap V(G')$ into $C_j \cap V(G')$. Clearly, $\delta(G', C_i, C_j) = -\delta(G', C_j, C_i)$.

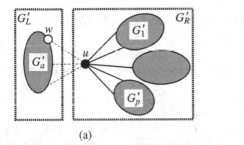 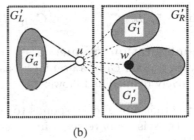

(a) (b)

Fig. 3. The bipartition (G'_L, G'_R) of a subgraph G' with a cut vertex u for the cases where (a) $u \in C_0$ and (b) $u \notin C_0$.

Let $G' = (V', E')$ be any induced subgraph of a graph $G = (V, E)$. Let u be an arbitrary cut vertex of G', and suppose that the induced subgraph $G[V' \setminus \{u\}]$ consists of p connected components $G'_1, G'_2, \ldots, G'_p$. Note that $p \geq 2$ since u is a cut vertex of G'. Let $G'_a = (V'_a, E'_a)$ be the connected component in $G[V' \setminus \{u\}]$ such that $\delta(G'_a, C_0, C_t) = \min\{\delta(G'_i, C_0, C_t) : 1 \leq i \leq p\}$. Then, the *bipartition* (G'_L, G'_R) of G' with a cut vertex u is to decompose G into two subgraphs G'_L and G'_R, defined as follows (see Fig. 3):

$$G'_L = \begin{cases} G'_a & \text{if } u \in C_0; \\ G[V'_a \cup \{u\}] & \text{if } u \notin C_0, \end{cases}$$

and $G'_R = G[V' \setminus V(G'_L)]$. Therefore, $V(G'_L)$ and $V(G'_R)$ form a partition of $V(G')$, that is, $V(G'_L) \cup V(G'_R) = V(G')$ and $V(G'_L) \cap V(G'_R) = \emptyset$.

Based on the bipartition (G'_L, G'_R) of a subgraph G' with a cut vertex u, we construct a reconfiguration sequence from $C_{0,G'}$ to $C_{t,G'}$, as follows:

(1) transform C_{0,G'_L} into C_{t,G'_L} without adding/deleting any vertex in G'_R; and

(2) transform C_{0,G'_R} into C_{t,G'_R} without adding/deleting any vertex in G'_L.

The following lemma will be used in Lemma 6 for proving that any vertex subset appeared in the reconfiguration sequence above is a vertex cover of G'. For notational convenience, let $C_{i,L} = C_{i,G'_L}$ and $C_{i,R} = C_{i,G'_R}$.

Lemma 4. *Let (G'_L, G'_R) be the bipartition of G' with a cut vertex u. Then,*

(a) *for any vertex cover C'_L of G'_L, the vertex subset $C'_L \cup C_{0,R}$ forms a vertex cover of G'; and*

(b) *for any vertex cover C'_R of G'_R such that $C_{0,R} \cap C_{t,R} \subseteq C'_R$, the vertex subset $C_{t,L} \cup C'_R$ forms a vertex cover of G'.*

Let $C'_q = C_{t,L} \cup C_{0,R}$. Lemma 4(a) implies that C'_q is a vertex cover of G'. Furthermore, $|C'_q| = |C_{t,L}| + |C_{0,R}|$ since $V(G'_L) \cap V(G'_R) = \emptyset$. Then, we give the following lemma.

Lemma 5. *Let $C'_q = C_{t,L} \cup C_{0,R}$. Then, $|C'_q| \leq \max\{|C_{0,G'}|, |C_{t,G'}|\}$.*

Reconfiguration Sequence. We now give our reconfiguration sequence between C_0 and C_t of a graph G, based on a *decomposition tree* T of G which is recursively defined as follows:

(A) the root r of T corresponds to the whole graph G; and

(B) if there is a cut vertex u in the subgraph G' corresponding to a node v of T, then v has two children v_L and v_R in T which correspond to the subgraphs G'_L and G'_R, respectively, where (G'_L, G'_R) is the bipartition of G' with u.

Then, each leaf of T corresponds to a 2-connected subgraph of G.

We now prove the key lemma.

Lemma 6. *Let α be a fixed integer, and T be a decomposition tree of a graph G. For every 2-connected subgraph G'' of G, suppose that*

$$f^*_{G''}(C_{0,G''}, C_{t,G''}; C_{0,G''} \cap C_{t,G''}) \leq \max\{|C_{0,G''}|, |C_{t,G''}|\} + \alpha. \qquad (4)$$

Then, for the subgraph G' corresponding to each node v of T, there is a reconfiguration sequence $\mathcal{C}' = \langle C'_0, C'_1, \ldots, C'_\ell \rangle$ such that

(a) $C'_0 = C_{0,G'}$ *and* $C'_\ell = C_{t,G'}$;

(b) $C_{0,G'} \cap C_{t,G'} \subseteq C'_i$ *for all vertex covers $C'_i \in \mathcal{C}'$; and*

(c) $f(\mathcal{C}') \leq \max\{|C_{0,G'}|, |C_{t,G'}|\} + \alpha$.

Proof. We prove the lemma by induction based on the decomposition tree T.

Base Step. Suppose that v is a leaf of T, and let G' be the subgraph corresponding to v. Then, G' is 2-connected, and hence Eq. (4) holds for G'. Therefore, Eqs. (2) and (4) imply that there exists a reconfiguration sequence between $C_{0,G'}$ and $C_{t,G'}$ satisfying all the three conditions (a)–(c).

Inductive Step. Let v be an internal node of T having two children v_L and v_R. Let G', G'_L and G'_R be the subgraphs corresponding to v, v_L and v_R, respectively, and hence (G'_L, G'_R) is a bipartition of G'. Suppose that the lemma holds for G'_L and G'_R. Then, G'_L has a reconfiguration sequence $\mathcal{C}^L = \langle C_0^L, C_1^L, \ldots, C_{\ell_L}^L \rangle$ such that

(a-L) $C_0^L = C_{0,L}$ and $C_{\ell_L}^L = C_{t,L}$;
(b-L) $C_{0,L} \cap C_{t,L} \subseteq C_i^L$ for all vertex covers $C_i^L \in \mathcal{C}^L$; and
(c-L) $f(\mathcal{C}^L) \leq \max\{|C_{0,L}|, |C_{t,L}|\} + \alpha$.

Similarly, G'_R has a reconfiguration sequence $\mathcal{C}^R = \langle C_0^R, C_1^R, \ldots, C_{\ell_R}^R \rangle$ such that

(a-R) $C_0^R = C_{0,R}$ and $C_{\ell_R}^R = C_{t,R}$;
(b-R) $C_{0,R} \cap C_{t,R} \subseteq C_j^R$ for all vertex covers $C_j^R \in \mathcal{C}^R$; and
(c-R) $f(\mathcal{C}^R) \leq \max\{|C_{0,R}|, |C_{t,R}|\} + \alpha$.

From the induction hypothesis, we construct a sequence $\mathcal{C}' = \langle C'_0, C'_1, \ldots, C'_\ell \rangle$ of vertex subsets of G', where $\ell = \ell_L + \ell_R$, defined as follows:

(i) $C'_i = C_i^L \cup C_{0,R}$ for all i, $0 \leq i \leq \ell_L$; and
(ii) $C'_i = C_{t,L} \cup C_{i-\ell_L}^R$ for all i, $\ell_L < i \leq \ell_L + \ell_R = \ell$.

Then, $C'_{\ell_L} = C_{t,L} \cup C_{0,R}$. In the following, we will show that $\mathcal{C}'$ is a reconfiguration sequence for G' satisfying all the three conditions (a)–(c). Let $\mathcal{C}'_{0,\ell_L} = \langle C'_0, C'_1, \ldots, C'_{\ell_L} \rangle$ and $\mathcal{C}'_{\ell_L,\ell} = \langle C'_{\ell_L}, C'_{\ell_L+1}, \ldots, C'_{\ell_L+\ell_R} \rangle$. Note that C'_{ℓ_L} is contained in both $\mathcal{C}'_{0,\ell_L}$ and $\mathcal{C}'_{\ell_L,\ell}$ for notational convenience.

We first show that $\mathcal{C}'$ satisfies the condition (a). By the construction (i) above and the condition (a-L), we have $C'_0 = C_0^L \cup C_{0,R} = C_{0,L} \cup C_{0,R} = C_{0,G'}$, as required. Similarly, by the construction (ii) above and the condition (a-R), we have $C'_\ell = C_{t,L} \cup C_{\ell_R}^R = C_{t,L} \cup C_{t,R} = C_{t,G'}$. Thus, $\mathcal{C}'$ satisfies the condition (a).

Before showing the conditions (b) and (c), we now prove that $\mathcal{C}'$ is a reconfiguration sequence between $C_{0,G'}$ and $C_{t,G'}$. It suffices to show that $\mathcal{C}'_{0,\ell_L}$ is a reconfiguration sequence from $C_{0,G'} = C_{0,L} \cup C_{0,R}$ to $C'_{\ell_L} = C_{t,L} \cup C_{0,R}$, and that $\mathcal{C}'_{\ell_L,\ell}$ is a reconfiguration sequence from $C'_{\ell_L} = C_{t,L} \cup C_{0,R}$ to $C_{t,G'} = C_{t,L} \cup C_{t,R}$.

Recall that $\mathcal{C}^L = \langle C_0^L, C_1^L, \ldots, C_{\ell_L}^L \rangle$ is a reconfiguration sequence between $C_0^L = C_{0,L}$ and $C_{\ell_L}^L = C_{t,L}$, and hence each $C_i^L \in \mathcal{C}^L$ is a vertex cover of G'_L. Since $C'_i = C_i^L \cup C_{0,R}$ for each vertex subset $C'_i \in \mathcal{C}'_{0,\ell_L}$, Lemma 4(a) implies that C'_i is a vertex cover of G'. Therefore, the sequence $\mathcal{C}'_{0,\ell_L}$ is a reconfiguration sequence from $C_{0,G'} = C_{0,L} \cup C_{0,R}$ to $C'_{\ell_L} = C_{t,L} \cup C_{0,R}$.

Recall also that $\mathcal{C}^R = \langle C_0^R, C_1^R, \ldots, C_{\ell_R}^R \rangle$ is a reconfiguration sequence between $C_0^R = C_{0,R}$ and $C_{\ell_R}^R = C_{t,R}$, and hence each $C_j^R \in \mathcal{C}^R$ is a vertex cover of G'_R. Furthermore, by the condition (b-R) we have $C_{0,R} \cap C_{t,R} \subseteq C_j^R$ for all $C_j^R \in \mathcal{C}^R$. Since $C'_i = C_{t,L} \cup C_{i-\ell_L}^R$ for each vertex subset $C'_i \in \mathcal{C}'_{\ell_L,\ell}$, Lemma 4(b) implies that C'_i is a vertex cover of G'. Therefore, the sequence $\mathcal{C}'_{\ell_L,\ell}$ is a reconfiguration sequence from $C'_{\ell_L} = C_{t,L} \cup C_{0,R}$ to $C_{t,L} \cup C_{t,R} = C_{t,G'}$.

In this way, $\mathcal{C}'$ is a reconfiguration sequence between $C_{0,G'}$ and $C_{t,G'}$.

We then show that $\mathcal{C}'$ satisfies the condition (b). Since $V(G'_L) \cap V(G'_R) = \emptyset$, we have $C_{0,G'} \cap C_{t,G'} = (C_{0,L} \cap C_{t,L}) \cup (C_{0,R} \cap C_{t,R})$. By the condition (b-L), we have $C_{0,L} \cap C_{t,L} \subseteq C_i^L$ for all i, $0 \le i \le \ell_L$. Therefore, for all vertex covers C'_i in $\mathcal{C}'_{0,\ell_L}$, by the construction (i) above we have

$$C_{0,G'} \cap C_{t,G'} = (C_{0,L} \cap C_{t,L}) \cup (C_{0,R} \cap C_{t,R}) \subseteq C_i^L \cup C_{0,R} = C'_i,$$

as required. Similarly, by the condition (b-R), we have $C_{0,R} \cap C_{t,R} \subseteq C_{i-\ell_L}^R$ for all i, $\ell_L \le i \le \ell_L + \ell_R$. Therefore, for all vertex covers C'_i in $\mathcal{C}'_{\ell_L,\ell}$, by the construction (ii) above we have

$$C_{0,G'} \cap C_{t,G'} = (C_{0,L} \cap C_{t,L}) \cup (C_{0,R} \cap C_{t,R}) \subseteq C_{t,L} \cup C_{i-\ell_L}^R = C'_i,$$

as required. In this way, $C_{0,G'} \cap C_{t,G'} \subseteq C'_i$ for all vertex covers $C'_i \in \mathcal{C}'$, and hence $\mathcal{C}'$ satisfies the condition (b).

We finally prove that $\mathcal{C}'$ satisfies the condition (c). Notice that

$$f(\mathcal{C}') = \max\{f(\mathcal{C}'_{0,\ell_L}), f(\mathcal{C}'_{\ell_L,\ell})\}. \tag{5}$$

Recall that $V(G'_L) \cap V(G'_R) = \emptyset$. Then, by the construction (i) above and the condition (c-L), we have

$$\begin{aligned}
f(\mathcal{C}'_{0,\ell_L}) &= f(\mathcal{C}^L) + |C_{0,R}| \le \max\{|C_{0,L}|, |C_{t,L}|\} + \alpha + |C_{0,R}| \\
&= \max\{|C_{0,L}| + |C_{0,R}|, |C_{t,L}| + |C_{0,R}|\} + \alpha \\
&= \max\{|C_{0,G'}|, |C'_{\ell_L}|\} + \alpha.
\end{aligned}$$

Since $C'_{\ell_L} = C_{t,L} \cup C_{0,R}$, by Lemma 5 we thus have

$$f(\mathcal{C}'_{0,\ell_L}) \le \max\{|C_{0,G'}|, |C_{t,G'}|\} + \alpha. \tag{6}$$

Similarly, by the construction (ii) above and the condition (c-R), we have

$$\begin{aligned}
f(\mathcal{C}'_{\ell_L,\ell}) &= |C_{t,L}| + f(\mathcal{C}^R) \le |C_{t,L}| + \max\{|C_{0,R}|, |C_{t,R}|\} + \alpha \\
&= \max\{|C_{t,L}| + |C_{0,R}|, |C_{t,L}| + |C_{t,R}|\} + \alpha \\
&= \max\{|C'_{\ell_L}|, |C_{t,G'}|\} + \alpha.
\end{aligned}$$

Therefore, by Lemma 5 we have

$$f(\mathcal{C}'_{\ell_L,\ell}) \le \max\{|C_{0,G'}|, |C_{t,G'}|\} + \alpha. \tag{7}$$

Equations (5), (6) and (7) prove that $\mathcal{C}'$ satisfies the condition (c). $\qquad\square$

Proof of Theorem 2. Recall that the root r of a decomposition tree T of a graph G corresponds to the whole graph G. Therefore, by Lemma 6 there exists a reconfiguration sequence $\mathcal{C}$ between $C_{0,G} = C_0$ and $C_{t,G} = C_t$ such that $C_{0,G} \cap C_{t,G} \subseteq C_i$ for all vertex covers $C_i \in \mathcal{C}$ and $f(\mathcal{C}) \le \max\{|C_{0,G}|, |C_{t,G}|\} + \alpha$. By Eq. (2) we thus have

$$f_G^*(C_0, C_t; C_0 \cap C_t) \le f(\mathcal{C}) \le \max\{|C_0|, |C_t|\} + \alpha,$$

as required. This completes the proof of Theorem 2. $\qquad\square$

5   Concluding Remarks

In this paper, we gave algorithmic results for the two reconfiguration problems on VERTEX COVER. We note again that our upper bound on the reconfiguration index gives an approximation algorithm with absolute performance guarantee.

Very recently, Mouawad et al. [10] proposed a linear-time algorithm to solve VERTEX COVER RECONFIGURATION for even-hole-free graphs and cacti. Their proof method is different from ours, and hence both results can coexist with each other. As one of interesting points of our paper, we proved that the reconfiguration index of a whole graph can be bounded only by the local computation, that is, the reconfiguration index of each 2-connected subgraph; this fact suggests that 2-connected subgraphs are essential for the problem.

Acknowledgment. We are grateful to Ryuhei Uehara for fruitful discussions. This work is partially supported by JSPS KAKENHI 25106504 and 25330003.

References

1. Bonamy, M., Johnson, M., Lignos, I., Patel, V., Paulusma, D.: Reconfiguration graphs for vertex colourings of chordal and chordal bipartite graphs. J. Comb. Optim. **27**, 132–143 (2014)
2. Bonsma, P., Cereceda, L.: Finding paths between graph colourings: PSPACE-completeness and superpolynomial distances. Theoret. Comput. Sci. **410**, 5215–5226 (2009)
3. Brandstädt, A., Le, V.B., Spinrad, J.P.: Graph Classes: A Survey. Society for Industrial and Applied Mathematics, Philadelphia (1999)
4. Conforti, M., Cornuéjols, G., Kapoor, A., Vušković, K.: Even-hole-free graphs part i: decomposition theorem. J. Graph Theory **39**, 6–49 (2002)
5. Garey, M.R., Johnson, D.S.: Computers and Intractability: A Guide to the Theory of NP-Completeness. Freeman, San Francisco (1979)
6. Gopalan, P., Kolaitis, P.G., Maneva, E.N., Papadimitriou, C.H.: The connectivity of Boolean satisfiability: computational and structural dichotomies. SIAM J. Comput. **38**, 2330–2355 (2009)
7. Hearn, R.A., Demaine, E.D.: PSPACE-completeness of sliding-block puzzles and other problems through the nondeterministic constraint logic model of computation. Theoret. Comput. Sci. **343**, 72–96 (2005)
8. Ito, T., Demaine, E.D., Harvey, N.J.A., Papadimitriou, C.H., Sideri, M., Uehara, R., Uno, Y.: On the complexity of reconfiguration problems. Theoret. Comput. Sci. **412**, 1054–1065 (2011)
9. Kamiński, M., Medvedev, P., Milanič, M.: Complexity of independent set reconfigurability problems. Theoret. Comput. Sci. **439**, 9–15 (2012)
10. Mouawad, A.E., Nishimura, N., Raman, V.: Vertex cover reconfiguration and beyond. In: Ahn, H.-K., Shin, C.-S. (eds.) ISAAC 2014. LNCS, vol. 8889, pp. 452–463. Springer, Heidelberg (2014)
11. Mouawad, A.E., Nishimura, N., Raman, V., Simjour, N., Suzuki, A.: On the parameterized complexity of reconfiguration problems. In: Gutin, G., Szeider, S. (eds.) IPEC 2013. LNCS, vol. 8246, pp. 281–294. Springer, Heidelberg (2013)
12. Pirzada, S., Dharwadker, A.: Applications of graph theory. J. Korean Soc. Ind. Appl. Math. **11**, 19–38 (2007)

Space Efficient Data Structures for Nearest Larger Neighbor

Varunkumar Jayapaul[1], Seungbum Jo[2], Venkatesh Raman[3],
and Srinivasa Rao Satti[2(✉)]

[1] Chennai Mathematical Institute, Chennai, India
[2] Seoul National University, Seoul, South Korea
sbcho@tcs.snu.ac.kr, ssrao@cse.snu.ac.kr
[3] The Institute of Mathematical Sciences, Chennai, India
vraman@imsc.res.in

Abstract. Given a sequence of n elements from a totally ordered set, and a position in the sequence, the nearest larger neighbor (NLN) query returns the position of the element which is closest to the query position, and is larger than the element at the query position. The problem of finding all nearest larger neighbors has attracted interest due to its applications for parenthesis matching and in computational geometry [1–3]. We consider a data structure version of this problem, which is to preprocess a given sequence of elements to construct a data structure that can answer NLN queries efficiently. We consider time-space tradeoffs for the problem in both the encoding (where the input is not accessible after the data structure has been created) and indexing model, and consider cases when the input is in a one or two dimensional array. We also consider the version when only the nearest larger right neighbor (NLRN) is sought (in one dimension). We initiate the study of this problem in two dimensions, and describe upper and lower bounds in the encoding and indexing models, distinguishing the two cases when all the elements are distinct or non-distinct.

1 Introduction and Motivation

Given a sequence of n elements from a totally ordered set, and a position in the sequence, the *nearest largest neighbor* (NLN) query asks for the position of an element which is closest to the query position, and is larger than the element at the query position. More formally, given an array A of length n containing elements from a totally ordered set, and a position i in A, we define the query: $NLN(i)$: return the index j such that $A[j] > A[i]$ and $|i-j| = \min\{k : A[i+k] > A[i]$ or $A[i - k] > A[i]$ for $k > 0\}$. Ties are broken to the left, and if there is no element greater than the query element, the query returns the answer ∞.

Seungbum Jo and Rao Satti—Research partly supported by Basic Science Research Program through the National Research Foundation of Korea (NRF) funded by the Ministry of Education, Science and Technology (Grant number 2012-0008241).

© Springer International Publishing Switzerland 2015
J. Kratochvíl et al. (Eds.): IWOCA 2014, LNCS 8986, pp. 176–187, 2015.
DOI: 10.1007/978-3-319-19315-1_16

In a similar way, we define NLRN (*right* nearest larger neighbor), and NLLN (*left* nearest larger neighbor) queries, which return the position of the nearest larger neighbor to the right and left, respectively, of the query position. In a symmetric way, one can also define nearest *smaller* neighbor problems. In this paper, we will stick to the version that seeks the larger neighbors.

2-dimensional NLN. We also consider a natural extension of the NLN problem to two-dimensional arrays. Here, we define the NLN of a query position as the closest position in the array, in terms of the L_1 distance, that contains an element larger than the element at the query position. More formally, given a position (i, j) in A, $NLN((i, j)) = (i', j')$ such that $A[i', j'] > A[i, j]$, and $|i - i'| + |j - j'| = \min\{(|x| + |y| : A[i + x, j + y] > A[i, j]\}$.

Encoding and Indexing Models. We consider the data structure versions of these problems in two different models that have been studied in the succinct data structures literature, namely the *indexing* and *encoding* models. In both these models, the data structure is created after preprocessing the input data. In the indexing model, the queries can be answered by probing the data structure as well as the input data, whereas in the encoding model, the query algorithm cannot access the input data. The size of the data structure in the encoding model is also referred to as the *effective entropy* [10] (of the input data, with respect to the problem).

Previous Work and Motivation. The problem of computing all (right) nearest larger neighbors has attracted much attention due to its importance as a preprocessing routine for answering range minimum queries, triangulation algorithms, reconstructing a binary tree from its traversal orders and matching a sequence of balanced parentheses [3]. While Berkman et al. [3] have given efficient parallel algorithms for the problem, recently Asano et al. [1] have given sequential time-space tradeoff results.

Fischer et al. [9] considered the problem of supporting NLRN and NLLN, and showed how a data structure supporting these two queries can be used in obtaining *entropy-bounded compressed suffix tree* representation. (They considered the min version of the problem instead of max, and named the operations NSV and PSV, for the next and previous smaller values, respectively.) They obtain two results:

(a) For any $f(n) = O(\lg n / \lg \lg n)$, one can construct an $O(n/f(n))$-bitindex that supports NSV and PSV queries in $O(f(n) \lg \lg n)$ time.
(b) Queries PSV and NSV can be answered in constant time, using an encoding of size $4n + o(n)$ bits.

The second result was later improved by Fischer [8] to $2.54n + o(n)$ bits. We show that this can be further improved to $2n + o(n)$ bits if all the elements are distinct. We also improve the first result by shaving a factor of $O(\lg \lg n)$ in the situation when only NLN is required and all the elements are distinct.

Given a two-dimensional array, Asano et al. [1] considered the *All Nearest Larger Neighbours* problem which asks for computing the NLN values for all the elements in the input array. They showed that this problem can be solved in $O(n^2 \lg n)$ time (and more generally, for any d-dimensional matrix in $O(n^d \lg n)$ time). To the best of our knowledge, the data structure version of the two-dimensional NLN problem, in which we are inteterested in constructing a data structure that answers online queries efficiently, has not been considered earlier.

For the case of binary sequences, the data structure version of the NLN problem can be solved by building an auxiliary structure to support *rank* and *select* queries on the bit vector. This uses $o(n)$ bits of extra space, in addition to the input array, and answers NLN (and also NLRN and NLLN) queries in $O(1)$ time.

We assume a standard word RAM model [11] with $O(1)$-time arithmetic and bitwise boolean operations, and we count space in terms of the number of bits used.

Our Results. For the case of one dimension, we first observe in Sect. 2, a $2n-o(n)$-bit lower bound for NLRN in the encoding model, by relating the problems to the well studied range minimum queries (RMQ). We also give an independent lower bound proof by directly counting the number of distinct configurations, which maybe of independent interest. In terms of upper bounds in the encoding model, a $2n+o(n)$ bits to answer NLRN queries and a $2.54n+o(n)$ bit structure to answer NLN queries exist through a structure of Fischer et al. [9]. We develop a $2n + o(n)$ bits structure for NLN if all elements are distinct.

Then in Sect. 3, we look at the problems in indexing model. We give a lower bound for time-space tradeoff for NLN and NLRN queries adapting the lower bound tradeoff proof of Brodal et al. [4] for RMQ. We also provide an algorithm that matches the tradeoff for NLN. For NLRN, our algorithm achieves the time-space product of $O(n \log c)$ (where the query takes $O(c)$ time) while the lower bound is $\Omega(n)$.

For the 2-dimensional NLN problem in the encoding model, we first show that $\Omega(n^2)$ bits are necessary to encode the array to support NLN queries, even when all the elements are distinct. We then describe an asymptotically optimal $\Theta(n^2)$-bit encoding that answers queries in $O(1)$ time, when all the elements in the input array are distinct. For the general case (without the restriction on the distinctness of the elements), we obtain an encoding that uses $O(n^2 \lg \lg n)$ bits and supports queries in $O(1)$ time. For indexing model in the general case, we construct an index of size $O(n^2)$ bits, which answers queries in $O(\lg \lg \lg n)$ time.

2   Data Structures for 1D in the Encoding Model

NLRN. We first show tight space bounds for NLRN encodings.

Theorem 1. *Any data structure which solves NLRN and NLN queries in the encoding model requires $2n - O(\lg n)$ bits.*

Proof. We reduce the RMQ (Range maximum query) problem [5] to the NLRN query problem. In the RMQ problem, we are given a static array, and the query comes with a pair of indices in the array, and the output is the index of position which has maximum value in the range between the pair of indices. Suppose we have a data structure to solve NLRN query, the algorithm to solve $RMQ(i, j)$ repeatedly finds NLRN(t) $\{i \leq t \leq j\}$ until the NLRN index goes out of the range (i, j) or there is no element larger than it to the right, at which point, it returns the last found NLRN value.

Now the lower bound follows from the lower bound for RMQ in encoding model [4]. □

The converse of the simulation in the proof of Theorem 1 – i.e., supporting NLRN queries using RMQ data structure also works – giving the following upper bound.

Lemma 1. *There exists a data structure in the encoding model that takes at most $2n + o(n)$ bits and supports NLRN queries in $O(\lg n)$ time.*

Proof. We use a $2n + o(n)$ bit RMQ structure [5]. We assume that the given sequence is in an array A, and $RMQ(i, j)$ returns the index of the maximum in the array between $A[i]$ and $A[j]$ ($A[i]$ and $A[j]$ inclusive). If there is a tie for the maximum, we assume that RMQ returns the smallest index that has the maximum in the range.

The idea is to do a 'doubling binary search' starting from i by calling RMQ on indices to the right of i, until we find the first occurrence where $RMQ(i, x)$ value differs from i, i.e. the first value larger than $A[i]$ appears in the doubling binary search; and this gives a bounded range to do another binary search to find the answer. ⊓

In fact, the query time can be improved to $O(1)$ using a 2d-Max-Heap as observed by Fischer [7]. We give the description for completeness.

Theorem 2 ([7]). *There exists a data structure in the encoding model that solves NLRN queries using $2n + o(n)$ bits in $O(1)$ time.*

Proof. A 2d-Max-Heap analogous to 2d-Min-Heap as described by Fischer [7] is constructed on array $A[1, n+1]$. The original structure in that paper solves the nearest smaller left neighbor problem.

The 2d-Max-Heap M_A of A is a labeled ordered tree with vertices $v_1, ..., v_{n+1}$ and each vertex corresponds to an index i. The root is labelled with $n+1$, and we treat $A[n+1]$ as ∞. The parent node of v_i is v_j if and only if $i < j$, $A[i] < A[j]$, and $A[k] \leq A[i]$ for all $i < k \leq j$. I.e. as we read the array from right to left, we make v_i the child of v_j if v_j is the nearest larger element (in the right) to v_i. It is clear that the sibling values are in increasing order from left to right, and that the parent of the node labelled v_i corresponds to the NLRN of $A[i]$.

As explained by Fischer [7], the labelled ordered tree can be represented using any of the succinct representation for such a tree, taking at most $2n + o(n)$ bits of space and can answer NLRN in constant time. □

NLN. Clearly an NLN query can be answered with two NLRN data structures –
one for the array A from left to right, and one for the reverse of the array – i.e.
reading the array from right to left. This results in a space of $4n + o(n)$ bits.
Fischer [8] has given a data structure that takes $2.54n + o(n)$ bits to answer both
NLRN and NLLN queries.

It is not clear whether NLN can be answered without explicitly answering NLLN and NLRN, and hence whether a representation using $(2.54 − \epsilon)n$
bits suffice to answer NLN queries. We conjecture that this is not possible, i.e.,
$2.54n + o(n)$ bits are necessary to answer NLN queries. However, when all elements are distinct, we show that one can do better.

Theorem 3. *There exists a data structure that uses $2n + o(n)$ bits and can
answer NLRN and NLLN queries in $O(1)$ time, when all elements are distinct.*

Proof. We show how to support NLLN and NLRN using a succinct representation of a Cartesian tree [5]. Given a node v in the Cartesian tree corresponding
to a position i in the input array, the position j corresponding to the parent
of v is either $NLLN(i)$ or $NLRN(i)$, depending on whether $j < i$ or $j > i$.
Suppose that $j < i$ (the other case is symmetric). Then the node corresponding to the position $NLRN(i)$ is the lowest ancestor of v which corresponds to
a position greater than i, which is in fact the next node in preorder after the
rightmost leaf of v. This can be computed in $O(1)$ time using any succinct binary
tree representation that supports rank and select of nodes in preorder and subtree size. The binary tree representation of [5] supports all these operations in
$O(1)$ time.   □

3   Data Structures for 1D in the Indexing Model

NLN. Recall that in the indexing model, the input can be accessed at query
time, and hence significant space saving should be possible. We first observe that
this can be extended to a time-space tradeoff for NLN.

Theorem 4. *For any parameter c, where $1 \leq c \leq n$, there exists a data structure
which solves NLN queries in the indexing model using $O(n/c)$ bits in $O(c)$ time.*

Proof. Break the input array A of n elements into chunks of size c. Extract the
maximal of each block and construct a conceptual array B of size n/c. Build
an encoding data structure for B that answers NLLN and NLRN queries, using
$2.54n/c + o(n/c)$ bits.

To solve an NLN query for index i, check in $O(c)$ time by scanning the
original input array, whether $A[i]$ is a maximal element of its block. If $A[i]$ is a
maximal element in its block, use the encoding structure for B to find the blocks
containing its NLLN and NLRN. Now, sequentially scan these two blocks to find
the NLN of the query element (in fact, if the two blocks are not at the same
distance from the block containing the query element, it is enough to scan one
of them).

If $A[i]$ is not the maximum of the block then in $O(c)$ time, check the block containing the query index, as well as the block before the query index and the block after the query index, by sequentially scanning the elements and find the NLN. The space used in addition to the input is $\frac{2.54n}{c} + o(\frac{n}{c})$ bits, and the time taken to solve a query is $O(c)$. $\qquad \square$

In the indexing model, Brodal, et al. [4] proved a lower bound for space-time tradeoff of the RMQ problem. The proof is in the non-uniform cell probe model [11]. In this model, computation is free, and time is counted as the number of cells accessed (probed) by the query algorithm. The algorithm is also allowed to be non-uniform, i.e., for different values of input parameter n, we can have different algorithms.

For n and any value of c, where $1 \le c \le n$, they define a set of arrays $C_{n,c}$ and a set of queries Q. They argue that for any RMQ algorithm which has access to an index of size n/c bits (in addition to the input array A), there exists an array in $C_{n,c}$ and a query in Q for which the algorithm is required to perform $\Omega(c)$ probes into A.

The following lower bound proof for NLRN and NLN follows along those lines. We defer the proof to the extended version.

Theorem 5. *Any data structure which stores $O(n/c)$ bits and answers NLRN (or NLN) queries in the indexing model, requires at least $\Omega(c)$ query time.*

NLRN. Interestingly, we find obtaining tradeoffs for NLRN harder than that for NLN, and we don't quite achieve the lower bound above. This is because, for an element that is not the maximum of a block, its NLRN can be quite far (unlike for NLN). Hence we use a different approach to find an upper bound for the trade-off.

Theorem 6. *There exists a data structure that supports NLRN queries in the indexing model in $O(c + \lg n)$ time using $O(\frac{n \lg n}{c^2} + \frac{n \lg c}{c})$ bits, for any integer $1 \le c \le n$.*

Proof. We construct a two-level index structure to answer the NLRN queries. At the top level, we break the given array $A[1, \ldots, n]$ into logical blocks of size b, for some parameter b to be determined later. We find the maximal value in each block, and store an array $C[1, \ldots, n/b]$ such that $C[i]$ stores the position of a maximal element in the ith block (if the maximal element occurs more than once, we store one of the positions that contains the maximal element). Thus $A[C[i]] \ge A[(i-1) * b + j]$, for $1 \le j \le b$. We now construct a complete binary tree with n/b leaves, and assign values with the nodes of this tree as follows. The i-th leaf from left-to-right in the tree is assigned the value $C[i]$. If v is a node with children v_1 and v_2, and v_1 and v_2 are assigned the values i_1 and i_2 respectively, then the value assigned to v is i_1 if $A[i_1] \ge A[i_2]$, and i_2 otherwise. (We essentially build a heap like structure with the maximal values of the blocks at the leaves, but instead of the values, we store the positions pointing to these values at the nodes of the tree.) The height of this tree is $\lg(n/b)$.

In the second level, we break each block into sub-blocks of size s, for some parameter $s < b$ to be chose later. For each sub-block, we store the position of a maximal element within the subblock, using $\lg s$ bits.

Thus the overall space usage of the structure is $O(\frac{n}{b} \lg n + \frac{n}{s} \lg s)$ bits.

To answer the NLRN query for an index i, we first scan the sub-block containing i (to the right of position i) to find its NLRN. If it is not found in this range, we scan the maximal values of each sub-block within the block containing i (again, to the right of position i) to look for a value that is greater than $A[i]$. If we find such a value, we then scan the corresponding sub-block to find the answer to the query. Otherwise, we use the complete binary tree structure to find the closest block to the right of the query position that whose maximal value is greater than $A[i]$. Once we find such a block, we again scan the maximal values of the sub-blocks within that block to find the sub-block that contains the answer, and finally scan that sub-block to find the answer.

The total query time is at most $s + b/s + O(\lg(n/b)) + b/s + s$ which is $O(s + b/s + \lg(n/b))$. By setting $b = c^2$ and $s = c$, we get the time and space bounds stated in the theorem.   □

In particular, by setting $c = \lg n$, we get an NLRN index that uses $O(n \lg \lg n / \lg n)$ bits and supports queries in $O(\lg n)$ time.

4   NLN on 2-Dimensional Arrays

Consider an $n \times n$ 2-dimensional (2D) matrix $A[1 \dots n][1 \dots n]$. Given two positions (i, j) and (i', j') in A, we define $dist((i, j), (i'j')) = |i - i'| + |j - j'|$. Given a 2D matrix A and a position (i, j) in A, we define $NLN((i, j))$ as a position (i', j') such that $A[i, j] < A[i', j']$, and $dist((i, j), (i'j'))$ is the least possible among all such (i', j').

A trivial solution to the NLN problem in 2D is to store $NLN((i, j))$, for $1 \le i, j \le n$. This requires $O(n^2 \lg n)$ bits, and supports queries in $O(1)$ time. In the following, we obtain improved results for the 2D NLN in the encoding and indexing models, and also describe some trade-off results.

4.1   2D NLN in the Encoding Model – Distinct Case

When there is no restriction on the elements of the matrix, one can show an n^2-bit lower bound for NLN encoding (follows easily for the case of a bit matrix). Using a simple encoding method, one can prove that the same asymptotic lower bound applies even when all the elements of the matrix are distinct (proof omitted), to obtain the following.

Theorem 7. *Any data structure which supports NLN queries on an $n \times n$ matrix A in encoding model requires at least $n^2/6$ bits, even when all the elements in A are distinct.*

We now obtain an asymptotically optimal upper bound for 2D NLN encoding for the distinct case.

Lemma 2. *There exists an encoding of an $n \times n$ matrix A that uses at most $O(n^2)$ bits, and supports NLN queries in $O(\lg n)$ time, provided all elements are distinct.*

Proof. The main idea is to divide the matrix recursively into blocks of geometrically increasing size, and store the NLN values of all elements, except the largest element and the elements whose answers are stored at a previous level, in each block explicitly. The following argument shows that this requires $O(n^2)$ bits overall.

In the first level, we divide A into $n^2/4$ blocks of size 2×2 each. Except for the largest element in each 2×2 block, the distance of NLN answer for the other three elements are bounded by 2. In general, at level k, we divide A into $n^2/4^k$ blocks of size $2^k \times 2^k$ each. In each of these $2^k \times 2^k$-sized blocks, there are four elements left for which we need to store the answer to their NLN queries. For three of these four elements, which do not correspond to the maximum value in the block, we store their answers at level k. Since the distance to the NLN answer for these three elements is bounded by 2^{k+2}, we can store these answers using $O(k)$ bits. Thus the total space usage is bounded by $\sum_{k=1}^{\lg n} (3n^2/4^k) * O(k) = O(n^2)$ bits.

One can support NLN queries in time proportional to the number of levels, i.e., in $O(\lg n)$ time, by going through each level, and checking in $O(1)$ time whether or not the answer is stored at that level, and if so, reading the answer from that level. □

We now describe another $O(n^2)$-bit encoding for the 2D NLN problem that supports queries in constant time.

Theorem 8. *There exists an encoding of an $n \times n$ matrix A that uses $O(n^2)$ bits and supports NLN queries in $O(1)$ time, provided all elements are distinct.*

Proof. The encoding is a small variant of the encoding described in the proof of Lemma 2. For each position in A, in some canonical order (say, row-major order), we write down the relative position (i.e., the distance in from the position to its answer in horizontal and vertical directions) of its NLN answer. We use a variable-length encoding, such as γ-code or δ-code [6], to write these answers. The proof of Lemma 2 implies that the sum of the lengths of all these answers is $O(n^2)$. We also store an indexable bit vector [12] indicating the starting positions of each code. This enables us to find the position where the answer to a given query starts and ends, in constant time. □

4.2   2D NLN in the Encoding Model – General Case

If there is no restriction of distinctness on the elements of the matirx, we divide the $n \times n$ matrix A into blocks of size $\lg n \times \lg n$. We use the following lemma that gives an upper bound on the number of distinct NLN positions of all the maximum elements in a block.

Lemma 3. *Given a block B of size $k \times k$, then maximum values of B have at most $4k - 4$ distinct NLN.*

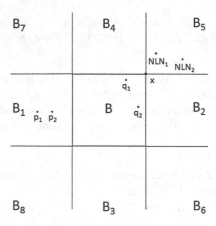

Fig. 1. Block B and its outside areas B_1 to B_8

Proof. Without loss of generality, suppose that B is defined as $A[ak \dots (a+1)k-1][bk \dots (b+1)k-1]$. Because of their maximality, all NLN of the maximum elements in B exist outside of the block. Divide the outside of B into 8 areas as $B_1 = \{A[i,j] | i < ak, bk < j < (b+1)k-1\}$, $B_2 = \{A[i,j] | i > (a+1)k-1, bk < j < (b+1)k-1\}$, $B_3 = \{A[i,j] | ak < i < (a+1)k-1, j < bk\}$, $B_4 = \{A[i,j] | ak < i < (a+1)k-1, j > (b+1)k-1\}$, $B_5 = \{A[i,j] | i \geq (a+1)k-1, j \geq (b+1)k-1\}$, $B_6 = \{A[i,j] | i \geq (a+1)k-1, j \leq bk\}$, $B_7 = \{A[i,j] | i \leq ak, j \geq (b+1)k-1\}$ and $B_8 = \{A[i,j] | i \leq ak, j \leq bk\}$ (see Fig. 1).

If there are two distinct NLN values in B_1, they cannot exist in the same row because all elements in B has smaller distance to the element that has smaller column value than the other one (see p_1 and p_2 in Fig. 1). So there exists at most $k-2$ distinct NLN in B_1. By the similar approach, there exists at most $4(k-2)$ distinct NLN elements in B_1, B_2, B_3 and B_4.

And there is at most one NLN element in B_5. To prove this, assmue that there are 2 elements at the different position q_1, q_2 in B and their NLN are in B_5 and distinct such as $NLN(q_1) = NLN_1$, $NLN(q_2) = NLN_2$ (see Fig. 1). If we define x as $((a+1)k-1, (b+1)k-1)$, then $dist(q_1, NLN_1) = dist(q_1, x) + dist(x, NLN_1) < dist(q_1, NLN_2) = dist(q_1, x) + dist(x, NLN_2)$ by the definition of NLN and we can derive $dist(x, NLN_1) < dist(x, NLN_2)$ from this inequality. But this contracts to the assumption that $NLN(q_2) = NLN_2$ because $dist(q_2, NLN_1) = dist(q_2, x) + dist(x, NLN_1) < dist(q_2, x) + dist(x, NLN_2) = dist(q_2, NLN_2)$. By similar approach, there are at most 4 distinct NLN in B_5, B_6, B_7 and B_8, totally $4k-4$ distinct NLN. □

Theorem 9. *There exists a data structure which solves NLN on an $n \times n$ matrix A in encoding model using $O(n^2 \lg \lg n)$ bits and $O(1)$ query time.*

Proof. As mentioned above, divide the A into blocks of size $\lg n \times \lg n$. We create an $n \times n$ bit-matrix B such $B[i,j] = 1$ if $A[i,j]$ is a maximal element in its block, and $B[i,j] = 0$ otherwise. We represent B by writing down its entries such that all the bits corresponding to block are consecutive (say, in row-major order), and store an auxiliary structure to support rank and select queries in the bit vector. This structure allows us to determine whether the query position contains a maximal element in its block or not, and if so, the rank among all the maximal or non-maximal elements within the block, in constant time.

If the query element is not a maximum element in the block, distance between the query and its NLN is less than $2 \lg n$. By encoding the difference of x and y co-odinates and direction, we can encode these elements using $O(n^2 \lg \lg n)$ bits and constant query time.

Suppose the query element is the maximum element in the block. In this case, we assign a color to each maximum element in the block such that all the elements with the same color have the same NLN. By Lemma 3, each block has at most $4 \lg n - 4$ distinct NLN for theses queries so we can store the colors of all the elements in A using at most $n^2 \lg (4 \lg n - 4) = O(n^2 \lg \lg n)$ bits. We also maintain the global table which matches the colors in each block and their corresponding positions in A using $n^2 / \lg^2 n \cdot (4 \lg n - 4) \cdot \lg n = O(n^2)$ bits. The query algorithm simply probes the look-up table and returns the answer, so it takes constant time.   □

4.3   NLN in the Indexing Model

In indexing model, we use 2-dimensional RMQ (range maximum query) structure which supports RMQ query in $n \times n$ matrix using $O(n^2)$ bits and $O(1)$ query time in indexing model [4] (one can also use the trade-off result in [4] to obtain a trade-off result for the NLN problem). We construct this RMQ structure on $2n - 1 \times 2n - 1$ ancillary matrix $A'[1 \ldots 2n - 1][1 \ldots 2n - 1$ which defined as $A'[i, j] = A[(i-j+n+1)/2, (i+j-n+1)/2]$ if both $(i-j+n+1)$ and $(i+j-n+1)$ are even and less than $2n+1$, $-\infty$ otherwise. Then if $A'[i, j]$ is not $-\infty$, for $k \geq 0$, RMQ in $A'[\max(i-k, 1) \ldots \min(i+k, 2n-1)][\max(i-k, 1) \ldots \min(i+k, 2n-1)]$ returns the maximum in all elements whose distance is smaller than k from the correspoding element to $A'[i, j]$ in A. (see Fig. 2). And we do not need to store A' itself for RMQ indexing structure on A' because all elements in A' can be easily derived from A using former relation. The following lemma shows that we can find NLN in A by performing a binary search on A' using RMQ queries.

Lemma 4. *Given $n \times n$ matrix $A[1 \ldots n][1 \ldots n]$. For $1 \leq i, j \leq n$, $0 \leq a \leq b$ and $k > 0$, if NLN of $A[i][j]$ is $A[i'][j']$ and $ak \leq dist(A[i][j], A[i'][j']) \leq bk$, We can find the NLN of $A[i][j]$, $A[i'][j']$ using $\lg ((b - a)k)$ RMQ queries on A', which defined as above in indexing model.*

Proof. We use a similar approach to Lemma 1. First set t as $(ak + bk)/2$. And we call a RMQ query with range $A'[\max(i - t, 1) \ldots \min(i + t, 2n - 1)][\max(i - t, 1) \ldots \min(i + t, 2n - 1)]$. If the value of the result position is less or equal than $A[i][j]$, we set t as $(t+bk)/2$, otherwise set t as $(t+ak)/2$ and call RMQ recursivly until find a smallest range whose RMQ is gearter than $A[i][j]$. Then the RMQ in this range gives the NLN of $A[i][j]$.   □

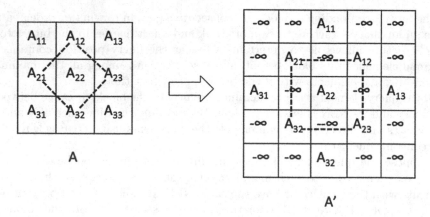

Fig. 2. 3×3 matrix A and its corresponding matrix A'. The dashed line shows that RMQ of the rhombus area in A matches to the RMQ of the rectangle area in A'.

Theorem 10. *There exists a data structure that solves NLN on an $n \times n$ matrix A in the indexing model using $O(n^2)$ bits and $O(\lg \lg \lg n)$ query time.*

Proof. Divide the A into blocks of size $\lg \lg n \times \lg \lg n$. And define the $(n/\lg \lg n) \times (n/\lg \lg n)$ matrix Q such that $Q]i][j]$ stores the maximum element in the block $A[i \lg \lg n \ldots (i+1) \lg \lg n - 1][j \lg \lg n \ldots (j+1) \lg \lg n - 1]$. Then we can define the data structure Q_{NLN} which supports NLN query on Q using $O(n^2/(\lg \lg n)^2 \times \lg \lg (n/\lg \lg n)) = o(n^2)$ bits with constant query time by encoding model in Theorem 9.

For query q, first find the nearest block B whose maximum element is larger than q using Q_{NLN}. If the distance from the block contains q to B is t, the possible minimum and maximum distance between q and NLN of q is $(t-2) \lg \lg n$ and $(t+2) \lg \lg n$ respectively. By Lemma 4, the total space contains RMQ on A' and Q_{NLN}, so they takes $O(n^2) + o(n) = O(n^2)$ bits and total query time is $O(\lg \lg \lg n)$. $\qquad \square$

5   Conclusions

Our main contribution is a systematic study of data structures for NLN and NLRN in one and two dimensional arrays in the encoding and indexing models.

More recently, for the two dimensional NLN problem, we obtained an encoding that supports NLN queries in constant time using $O(n^2)$ bits in the general case. And for $0 < c \leq n$, we also have a data structure for NLN queries on two-dimensional binary array that supports queries in $O(c)$ time using an index of $O(n/c)$ additional bits. Also, for the NLRN problem in 1D, we obtained an index that takes $O((n/c) \lg c)$ bits and answers queries in $O(c)$ time.

We end with specific open problems that can trigger further work.

- Is there a data structure that takes less than $2.54n + o(n)$ bits and can answer NLN queries in a one dimensional array in constant time in the general case (when elements may repeat) in the encoding model?
- In the indexing model, is it possible to achieve $O(c)$ query time for NLRN using $O(n/c)$ bits of space for one dimensional arrays?

References

1. Asano, T., Bereg, S., Kirkpatrick, D.: Finding nearest larger neighbors. In: Albers, S., Alt, H., Näher, S. (eds.) Efficient Algorithms. LNCS, vol. 5760, pp. 249–260. Springer, Heidelberg (2009)
2. Asano, T., Kirkpatrick, D.: Time-space tradeoffs for all-nearest larger neighbors problems. In: Dehne, F., Solis-Oba, R., Sack, J.-R. (eds.) WADS 2013. LNCS, vol. 8037, pp. 61–72. Springer, Heidelberg (2013)
3. Berkman, O., Schieber, B., Vishkin, U.: Optimal doubly logarithmic parallel algorithms based on finding all nearest smaller values. J. Algorithms **14**(3), 344–370 (1993)
4. Brodal, G.S., Davoodi, P., Rao, S.S.: On space efficient two dimensional range minimum data structures. Algorithmica **63**(4), 815–830 (2012)
5. Davoodi, P., Raman, R., Satti, S.R.: Succinct representations of binary trees for range minimum queries. In: Gudmundsson, J., Mestre, J., Viglas, T. (eds.) COCOON 2012. LNCS, vol. 7434, pp. 396–407. Springer, Heidelberg (2012)
6. Elias, P.: Universal codeword sets and representations of the integers. IEEE Trans. Inf. Theory **21**(2), 194–203 (1975)
7. Fischer, J.: Optimal succinctness for range minimum queries. In: López-Ortiz, A. (ed.) LATIN 2010. LNCS, vol. 6034, pp. 158–169. Springer, Heidelberg (2010)
8. Fischer, J.: Combined data structure for previous- and next-smaller-values. Theor. Comput. Sci. **412**(22), 2451–2456 (2011)
9. Fischer, J., Mäkinen, V., Navarro, G.: Faster entropy-bounded compressed suffix trees. Theor. Comput. Sci. **410**(51), 5354–5364 (2009)
10. Golin, M., Iacono, J., Krizanc, D., Raman, R., Rao, S.S.: Encoding 2D range maximum queries. In: Asano, T., Nakano, S., Okamoto, Y., Watanabe, O. (eds.) ISAAC 2011. LNCS, vol. 7074, pp. 180–189. Springer, Heidelberg (2011)
11. Peter, B.: Miltersen Cell probe complexity - a survey. In: FSTTCS (1999)
12. Raman, R., Raman, V., Satti, S.R.: Succinct indexable dictionaries with applications to encoding k-ary trees, prefix sums and multisets. ACM Trans. Algorithms **3**(4), 510–534 (2007). Article 43

Playing Several Variants of Mastermind with Constant-Size Memory is not Harder than with Unbounded Memory

Gerold Jäger[1] and Marcin Peczarski[2]([✉])

[1] Department of Mathematics and Mathematical Statistics,
University of Umeå, 901-87 Umeå, Sweden
gerold.jaeger@math.umu.se
[2] Institute of Informatics, University of Warsaw,
ul. Banacha 2, 02-097 Warszawa, Poland
marpe@mimuw.edu.pl

Abstract. We investigate a version of the Mastermind game, where the codebreaker may only store a constant number of questions and answers, called Constant-Size Memory Mastermind, which was recently introduced by Doerr and Winzen. We concentrate on the most difficult case, where the codebreaker may store only one question and one answer, called Size-One Memory Mastermind. We consider two variants of the game: the original one, where the answer is coded with white and black pegs, and the simplified one, where only black pegs are used in the answer. We show that for two pegs and an arbitrary number of colors, the number of questions needed by the codebreaker in an optimal strategy in the worst case for these games is equal to the corresponding number of questions in the games without a memory restriction. In other words, for these cases restricting the memory size does not make the game harder for the codebreaker. This is a continuation of a result of Doerr and Winzen, who showed that this holds asymptotically for a fixed number of colors and an arbitrary number of pegs. Furthermore, by computer search we determine additional pairs (p, c), where again the numbers of questions in an optimal strategy in the worst case for Size-One Memory Mastermind and original Mastermind are equal.

Keywords: Game theory · Logic game · Mastermind · Space complexity

1 Introduction

Mastermind is a two-player board game invented by Meirowitz in 1970. The players are usually called *codemaker* and *codebreaker*. The codemaker chooses a secret code consisting of 4 pegs and 6 possible colors for each peg. The codebreaker has to give several questions about the code, until he or she has found the correct secret. The codemaker has to evaluate each question by giving an answer consisting of black and white pegs, where each black peg corresponds to a peg of the codebreaker's question which is correct in position and color, and each white peg corresponds to another peg which is correct only in color.

J. Kratochvíl et al. (Eds.): IWOCA 2014, LNCS 8986, pp. 188–199, 2015.
DOI: 10.1007/978-3-319-19315-1_17

The codebreaker's aim is to minimize the number of questions needed to find the secret. There are two ways of optimizing the codebreaker's strategy: minimizing the number of questions on *average* and minimizing the number of questions in the *worst case*. Both optimum strategies have been found for original Mastermind with 4 pegs and 6 colors using $5625/1296 \approx 4.34$ questions [16] and 5 questions [15], respectively.

Original Mastermind can easily be extended to *Generalized Mastermind* with p pegs and c colors. Let $f(p,c)$ be the number of questions needed by the codebreaker in an optimal strategy in the worst case for this game. Chen and Lin. theoretically determined that $f(2,c) = \lfloor c/2 \rfloor + 2$ for $c \geq 2$ [2]. We determined a similar formula for $f(3,c)$ and proved a tight upper and lower bound for $f(4,c)$ and an upper and a lower bound for $f(p,2)$ [12]. Furthermore, we obtained additional values $f(p,c)$ by a computer search. Recently, Doerr et al. have shown that $f(c,c) = \mathcal{O}(c \log \log c)$ [4], improving the classical result $f(c,c) = \mathcal{O}(c \log c)$ of Chvátal [3]. Also many variants of Mastermind have been investigated in the literature. One important variant is black-peg Mastermind, where the codemaker only gives black-peg information and not white-peg information [11,13]. Let $b(p,c)$ be the number of questions needed by the codebreaker in an optimal strategy in the worst case for this variant of the game. We proved, among others, the exact formula $b(p,c) = c + p - 1$ for $p = 2,3$ and $c \geq 2$ [13]. Other variants are static Mastermind [10] and the AB game [1,14].

In this work we investigate another version of the Mastermind game, which has recently been introduced by Doerr and Winzen [6,7]. They add to original Generalized Mastermind a so-called size-m memory restriction for the codebreaker. Then the codebreaker can store only up to m questions and the codemaker's corresponding answers and he or she can decide about the next question based only on this information. However, after receiving the answer, the codebreaker can decide which m of the last $m+1$ questions and corresponding answers he or she can store. For $m = 1$ this means that the codebreaker has a memory for storing only one pair of question and answer. Based only on this information he or she chooses the next question. After receiving the answer, he or she has two pairs of question and answer, and he or she decides which one to keep. The other one is discarded. Our codebreaker's strategy is even more restrictive. The codebreaker asks the question Q_i and remembers it. After receiving the answer A_i he or she remembers it and then decides about the next question Q_{i+1}. He or she asks and remembers Q_{i+1} and immediately forgets the question Q_i and the answer A_i. After receiving the answer A_{i+1} he or she remembers also A_{i+1}.

It should be noted that the memory, as defined in the work of Doerr and Winzen and in this paper, is in fact not constant. Memorizing a question requires $\theta(p \log c)$ bits, and memorizing an answer requires $\theta(\log p)$ bits. Hence, the memory increases linearly in p and logarithmically in c.

Let $f_m(p,c)$ and $b_m(p,c)$ be the number of questions needed by the codebreaker in an optimal strategy in the worst case for Size-m Memory Mastermind and its black-peg variant, respectively. Doerr and Winzen proved that for a fixed number of colors c the codebreaker has a size-one memory strategy winning the

Mastermind game with p pegs using $\mathcal{O}(p/\log p)$ questions in the worst case, in other words, $f_1(p,c) = \mathcal{O}(p/\log p)$. A lower bound of $\Omega(p/\log p)$ was previously known, even without memory restriction, and it follows from an information theoretic argument [3,9]. Doerr and Winzen considered also the black-peg variant of this game and described that for two colors this game is equivalent to the memory restricted black box complexity of ONEMAX_n (for details see [5–7]). Furthermore, they proved $b_1(p,c) = \mathcal{O}(p/\log p)$ for a fixed number of colors. Interestingly, this disproves the conjecture of Droste, Jansen, Wegener that a memory restriction of size one leads to a black box complexity of ONEMAX_n of order $\theta(n \log n)$ [8].

In this work we show an *exact* formula for the number of questions needed by the codebreaker in an optimal strategy in the worst case for Size-m Memory Mastermind with two pegs. In other words, we fix the number of pegs and not the number of colors. Our main result can be summarized in the following theorem.

Theorem 1. *The codebreaker has a size-one memory strategy winning the Mastermind game with two pegs and $c \geq 2$ colors using exactly $\lfloor c/2 \rfloor + 2$ questions in the worst case. Thus, it holds that $f_1(2,c) = f(2,c)$ for all c.*

We obtain also a corresponding result for the black-peg variant, for both, two and three pegs.

Theorem 2. *The codebreaker has a size-one memory strategy winning the black-peg variant of the Mastermind game with two or three pegs and $c \geq 2$ colors using exactly $c + p - 1$ questions in the worst case. Thus, it holds that $b_1(p,c) = b(p,c)$ for $p = 2, 3$ and all c.*

This paper is organized as follows. In Sect. 2 we present the algorithm for Theorem 1, and we prove its correctness in Sect. 3. Then we present the proof of Theorem 2 for $p = 2$ in Sect. 4. We describe additional results with more than two pegs in Sect. 5, and make some conclusions and suggest future work in Sect. 6. Regarding the proof of Theorem 2 for $p = 3$ we refer to the full version of this paper.

Let $x\text{B}$ denote the answer with x black pegs and no white pegs, and let $x\text{W}$ denote the answer with x white pegs and no black pegs. Furthermore, let $\emptyset$ and 0B denote the zero (empty) answer. Observe that for the two pegs game an answer with one black peg and one white peg is impossible.

2   Algorithm

A strategy for two-pegs Mastermind can be found in [12]. It starts with the questions $(0,1)$, $(2,3)$, $(4,5)$, i.e., we use questions of the pattern $(2i, 2i + 1)$. At most two of them can receive a non-empty answer. The codebreaker needs to memorize these two questions and the associated answers. If the codebreaker is allowed to memorize only one question and answer, we use the following idea. We start asking questions as in [12], but after the first non-empty answer we change the pattern of asked questions so that it codes this question and its corresponding answer. Further questions are chosen such that one question codes all information needed to guess the secret. The key idea is to code and recode a history of questions and answers into the current question, leading to a sequence of further questions. The algorithm is divided into phases. Let $k = \lfloor c/2 \rfloor$. Note that in the following all values i and j can always be computed from the previous question and the number of colors c.

Phase 1. We ask successively at most $k - 2$ questions of the form $(2i, 2i + 1)$ for $0 \leq i \leq k - 3$, in increasing order of i. If the answers to all these questions are empty, then we go to Phase 3.

Otherwise, let i be the integer such that the first question with a non-empty answer is $(2i, 2i + 1)$. We consider several cases:

- $(2i, 2i + 1)$ receives the answer 2B: The game ends.
- $(2i, 2i + 1)$ receives the answer 2W: We ask the question $(2i + 1, 2i)$, which has to be answered with 2B, and the game ends.
- $(2i, 2i + 1)$ receives the answer 1W: We go to Phase 2a.
- $(2i, 2i + 1)$ receives the answer 1B: We go to Phase 2b.

Phase 2a. Up to now $i+1$ questions have been asked, and the question $(2i, 2i+1)$ has been answered with 1 W in Phase 1. Now we ask at most $k - i - 1$ questions of the pattern

$$(2i + 2, l), (2i + 3, l + 1), \ldots, (l - 1, 2k - 1),$$

where l will be chosen in the following. Note that the second color in the first question is the next color to the first color in the last question. It must hold $(l - 1) - (2i + 2) = (2k - 1) - l$. Hence, we have $l = k + i + 1$. Thus, we ask the questions

$$(2i + 2, k + i + 1), (2i + 3, k + i + 2), \ldots, (k + i, 2k - 1).$$

Observe that the answers 2B and 2 W are impossible in this phase. Hence, we consider three cases:

- If the first j questions have received the empty answer and the question $(2i + j + 2, k + i + j + 1)$ is answered with 1W, where $j = 0, 1, 2, \ldots, k - i - 2$, we go to Phase 4a.
- If the first j questions have received the empty answer and the question $(2i + j + 2, k + i + j + 1)$ is answered with 1B, where $j = 0, 1, 2, \ldots, k - i - 2$, we go to Phase 4b.
- If all $k - i - 1$ questions receive the empty answer, we go to Phase 5a.

Phase 2b. Up to now $i+1$ questions have been asked, and the question $(2i, 2i+1)$ has been answered with 1B in Phase 1. Now we ask at most $k - i - 1$ questions of the pattern

$$(k + i + 1, 2i + 2), (k + i + 2, 2i + 3), \ldots, (2k - 1, k + i).$$

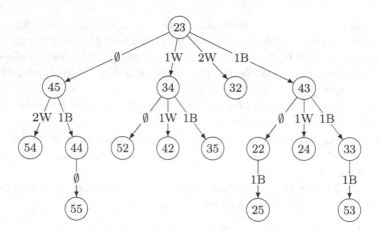

Fig. 1. Phase 3 for $c = 6$ colors

Note that the above questions differ from the questions of Phase 2a only by reversing the order of the colors. As previously, the answers 2B and 2W are impossible in this phase, and we consider three cases:

- If the first j questions have received the empty answer and the question $(k + i + j + 1, 2i + j + 2)$ is answered with 1W, where $j = 0, 1, 2, \ldots, k - i - 2$, we go to Phase 4c.
- If the first j questions have received the empty answer and the question $(k + i + j + 1, 2i + j + 2)$ is answered with 1B, where $j = 0, 1, 2, \ldots, k - i - 2$, we go to Phase 4d.
- If all $k - i - 1$ questions receive the empty answer, we go to Phase 5b.

Phase 3. All $k - 2$ first questions have received the empty answer. If c is even, then there are four possible colors: $c - 4$, $c - 3$, $c - 2$, $c - 1$. They can be denoted also as $2k - 4$, $2k - 3$, $2k - 2$, $2k - 1$. If c is odd, then there are five possible colors: $c - 5$, $c - 4$, $c - 3$, $c - 2$, $c - 1$. They can be denoted also as $2k - 4$, $2k - 3$, $2k - 2$, $2k - 1$, $2k$. In both cases we can find the secret with four questions using only colors from the set of possible colors. Figure 1 shows the strategy for Phase 3 when the game was started with $c = 6$ colors, and 4 colors (namely the colors 2, 3, 4, 5) remain possible for that phase. Figure 2 shows the strategy for Phase 3, when the game was started with $c = 7$ colors, and 5 colors (namely the colors 2, 3, 4, 5, 6) remain possible for that phase. The nodes represent questions and edges represent answers. Questions which can be answered with 2B are drawn in a circle, whereas questions which cannot be answered with 2B are drawn in a square. Note that questions in leafs are always answered with 2B. If c is even, then the strategy from Fig. 1 can be transformed into a general strategy by the color mapping $2 \mapsto 2k - 4$, $3 \mapsto 2k - 3$, $4 \mapsto 2k - 2$, $5 \mapsto 2k - 1$. If c is odd, then the strategy from Fig. 2 can be transformed into a general strategy by the above color mapping with the additional mapping $6 \mapsto 2k$.

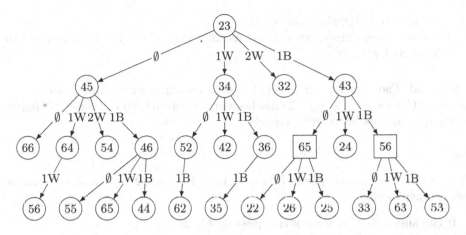

Fig. 2. Phase 3 for $c = 7$ colors

Phase 4a. The question $(2i, 2i + 1)$ has been answered with $1\,\mathrm{W}$ in Phase 1. The question $(2i + j + 2, k + i + j + 1)$ has been answered with $1\,\mathrm{W}$ in Phase 2a. Hence, there are two possible secrets, namely

$$(2i + 1, 2i + j + 2), (k + i + j + 1, 2i).$$

We ask the question $(2i + 1, 2i + j + 2)$, and we consider two cases:

- If the answer is 2B, the game ends.
- If the answer is empty, we ask the question $(k + i + j + 1, 2i)$, which has to be answered with 2B.

Note that all four colors used in this phase (and all Phases 4) are distinct, because $2i \neq k + i + j + 1$, $2i + 1 \neq k + i + j + 1$ and $2i + j + 2 \neq k + i + j + 1$.

Phase 4b. The question $(2i, 2i + 1)$ has been answered with $1\,\mathrm{W}$ in Phase 1. The question $(2i + j + 2, k + i + j + 1)$ has been answered with 1B in Phase 2a. Hence, there are two possible secrets, namely

$$(2i + 1, k + i + j + 1), (2i + j + 2, 2i).$$

We ask the same question as in Phase 4a, i.e., $(2i + 1, 2i + j + 2)$, and we consider two further cases:

- If the answer is 1W, we ask the question $(2i + j + 2, 2i)$.
- If the answer is 1B, we ask the question $(2i + 1, k + i + j + 1)$.

In both cases the latter question has to be answered with 2B.

Phase 4c. The question $(2i, 2i + 1)$ has been answered with 1B in Phase 1. The question $(k + i + j + 1, 2i + j + 2)$ has been answered with $1\,\mathrm{W}$ in Phase 2b. Hence, there are two possible secrets, namely

$$(2i, k + i + j + 1), (2i + j + 2, 2i + 1).$$

We ask the question $(2i + j + 2, 2i + 1)$, and we consider two cases:

- If the answer is 2B, the game ends.
- If the answer is empty, we ask the question $(2i, k + i + j + 1)$, which has to be answered with 2B.

Phase 4d. The question $(2i, 2i + 1)$ has been answered with 1B in Phase 1. The question $(k + i + j + 1, 2i + j + 2)$ has been answered with 1B in Phase 2b. Hence, there are two possible secrets, namely

$$(2i, 2i + j + 2), (k + i + j + 1, 2i + 1).$$

We ask the same question as in Phase 4c, i.e., $(2i + j + 2, 2i + 1)$, and we consider two further cases:

- If the answer is 1W, we ask the question $(2i, 2i + j + 2)$.
- If the answer is 1B, we ask the question $(k + i + j + 1, 2i + 1)$.

In both cases the latter question has to be answered with 2B.

Phase 5a. The question $(2i, 2i + 1)$ has been answered with 1 W in Phase 1. All other questions have received the empty answer. It is impossible to reach this phase if c is even, because all colors except $2i$ and $2i + 1$ are impossible.

If c is odd, then there are two possible secrets, namely

$$(c - 1, 2i), (2i + 1, c - 1).$$

We ask the question $(c - 1, 2i)$, and we consider two cases:

- If the answer is 2B, the game ends.
- If the answer is 1W, we ask the question $(2i + 1, c - 1)$, which has to be answered with 2B.

Phase 5b. The question $(2i, 2i + 1)$ has been answered with 1B in Phase 1. All other questions have received the empty answer. If c is even, then there are two possible secrets, namely

$$(2i, 2i), (2i + 1, 2i + 1).$$

We ask the question $(2i, 2i)$, and we consider two cases:

- If the answer is 2B, the game ends.
- If the answer is empty, we ask the question $(2i + 1, 2i + 1)$, which has to be answered with 2B.

If c is odd, then there are four possible secrets, namely

$$(2i, 2i), (2i + 1, 2i + 1), (2i, c - 1), (c - 1, 2i + 1).$$

We ask the question $(2i, c - 1)$, and we consider four cases:

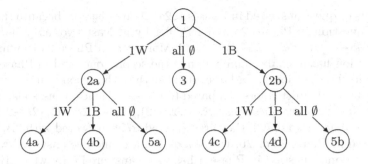

Fig. 3. Phase transition diagram

- If the answer is 1B, we ask the question $(2i, 2i)$, which has to be answered with 2B.
- If the answer is empty, we ask the question $(2i + 1, 2i + 1)$, which has to be answered with 2B.
- If the answer is 2B, the secret is $(2i, c - 1)$, and the game ends.
- If the answer is 1W, we ask the question $(c - 1, 2i + 1)$, which has to be answered with 2B.

Small Number of Colors. If the number of colors is 4 or 5, then Phase 1 reduces to zero questions and thus there is also no Phase 2, 4 and 5. The whole algorithm reduces to Phase 3 only, using at most 4 questions. If the number of colors is 2 or 3, then we use the part of Phase 3, which does not ask the first question, and assume that it has received the empty answer. In those cases we find a secret in at most 3 questions. It is easy to see that $f_1(2, 1) = f(2, 1) = 1$.

3   Proof of Correctness

First, we count the number of questions needed to find a secret. Phase 1 uses $i + 1 \leq k - 2$ questions. Phase 2 uses at most $k - i - 1$ questions. Phase 3 uses at most four questions. Phases 4 and 5 use at most two questions. Figure 3 shows all possible phase transitions. This leads to the following three possible scenarios:

- Phase 1, Phase 2, Phase 4;
- Phase 1, Phase 2, Phase 5;
- Phase 1, Phase 3.

It is easy to see that we find a secret in at most $k + 2$ questions.

Next, we need to check if the above algorithm states a strategy for the codebreaker with size-one memory. It is sufficient to prove that the asked question depends only on the previous question and the received answer. This will be implied by the property that either a question is used only in one game state or if a question is used in two game states, then the sets of answers in these states are disjoint.

The sets of questions asked in Phases 1, 2a, 2b are disjoint, because the colors used in a question in Phases 2a and 2b differ by at least two, as $(k + i + 1) - (2i + 2) = k - i - 1 \geq k - (k - 3) - 1 = 2$. Moreover, in Phase 2a the first color in the question has a smaller number than the second one, and in Phase 2b the first color in the question has a larger number than the second one.

Phase 3 uses distinct colors compared to Phase 1. Hence, it uses also distinct questions. Note that the questions $(2k-4, 2k-2)$, $(2k-3, 2k-1)$, $(2k-2, 2k-4)$, $(2k-1, 2k-3)$ are used only if they are expected to be answered with 2B, as they are also used in Phase 2a or 2b, where a question cannot be answered with 2B.

As the second question in Phase 4 has to be answered only with 2B, it can be used in all other phases. The first question of Phase 4 is distinct from all questions of Phase 1, because:

– the first color has a smaller number than the second one and the first one is odd, e.g., the question $(2i + 1, 2i + j + 2)$, or
– the first color has a larger number than the second one and the second one is odd, e.g., the question $(2i + j + 2, 2i + 1)$.

Phase 4 uses distinct questions compared to Phase 3, because one color in each question of Phase 4 has a number smaller than $2k - 4$ (this is the color which was used in Phase 1).

In the following we will prove that the first question used in Phase 4 is not used in Phase 2. The first question used in Phase 4a or 4b cannot be used in Phase 2b, because it has an opposite color number order. Now assume that the first question used in Phase 4a or 4b is also used in Phase 2a. Let $(2i' + j' + 2, k + i' + j' + 1)$ be a question asked in Phase 2a. Let $(2i + 1, 2i + j + 2)$ be the first question asked in Phase 4a or 4b. This implies that

$$2i' + j' + 2 = 2i + 1, \tag{1}$$
$$k + i' + j' + 1 = 2i + j + 2. \tag{2}$$

Hence, we have $2(k + i' + j' + 1) - (2i' + j' + 2) = 2(2i + j + 2) - (2i + 1)$ and $2k + j' = 2i + 2j + 3$. However, $i + j \leq k - 2$ and thus $j' \leq -1$, which is a contradiction.

A similar result can be proved for Phases 4c, 4d and Phases 2a, 2b as follows. The first question used in Phase 4c or 4d cannot be used in Phase 2a, because it has an opposite color number order. Now assume that the first question used in Phase 4c or 4d is also used in Phase 2b. Let $(k + i' + j' + 1, 2i' + j' + 2)$ be a question asked in Phase 2b. Let $(2i + j + 2, 2i + 1)$ be the first question asked in Phase 4c or 4d. This implies exactly the same Eqs. (1) and (2) as previously, and we have a contradiction again.

The questions used in Phase 5a or 5b are not used in Phase 3, because $2i < 2k - 4$ and $2i + 1 < 2k - 4$ holds. They are also not used in any other phase, because only in Phases 3 and 5, questions can contain the color $2k$ (i.e., color $c - 1$ for odd c) or twice the same color.

This finishes the proof of correctness and of Theorem 1.

4 Algorithm for Black-Peg Variant

Let us consider a Mastermind strategy, where we choose only questions for which the answer pB (p black pegs) is possible. In other words, we choose questions only from the set of secrets which are still possible based on all answers given yet by the codemaker. We call such a strategy a P-strategy.

Now we prove that every P-strategy is a size-one memory strategy. Every strategy-tree for Mastermind has to contain as nodes all possible questions, because for every possible secret there must be a question equal to this secret and answered with p black pegs. Every question appears at most once in the P-strategy-tree, because distinct paths in the strategy-tree lead to disjoint sets of possible secrets. Hence, every P-strategy-tree contains exactly c^p nodes, where each node is labeled with another question. Therefore, each question appears exactly once. Moreover, a P-strategy-tree has exactly $c^p - 1$ edges. In general, to store the game state, the codebreaker has to remember the whole path from the root to the present node, but in the P-strategy the label of the node (a question) uniquely determines the game state. Hence, the codebreaker has to memorize only one question to store a position in the strategy-tree, and the P-strategy-tree immediately defines a strategy for Size-One Memory Mastermind.

A strategy for the black-peg variant of two-pegs Mastermind can be found in [13]. We adapt this strategy to a size-one memory P-strategy. Note that in this game we have only three possible answers, namely 0B, 1B and 2B.

Consider the following strategy for $c \geq 2$. We ask successively at most $c - 1$ questions of the form (i, i) for $0 \leq i \leq c-2$, in increasing order of i. If the answers of all these questions are empty, then the only possible secret is $(c - 1, c - 1)$. We ask the question $(c - 1, c - 1)$, which has to be answered with 2B. Thus, we are done in c questions. Otherwise, let i be the integer such that the first question with a non-empty answer is (i, i). We consider three cases:

1. (i, i) receives the answer 2B: We are done in $i + 1 \leq c - 1$ questions.
2. (i, i) receives the answer 1B: We ask the question $(i, i + 1)$.
3. (i, i) receives the answer 0B: By construction, this case is not possible.

Regarding the question $(i, i + 1)$ (Case 2) we have three next cases:

4. $(i, i + 1)$ receives the answer 2B: We are done in $i + 2 \leq c$ questions.
5. $(i, i + 1)$ receives the answer 1B: Then $i < c - 2$ and we have $c - i - 2$ possible secrets: $(i, i + 2), (i, i + 3), \ldots, (i, c - 1)$, which we can test in $c - i - 2$ further questions. Then totally we have at most $i + 2 + c - i - 2 = c$ questions.
6. $(i, i + 1)$ receives the answer 0B: Then we ask the question $(i + 1, i)$.

Regarding the question $(i + 1, i)$ (Case 6) we have three further cases:

7. $(i + 1, i)$ receives the answer 2B: We are done in $i + 3 \leq c + 1$ questions.
8. $(i + 1, i)$ receives the answer 1B: Then $i < c - 2$ and we have $c - i - 2$ possible secrets: $(i + 2, i), (i + 3, i), \ldots, (c - 1, i)$, which we can test in $c - i - 2$ further questions. Totally, we have at most $i + 3 + c - i - 2 = c + 1$ questions.
9. $(i + 1, i)$ receives the answer 0B: This case is not possible.

We conclude that $b_1(2, c) \leq c+1$. By [13], it holds that $b_1(2, c) \geq b(2, c) = c+1$ for $c \geq 2$. It is easy to see that $b_1(2, 1) = b(2, 1) = 1$. This finishes the proof of Theorem 2 for $p = 2$.

For a better understanding of the strategy, below we present a function for the codebreaker. The function takes a question and an answer (0B or 1B) and returns the next question. We omit the answer 2B, because after that the game ends.

$$(i, i), 0B \mapsto (i + 1, i + 1) \text{ for } i = 0, 1, \ldots, c - 2,$$
$$(i, i), 1B \mapsto (i, i + 1) \qquad \text{for } i = 0, 1, \ldots, c - 2,$$
$$(i, i + 1), 0B \mapsto (i + 1, i) \qquad \text{for } i = 0, 1, \ldots, c - 2,$$
$$(i, i + j + 1), 1B \mapsto (i, i + j + 2) \text{ for } i = 0, 1, \ldots, c - 3, j = 0, 1, \ldots, c - i - 3,$$
$$(i + j + 1, i), 1B \mapsto (i + j + 2, i) \text{ for } i = 0, 1, \ldots, c - 3, j = 0, 1, \ldots, c - i - 3.$$

Note that the questions $(i, i + j + 1)$ for $j > 0$ and $(i + j + 1, i)$ for $j \geq 0$ cannot be answered with 0B.

5   More Pegs

Using computer search, we can show that there are additional pairs (p, c), where the numbers of questions needed by the codebreaker in an optimal strategy in the worst case for Size-One Memory Mastermind and original Mastermind are equal, i.e., it holds that $f(p, c) = f_1(p, c)$ or $b(p, c) = b_1(p, c)$. For obtaining these results we have modified the program from [12] for computing the number of questions $f(p, c)$ needed by the codebreaker in an optimal strategy in the worst case for original Mastermind. As the program was not able to compute $f_1(p, c)$ or $b_1(p, c)$ directly, we have restricted the computations to find only P-strategies. Let $f_0(p, c)$ and $b_0(p, c)$ be the worst case number of questions in a P-strategy for the original and black-peg variant, respectively. Clearly, it holds that $f_0(p, c) \geq f_1(p, c) \geq f(p, c)$ and $b_0(p, c) \geq b_1(p, c) \geq b(p, c)$. Hence, if we obtain that $f_0(p, c) = f(p, c)$ or $b_0(p, c) = f(p, c)$, we can conclude the value of $f_1(p, c)$ or $b_1(p, c)$. The computed values are listed in the following, where the values of $f(p, c)$ and $b(p, c)$ are taken from [12,13].

$$
\begin{array}{ll}
f_0(2, 2) = f(2, 2) = 3 & f_0(4, 2) = f(4, 2) = 4 \\
f_0(2, 3) = f(2, 3) = 3 & f_0(4, 3) = f(4, 3) = 4 \\
f_0(2, 4) = f(2, 4) = 4 & f_0(4, 4) = 5 > f(4, 4) = 4 \\
f_0(2, 5) = 5 > f(2, 5) = 4 & f_0(4, 5) = f(4, 5) = 5 \\
f_0(2, 6) = 6 > f(2, 6) = 5 & f_0(4, 6) = 6 > f(4, 6) = 5 \\
f_0(3, 2) = f(3, 2) = 3 & f_0(5, 2) = f(5, 2) = 4 \\
f_0(3, 3) = f(3, 3) = 4 & f_0(5, 3) = f(5, 3) = 4 \\
f_0(3, 4) = f(3, 4) = 4 & f_0(5, 4) = f(5, 4) = 5 \\
f_0(3, 5) = f(3, 5) = 5 & f_0(6, 2) = f(6, 2) = 5 \\
f_0(3, 6) = 6 > f(3, 6) = 5 & f_0(6, 3) = f(6, 3) = 5 \\
\end{array}
$$

$$
\begin{array}{ll}
b_0(4, 2) = 5 = b(4, 2) = 5 & b_0(5, 2) = 6 > b(5, 2) = 5 \\
b_0(4, 3) = 9 > b(4, 3) = 5 & b_0(5, 3) = 9 > b(5, 3) = 6 \\
b_0(4, 4) = 10 > b(4, 4) = 6 & b_0(6, 2) = 7 > b(6, 2) = 6 \\
\end{array}
$$

6   Conclusions and Future Work

In this work we continued the work of Doerr and Winzen [6,7], who showed that in original Mastermind for a fixed number of colors, bounding memory for the codebreaker does not increase the asymptotic number of questions in the worst case. We show that similar holds for two pegs. However, our result is not only an asymptotic bound, but an exact one.

The computational results from Sect. 5 show that the approach used for the black-peg variant of Size-One Memory Mastermind with two or three pegs cannot be extended to a larger number of pegs, as for $p = 4, 5, 6$ there exists a c such that $b_0(p, c) > b(p, c)$, i.e., there is no appropriate P-strategy.

For future research we suggest to search for a pair (p, c), where the size-one memory version needs more questions than the version with unbounded memory.

References

1. Chen, S.T., Lin, S.S.: Optimal algorithms for $2 \times n$ AB games–a graph-partition approach. J. Inform. Sci. Eng. **20**(1), 105–126 (2004)
2. Chen, S.T., Lin, S.S.: Optimal algorithms for $2 \times n$ Mastermind games–a graph-partition approach. Comput. J. **47**(5), 602–611 (2004)
3. Chvátal, V.: Mastermind. Combinatorica **3**(3–4), 325–329 (1983)
4. Doerr, B., Spöhel, R., Thomas, H., Winzen, C.: Playing mastermind with many colors. In: Proceedings of the 24th Annual ACM-SIAM Symposium on Discrete Algorithms (SODA 2013), SIAM, pp. 695–704 (2013)
5. Doerr, B., Winzen, C.: Memory-restricted black-box complexity of OneMax. Inform. Process. Lett. **112**(1–2), 32–34 (2012)
6. Doerr, B., Winzen, C.: Playing Mastermind with constant-size memory. In: Proceedings of 29th International Symposium on Theoretical Aspects of Computer Science (STACS 2012). Leibniz International Proceedings in Informatics (LIPIcs), vol. 14, pp. 441–452. Schloss Dagstuhl - Leibniz-Zentrum für Informatik (2012)
7. Doerr, B., Winzen, C.: Playing Mastermind with constant-size memory. Theory Comput. Syst. **55**(4), 658–684 (2014)
8. Droste, S., Jansen, T., Wegener, I.: Upper and lower bounds for randomized search heuristics in black-box optimization. Theory Comput. Syst. **39**(4), 525–544 (2006)
9. Erdős, P., Rényi, A.: On two problems of information theory. Magyar Tud. Akad. Mat. Kutató Int. Közl. 8, 229–243 (1963)
10. Goddard, W.: Static Mastermind. J. Combin. Math. Combin. Comput. **47**, 225–236 (2003)
11. Goodrich, M.T.: On the algorithmic complexity of the Mastermind game with black-peg results. Inform. Process. Lett. **109**(13), 675–678 (2009)
12. Jäger, G., Peczarski, M.: The number of pessimistic guesses in Generalized Mastermind. Inform. Process. Lett. **109**(12), 635–641 (2009)
13. Jäger, G., Peczarski, M.: The number of pessimistic guesses in Generalized Black-peg Mastermind. Inform. Process. Lett. **111**(19), 933–940 (2011)
14. Jäger, G., Peczarski, M.: The worst case number of questions in Generalized AB game with and without white-peg answers. Discrete Appl. Math. **184**, 20–31 (2015)
15. Knuth, D.E.: The computer as Mastermind. J. Recr. Math. **9**(1), 1–6 (1976–1977)
16. Koyama, K., Lai, T.W.: An optimal Mastermind strategy. J. Recr. Math. **25**(4), 251–256 (1993)

On Maximum Common Subgraph Problems in Series-Parallel Graphs

Nils Kriege, Florian Kurpicz$^{(\boxtimes)}$, and Petra Mutzel

Department of Computer Science, Technische Universität Dortmund,
Dortmund, Germany
{nils.kriege,florian.kurpicz,petra.mutzel}@tu-dortmund.de

Abstract. The complexity of the maximum common connected subgraph problem in partial k-trees is still not fully understood. Polynomial-time solutions are known for degree-bounded outerplanar graphs, a subclass of the partial 2-trees. On the contrary, the problem is known to be **NP**-hard in vertex-labeled partial 11-trees of bounded degree. We consider series-parallel graphs, i.e., partial 2-trees. We show that the problem remains **NP**-hard in biconnected series-parallel graphs with all but one vertex of degree bounded by three. A positive complexity result is presented for a related problem of high practical relevance which asks for a maximum common connected subgraph that preserves blocks and bridges of the input graphs. We present a polynomial time algorithm for this problem in series-parallel graphs, which utilizes a combination of BC- and SP-tree data structures to decompose both graphs.

Keywords: Maximum Common Subgraph · Block and Bridge Preserving · Series-parallel graphs

1 Introduction

Finding a maximum common subgraph (MCS) of two input graphs is an important task in many application domains like pattern recognition and cheminformatics [18]. MCS is well known to be **NP**-hard. Since practically relevant graphs, e.g., derived from small molecules, often have small treewidth [9], it is highly relevant to develop polynomial time algorithms for tractable graph classes and to clearly identify graph classes, where MCS remains **NP**-hard. For the related subgraph isomorphism problem such a clear demarcation for partial k-trees is known. Subgraph isomorphism is solvable in polynomial time in partial k-trees if the smaller graph either is k-connected or has bounded degree [7,13]. However, it is **NP**-complete when the smaller graph is not k-connected or has more than k vertices of unbounded degree [8]. MCS apparently is at least as hard as subgraph isomorphism; two recent results show that it actually is considerably harder: Akutsu [2] has shown that finding a connected MCS is **NP**-hard in

---

This work was supported by the German Research Foundation (DFG), priority programme "Algorithms for Big Data" (SPP 1736).

© Springer International Publishing Switzerland 2015
J. Kratochvíl et al. (Eds.): IWOCA 2014, LNCS 8986, pp. 200–212, 2015.
DOI: 10.1007/978-3-319-19315-1_18

vertex-labeled partial 11-trees of bounded degree. Furthermore it was believed that the problem of finding a maximum common k-connected subgraph of k-connected partial k-trees (k-MCS) can be solved with the same technique that was successfully used for subgraph isomorphism. Recently, it was shown that these techniques are insufficient even for series-parallel graphs, for which a new approach based on SP-trees was devised [10]. Further polynomial time algorithms were proposed for connected MCS of almost trees and outerplanar graphs of bounded degree [1,3].

Motivated by the fact that even subgraph isomorphism is **NP**-hard when the smaller graph is a tree and the other is outerplanar, a problem variation referred to as BBP-MCS was considered [17,18]. Here, the common subgraph is required to preserve blocks, i.e., maximal biconnected subgraphs, and bridges of the input graphs, which renders efficient algorithms for outerplanar graphs possible [17]. Notably, BBP-MCS yields meaningful results for molecular graphs in practice and even compares favorably to ordinary MCS in empirical studies [16,18].

Our Contribution. On the theoretical side, we prove that finding a connected MCS of two biconnected series-parallel graphs with all but one vertex of degree bounded by three is **NP**-hard. We obtain this result by a polynomial-time reduction of the *Numerical Matching with Target Sums* problem. Furthermore, we consider BBP-MCS in series-parallel graphs and propose a polynomial time solution, thus, generalizing the known result for outerplanar graphs. Employing BC- and SP-tree decompositions of the input graphs allows us to identify subproblems closely related to k-MCS, $k \in \{1, 2\}$. We make use of a classical approach for the maximum common subtree problem [14], i.e., 1-MCS, and a recently proposed algorithm for 2-MCS [10] to obtain our main result. Our approach yields a running time of $\mathcal{O}(n^6)$ in series-parallel and $\mathcal{O}(n^6)$ in outerplanar graphs, where n is the maximum number of vertices in one of the input graphs.

2 Preliminaries

Let G be a simple graph. We denote the set of *vertices* by $V(G)$ and the set of *edges* by $E(G)$. A graph is *connected* if there is a path between any two vertices. Each maximal connected subgraph $G' \subseteq G$ is called *connected component*. Let $V \subseteq V(G)$, then $G[V]$ denotes the *induced subgraph* $G' \subseteq G$ with $V(G') = V$ and $E(G') = \{(u, v) \in V \times V : (u, v) \in E(G)\}$. A set $S \subseteq V(G)$ is called $|S|$-*separator* or *separator* of a connected graph G if $G \setminus S := G[V(G) \setminus S]$ consists of at least two connected components. If $S = \{v\}$ is a separator then v is called *cutvertex*. A graph G with $|V(G)| > k$ is called k-*connected* if there is no j-separator of G with $j < k$ and *biconnected* if $k = 2$. We define $[n] := \{1, \ldots, n\}$ for all $n \in \mathbb{N}$. A *path* is a sequence of vertices $(v_0, v_1, \ldots, v_n)$ such that $(v_{i-1}, v_i) \in E(G)$ for all $i \in [n]$. A path with $v_n = v_0$ is called *cycle*. The *length* of a path or cycle is the number of edges contained in it. Let $(v_0, v_1, \ldots, v_n)$ be a cycle, an edge (v_i, v_j) such that $1 \neq |i - j| < n$ is called *chord*. Cycles without chords are *chordless*.

A graph G is *bipartite* if there are two disjoint sets $U, U' \subseteq V(G)$ such that $U \cup U' = V(G)$ and for all $(u, v) \in E(G)$ neither $u, v \in U$ nor $u, v \in U'$.

A *matching* of G is a set of edges $M \subseteq E(G)$ such that $u = u' \iff v = v'$ for all $((u, v), (u', v')) \in M \times M$. The *maximum weighted bipartite matching* problem (MwbM) asks for the maximum weight of a matching of a weighted bipartite graph and is solvable in $\mathcal{O}(n^3)$, e.g., with the Hungarian method [11].

K_n denotes the complete graph with n vertices and $K_2^{s,t}$ denotes an instance of the K_2 where one vertex is called s- and the other t-vertex. A graph is *series-parallel* if each maximal biconnected subgraph can be constructed starting with a finite set of $K_2^{s,t}$ by performing a sequence of the following two operations.

S-Operation: Merge the s-vertex of one component with the t-vertex of a different component. The vertex created by merging remains unnamed.

P-Operation: Merge the s- and t-vertices of two different components of the set. The resulting vertices are called s- and t-vertex.

By definition, series-parallel graphs are at most biconnected and equivalent to partial 2-trees [4], i.e., graphs with treewidth at most 2. We use the notation and definition introduced in [5] to define the SP-tree decomposition of series-parallel graphs.

Definition 1 (SP-tree). *Let G be a biconnected series-parallel graph with at least three vertices. Then the SP-tree of G, denoted by $\mathrm{SP}(G) = T^{\mathrm{SP}}$, is the smallest tree such that the following conditions are satisfied:*

SP1 *each node[1] λ of T^{SP} is associated with a skeleton graph $S_\lambda = (V_\lambda, E_\lambda)$. Each edge $e = (u, v) \in E_\lambda$ is either a real or a virtual edge. If e is a virtual edge, then $S = \{u, v\}$ is a separator of G.*

SP2 *T^{SP} has two different types of nodes. S-nodes where the skeleton graph is a chordless cycle and P-nodes which have a skeleton graph consisting of multiple parallel edges between exactly two vertices.*

SP3 *for two adjacent nodes λ and η in T^{SP}, the skeleton graph S_λ contains a virtual edge e_η representing S_η and vice versa. The node η is called* pertinent *to the edge e_η.*

SP4 *The graph resulting by merging all skeleton graphs in a way that each virtual edge is replaced by the skeleton of its pertinent node in T^{SP} is exactly G.*

The sets of S-nodes and P-nodes in T^{SP} are denoted by $V_S(T^{\mathrm{SP}})$ or $V_P(T^{\mathrm{SP}})$ and T^{SP} is bipartite regarding these two sets. Let $r \in E(G)$, the *rooted SP-tree* is obtained by rooting T^{SP} at the node λ with $r \in V(S_\lambda)$. A rooted SP-tree induces a parent-child relation where a node λ is the parent of an adjacent node η if the path from the root node to λ is shorter than the path from the root node to η. If a node λ is the parent of a node η and $e_\lambda \in E(S_\eta)$ is the virtual edge pertinent to λ in η, then e_λ is called *reference edge* of λ and denoted by $\mathrm{ref}(\lambda)$.

Let G be a graph. Each maximal connected subgraph without a cutvertex with respect to that component is called a *block*. There are two different types of blocks: A maximal biconnected subgraph and a *bridge*, i.e., a K_2. Any two blocks

---

[1] We call vertices of SP- and BC-trees nodes and vertices of the input graphs vertices.

of G may have at most one vertex in common, which must be a cutvertex. Blocks that are not bridges are called *non-bridge* block. Let B denote the set of blocks of G and C the set of cutvertices of G. The graph with vertices $B \cup C$ and edges between each $b \in B$ and $c \in C$ iff $V(c) \in V(b)$ is called *block graph* of G and denoted by $BC(G)$. If G is connected, the block graph is a tree and referred to as *BC-tree*. Each node Λ in a BC-tree has a skeleton graph S_Λ consisting of the vertices and edges represented by the node. Let $T^{BC} = BC(G)$ and $r \in E(G)$, the *rooted BC-tree* $T^{BC,r}$ is obtained by rooting T^{BC} at the B-node Λ such that $r \in V(S_\Lambda)$. It induces a parent-child relation as defined above in $T^{BC,r}$ and also a parent-child relation between the nodes of the SP-trees of the skeleton graphs. Since those only exists for non-bridge nodes, denoted by $V_{Bl}(T^{BC,r})$, there are two cases: Let T^{SP}_Λ be the SP-tree of the skeleton graph of $\Lambda \in V_B(T^{BC})$. First, Λ is the root of T^{BC}, then T^{SP}_Λ is rooted at r. Otherwise, let Ξ be the parent of Λ hence Ξ is a cutvertex with $v = V(S_\Xi)$. Then T^{SP}_Λ is rooted at the P- or S-node such that v is in the skeleton graph of this node (P-node if existing). $V_{Br}(T^{BC,r})$ denotes the bridges of the BC-tree. Greek upper- and lowercase letters denote B-, C- and S-, P-nodes, resp. Latin letters denote vertices of the input graphs.

Let G and H be graphs. If $V(H) \subseteq V(G)$ and $E(H) \subseteq E(G)$, then H is called *subgraph* of G. The graphs G and H are *isomorphic*, if there is a bijection $\phi: V(G) \rightarrow V(H)$, such that $(u,v) \in E(G) \iff (\phi(u), \phi(v)) \in E(H) \; \forall u, v \in V(G)$ and H is *subgraph isomorphic* to G, if H is isomorphic to a subgraph of G. We say u is *mapped* to v' if $\phi(u) - v'$. There is a *common subgraph isomorphism* between G and H, if there are sets $R \subseteq V(G)$ and $S \subseteq V(H)$ such that the induced subgraphs $G[R]$ and $H[S]$ are isomorphic. Let ϕ be the common subgraph isomorphism, then ϕ is a *maximum common subgraph isomorphism* if there is no common subgraph isomorphism ϕ' with $|\text{dom}(\phi')| > |\text{dom}(\phi)|$, where $\text{dom}(\phi)$ denotes the *domain* of ϕ. A common subgraph is called *maximum common subgraph* (MCS) if there is no common subgraph containing more vertices.

Definition 2 (Maximum Common Subgraph Problem (MCS)). *Given two graphs G and H, find the order of a maximum common connected subgraph.*

Please notice, that MCS can denote both: the problem and a subgraph. In the following we assume that the input graphs are connected series-parallel graphs and common subgraphs must be induced subgraphs of both input graphs.

3 MCS in Series-Parallel Graphs with Bounded Degree

In this section, we consider MCS where both input graphs are biconnected and have degree at most 3 for all but 1 vertex ($MCS^{\leq 3,1}$). We prove that this problem is **NP**-hard and improve the result for subgraph isomorphism that, transferred to MCS, states that $MCS^{\leq 4,2}$ is **NP**-hard [8].

Since the running time of an algorithm is given with respect to the size of the input, a reasonable encoding is demanded, e.g., an integer n can be encoded in $\log n$ bits. An **NP**-complete problem may no longer be **NP**-complete if the instances are encoded unary. *Strongly* **NP**-complete problems are **NP**-complete

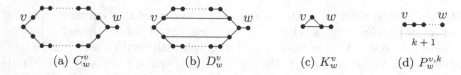

(a) C_w^v (b) D_w^v (c) K_w^v (d) $P_w^{v,k}$

Fig. 1. Gadgets used to create the graphs G and H for a NMwTS instance.

even if the input is encoded unary [6]. Hence even the values of numbers can be used. To prove that MCS$^{\leq 3,1}$ is **NP**-hard we show that there is a polynomial-time reduction from the following problem which is strongly **NP**-complete [6].

Definition 3 (Numerical Matching with Target Sums (NMwTS)).
Given two disjoint sets X and Y with $|X| = |Y| = n$, a size function $s\colon X \cup Y \to \mathbb{Z}^+$ and a vector $\boldsymbol{b} = \langle b_1, b_2, \ldots, b_n \rangle$ with $b_i \in \mathbb{Z}^+$ for all $i \in [n]$. Can $X \cup Y$ be partitioned into disjoint sets $A_1, A_2, \ldots, A_n$ each containing one element from each of X and Y, such that $\sum_{a \in A_i} s(a) = b_i$ for all $i \in [n]$?

3.1 Construction of the Polynomial-Time Reduction

For an instance $(X, Y, s, \boldsymbol{b})$ of NMwTS we construct two graphs, G and H to represent the values of the elements in X, Y and $\boldsymbol{b}$. Let $\Sigma_s := \sum_{z \in X \cup Y} s(z)$ and $\Sigma_b := \sum_{i=1}^n b_i$. B_w^v denotes a cycle with $2\Sigma_s + 2$ vertices such that each path from v to w has length Σ_s. C_w^v is an instance of $B_{v'}^v$ with an additional vertex w called *anchor vertex* and an edge (v', w). D_w^v is an extension of C_w^v with two chords such that it is still outerplanar and there are two edge disjoint paths of length 4 from v to w.[2] K_w^v is an instance of K_3, where two vertices are denoted by v and v', with additional vertex w and an edge (v', w). Last, $P_w^{v,k}$ is a path of length k, where the vertices of degree 1 are denoted by v and w, see Fig. 1.

$$B_k = \bigcup_{i=1}^n \left(C_{c_i}^{\bar{x}} \cup D_{c_{i+n}}^{\bar{x}} \right) \cup \bigcup_{i=1+k}^{n+k-1} \left(P_{c_{i+1}}^{c_i,2} \right) \tag{1}$$

$$G = \bigcup_{i=1}^n \left(K_{\bar{x}_{1,i}}^{c_i} \cup K_{\bar{y}_{1,i}}^{c_{i+n}} \cup \bigcup_{j=2}^{s(x_i)} K_{\bar{x}_{j,i}}^{\bar{x}_{j-1,i}} \cup \bigcup_{j=2}^{s(y_i)} K_{\bar{y}_{j,i}}^{\bar{y}_{j-1,i}} \cup P_{\bar{y}_{s(y_i)}}^{\bar{x}_{s(x_i)},3} \right) \cup B_0 \tag{2}$$

$$H = \bigcup_{i=1}^n \left(K_{v_{1,i}}^{c_i} \cup K_{c_{i+n}}^{v_{b_i,i}} \cup \bigcup_{j=2}^{b_i} K_{v_j}^{v_{j-1}} \right) \cup B_n \tag{3}$$

The graphs G and H contain a subgraph which we call *base-gadget*, see Eq. 1. It consists of $2n$ cycles, $C_{c_i}^{\bar{x}}$ and $D_{c_{i+n}}^{\bar{x}}$ for $i \in [n]$. All cycles share the vertex $\bar{x}$, which is the only vertex with unbounded degree. The subgraphs $\bigcup_{i=1+k}^{n+k-1} P_{c_{i+1}}^{c_i,2}$,

---

[2] If an instance of NMwTS does not allow the construction of D_w^v, all values are multiplied by 3.

$k \in \{0, n\}$ are called *anchor paths* and are required to assure that G and H are biconnected. The index k in Eq. 1 is used to connect either the anchor vertices of cycles containing chords ($k = 0$) or of the chordless cycles ($k = n$). The graph G represents the values of the elements in X and Y, see Eq. 2. There is an *xy-path* between the anchor vertices c_i and c_{i+n} representing the values of x_i and y_i. The i-th xy-path consists of $s(x_i)$ connected K_w^v's (x-path) and $s(y_i)$ connected K_w^v's (y-path). The x- and the y-path are connected by one $P_w^{v,3}$ called *separating path*. Analogously, H represents the values in the vector b, see Eq. 3. There is a *b-path* between the anchor vertices c_i and c_{i+n} representing the value b_i. The i-th b-path consists of $b_i + 1$ K_w^v's.

Both, G and H, are series-parallel and can be computed in polynomial time with respect to the input size of NMwTS since the problem is **NP**-complete in the strong sense.

Lemma 1. *G and H are biconnected series-parallel graphs and can be constructed in polynomial time with respect to the values of the NMwTS instance.*

Proof (Sketch). Consider either G or H without $\bar{x}$, due to the anchor paths, the graph is connected, the same is true for G and H without any other vertex. Paths and cycles are series-parallel. Hence, K_w^v's are series-parallel and thus the xy-, b-paths and base-gadgets are series-parallel, too. They can be merged with P-operations such that $\bar{x}$ and an anchor vertex are the s- and t-nodes. Also G and H contain $|V(G)| = n(4\Sigma_s + 3) + 3(\Sigma_s + 1), |V(H)| = n(4\Sigma_s + 3) + 3(\Sigma_b + 1), |E(G)| = 4(\Sigma_s n + \Sigma_s + n) + 3n - 2$ and $|E(H)| = 4(\Sigma_s n + \Sigma_b + 2n) + n - 2$ vertices and edges, which is polynomial regarding the instance size of NMwTS. □

Due to their construction, all MCS of G and H have common characteristics regarding their size and the vertices contained in them. First we show, that not all vertices in the xy- and b-paths can be contained in an MCS.

Lemma 2. *Let P be an xy-path and P' be a b-path each with an additional edge incident to the vertices with degree one, then an MCS of P and P' has size $\min(|V(P)|, |V(P')|) - 1$.*

Proof (Sketch). Due to their construction there are $k, l \in \mathbb{N}$ such that $3k = |V(P)|$ and $3l = |V(P')|$. If $k \le l$, then the xy-path contains more than one K_3 less than the b-path. Since the separating path cannot be mapped to a K_3 there is at least one vertex which cannot be contained in an MCS. If $k > l$, then the xy-path contains at least two more K_3's than the b-path, hence each vertex except one in the b-path can be contained in the MCS. □

We can also prove, that all vertices in the base gadgets are contained in the MCS except for the vertices only contained in the anchor paths.

Lemma 3. *Let B_0 and B_n be two base-gadgets, then an MCS of B_0 and B_n has size $|V(B_0)| - n + 1$.*

Proof (Sketch). The vertices with unbounded degree are mapped to each other, as otherwise not all cycles can be contained in the MCS. In B_0 the anchor paths are between the chordless cycles and in B_n the anchor paths are between the cycles with chords. If vertices of cycles of different types are mapped, then one vertex of each cycle and the adjacent anchor vertex cannot be contained in the MCS. Hence, only the $n - 1$ vertices contained in the anchor paths cannot be contained in the MCS.   □

3.2   Correctness of the Polynomial-Time Reduction

For the reduction, we show that an instance of NMwTS has a numerical matching if and only if an MCS of the corresponding graphs G and H has a specific size.

Lemma 4. *An instance (X, Y, s, b) of NMwTS has a numerical matching if and only if $|V(G)| = |V(H)|$ and an MCS of G and H has size $|V(G)| - 2n + 1$.*

Proof (Sketch). Let (X, Y, s, b) be an instance of NMwTS and G, H graphs constructed as described above. Assume that there is a numerical matching. Hence, $\Sigma_s = \Sigma_b$ and thus $|V(G)| = |V(H)|$. An MCS of all xy-paths, b-paths and the base-gadgets has size $|V(G)| - n$ and $|V(G)| - n + 1$ (Lemmas 2 and 3). Even though they have been considered separately, the results can be combined, since all relevant vertices, the ones adjacent to the base-gadget or the xy-paths and b-paths, are contained in each MCS.

Now assume $|V(G)| = |V(H)|$ and there is an MCS with size $|V(G)| - 2n + 1$. Since we only consider connected MCSs, the vertex with unbounded degree must be contained in this MCS. For each xy-path and b-path there has to be one vertex which cannot be contained in an MCS (Lemma 2). The same is true for the base-gadgets, since the vertices of the anchor paths cannot be contained (Lemma 3). The vertices of the separating paths are not contained in an MCS. Thus the values of the elements of X and Y are correctly bipartitioned. Due to the size of the graphs for each b_i there is an x_j and $y_{j'}$ such that $b_i = s(x_j) + s(y_{j'})$. □

Since G and H both have a maximum degree bounded by 3 for all but one vertex, the next result follows accordingly.

Theorem 1. $\text{MCS}^{\leq 3,1}$ *in biconnected series-parallel graphs is **NP**-hard.*

4   The Block-and-Bridge Preserving Maximum Common Subgraph Problem in Series-Parallel Graphs

In this section we consider the *block-and-bridge preserving* MCS (BBP-MCS) which has been introduced in [17] and also is used in [3]. An MCS is a BBP-MCS if it satisfies the following two conditions:

(BBP1) Any two vertices in different blocks in a common subgraph must not be contained in the same block of an input graph.

(BBP2) Each bridge in a common subgraph is a bridge in both input graphs.

We use the polynomial time algorithm for computing the size of a biconnected MCS of two biconnected series-parallel graphs (2-MCS) [10] to obtain an algorithm which solves BBP-MCS in arbitrary series-parallel graphs. To do so, we make use of a characteristic of an MCS: Every two vertices in an input graph which are not in the same block cannot be in the same block in any common subgraph. Hence, vertices in one block can only be mapped to vertices contained in exactly one block due to condition (BBP1). With respect to BBP-MCS, cutvertices have only to be considered if two of them are mapped. Since in biconnected graphs there are no cutvertices, BBP-MCS and 2-MCS are equivalent in those.

4.1 The Algorithm

We present an algorithm which solves BBP-MCS in polynomial time. The algorithm uses the BC-trees of the input graphs as underlying data structure. We apply the idea presented in [14] for MCS in trees, to the BC-trees. We decompose the BC-trees in rooted subtrees and compute the BBP-MCS for those. These results are then combined with MwbM. Since we want to solve BBP-MCS we do not need to compare all combination of subtrees, hence we use block split graphs to define the ones that must be considered. Let G be a series-parallel graph and $T_G^{BC,r}$ the BC-tree of G rooted at $r \in E(G)$. Let $S \subseteq V(G)$ be a 1- or 2-separator of G and $\{C_1, \ldots, C_n\}$ the connected components of $G \setminus S$ such that $r \in E(C_1)$. Then $\overline{T_{G,S}^{BC,r}}$ denotes the induced subgraph $G[V(C_1) \cup S]$ and $T_{G,S}^{BC,r}$ denotes the induced subgraph $G[\bigcup_{i=2}^{n} C_i \cup S]$, called the *block split graphs* of G.   BBP-MCS is the main procedure of the algorithm. Given two

---

Algorithm 1. BBP-MCS(G, H)

Input: Two series-parallel graphs G and H.
Output: Size of a BBP-MCS of the series-parallel graphs G and H.
1: $T_G^{BC} \leftarrow \mathrm{BC}(G)$; $T_H^{BC} \leftarrow \mathrm{BC}(H)$; $z \leftarrow 0$
2: **for all** $(\Lambda, \Lambda') \in V_{Bl}(T_G^{BC}) \times V_{Bl}(T_H^{BC})$ **do**
3:      $T_\Lambda^{SP} \leftarrow \mathrm{SP}(S_\Lambda^{BC})$; $T_{\Lambda'}^{SP} \leftarrow \mathrm{SP}(S_{\Lambda'}^{BC})$
4:      **for all** $(\lambda, \lambda') \in V_S(T_\Lambda^{SP}) \times V_S(T_{\Lambda'}^{SP})$ **do**
5:          $r \leftarrow$ arbitrary $(u, v) \in E(S_\lambda) \cap E(G)$; $\mathrm{Root}(T_G^{BC}, r)$
6:          **for all** $r' = (u', v') \in E(S_{\lambda'}) \cap E(H)$ **do**
7:              $\mathrm{Root}(T_H^{BC}, r')$
8:              $p_1 \leftarrow$ BBP-MCS-S$(u, v, \lambda, u', v', \lambda')$
9:              $p_2 \leftarrow$ BBP-MCS-S$(u, v, \lambda, v', u', \lambda')$
10:             $z \leftarrow \max(z, p_1, p_2)$
11: **for all** $(\Lambda, \Lambda') \in V_{Br}(T_G^{BC}) \times V_{Br}(T_H^{BC})$ **do**
12:     $r = (u, v) \leftarrow E(S_\Lambda^{BC})$; $\mathrm{Root}(T_G^{BC}, r)$
13:     $r' = (u', v') \leftarrow E(S_{\Lambda'}^{BC})$; $\mathrm{Root}(T_H^{BC}, r')$
14:     $p_1 \leftarrow$ BBP-MCS-C(u, u') + BBP-MCS-C(v, v')
15:     $p_2 \leftarrow$ BBP-MCS-C(u, v') + BBP-MCS-C(v, u')
16:     $z \leftarrow \max(z, p_1, p_2)$
17: **return** $z + 2$

---

series-parallel graphs it computes the size of a BBP-MCS. To do so, first the BC-trees and SP-trees of the non-bridge nodes are computed. Then the 2-MCS for each combination of bridges or non-bridge nodes is computed. For each of those combinations, the BC-trees are rooted at an edge r and r' in the skeleton graphs of these nodes. If the two nodes are non-bridge nodes, the BBP-MCS of those is computed using a 2-MCS algorithm modified to handle cutvertices, see Procedure 2.

Procedure 2. BBP-MCS-S$(u, v, \lambda, u', v', \lambda')$

Input: Vertices $u, v \in V(G), u', v' \in V(H)$ and S-nodes $\lambda \in V_S(T^{\mathrm{SP}}_{\cdot, G}), \lambda' \in V_S(T^{\mathrm{SP}}_{\cdot, H})$.
Output: Size of a BBP-MCS of $T^{\mathrm{BC},r}_{G, \{u,v\}}$ and $T^{\mathrm{BC},r}_{G, \{u', v'\}}$ such that $u \mapsto u'$ and $v \mapsto v'$.
 1: $e = (v, w) \leftarrow \mathrm{NEXT}(v, \lambda);$ $e' = (v', w') \leftarrow \mathrm{NEXT}(v', \lambda')$
 2: **if** $e = \mathrm{ref}(\lambda)$ **then return** BBP-MCS-S$(u, v, \mathrm{pS}(\lambda), u', v', \lambda')$
 3: **if** $e' = \mathrm{ref}(\lambda')$ **then return** BBP-MCS-S$(u, v, \lambda, u', v', \mathrm{pS}(\lambda'))$
 4: **if** $w = u$ and $w' = u'$ **then return** MCS-E$(e, \lambda, e', \lambda')$ + BBP-MCS-C(w, w')
 5: **if** $w = u$ xor $w' = u'$ **then return** $-\infty$
 6: $z \leftarrow$ MCS-E$(e, \lambda, e', \lambda')$+BBP-MCS-S$(u, w, \lambda, u', w', \lambda')$+BBP-MCS-C$(w, w')$+1
 7: **if** $e \notin E(G)$ or $e' \notin E(H)$ **then**
 8:     **if** $e \in E(G)$ **then** $M \leftarrow \{\lambda\}$ **else** $M \leftarrow \mathrm{cS}(e)$
 9:     **if** $e' \in E(H)$ **then** $M' \leftarrow \{\lambda'\}$ **else** $M' \leftarrow \mathrm{cS}(e')$
10:     **for all** $(\eta, \eta') \in M \times M'$ **do**
11:         $p \leftarrow$ BBP-MCS-S$(u, v, \eta, u', v', \eta')$
12:         $z \leftarrow \max(z, p)$
13: **return** z

BBP-MCS-S computes the 2-MCS of two non-bridge blocks [10, MCS-S]. It utilizes the rooted SP-trees given by the BC-tree decomposition. To obtain a BBP-MCS each common subgraph of a non-bridge block must be biconnected. Hence we traverse through the skeleton graph of the SP-trees and in the end have to return to the first visited vertex of the non-bridge block as otherwise the computed subgraph of the block is not biconnected. Whenever BBP-MCS-S is called, there are three cases regarding the edges incident to the considered vertices: If both are real, the extension of the mapping is straightforward. If both are virtual, the block split graphs $T^{\mathrm{BC},r}_{G, \{v,w\}}$ and $T^{\mathrm{BC},r'}_{H, \{v',w'\}}$ must be mapped, where v, w, v' and w' are the considered vertices. If one is real while the other is virtual, the real edges in an S-node pertinent to the virtual edge have to be considered. In addition to these cases, whenever two cutvertices w, w' are mapped, the block split graphs $T^{\mathrm{BC},r}_{G, \{w\}}$ and $T^{\mathrm{BC},r'}_{H, \{w'\}}$ must be mapped.

BBP-MCS-C computes the size of a BBP-MCS of two block split graphs obtained from cutvertices. Therefore, 0 is returned if the given vertices u and u' are not both cutvertices. Otherwise, we consider their child nodes $\mathrm{cB}(u)$ and $\mathrm{cB}(u')$ in the BC-trees rooted at r and r', respectively. To this end, we create a weighted complete bipartite graph C with vertex partition $\mathrm{cB}(u) \cup \mathrm{cB}(u')$. The weight $w \colon E(C) \to \mathbb{N} \cup \{-\infty\}$ of an edge is the size of a BBP-MCS of the two block split graphs associated with its endpoints. All edges incident to nodes

Procedure 3. BBP-MCS-C(u, u')

Input: Two cutvertices $u \in V(G)$, $u' \in V(H)$.
Output: Size of a BBP-MCS of $T_{G,\{u\}}^{\mathrm{BC},r}$ and $T_{G,\{u'\}}^{\mathrm{BC},r}$ such that $u \mapsto u'$.

1: **if** $\nexists \lambda \in V_{\mathrm{C}}(T_G^{\mathrm{BC}})$: $u \in V(S_\lambda)$ or $\nexists \lambda' \in V_{\mathrm{C}}(T_H^{\mathrm{BC}})$: $u' \in V(S_\lambda)$ **then return** 0
2: $M \leftarrow \mathrm{cB}(u)$; $M' \leftarrow \mathrm{cB}(u')$; $w \leftarrow \emptyset$
3: **for all** $d = (\Lambda, \Lambda') \in V_{\mathrm{Bl}}(M) \times V_{\mathrm{Bl}}(M')$ **do**
4: $\quad$ $T_\Lambda^{\mathrm{SP}} \leftarrow \mathrm{SP}(S_\Lambda^{\mathrm{BC}})$; $T_{\Lambda'}^{\mathrm{SP}} \leftarrow \mathrm{SP}(S_{\Lambda'}^{\mathrm{BC}})$
5: $\quad$ **if** $\exists \lambda \in V_P(T_\Lambda^{\mathrm{SP}})$: $u \in V(S_\lambda)$ **then** $N \leftarrow \mathrm{cS}(\lambda)$ **else** $N \leftarrow \{\lambda\}$
6: $\quad$ **if** $\exists \lambda' \in V_P(T_{\Lambda'}^{\mathrm{SP}})$: $u' \in V(S_{\lambda'})$ **then** $N' \leftarrow \mathrm{cS}(\lambda')$ **else** $N' \leftarrow \{\lambda'\}$
7: $\quad$ **for all** $(\lambda, \lambda') \in N \times N'$ **do**
8: $\quad\quad$ $(s,t) \leftarrow$ arbitrary $(s,t) \in E(S_\lambda) \cap E(G)$: $s = u$
9: $\quad\quad$ **for all** $(s',t') \in E(S_{\lambda'}) \cap E(H)$: $s' = u'$ **do**
10: $\quad\quad\quad$ $w(d) \leftarrow \max(w(d), \text{BBP-MCS-S}(s,t,\lambda,s',t',\lambda'))$
11: **for all** $d = (\Lambda, \Lambda') \in V_{\mathrm{Br}}(M) \times V_{\mathrm{Br}}(M)$ **do**
12: $\quad$ $(u,v) \leftarrow \text{NEXT}(u,\Lambda)$; $(u',v') \leftarrow \text{NEXT}(u',\Lambda')$
13: $\quad$ $w(d) \leftarrow \text{BBP-MCS-C}(v,v') + 1$
14: **return** $\text{MwBMatching}(M, M', w)$

Procedure 4. MCS-E$(e, \lambda, e', \lambda')$

Input: Edges $e = (u,v) \in E(S_\lambda)$ and $e' = (u',v') \in E(S'_\lambda)$
Output: Size of a BBP-MCS of $T_{G,\{u,v\}}^{\mathrm{BC},r}$ and $T_{G,\{u',v'\}}^{\mathrm{BC},r}$ such that $u \mapsto u'$ and $v \mapsto v'$.

1: **if** $e \in E(G)$ xor $e' \in E(H)$ **then return** $-\infty$
2: **if** $e \in E_r(S_\lambda)$ or $e' \in E_r(S_{\lambda'})$ **then return** 0
3: $M \leftarrow \mathrm{cS}(e)$; $M' \leftarrow \mathrm{cS}(e')$; $w \leftarrow \emptyset$
4: **for all** $d = (\eta, \eta') \in M \times M$ **do**
5: $\quad$ $w(d) \leftarrow \text{BBP-MCS-S}(u,v,\eta,u',v',\eta')$
6: $p \leftarrow \text{MwBMatching}(M, M', w)$
7: **if** $p = 0$, $e \notin E(G)$ or $e' \notin E(H)$ **then return** $-\infty$ **else return** p

associated with a block and a bridge have weight $-\infty$ as a mapping of those contradicts restriction (BBP1). It is important to notice, that the computation of the BBP-MCS of two blocks is not the same as in the main procedure, since the cutvertices must be mapped. Hence, we only consider mappings where these vertices are mapped, see Lines 8 and 9, Procedure 3. The child S-nodes of a P-node λ are denoted by $\mathrm{cS}(\lambda)$ and $\mathrm{pS}(\lambda)$ refers to its parent.

MCS-E is also called whenever the considered subgraph is extended by adding a new vertex to it. If both edges between the newly mapped vertex and the vertex added before are virtual, then the vertices are a separator, see (SP1) and the BBP-MCS of the block split graph has to be added to the result. $\text{NEXT}(u,\lambda)$ and $\text{NEXT}(u,\Lambda)$ return the vertex adjacent to u which yet has not been considered in the skeleton graph of λ and Λ, respectively. ROOT roots the BC-tree at the given edge and induces a rooting in all considered SP-trees. A more detailed description of the algorithm can be found in [12].

4.2   Analysis

We argue that Algorithm 1 solves BBP-MCS in polynomial time and show that if both input graphs are outerplanar then the running time can be improved.

Theorem 2. *Algorithm 1 solves BBP-MCS in series-parallel graphs in time* $\mathcal{O}(n^6)$.

Proof. The correctness of the algorithm is based on the argumentation above and [10]. To prove the running time, we transform the algorithm in a dynamic programming approach. In [10, Th.1] it is shown, that 2-MCS can be solved in time $\mathcal{O}(n^6)$ while storing the 2-MCS of two split graphs in a table of size $\mathcal{O}(n^4)$.

We assume w.l.o.g. that for each smaller block split graph the BBP-MCS has been computed whenever BBP-MCS-S is called. The size of BBP-MCS for each pair of child blocks has already been computed, hence the BBP-MCS of the block split graphs can be obtained with MwbM in $\mathcal{O}(n^3)$ (BBP-MCS-C). The tables have size $\mathcal{O}(n^4)$ and the only loop requires total time $\mathcal{O}(n^2)$ because the results have already been computed. Thus one call of BBP-MCS-S has running time $\mathcal{O}(n^4)$. Since there are only $\mathcal{O}(n^2)$ possible combinations of block split graphs the total running time of BBP-MCS-S is $\mathcal{O}(n^6)$.

BBP-MCS-C can be computed in time $\mathcal{O}(n^5)$ since the size of a BBP-MCS of the smaller block split graphs has already been computed. Again MwbM can be solved in time $\mathcal{O}(n^3)$ since at most $\mathcal{O}(n^2)$ of these matching problems must be solved, the total running time of BBP-MCS-C is $\mathrm{O}(n^5)$.

As MCS-E has not been changed with respect to [10], its running time is $\mathcal{O}(n^5)$, resulting in a total running time of $\mathcal{O}(n^6)$.   □

Even though MwbM can be solved in $\mathcal{O}(n^3)$ it is a limiting factor regarding the running time. If we consider outerplanar graphs each P-node in the SP-trees has degree two which concludes in the following theorem.

Theorem 3. *BBP-MCS in outerplanar graphs can be solved in time* $\mathcal{O}\left(n^5\right)$.

Proof. The proof is similar to the proof of Theorem 2. Since all P-nodes in SP-trees of outerplanar graphs have degree two, the total running time of BBP-MCS-S reduces to $\mathcal{O}(n^4)$. Moreover, there is no need to use MwbM, as the bipartite graphs are K_2's. Consequently the running time of MCS-E is $\mathcal{O}(n^3)$.

BBP-MCS-C considers all adjacent nodes in the BC-tree whose number is not restricted if the graph is outerplanar and therefore still unbounded. Therefore the total running time is $\mathcal{O}(n^5)$.   □

It was known that BBP-MCES in outerplanar graphs can be solved in $\mathcal{O}(n^7)$, where MCES refers to a variation of the problem that asks for edge-induced common subgraphs with maximum number of edges [17]. Note that — in contrast to the variant we consider — a subgraph where in one input graph two vertices are adjacent while the vertices in the other are not, is a feasible MCES.

5 Concluding Remarks

We have shown that MCS in series-parallel graphs with degree bounded by 3 for all but one vertex is **NP**-hard by reduction of NMwTS. Then we have extended a 2-MCS algorithm [10] to solve BBP-MCS with running time $\mathcal{O}(n^6)$. In outerplanar graphs, it can solve BBP-MCS in $\mathcal{O}(n^5)$ which is an improvement regarding the algorithm solving BBP-MCES in outerplanar graphs in $\mathcal{O}(n^7)$ [17]. BBP-MCES in outerplanar graphs was taken as basis to obtain polynomial time solutions for MCES in outerplanar graphs of bounded degree [3] and it has yet to be decided whether MCES in series-parallel graphs of bounded degree can be solved in polynomial time. To the author's best knowledge, there is only one problem which is known to be solvable in polynomial time in outerplanar graphs, but is **NP**-complete in series-parallel graphs: the *edge-disjoint paths problem* [15]. It still is unknown, whether MCS in series-parallel graphs is solvable in polynomial time if all vertices have bounded degree. Since the series-parallel graphs are equivalent to the partial 2-trees, there is a parameterized class of graphs, i.e., the partial k-trees, for which it is known that MCS is **NP**-complete for $k \geq 11$ even when the degree is bounded [2]. For all other $k > 1$, the complexity has yet to be decided.

References

1. Akutsu, T.: A polynomial time algorithm for finding a largest common subgraph of almost trees of bounded degree. IEICE Trans. Fundam. **E76–A**(9), 1488–1493 (1993)
2. Akutsu, T., Tamura, T.: On the complexity of the maximum common subgraph problem for partial k-trees of bounded degree. In: Chao, K.-M., Hsu, T., Lee, D.-T. (eds.) ISAAC 2012. LNCS, vol. 7676, pp. 146–155. Springer, Heidelberg (2012)
3. Akutsu, T., Tamura, T.: A polynomial-time algorithm for computing the maximum common connected edge subgraph of outerplanar graphs of bounded degree. Algorithms **6**(1), 119–135 (2013)
4. Brandstädt, A., Le, V.B., Spinrad, J.P.: Graph Classes: A Survey. Society for Industrial and Applied Mathematics, Philadelphia (1999)
5. Chimani, M., Hliněný, P.: A tighter insertion-based approximation of the crossing number. In: Aceto, L., Henzinger, M., Sgall, J. (eds.) ICALP 2011, Part I. LNCS, vol. 6755, pp. 122–134. Springer, Heidelberg (2011)
6. Garey, M.R., Johnson, D.S.: Computers and Intractability: A Guide to the Theory of NP-completeness. WH Freeman and Company, New York (1979)
7. Gupta, A., Nishimura, N.: Sequential and parallel algorithms for embedding problems on classes of partial k-trees. In: Schmidt, E.M., Skyum, S. (eds.) SWAT 1994. LNCS, vol. 824. Springer, Heidelberg (1994)
8. Gupta, A., Nishimura, N.: The complexity of subgraph isomorphism for classes of partial k-trees. Theoret. Comput. Sci. **164**(1–2), 287–298 (1996)
9. Horvth, T., Ramon, J.: Efficient frequent connected subgraph mining in graphs of bounded tree-width. Theoret. Comput. Sci. **411**(3133), 2784–2797 (2010)
10. Kriege, N., Mutzel, P.: Finding maximum common biconnected subgraphs in series-parallel graphs. In: Csuhaj-Varjú, E., Dietzfelbinger, M., Ésik, Z. (eds.) MFCS 2014, Part II. LNCS, vol. 8635, pp. 505–516. Springer, Heidelberg (2014)

11. Kuhn, H.W.: The hungarian method for the assignment problem. Naval Res. Logistics Q. **2**(1–2), 83–97 (1955)
12. Kurpicz, F.: Efficient algorithms for the maximum common subgraph problem in partial 2-trees. Master's thesis, TU Dortmund (2014)
13. Matouek, J., Thomas, R.: On the complexity of finding iso- and other morphisms for partial k-trees. Discrete Math. **108**(1–3), 343–364 (1992)
14. Matula, D.W.: Subtree isomorphism in $O(n^{5/2})$. In: Algorithmic Aspects of Combinatorics, Ann. Discrete Math., vol. 2, pp. 91–106 (1978)
15. Nishizeki, T., Vygen, J., Zhou, X.: The edge-disjoint paths problem is np-complete for series-parallel graphs. Discrete Appl. Math. **115**(1), 177–186 (2001)
16. Schietgat, L., Costa, F., Ramon, J., De Raedt, L.: Effective feature construction by maximum common subgraph sampling. Mach. Learn. **83**(2), 137–161 (2011)
17. Schietgat, L., Ramon, J., Bruynooghe, M.: A polynomial-time metric for outerplanar graphs. In: Mining and Learning with Graphs (MLG) (2007)
18. Schietgat, L., Ramon, J., Bruynooghe, M., Blockeel, H.: An efficiently computable graph-based metric for the classification of small molecules. In: Boulicaut, J.-F., Berthold, M.R., Horváth, T. (eds.) DS 2008. LNCS (LNAI), vol. 5255, pp. 197–209. Springer, Heidelberg (2008)

Profile-Based Optimal Matchings in the Student/Project Allocation Problem

Augustine Kwanashie[1]($\boxtimes$), Robert W. Irving[1], David F. Manlove[1],
and Colin T.S. Sng[2]

[1] School of Computing Science, University of Glasgow, Glasgow, UK
a.kwanashie.1@research.gla.ac.uk
[2] EBay Inc., Austin, TX, USA

Abstract. In the *Student/Project Allocation problem* (SPA) we seek to
assign students to individual or group projects offered by lecturers. Students provide a list of projects they find acceptable in order of preference.
Each student can be assigned to at most one project and there are constraints on the maximum number of students that can be assigned to each
project and lecturer. We seek matchings of students to projects that are
optimal with respect to *profile*, which is a vector whose rth component
indicates how many students have their rth-choice project. We present
an efficient algorithm for finding a *greedy maximum matching* in the SPA
context – this is a maximum matching whose profile is lexicographically
maximum. We then show how to adapt this algorithm to find a *generous
maximum matching* – this is a matching whose reverse profile is lexicographically minimum. Our algorithms involve finding optimal flows in
networks. We demonstrate how this approach can allow for additional
constraints, such as lecturer lower quotas, to be handled flexibly.

1 Introduction

In most academic programmes students are usually required to take up individual or group projects offered by lecturers. Students are required to rank a
subset of the projects they find acceptable in order of preference. Each project
is offered by a unique lecturer who may also be allowed to rank the projects
she offers or the students who are interested in taking her projects in order of
preference. Each student can be assigned to at most one project and there are
usually constraints on the maximum number of students that can be assigned
to each project and lecturer. The problem then is to assign students to projects
in a manner that satisfies these capacity constraints while taking into account
the preferences of the students and lecturers involved. This problem has been
described in the literature as the *Student-Project Allocation problem* (SPA). In

---

D.F. Manlove—Supported by Engineering and Physical Sciences Research Council
grant EP/K010042/1.
C.T.S. Sng—Work done while at the School of Computing Science, University of
Glasgow.

© Springer International Publishing Switzerland 2015
J. Kratochvíl et al. (Eds.): IWOCA 2014, LNCS 8986, pp. 213–225, 2015.
DOI: 10.1007/978-3-319-19315-1_19

some cases, lecturer *lower quotas*, indicating the minimum number of students to be assigned to each lecturer, may also be specified.

Although described in an academic context, applications of SPA need not be limited to assigning students to projects but may extend to other scenarios, such as the assignment of employees to posts in a company where available posts are offered by various departments. It is widely accepted that matching problems (like SPA) are best solved by centralised matching schemes where agents submit their preferences and a central authority computes an optimal matching that satisfies all the specified criteria [5]. Moreover the potentially large number of students and projects involved in these schemes motivates the need to discover efficient algorithms for finding optimal matchings.

In SPA, students are always required to provide preference lists over projects. However, variants of the problem may be defined depending on the presence and nature of lecturer preference lists. Some variants of SPA require both students and lecturers to provide preference lists. These variants include: (i) the *Student/Project Allocation problem with lecturer preferences over Students* (SPA-S) [2] which requires each lecturer to rank the students who find at least one of her offered projects acceptable, in order of preference, (ii) the *Student/Project Allocation problem with lecturer preferences over Projects* (SPA-P) [11,14] which involves lecturers ranking the projects they offer in order of preference and (iii) the *Student/Project Allocation problem with lecturer preferences over Student-Project pairs* (SPA-(S,P)) [2,3] where lecturers rank student-project pairs in order of preference. These variants of SPA have been studied in the context of the well-known *stability* solution criterion for matching problems [5]. The general stability objective is to produce a matching M in which no student-project pair that are not currently matched in M can simultaneously improve by being paired together (thus in the process potentially abandoning their partners in M). A full description of the results relating to these SPA variants can be found in [13].

1.1   One-Sided Preferences and Profile-Based Optimality

In many practical SPA applications it is considered appropriate to allow only students to submit preferences over projects. When preferences are specified by only one set of agents in a two-sided matching problem, the notion of stability becomes irrelevant. This motivates the need to adopt alternative solution criteria when lecturer preferences are not allowed. In this subsection we mention some of these solution criteria and briefly present results relating to them. These criteria consider the size of the matchings produced as well as the satisfaction of the students involved.

When the preference lists of the lecturers are ignored, the SPA problem becomes a two-sided matching problem with one-sided preferences. Various optimality criteria for such problems have been studied in the literature [13]. Some of these criteria depend on the *profile* or the *cost* of a matching. In the SPA context, the profile of a matching is a vector whose rth component indicates the number of students obtaining their rth-choice project in the matching. The cost of a matching (w.r.t. the students) is the sum of the ranks of the assigned projects in the students' preference lists (that is, the sum of rx_r taken over all components r of the profile, where x_r is the rth component value). A *minimum cost maximum matching* is a maximum cardinality matching with minimum cost. A *rank-maximal matching* is a matching that has lexicographically maximum profile [8,10]. That is the maximum number of students are assigned to their first-choice project and subject to this, the maximum number of students are assigned to their second choice project and so on. However a rank maximal matching need not be a maximum matching in the given instance (see, e.g., [13, p. 43]). Since it is usually important to match as many students as possible, we may first optimise the size of the matching before considering student satisfaction. Thus we define a *greedy maximum matching* [6,9,15] as a maximum matching which has lexicographically maximum profile. The intuition behind both rank-maximal and greedy maximum matchings is to maximize the number of students matched with higher ranked projects. This may lead to some students being matched to projects that are relatively low on their preference lists. An alternative approach is to find a *generous maximum matching* which is a maximum matching in which the minimum number of students are matched to their Rth-choice project (where R is the maximum length of any students' preference list) and subject to this, the minimum number of students are matched to their $(R-1)$th-choice project and so on. Greedy and generous maximum matchings have been used to assign students to projects in the School of Computing Science, and students to elective courses in the School of Medicine, both at the University of Glasgow, since 2007.

A special case of SPA, where each project is offered by a unique lecturer with an infinite upper quota and zero lower quota, can be modelled as the *Capacitated House Allocation problem* (CHA). This is a variant of the well-studied *House Allocation problem* (HA) [7,19] which involves the allocation of a set of indivisible goods (which we call houses) to a set of applicants. In CHA, each applicant is required to rank a subset of the houses in order of preference with the houses having no preference over applicants. The applicants play the role of students and the houses play the role of projects and lecturers. As in the case of SPA, we seek to find a many-to-one matching comprising applicant-house pairs. Efficient algorithms for finding profile-based optimal matchings in CHA have been studied in the literature [6,9,15,17]. The most efficient of these is the $O(R^*m\sqrt{n})$ algorithm for finding rank-maximal, greedy maximum and generous maximum matchings in CHA problems due to Huang et al. [6] where R^* is the maximum rank of any applicant in the matching, m is the sum of all the preference list lengths and n is the total number of applicants and houses. These models however fail to address the issue of load balancing among lecturers. In order to keep the assignment of students fair each lecturer will typically have a minimum (lower quota) and maximum (capacity/upper quota) number of students they are expected to supervise. These numbers may vary for different lecturers according to other administrative and academic commitments. Finding efficient algorithms for profile-based optimal matchings when considering these lecturer upper and lower quotas is the main motivation of this paper.

The CHA algorithms mentioned above are based on modelling the problem in terms of a bipartite graph with the aim of finding a matching in the graph which satisfies the stated criteria. However a more flexible approach would be to model the problem as a network with the aim of finding a flow that can be converted to a matching which satisfies the stated criteria. SPA has also been investigated in the network flow context [1,18] where a *minimum cost maximum flow algorithm* is used to find a minimum cost maximum matching and other profile-based optimal matchings. The model presented in [18] allows for lower quotas on lecturers and projects as well as alternative lecturers to supervise each project. By an appropriate assignment of edge weights in the network it is shown that a minimum cost maximum flow algorithm (due to Orlin [16]) can find rank maximal, generous maximum and greedy maximum matchings in a SPA instance. This takes $O(m \log n(m + n \log n))$ time in the worst case, where m and n are the number of vertices and edges in the network respectively. In the SPA context this takes $O(m_2^2 \log(n_1 + n_2) + m_2(n_1 + n_2) \log^2(n_1 + n_2))$ time, where n_1 is the number of students, n_2 is the number of projects and m_2 is the sum of all the students' preference list lengths. However this approach involves assigning exponentially large edge weights (see, e.g., [13, p. 405]), which may be computationally infeasible for larger problem instances due to floating point inaccuracies in dealing with such high numbers. For example given a large SPA instance involving say, $n_1 = 100$ students each ranking $R = 10$ projects in order of preference, edge weights could potentially be of the order $n_1^R = 100^{10} = 10^{20}$ (and arithmetic involving such weights could easily require more than the 15–17 significant figures available in a 64-bit double-precision floating representation). Since the flow algorithms involve comparing these edge weights, floating point precision errors could easily cause them to fail in practice. Moreover using the standard assumption that arithmetic on numbers of magnitude $O(n_1)$ takes constant time, arithmetic on edge weights of magnitude $O(n_1^R)$ would add an additional factor of $O(R)$ onto the running time of Orlin's algorithm.

1.2 Our Contribution

In this paper we present efficient algorithms for finding optimal matchings to SPA problems based on the profile-based greedy maximum and generous maximum optimality criteria. Our model allows for lecturer upper and lower quotas and finds these profile-based optimal matchings without the need for exponentially-large edge weights.

We model SPA as a network flow problem and describe a modified augmenting path algorithm for finding a maximum flow which can then be transformed to an optimal SPA matching. This approach introduces greater flexibility by allowing side constraints like lecturer lower quotas to be added to the model. Our algorithms run in $O(n_1^2 R m_2)$ time. The elimination of large edge weights comes at the expense of a slightly slower running time than that of Orlin's algorithm in the worst case (i.e. slower by a factor of $O(\max\{\frac{n_1}{\log n_1}, \frac{n_2}{\log n_2}\})$ in all cases. See [12] for further details).

The remainder of this paper is organised as follows. In Sect. 2 we formally define the model. In Sect. 3 we describe an efficient algorithm for finding a greedy maximum matching given a SPA instance. In Sect. 4 we show how this algorithm can be modified in order to find a generous maximum matching. Finally in Sect. 5 we explain how the approach can be extended to allow lecturer lower quotas. All proofs for this paper can be found in [12].

2 Preliminary Definitions

An instance I of the SPA problem consists of a set S of students, a set $\mathcal{P}$ of projects and a set $\mathcal{L}$ of lecturers. Each student s_i ranks a set $A_i \subseteq \mathcal{P}$ of projects that she considers acceptable in order of preference. This *preference list* of projects may contain ties. Each project $p_j \in \mathcal{P}$ has an upper quota c_j indicating the maximum number of students that can be assigned to it. Each lecturer $l_k \in \mathcal{L}$ offers a set of projects $P_k \subseteq \mathcal{P}$ and has an upper quota d_k^+ indicating the maximum number of students that can be assigned to l_k. Unless explicitly mentioned, we assume that all lecturer lower quotas are equal to 0. The sets $\{P_1, \ldots, P_k\}$ partition $\mathcal{P}$. If project $p_j \in P_k$, then we denote $l_k = l(p_j)$.

An *assignment* M in I is a subset of $S \times \mathcal{P}$ such that:

1. Student-project pair $(s_i, p_j) \in M$ implies $p_j \in A_i$.
2. For each student $s_i \in S$, $|\{(s_i, p_j) \in M : p_j \in A_i\}| \leq 1$.

If $(s_i, p_j) \in M$ we denote $M(s_i) = p_j$. For a project p_j, $M(p_j)$ is the set of students assigned to p_j in M. Also if $(s_i, p_j) \in M$ and $p_j \in P_k$ we say student s_i is assigned to project p_j and to lecturer l_k in M. We denote the set of students assigned to a lecturer l_k as $M(l_k)$. A *matching* in this problem is an assignment M that satisfies the capacity constraints of the projects and lecturers. That is, $|M(p_j)| \leq c_j$ for all projects $p_j \in \mathcal{P}$ and $|M(l_k)| \leq d_k^+$ for all lecturers $l_k \in \mathcal{L}$.

Given a student s_i and a project $p_j \in A_i$, we define $rank(s_i, p_j)$ as $1 +$ the number of projects that s_i prefers to p_j. Let R be the maximum rank of a project in any student's preference list. We define the *profile* $\rho(M)$ of a matching M in I as an R-tuple $(x_1, x_2, ..., x_R)$ where for each r $(1 \leq r \leq R)$, x_r is the number of students s_i assigned in M to a project p_j such that $rank(s_i, p_j) = r$. Let $\alpha = (x_1, x_2, ..., x_R)$ and $\sigma = (y_1, y_2, ..., y_R)$ be any two profiles. We define the *empty profile* $O_R = (o_1, o_2, ..., o_R)$ where $o_r = 0$ for all r $(1 \leq r \leq R)$. We also define the *negative infinity profile* $B_R^- = (b_1, b_2, ..., b_R)$ where $b_r = -\infty$ $(1 \leq r \leq R)$ and the *positive infinity profile* $B_R^+ = (b_1, b_2, ..., b_R)$ where $b_r = +\infty$ $(1 \leq r \leq R)$. We define the sum of two profiles α and σ as $\alpha + \sigma = (x_1 + y_1, x_2 + y_2, ..., x_R + y_R)$. Given any q $(1 \leq q \leq R)$, we define $\alpha + q = (x_1, ..., x_{q-1}, x_q + 1, x_{q+1}, ..., x_R)$. We define $\alpha - q$ in a similar way.

We define the total order $\succ_L$ on profiles as follows. We say α *left dominates* σ, denoted by $\alpha \succ_L \sigma$ if there exists some r $(1 \leq r \leq R)$ such that $x_{r'} = y_{r'}$ for $1 \leq r' < r$ and $x_r > y_r$. We define *weak left domination* as follows. We say $\alpha \succeq_L \sigma$ if $\alpha = \sigma$ or $\alpha \succ_L \sigma$. We may also define an alternative total order $\prec_R$ on profiles as follows. We say α *right dominates* σ $(\alpha \prec_R \sigma)$ if there exists some

r $(1 \leq r \leq R)$ such that $x_{r'} = y_{r'}$ for $r < r' \leq R$ and $x_r < y_r$. We also define *weak right domination* as follows. We say $\alpha \preceq_R \sigma$ if $\alpha = \sigma$ or $\alpha \prec_R \sigma$.

The SPA problem can be modelled as a network flow problem. Given a SPA instance I, we construct a flow network $N(I) = \langle G, c \rangle$ where $G = (V, E)$ is a directed graph and c is a non-negative capacity function $c : E \to \mathbb{R}^+$ defining the maximum flow allowed through each edge in E. The network consists of a single source vertex v_s and sink vertex v_t and is constructed as follows. Let $V = \{v_s, v_t\} \cup \mathcal{S} \cup \mathcal{P} \cup \mathcal{L}$ and $E = E_1 \cup E_2 \cup E_3 \cup E_4$ where $E_1 = \{(v_s, s_i) : s_i \in \mathcal{S}\}$, $E_2 = \{(s_i, p_j) : s_i \in \mathcal{S}, p_j \in A_i\}$, $E_3 = \{(p_j, l_k) : p_j \in \mathcal{P}, l_k = l(p_j)\}$ and $E_4 = \{(l_k, v_t) : l_k \in \mathcal{L}\}$. We set the capacities as follows: $c(v_s, s_i) = 1$ for all $(v_s, s_i) \in E_1$, $c(s_i, p_j) = 1$ for all $(s_i, p_j) \in E_2$, $c(p_j, l_k) = c_j$ for all $(p_j, l_k) \in E_3$ and $c(l_k, v_t) = d_k^+$ for all $(l_k, v_t) \in E_4$.

We call a path P' from v_s to some project p_j a *partial augmenting path* if P' can be extended by adding the edges $(p_j, l(p_j))$ and $(l(p_j), v_t)$ to form an augmenting path with respect to flow f. Given a partial augmenting path P' from v_s to p_j, we define the *profile* of P', denoted $\rho(P')$, as follows:

$$\rho(P') = O_R + \sum \{rank(s_i, p_j) : (s_i, p_j) \in P' \wedge f(s_i, p_j) = 0\}$$
$$- \sum \{rank(s_i, p_j) : (p_j, s_i) \in P' \wedge f(s_i, p_j) = 1\}$$

where additions are done with respect to the $+$ and $-$ operations on profiles. Unlike the profile of a matching, the profile of an augmenting path may contain negative values. Also if P' can be extended to a full augmenting path P with respect to flow f by adding the edges $(p_j, l(p_j))$ and $(l(p_j), v_t)$ where v_s and p_j are the endpoints of P', then we define the profile of P, denoted by $\rho(P)$, to be $\rho(P) = \rho(P')$. Multiple partial augmenting paths may exist from v_s to p_j, thus we define the *maximum profile of a partial augmenting path* from v_s to p_j with respect to $\succ_L$, denoted $\Phi(p_j)$, as follows:

$$\Phi(p_j) = \max_{\succ_L} \{\rho(P') : P' \text{ is a partial augmenting path from } v_s \text{ to } p_j\}.$$

An augmenting path P is called a *maximum profile augmenting path* if $\rho(P) = \max_{\succ_L} \{\Phi(p_j) : p_j \in \mathcal{P}\}$. Let f be an integral flow in N. We define the matching $M(f)$ in I induced by f as follows: $M(f) = \{(s_i, p_j) : f(s_i, p_j) = 1\}$. Clearly by construction of N, $M(f)$ is a matching in I, such that $|M(f)| = |f|$. If f is a flow and P is an augmenting path with respect to f then $\rho(M') = \rho(M) + \rho(P)$ where $M = M(f), M' = M(f')$ and f' is the flow obtained by augmenting f along P. Also given a matching M in I, we define a flow $f(M)$ in N corresponding to M as follows:

$\forall$ $(v_s, s_i) \in E_1, f(v_s, s_i) = 1$ if s_i is matched in M and $f(v_s, s_i) = 0$ otherwise.
$\forall$ $(s_i, p_j) \in E_2, f(s_i, p_j) = 1$ if $(s_i, p_j) \in M$ and $f(s_i, p_j) = 0$ otherwise.
$\forall$ $(p_j, l_k) \in E_3, f(p_j, l_k) = c_j'$ where $c_j' = |M(p_j)|$
$\forall$ $(l_k, v_t) \in E_4, f(l_k, v_t) = d_k'$ where $d_k' = |M(l_k)|$

students' preferences: lecturers' offerings:

$s_1 : p_1 \quad p_2 \quad p_3$ $l_1 : \{p_1, p_2\}$

$s_2 : p_1$ · $l_2 : \{p_3\}$

$s_3 : p_2 \quad p_3$ project capacities: $c_1 = 1, c_2 = 1, c_3 = 1$

                                     lecturer capacities: $d_1 = 2, d_2 = 1$

Fig. 1. A SPA instance I

We define a student s_i to be *exposed* if $f(v_s, s_i) = 0$ meaning that there is no flow through s_i. Similarly we define a project p_j to be *exposed* if $f(p_j, l_k) < c_j$ and $f(l_k, v_t) < d_k^+$ where $l_k = l(p_j)$.

Let M be a matching of size k in I. We say that M is a *greedy k-matching* if there is no other matching M' such that $|M'| = k$ and $\rho(M') \succ_L \rho(M)$. If k is the size of a maximum cardinality matching in I, we call M a *greedy maximum matching* in I. Also we say that M is a *generous k-matching* if there is no other matching M' such that $|M'| = k$ and $\rho(M') \prec_R \rho(M)$. If k is the size of a maximum cardinality matching in I, we call M a *generous maximum matching* in I.

Figure 1 shows a sample SPA instance with greedy and generous maximum matchings $M_1 = \{(s_1, p_3), (s_2, p_1), (s_3, p_2)\}$ and $M_2 = \{(s_1, p_2), (s_2, p_1), (s_3, p_3)\}$ respectively.

3   Greedy Maximum Matchings in SPA

In this section we present the algorithm GREEDY-MAX-SPA for finding a greedy maximum matching given a SPA instance. The algorithm is based on the general Ford-Fulkerson algorithm for finding a maximum flow in a network [4]. We obtain maximum profile augmenting paths by adopting techniques used in the bipartite matching approach for finding a greedy maximum matching in HA [9] and CHA [17].

The GREEDY-MAX-SPA algorithm shown in Algorithm 1 takes in a SPA instance I as input and returns a greedy maximum matching M in I. A flow network $N(I) = \langle G, c \rangle$ is constructed as described in Sect. 2. Given a flow f in $N(I)$ that yields a greedy k-matching $M(f)$ in I, if k is not the size of a maximum flow in $N(I)$, we seek to find a maximum profile augmenting path P with respect to f in $N(I)$ such that the new flow f' obtained by augmenting f along P yields a greedy $(k+1)$-matching $M(f')$ in I. Lemmas 1 and 2 show the correctness of this approach. We firstly show that if k is smaller than the size of a maximum flow in $N(I)$ then such a path is bound to exist.

Lemma 1. *Let I be an instance of SPA and let η denote the size of a maximum matching in I. Let k ($1 \leq k < \eta$) be given and suppose that M_k is a greedy k-matching in I. Let $N = N(I)$ and $f = f(M_k)$. Then there exists an augmenting path P with respect to f in N such that if f' is the result of augmenting f along P then $M_{k+1} = M(f')$ is a greedy $(k+1)$-matching in I.*

Algorithm 1. GREEDY-MAX-SPA

Require: SPA instance I;
Ensure: return matching M;
 1: define flow network $N(I) = \langle G, c \rangle$;
 2: define empty flow f;
 3: **loop**
 4:    $P = $ GET-MAX-AUG$(N(I), f)$;
 5:    **if** $P \neq null$ **then**
 6:       augment f along P;
 7:    **else**
 8:       return $M(f)$;

Lemma 2. *Let f be a flow in N and let $M_k = M(f)$. Suppose that M_k is a greedy k-matching. Let P be a maximum profile augmenting path with respect to f. Let f' be the flow obtained by augmenting f along P. Now let $M_{k+1} = M(f')$. Then M_{k+1} is a greedy $(k+1)$-matching.*

The GET-MAX-AUG algorithm shown in Algorithm 2 accepts a flow network $N(I)$ and flow f as input and finds an augmenting path of maximum profile relative to f or reports that none exists. The latter case implies that $M(f)$ is already a greedy maximum matching. The method consists of three phases: an initialisation phase (lines 1–11), the main phase which is a loop containing two other loops (lines 12–27) and a final phase (lines 28–35) where the augmenting path is generated and returned.

For each project p_j the GET-MAX-AUG method maintains a variable $\rho(p_j)$ describing the profile of a partial augmenting path P' from some exposed student to p_j. It also maintains, for every project $p_j \in \mathcal{P}$, a pointer $pred(p_j)$ to the student or lecturer preceding p_j in P'. For every lecturer $l_k \in \mathcal{L}$ a pointer $pred(l_k)$ is also used to refer to any project preceding l_k in P'. Thus the final augmenting path produced will pass through each lecturer or project at most once. The initialisation phase of the method involves setting all $pred$ pointers to **null** and ρ profiles to B_R^-. Next, the method seeks to find, for each project p_j, a partial augmenting path $((v_s, s_i), (s_i, p_j))$ from the source, through an exposed student s_i to p_j should one exist. In the presence of multiple paths satisfying this criterion, the path with the best profile (w.r.t. $\succ_L$) is selected. The variables $pred(p_j)$ and $\rho(p_j)$ are updated accordingly. Thus at the end of this phase $\rho(p_j)$ indicates the maximum profile of an augmenting path of length 2 via some exposed student to p_j should one exist. If such a path does not exist then $\rho(p_j)$ and $pred(p_j)$ retain their initial values of B_R^- and **null** respectively.

In the main phase, the algorithm then runs $|f|$ iterations, at each stage attempting to increase the quality (w.r.t. $\succ_L$) of the augmenting paths described by the ρ profiles. Each iteration runs two loops. Each loop identifies cases where the flow through one edge in the network can be reduced in order to allow the flow through another to be increased while improving the profile of the projects involved. In both loops, the decision on whether to switch the flow between candidate edges is made based on an edge relaxation operation similar to that used in the Bellman-Ford algorithm for solving the single source shortest path problem in which edge weights may be negative. In the first loop, we seek to evaluate the gain that may be derived from switching the flow through a student from one project to another. Given an edge (s_i, p_k) with a flow of 1 in f and edge (s_i, p_j) with no flow in f, we define σ to be the resulting profile of p_j if the partial augmenting path ending at p_k is to be extended (via s_i) to p_j. Thus σ will become the new value of $\rho(p_j)$ should this extension take place. If $\sigma \succ_L \rho(p_j)$ (i.e. if the proposed profile is better than the current one), we extend the augmenting path to p_j and update $\rho(p_j) = \sigma$ and $pred(p_j) = s_i$.

In the second loop, we seek to evaluate the gain that may be derived from switching flow to some lecturer from one project to another. Given a lecturer l_k, let $P_k' \subseteq P_k$ be the set of projects offered by l_k with positive outgoing flow and $P_k'' \subseteq P_k$ be the set of projects offered by l_k that are undersubscribed in $M(f)$. Then we seek to determine if an improvement can be obtained by switching a unit of flow from some project $p_j \in P_k'$ to some other project $p_m \in P_k''$. This is achieved by comparing the $\rho(p_j)$ and $\rho(p_m)$ profiles and updating $\rho(p_j) = \rho(p_m)$, $pred(p_j) = l_k$ and $pred(l_k) = p_m$ if $\rho(p_m) \succ_L \rho(p_j)$ where $\rho(p_m)$ represents the profile of a partial augmenting path that does not already pass through l_k (i.e., $pred(p_m) \neq l_k$). This means that the partial augmenting path ending at p_m can be extended further (via l_k) to p_j while improving its profile. The intuition is that, after augmenting along such a path, p_m gains an extra student while p_j loses one.

During the final phase, we iterate through all exposed projects and find the one with the largest profile with respect to $\succ_L$ (say p_q). An augmenting path is then constructed through the network using the $pred$ values of the projects and lecturers and the matched edges in $M(f)$ starting from p_q. The generated path is returned to the calling algorithm. If no exposed project exists, the method returns **null**. We next show that GET-MAX-AUG method produces such a maximum profile augmenting path in N with respect to f should one exist.

Lemma 3. *Given a SPA instance I, let f be a flow in $N = N(I)$ where $k = |f|$ is not the size of a maximum matching in I and $M(f)$ is a greedy k-matching in I. Algorithm GET-MAX-AUG finds a maximum profile augmenting path in N with respect to f.*

From Lemmas 1, 2 and 3, we can conclude that the algorithm GREEDY-MAX-SPA finds a greedy maximum matching given a SPA instance. Concerning the complexity of the algorithm, the main loop calls GET-MAX-AUG η times where η is the size of a maximum cardinality matching in I. The first phase of GET-MAX-AUG performs $O(m_2)$ profile comparison operations and $O(n_3)$ initialisation steps for the lecturer $pred$ values where $m_2 = |E_2|$, $n_3 = |\mathcal{L}|$, and each profile comparison step requires $O(R)$ time. The loop in the main phase of GET-MAX-AUG runs k times where k is the value of the flow obtained at that time. The first and second loops perform $O(m_2)$ and $O(n_2)$ relaxation steps respectively where $n_2 = |\mathcal{P}|$ and each relaxation step requires $O(R)$ time to compare profiles. The final phase of the algorithm performs $O(n_2)$ profile comparisons, each also taking

---

Algorithm 2. GET-MAX-AUG (method for GREEDY-MAX-SPA)

Require: flow network $N(I) = \langle G, c \rangle$ where $G = (V, E)$, flow f where $M(f)$ is a greedy $|f|$-matching;
1: /* initialisation */
2: **for** project $p_j \in \mathcal{P}$ **do**
3:　　$\rho(p_j) = B_R^-$;
4:　　$pred(p_j) = \textbf{null}$;
5:　　**for each** exposed student $s_i \in S$ such that $p_j \in A_i$ **do**
6:　　　　$\sigma = O_R + rank(s_i, p_j)$;
7:　　　　**if** $\sigma \succ_L \rho(p_j)$ **then**
8:　　　　　　$\rho(p_j) = \sigma$;
9:　　　　　　$pred(p_j) = s_i$;
10: **for** lecturer $l_k \in \mathcal{L}$ **do**
11:　　$pred(l_k) = \textbf{null}$;
12: /* main phase */
13: **for** $1...|f|$ **do**
14:　　/* first loop */
15:　　**for each** $(s_i, p_j) \in E$ where $f(s_i, p_j) = 0$ and $f(s_i, p_k) = 1$ for some $p_k \in A_i$ **do**
16:　　　　$\sigma = \rho(p_k) - rank(s_i, p_k) + rank(s_i, p_j)$;
17:　　　　**if** $\sigma \succ_L \rho(p_j)$ **then**
18:　　　　　　$\rho(p_j) = \sigma$;　　$pred(p_j) = s_i$;
19:　　/* second loop */
20:　　**for each** lecturer $l_k \in \mathcal{L}$ **do**
21:　　　　$\sigma = B_R^-$;　　$p_z = \textbf{null}$;
22:　　　　**for each** project $p_m \in P_k$ such that $l(p_m) = l_k \wedge f(p_m, l_k) < c_m$ **do**
23:　　　　　　**if** $\rho(p_m) \succ_L \sigma$ **then**
24:　　　　　　　　$\sigma = \rho(p_m)$;　　$p_z = p_m$;
25:　　　　**for each** project $p_j \in P_k$ such that $l(p_j) = l_k \wedge f(p_j, l_k) > 0 \wedge p_j \neq p_z$ **do**
26:　　　　　　**if** $\sigma \succ_L \rho(p_j)$ **then**
27:　　　　　　　　$\rho(p_j) = \sigma$;　　$pred(p_j) = l_k$;　　$pred(l_k) = p_z$;
28: /* final phase */
29: $\rho = \max_{\succ_L}(\{B_R^-\} \cup \{\rho(p_j) : p_j \in P \text{ is exposed}\})$;
30: **if** $\rho \succ_L B_R^-$ **then**
31:　　$p_q = \arg\max_{\succ_L}(\{B_R^-\} \cup \{\rho(p_j) : p_j \in P \text{ is exposed}\})$;
32:　　$Q = $ path obtained by following $pred$ values and matched edges in $M(f)$ from p_q to an exposed student;
33:　　**return** $\langle v_s \rangle$ ++ $reverse(Q)$ ++ $\langle l(p_q), v_t \rangle$; /*++ denotes concatenation*/
34: **else**
35:　　**return null**;

$O(R)$ time. Thus the overall time complexity of the GET-MAX-AUG method is $O(m_2 R + n_3 + kR(m_2 + n_2) + n_2 R) = O(kR(m_2))$. Thus the overall time complexity of the GREEDY-MAX-SPA algorithm is $O(n_1^2 m_2 R)$. A straightforward refinement of the algorithm can be made by observing that if no profile is updated during an iteration of the main loop, then no further profile improvements can be made and we can terminate the main loop at this point. We conclude with the following theorem.

Theorem 4. *Given a* SPA *instance* I, *a greedy maximum matching in* I *can be obtained in* $O(n_1^2 R m_2)$ *time.*

4   Generous Maximum Matchings in SPA

Analogous to the case for greedy maximum matchings, generous maximum matchings can also be found by modelling SPA as a network flow problem. Given a SPA instance I we define the following terms relating to partial augmenting paths in $N(I)$. For each project $p_j \in \mathcal{P}$, we define the *minimum profile of a partial augmenting path* from v_s through an exposed student to p_j with respect to $\prec_R$, denoted $\Phi'(p_j)$, as follows: $\Phi'(p_j) = \min_{\prec_R}\{\rho(P') : P' \text{ is a partial augmenting path from } v_s \text{ to } p_j\}$.

If a partial augmenting path P' ending at project p_j can be extended to an augmenting path P by adding edges $(p_j, l(p_j))$ and $(l(p_j), v_t)$ then such an augmenting path is called a *minimum profile augmenting path* if $\rho(P) = \min_{\prec_R}\{\Phi'(p_j) : p_j \in \mathcal{P}\}$. A similar approach to that used to find a greedy maximum matching can be adopted in order to find a generous maximum matching. The main GREEDY-MAX-SPA algorithm will remain unchanged (we will call it GENEROUS-MAX-SPA for convenience) as the intuition remains to successively find larger generous k-matchings until a generous maximum matching is obtained. We however make slight changes to the GET-MAX-AUG algorithm in order to find a minimum profile augmenting path in the network should one exist (the resulting algorithm is then known as GET-MIN-AUG). The changes are as follows. (i) We replace all occurrences of left domination $\succ_L$ with right domination $\prec_R$. (ii) We also replace all occurrences of negative infinity profile B_R^- with a positive infinity profile $B_R^|$. (iii) Finally we replace both max functions (in lines 29 and 31) with the min function. Analogous statements and proofs of Lemmas 1, 2 and 3 exist in this context. Thus we may conclude with the following theorem concerning the GENEROUS-MAX-SPA algorithm.

Theorem 5. *Given a* SPA *instance* I, *a generous maximum matching in* I *can be obtained in* $O(n_1^2 R m_2)$ *time.*

5   Lecturer Lower Quotas

We have so far considered a SPA model in which each lecturer l_k has an upper quota. In this section we discuss how the algorithm presented above can be modified to allow lecturer lower quotas. We call this extension the *Student/Project problem with Lecturer lower quotas* (SPA-L). In an instance I of SPA-L, each lecturer l_k now additionally has a lower quota $d_k^-(I)$ (it will be helpful to indicate specific instances to which these lower bounds refer within the notation). We assume that $d_k^-(I) \geq 0$ and $d_k^+(I) \geq \max\{d_k^-(I), 1\}$. In the SPA-L context, our definition of a matching as presented in Sect. 2 needs to be tightened slightly. A *constrained matching* is a matching M in the SPA context with the additional property that, for each lecturer l_k, $|M(l_k)| \geq d_k^-(I)$. We seek to find greedy and generous maximum constrained matchings should they exist.

Let I be a SPA-L instance. Also let I' be a SPA instance constructed from I by setting $d_k^-(I') = 0$ and $d_k^+(I') = d_k^-(I)$ for each lecturer l_k. Firstly we find a greedy maximum matching M' in I' using the GREEDY-MAX-SPA algorithm. If $f' = f(M')$ is not a saturating flow (i.e., one in which all edges $(l_k, v_t) \in E_4$ are saturated), then I admits no constrained matching. Otherwise we augment f' in $N(I)$ by calling GREEDY-MAX-SPA on I, changing line 2 so that flow f is assigned to be f' initially. We continuously augment the flow until no augmenting path exists. The matching $M = M(f)$ obtained from the resulting flow f is a greedy maximum constrained matching in I. Generous maximum constrained matchings can also be found by using GENEROUS-MAX-SPA and GET-MIN-AUG instead of GREEDY-MAX-SPA and GET-MAX-AUG respectively.

Theorem 6. *Let I be a SPA-L instance. Each of the problems of finding a greedy or generous maximum constrained matching, or reporting that no such matching exists, can be solved in $O(n_1^2 R m_2)$ time.*

References

1. Abraham, D.J.: Algorithmics of two-sided matching problems. Master's thesis, University of Glasgow, Department of Computing Science (2003)
2. Abraham, D.J., Irving, R.W., Manlove, D.F.: Two algorithms for the Student-Project allocation problem. J. Discrete Algorithms **5**(1), 79–91 (2007)
3. El-Atta, A.H.A., Moussa, M.I.: Student project allocation with preference lists over (student, project) pairs. In: Proceedings of ICCEE 09: The 2nd International Conference on Computer and Electrical Engineering, pp. 375–379 (2009)
4. Ford, L.R., Fulkerson, D.R.: Flows in Networks. Princeton University Press, Princeton (1962)
5. Gusfield, D., Irving, R.W.: The Stable Marriage Problem: Structure and Algorithms. MIT Press, Cambridge (1989)
6. Huang, C.-C., Kavitha, T., Mehlhorn, K., Michail, D.: Fair matchings and related problems. In: IARCS Annual Conference on Foundations of Software Technology and Theoretical Computer Science (FSTTCS 2013), vol. 24, pp. 339–350 (2013)
7. Hylland, A., Zeckhauser, R.: The efficient allocation of individuals to positions. J. Polit. Econ. **87**(2), 293–314 (1979)
8. Irving, R.W.: Greedy matchings. Technical Report TR-2003-136, University of Glasgow, Department of Computing Science (2003)
9. Irving, R.W.: Greedy and generous matchings via a variant of the Bellman-Ford algorithm (2006) (Unpublished manuscript)
10. Irving, R.W., Kavitha, T., Mehlhorn, K., Michail, D., Paluch, K.: Rank-maximal matchings. ACM Trans. Algorithms **2**(4), 602–610 (2006)
11. Iwama, K., Miyazaki, S., Yanagisawa, H.: Improved approximation bounds for the student-project allocation problem with preferences over projects. J. Discrete Algorithms **13**, 59–66 (2012)
12. Kwanashie, A., Irving, R.W., Manlove, D.F., Sng, C.T.S.: Profile-based optimal matchings in the Student/Project Allocation problem. CoRR Technical Report 1403.0751 (2014). http://arxiv.org/abs/1403.0751

13. Manlove, D.F.: Algorithmics of Matching Under Preferences. World Scientific, Singapore (2013)
14. Manlove, D.F., O'Malley, G.: Student project allocation with preferences over projects. J. Discrete Algorithms **6**, 553–560 (2008)
15. Mehlhorn, K., Michail, D.: Network problems with non-polynomial weights and applications (2006) (Unpublished manuscript)
16. Orlin, J.B.: A faster strongly polynomial minimum cost flow algorithm. Oper. Res. **41**(2), 338–350 (1993)
17. Sng, C.T.S.: Efficient Algorithms for Bipartite Matching Problems with Preferences. Ph.D. thesis, University of Glasgow, Department of Computing Science (2008)
18. Zelvyte, M.: The Student-Project Allocation problem: a network flow model. Honours project dissertation, University of Glasgow, School of Mathematics and Statistics (2014)
19. Zhou, L.: On a conjecture by Gale about one-sided matching problems. J. Econ. Theor. **52**(1), 123–135 (1990)

The Min-max Edge q-Coloring Problem

Tommi Larjomaa[1] and Alexandru Popa[2]($\boxtimes$)

[1] Department of Communications and Networking, Aalto University School of
Electrical Engineering, Aalto, Finland
tommi.larjomaa@gmail.com
[2] Faculty of Informatics, Masaryk University, Brno, Czech Republic
popa@fi.muni.cz

Abstract. In this paper we introduce and study a new problem named *min-max edge q-coloring* which is motivated by applications in wireless mesh networks. The input of the problem consists of an undirected graph and an integer q. The goal is to color the edges of the graph with as many colors as possible such that: (a) any vertex is incident to at most q different colors, and (b) the maximum size of a color group (i.e. set of edges identically colored) is minimized. We show the following results:

1. Min-max edge q-coloring is NP-hard, for any $q \geq 2$.
2. A polynomial time exact algorithm for min-max edge q-coloring on trees.
3. Exact formulas of the optimal solution for cliques.
4. An approximation algorithm for planar graphs.

1 Introduction

Traditionally, backbone connectivity in networks of various sizes has been built using wired infrastructure. Even though the bandwidth that modern wired networking technology offers is no doubt better than that of wireless alternatives, the material and installation costs of wired networks is a significant drawback. Therefore, the concept of wireless mesh networks (WMNs) has received a lot of attention and has been researched actively during the past decade [2,3,7].

In a multi-channel WMN, each node is able to use multiple non-overlapping frequency channels. The use of many channels inside the same network can significantly improve overall performance; interference from neighboring nodes can be decreased substantially, when nodes do not need to use the same radio channel for every link. Multiple radio channels in the network means, that at least some of the nodes need to handle more than one channel at a time. In many proposed designs the multi-channel feature is achieved by packet-by-packet reconfiguration of the radio [8,11,15]. However, one of the drawbacks of this kind of continuous channel switching of a single radio interface is that it requires precise synchronization throughout the network.

An alternative approach would be to fit multiple radio interfaces to each node, thus allowing a more persistent channel allocation per interface. A couple of such multi-NIC (network interface card) architectures have been proposed

© Springer International Publishing Switzerland 2015
J. Kratochvíl et al. (Eds.): IWOCA 2014, LNCS 8986, pp. 226–237, 2015.
DOI: 10.1007/978-3-319-19315-1_20

by Raniwala et al. [12,13]. Their simulation and testbed experiments show a promising improvement with only two NICs per node, compared to a single-channel WMN. Another appealing feature of these architectures is that they are based on readily available, commodity IEEE 802.11 interfaces, requiring only systems software modification.

The scenario of two or more NICs per node with fixed channels imposes some limitations to the assignment of channels on each interface. In order to set up a link between two nodes, both of them have to have at least one of their interfaces set to the same channel. On the other hand, links inside an interference range should use as many different channels as possible. Thus, the channels need to be assigned carefully in order to both keep every required link possible and maximize useful bandwidth throughout the network.

The channel assignment problem can be modeled as a type of edge coloring problem: given a graph G, the edges have to be colored so that there are at most q different colors incident to each vertex. Here, vertices, edges and colors represent network nodes, links and channels, respectively. A coloring that satisfies this constraint, is called an edge q-coloring. Note, that the coloring constraint differs from the traditional coloring problems, where adjacent items are not allowed to have the same color. Also the goal is different; instead of minimizing the amount of colors, a large amount of different colors in an edge q-coloring is often a desired state of things.

Previously, the channel assignment was formulated as the max edge q-coloring problem, where the goal was to maximize the total number of colors. The drawback of this model is that in an optimal solution the same color is assigned to many edges while other colors are used only once. We remind the reader that in the wireless mesh network setting, having the same color assigned to many edges is equivalent to having the same frequency used many times, and therefore, having interference. Since the goal of the application is to minimize the interference, max edge q-coloring is perhaps not the ideal theoretical formulation (although max edge q-coloring is still interesting as a combinatorial problem). Instead, we aim to have the color components as balanced as possible. Therefore, we newly introduce the min-max edge q-coloring where the goal is to minimize the maximum size of a color group. It is true that the interference cannot be reduced completely under this model, since the same frequency is assigned to edges of connected subgraphs, but we believe that it is more realistic. Also, we believe that the min-max edge q coloring problem is an interesting combinatorial problem by itself.

We would like to emphasize that the approximation algorithms proposed for the max edge q-coloring [1, 4–6] indeed use the same color many times, since they select first a maximum matching in the graph, color the edges of the matching with distinct colors, remove them from the graph, and, finally, color all the edges in the same connected component using an identical color. There exist instances where the maximum color class returned by the 2-approximation algorithm for max edge q coloring [1] has size $O(|V|)$, but the optimal solution of the min-max q coloring is constant. Thus, the situation presented in the previous paragraph is definitely not an artificial example and it reflects precisely the behavior of known algorithms.

Related Work. The problem of finding a maximum edge q-coloring of a given graph has been studied by Feng et al. [4–6]. They provide a 2-approximation algorithm for $q = 2$ and a $(1 + \frac{4q-2}{3q^2-5q+2})$-approximation for $q > 2$. They show that the problem is solvable in polynomial time for trees and complete graphs in the case $q = 2$, but the complexity for general graphs has been left as an open problem. Adamaszek and Popa [1] show that the problem is APX-hard and present a 5/3-approximation algorithm for graphs which have a perfect matching. The maximum edge q-coloring is also considered in combinatorics and is a particular case of the anti-Ramsey number. For a brief description of the connection of the two problems, the reader can refer to [1].

To the best of our knowledge there has been no prior research on the min-max edge q-coloring problem.

Our Contributions. In this paper we introduce and study the min-max edge q-coloring problem. First, in Sect. 2 we prove that the problem is NP-hard for any $q \geq 2$. The proof is split into two parts. We first show the NP-hardness for a more general version in which each vertex is allowed to have an independent value of q. In the second part we show how to introduce extra gadgets in order to force all the vertices to have the same value of q.

Then, in Sect. 3 we show an exact polynomial time algorithm for trees, for $q = 2$. We first show that the optimal solution in a tree is at least $\Delta/2$ and at most Δ, where Δ is the maximum degree of the tree. Then, the algorithm uses binary search to find a value c, such that the input admits a coloring in which the largest color group is at most c. Given a value c, we select in turn each vertex as the root of the tree and try to construct a solution in a bottom up fashion. This is not straightforward as for each vertex we have to solve a knapsack instance (fortunately, these instances are solvable in polynomial time).

In Sect. 4 we analyze the value of the optimal solution on complete graphs. Since the problem on general graphs seems difficult, the study of special classes of graphs helps us to understand better the structure of the problem. We provide the exact formulas of the optimal solutions for cliques.

Section 5 summarizes the results and briefly discusses possible future research directions. Due to space limitations we omitted some of the proofs. Detailed proofs and several additional results are presented in [9].

2   NP-hardness of Min-max Edge q-Coloring

In this section we prove that the min-max edge q-coloring problem is NP-hard for $q \geq 2$, giving little hope of finding a general exact polynomial time algorithm for it. The proof is split into two steps. First we prove NP-hardness for a more general version of the problem, defined next, where each vertex is assigned a value for q individually.

Problem 1 (General min-max edge q-coloring problem). The input is a graph $G = (V, E)$, and for each vertex v_i there is a positive integer q_i. A feasible solution is a coloring of edges, such that for each vertex v_i, there are at most q_i different colors incident to it. The goal is to find a coloring σ such that the size of the largest color group, $\max_c |\{e \in E | \sigma(e) = c\}|$, is minimized.

The reduction is made from monotone one-in-three SAT (Definition 1), which is known to be NP-complete [14]. By modifying this reduction slightly we can prove NP-hardness for the min-max edge q-coloring problem with a constant value of q.

Definition 1 (Monotone one-in-three SAT Problem). *The input is a Boolean 3CNF-formula ϕ, where each literal is simply a variable; there is no negation. Determine whether a truth assignment for the variables exists, such that for each clause, there is exactly one literal that is true, while the other two literals are false.*

Now we state the NP-hardness for the general edge q-coloring problem.

Theorem 1. *Problem 1 is NP-hard.*

Proof. We use a reduction from monotone one-in-three SAT (Definition 1), which goes as follows. There are m clauses and n variables in the formula ϕ. For each clause, there is a single vertex c_j with $q = 2$ (we use this notation as a shorthand for "at most 2 different colors can be incident to c_j"). For each variable x_i, there are three vertices: a_i, b_i and v_i having $q = 1$, $q = 1$ and $q = 2$, respectively. Each vertex v_i is adjacent to vertices a_i and b_i. If a variable is present in a clause, the corresponding variable vertex a_i is adjacent to the clause vertex c_j. For each a_i, there are additional leaves adjacent to it, so that $\deg(a_i) = 2m_i$, where m_i is the number of clauses the variable is present in.[1] Moreover, for each b_i, there are additional leaves so that $\deg(b_i) = L - 2m_i$, where $L = 4m + n$. Finally, there is a vertex f with $q = 1$, that is adjacent to each v_i. The resulting graph is of polynomial size in m. Figure 1 illustrates the idea of the reduction by showing the full gadget of a single variable.

Next we show that if ϕ is satisfiable, there is a feasible coloring for the reduction, whose largest color group is L. For each variable x_i that is false in the satisfying truth assignment, color the edges incident to a_i with the color F, which is the color incident to the vertex f. There are two edges incident to some a-vertex per each literal, and $2m$ false literals, so there are in total $4m + n = L$ edges colored with F. Since v_i is incident to only one color at this point, we give a distinct color for the edges of b_i, of which there are less than L.

For each true variable x_i we choose a distinct color and use it to color edges incident to both a_i and b_i. These color groups have thus $L - 2m_i + 2m_i = L$ edges. Since the truth assignment is satisfying, there is one color representing a true variable and the color F representing false variables incident to each clause vertex, which makes the coloring feasible.

---

[1] We can safely assume that each variable has at most one literal in any clause.

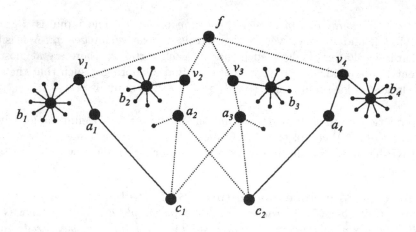

Fig. 1. The reduction from formula $(x_1 \lor x_2 \lor x_3) \land (x_2 \lor x_3 \lor x_4)$. The dotted edges are assigned the color F in a coloring where none of the colors have more than $L = 12$ edges.

Finally, we show that if the formula is not satisfiable, the optimum of the reduction is more than L (in other words, if the optimum of the reduction is less or equal to L, the formula is satisfiable). In a feasible coloring of a reduction from an unsatisfiable formula, there are two possibilities, because otherwise, the coloring gives a truth assignment. Either there are clauses in which two or more variables and their a-vertices have a color different from F, or there are clauses in which all variables are using color F (or both).

In the first case, two variable vertices a_i and a_j necessarily share a color, which we denote by C. Consequently, the vertices v_i and v_j are both saturated with colors F and C. Note, that for any variable x_k, $\deg(b_k) \geq L - 2m > L/2$. Thus, if the vertices b_i and b_j are assigned the same color, the limit L is immediately exceeded. On the other hand, if one of those vertices, say, b_i takes the color F, and b_j takes the color C, there are already L edges colored with C due to the variable x_j plus the edges incident to a_i.

In the second case we can assume that the clauses that have not only false literals in them, have exactly one true literal, since the other case was already discussed. Now, there are more than $2m$ false literals, and, as observed before, there are two edges per literal incident to the a-vertices. Thus, there must be more than $4m + n = L$ edges colored with F. □

As we go on to prove NP-hardness for min-max edge q-coloring, where each vertex has the same value for q, we use a slightly modified version of the previous reduction. The idea is to mimic vertices with $q = 1$ or $q = 2$. This is done by saturating vertices with an appropriate number of different colors that already have L edges. We proceed with the theorem and proof.

Theorem 2. *The min-max edge q-coloring problem is NP-hard for $q \geq 2$.*

Proof. We begin by showing how to force a vertex with any value of q to allow only one or two new colors for its additional edges, given the upper bound L for color group size. Observe that the optimum for a $(qL + 1)$-star, namely a star with qL leaves, is exactly L. We take $q - 1$ such stars, pick one leaf from each star and identify them as one vertex. In an optimal coloring of the acquired gadget, the contracted vertex v is incident to $q - 1$ different colors of size L. As we add edges to v, they can be colored with only one color in order to keep color group sizes below L. If we want a vertex that allows two colors, we pick $q - 2$ leaves from different qL-stars (we can use the same stars as before, since there are plenty of leaves left) and identify them as one.

Using such gadgets that mimic vertices with $q = 1$ and $q = 2$, we straight-forwardly construct a reduction equivalent to the one used in the proof of Theorem 1. Now it remains to show that the number of additional vertices and edges in the new reduction is polynomially bounded in the size of the formula.

We show that we need only $(q-1)$ stars to be able to mimic enough vertices. In the original reduction, there is one vertex per clause, three vertices per variable[2] and the vertex f. In total we have $M = m + 3n + 1$ vertices that need to be mimicked. We need at most $q - 1$ leaves to mimic one vertex, so having $qL \geq 2L \geq M$ will suffice. Assume the opposite, which yields

$$M > 2L \Leftrightarrow m + 3n + 1 > 8m + 2n \Leftrightarrow n > 7m - 1.$$

This contradicts with the fact that there can be at most $3m$ variables in a 3CNF-formula, that is, $n \leq 3m$. So, the number of additional edges needed for the modified reduction is $(q - 1)qL = O(m + n)$, since q is constant.   □

3   Exact Polynomial Time Algorithm for Trees

In this section we present an exact polynomial time algorithm for solving the min-max edge 2-coloring problem on trees. First of all we give the following bounds of the optimal solution.

Lemma 1. *For an instance of the min-max edge 2-coloring problem, where the graph is a tree T, $OPT \in \left[\frac{\Delta}{2}, \Delta - 1\right]$, where Δ is the maximum degree of T.*

Proof. The lower bound follows from the fact that there is a vertex with Δ edges incident to it, and only two distinct colors can be assigned to these edges. The upper bound can always be achieved with the following coloring. Choose an arbitrary vertex v_r as the root vertex, and color its edges evenly with two colors. For each child v of v_r, there are $deg(v) - 1$ uncolored edges that can be colored with a new color, since v had only one edge colored previously. The same is repeated iteratively for each child vertex of a visited vertex. No more than $\Delta - 1$ edges are colored with any color.   □

---

[2] We do not need to take into account the leaves of the variable vertices; a leaf allows only one color incident to it, no matter what value q has.

Next, we describe informally the polynomial time algorithm for trees (the algorithm is presented formally as Algorithm 1). The idea of the algorithm is to try to color the tree with different candidate values for optimum from the interval $\left[\frac{\Delta}{2}, \Delta - 1\right]$, until candidates c and $c - 1$ are found so that c leads to a feasible coloring whereas $c - 1$ does not. This is repeated for each vertex as the root vertex, and the smallest successful value of c is the optimum. Using binary search we only need to test $O(\log \Delta)$ different candidates per root.

Once we have fixed a root and a value c for the maximum color class, the algorithm proceeds in a bottom up fashion, starting from the leaves. For each vertex v, we have to color the edges incident to v with two colors, each color class having size at most c. We proceed as follows: for each vertex v we color with one color as many edges incident to v and then we transmit to the parent of v the number of edges that are left uncolored (this number is termed *residual number* in Algorithm 1). Thus, for each vertex v, we have to solve a knapsack instance, where the size of the knapsack is c, the maximum size of a color class, and the items are the residual values of the children. The residual value of v is the sum of the residual values of its children minus the optimal solution of the knapsack instance plus one (this is the edge that connects v to its parent).

Essentially, one run through the loop starting at step 8 minimizes the residual number of the root vertex with respect to an optimum candidate c. If a residual number exceeds c, the combination of the root vertex and the optimum candidate does not lead to a feasible coloring. Changing the root vertex, however, changes the parental relationships between the vertices, and consequently the residual numbers, even if the optimum candidate was the same. This is why we need to iterate the minimization process with all combinations of root vertices and optimum candidates to be sure. Since there are merely $O(n\Delta)$ of such combinations, this does not compromise the algorithm running in polynomial time.

As a final note, since the knapsack problem is known to be NP-hard, it might give reason to believe that step 9 of the adlgorithm does not run in polynomial time in general. Fortunately, it is also well known that knapsack instances are solvable in $O(nW)$ time, where n is the number of items and W is the size of the knapsack. Since at any vertex there are at most Δ items (children) and the knapsack size is also at most Δ, any knapsack instance encountered in step 9 is solvable in $O(\Delta^2)$ time.

4   Complete Graphs

In this section we present formulas for the optimal solution of the min-max edge 2-coloring problem in the case of complete graphs. More precisely, we show that an optimal min-max edge 2-coloring of an n-clique K_n achieves $\mathrm{OPT}(K_n) \geq \left\lceil \frac{1}{3}|E(K_n)| \right\rceil$. The proof is split in parts. Also, we show that the bound is tight in most cases and present exact formulas for the optimum in all cases. Before we begin, we define a term used frequently later on.

Input: A tree graph T

1. $m \longleftarrow \Delta - 1$
2. For each vertex v^r
    3. Label each vertex of T with its distance from the root vertex (via e.g. BFS)
    4. $l \longleftarrow \left\lceil \frac{\Delta}{2} \right\rceil, u \longleftarrow \Delta - 1$
    5. Repeat
        6. Assign for each non-root vertex v a residual number $v_l \longleftarrow 1$, and for the root $v_l^r \longleftarrow 0$
        7. $c \longleftarrow \lfloor \frac{1}{2}(l+u) \rfloor$
        8. For each non-leaf vertex in descending order of distance from root
            9. Solve the following knapsack instance:
                Denote the children of v by v^i. Size of the knapsack is c, and the item sizes are the residual numbers v_l^i of the children.
            10. Store the set of indices of the children in the knapsack solution to S
            11. If $\sum_i v_l^i - \sum_{j \in S} v_l^j + v_l > c$: $l \longleftarrow c+1$ and go to step 16
            12. Color the uncolored edges incident to $v^i, i \in S$, and all their successors with a new color
            13. $v_l \longleftarrow v_l + \sum_i v_l^i - \sum_{j \in S} v_l^j$
        14. Color the remaining uncolored edges connected to the root with one color
        15. Store the current coloring to U, and set $u \longleftarrow c$
        16. If $l = u$, revert to the coloring U, jump out of the loop to step 17
    17. if $u < m$: $m \longleftarrow u$ and $M \longleftarrow U$
18. Revert to the coloring M

Output: m

Algorithm 1. Tree 2-coloring algorithm

Definition 2 (Color Subgraph). *For a given feasible edge q-coloring of a graph G and a color c, a color subgraph G^c is an edge induced subgraph of G, induced by all edges with the color c.*

The first observation concerns a color that is not incident to every vertex of the clique. Such a color can share vertices with only a limited number of other colors. This and the forthcoming lemmas help narrow down the different ways of how a clique can be colored.

Lemma 2. *In a feasible edge 2-coloring of a clique K_n and for any color c, a color subgraph K_n^c cannot share vertices with more than two other color subgraphs, if $V(K_n^c) \subset V(K_n)$.*

Proof. Assume the opposite. In a feasible coloring of K_n, let K_n^c be a color subgraph that shares vertices with $k \geq 3$ other color subgraphs $K_n^{c_1}, \ldots, K_n^{c_k}$, and $V(K_n^c) \subset V(K_n)$. Now, a vertex v in $V(K_n^c)$ is incident to two colors: c and c_i, the latter being assigned to the edges going from v to vertices not in $V(K_n^c)$. Formally, $V(K_n) \setminus V(K_n^c) \subset V(K_n^{c_i})$ for each $i = 1, \ldots, k$. Thus, we have a set of vertices $V(K_n^{c_i})$ that is incident to k colors, which makes the coloring not feasible, a contradiction. $\square$

Next, we look at a more specific case of the situation described in the above lemma. When a color is not incident to all vertices and shares vertices with exactly two other colors, there are exactly three colors, all of which are necessarily incident to the other two. This coloring strategy actually turns out to be the best in the end.

Lemma 3. *Given a feasible edge 2-coloring of K_n, for which there is a color subgraph K_n^c that shares vertices with exactly two other color subgraphs, and $V(K_n^c) \subset V(K_n)$, the coloring has exactly three colors, whose color subgraphs have these same properties.*

Proof. Let K_n^c be a color subgraph of K_n that shares vertices with exactly two other color subgraphs $K_n^{c_1}$ and $K_n^{c_2}$. As in the proof of Lemma 2, $V(K_n) \setminus V(K_n^c) \subset V(K_n^{c_i}), i = 1, 2$. Thus, all vertices $V(K_n) \setminus V(K_n^c)$ are saturated with 2 colors. Since $V(K_n^c)$ was assumed to be incident to only the three colors, there cannot be any other colors. Furthermore, none of the colors is incident to all vertices. □

In the following lemma we cover the remaining non-trivial alternative, that is, there is a color that shares vertices with exactly one other color. This implies the presence of a color incident to all vertices. From now on we call such a color *global*.

Lemma 4. *Given a feasible edge 2-coloring of K_n and a color subgraph K_n^c that shares vertices with exactly one other color subgraph K_n^F, and $V(K_n^c) \subset V(K_n)$, the color F is incident to all vertices of K_n.*

Proof. The edges between $V(K_n^c)$ and the rest of the vertices must be colored with some other color than c. Since c is incident only to $V(K_n^c)$, the edges between $V(K_n^c)$ and the rest of the vertices must be colored with F. Thus, F is incident to all vertices of K_n. □

We now have enough tools to provide the actual lower bound. First we show that if there are more than four colors, one of them must be global. This, in turn, yields that one of the colors has over one third of all edges. Since the alternative is to have three or less different colors, the lower bound follows.

Theorem 3. *For min-max edge 2-coloring, the following holds:*

$$OPT(K_n) \geq \left\lceil \frac{1}{3}|E(K_n)| \right\rceil = \left\lceil \frac{n(n-1)}{6} \right\rceil \tag{1}$$

Proof. First of all, we observe that in order to have $OPT(K_n) < \left\lceil \frac{1}{3}|E(K_n)| \right\rceil$, at least four different colors must be used in an optimal coloring. Assume this is possible. With at least four colors, Lemmas 2 and 3 imply that the colors not incident to all vertices can share vertices with only one other color. By Lemma 4, that other color is the global color F. Now, let K_n^c be the color subgraph with the largest proper subset of vertices of K_n, and let $k_c = |V(K_n^c)|$. Edges of only the global color fill the cut (i.e. the set of edges between two groups of vertices) between $V(K_n^c)$ and the rest of the $n - k_c$ vertices, thus

$$k_c(n - k_c) \leq \frac{1}{3}|E(K_n)| = \frac{n(n-1)}{6}.$$

With the help of basic calculus, this yields

$$k_c \leq \frac{n}{2} - \sqrt{\frac{n^2 + 2n}{12}} < \left(\frac{1}{2} - \frac{1}{\sqrt{12}}\right)n < \frac{1}{3}n \tag{2}$$

or

$$k_c \geq \frac{n}{2} + \sqrt{\frac{n^2 + 2n}{12}} > \left(\frac{1}{2} + \frac{1}{\sqrt{12}}\right)n > \frac{2}{3}n. \tag{3}$$

If (2) is true, there are two possibilities: either all non-global colors are incident to a total of less than a third of all vertices, or there is a set of non-global colors that are incident to a total of k vertices, so that $\frac{1}{3}n \leq k \leq \frac{2}{3}n$. In the former case, $|E(K_n^F)| > \frac{1}{3}|E(K_n)|$, a contradiction. In the latter case, the cut between the k and the other $n - k$ vertices are again filled with edges of the global color, as in the case of k_c, but this time k fails to satisfy (2) or (3), leading to a contradiction.

If (3) is true, there are $k < \frac{1}{3}n$ vertices for the rest of the colors to occupy. In total, these vertices have at most the following amount of edges between them:

$$|E(K_k)| = \frac{k(k-1)}{2} < \frac{\frac{1}{3}n^2 - n}{6} < \frac{n^2 - n}{6} = \frac{1}{3}|E(K_n)|.$$

Thus, over two thirds of the edges are left for the two other colors to share, leaving the lower bound out of reach.

Now, the only way to achieve the suggested lower bound is by using three colors, in which case the bound is trivial. □

Most of the time, the lower bound is actually tight, and it is achievable only with a coloring described in Lemma 3 (i.e. every vertex incident to exactly two colors, no global color) for two reasons. First, as we saw in the above proof, the lower bound is out of reach using four colors. Second, if one of the three colors is global, the other colors need to satisfy either (2) or (3), leaving at least one of them too small.

The most even way to distribute the edges between three colors is to first divide the vertices of K_n to three groups of sizes $k = \lfloor \frac{n}{3} \rfloor$ and $k + 1$, depending on the remainder of the division. Each color is then incident to the vertices of two of the groups, each group is incident to two colors.

If the remainder is 1, then $k = \frac{n-1}{3}$. One color is incident to $2k$ vertices, while two other colors are incident to $2k + 1$ vertices each. If the "smaller" color subgraph with $2k$ vertices can accommodate one third of the edges, distributing the rest of the edges evenly to the two "bigger" color subgraphs is trivial. Otherwise, it is not possible to color the edges quite evenly, and the remaining over two thirds of edges still need to be shared between the bigger color groups. More precisely, the exact optimum can be written as

$$\text{OPT}(K_n) = \max\left(\left\lceil\frac{1}{3}|E(K_n)|\right\rceil, \left\lceil\frac{5}{4}k(k+1)\right\rceil\right).$$

If the remainder is 2, then $k = \frac{n-2}{3}$. There are two colors incident to $2k+1$ vertices and one color incident to $2k+2$ vertices. Achieving the lower bound is possible, if the bigger color subgraph can avoid coloring more than one third of all edges. If not, the minimum size of the bigger color group is the optimum, that is

$$\text{OPT}(K_n) = \max\left(\left\lceil\frac{1}{3}|E(K_n)|\right\rceil, (k+1)^2\right).$$

5 Conclusions and Future Work

The goal of this paper is to analyze the problem of efficiently allocating channels in wireless mesh networks from a theoretic point of view and to design and analyze some basic approximation algorithms. The analysis is simplified by modelling the channel allocation problem as a graph coloring problem, namely min-max edge q-coloring. The concept of edge q-coloring captures the restriction in some proposed WMN architectures, where each network node can use at most a number of different frequency channels at once. Furthermore, we give the most attention to the case $q = 2$, since it has been considered important from a practical perspective.

For the min-max edge q-coloring problem, we prove NP-hardness, both in a more general case (see Problem 1), where each vertex has its individual value for q, and in the case where the value of $q \geq 2$ is constant for each vertex. We show lower bounds for the optimum in terms of maximum and average degree. We also introduce an exact polynomial time algorithm for trees and provide the exact formulas of the optimal solutions for cliques.

In [9] we present some extra results. First, for bicliques, we present a lower bound which is tight when both parts of the graph have an even number of vertices (and almost tight for the other cases). For a hypercube Q_n we give a lower bound which is tight for even n, and similarly, almost tight for odd n. Although these classes of graphs have a very simple structure, finding lower bounds is much more difficult than in the case of the max edge q-coloring problem.

A good lower bound of the optimal solution is necessary in order to design approximation algorithms. For the min-max edge q-coloring problem, a trivial lower bound is half of the maximum degree. In [9], we also show another lower bound in terms of the average degree of the graph, namely that $\text{OPT} \geq \frac{d^2}{2q^2}$, where d is the average degree.

Moreover, we design approximation algorithm for planar graphs which achieves a sublinear approximation ratio. The algorithm uses a theorem of Lipton and Tarjan [10] which says that a planar graph admits a small balanced separator.

Interesting directions for future research include finding hardness of approximation results and better algorithms, especially for min-max edge q-coloring on general graphs. Also it might be interesting to see how the proposed algorithms would affect performance, if applied to actual Wireless Mesh Networks.

Acknowledgements. We would like to thank anonymous reviewers for their useful comments.

References

1. Adamaszek, A., Popa, A.: Approximation and hardness results for the maximum edge q-coloring problem. In: Cheong, O., Chwa, K.-Y., Park, K. (eds.) ISAAC 2010, Part II. LNCS, vol. 6507, pp. 132–143. Springer, Heidelberg (2010)
2. Akyildiz, I., Wang, X., Wang, W.: Wireless mesh networks: a survey. Comput. Netw. **47**(4), 445–487 (2005)
3. Draves, R., Padhye, J., Zill, B.: Routing in multi-radio, multi-hop wireless mesh networks. In: MobiCom 2004, pp. 114–128. ACM (2004)
4. Feng, W., Chen, P., Zhang, B.: Approximate maximum edge coloring within factor 2: a further analysis. In: ISORA, pp. 182–189 (2008)
5. Feng, W., Zhang, L., Qu, W., Wang, H.: Approximation algorithms for maximum edge coloring problem. In: Cai, J.-Y., Cooper, S.B., Zhu, H. (eds.) TAMC 2007. LNCS, vol. 4484, pp. 646–658. Springer, Heidelberg (2007)
6. Feng, W., Zhang, L., Wang, H.: Approximation algorithm for maximum edge coloring. Theor. Comput. Sci. **410**(11), 1022–1029 (2009)
7. Gupta, B., Acharya, B., Mishra, M.: Optimization of routing algorithm in wireless mesh networks. In: NaBIC 2009, pp. 1150–1155. IEEE (2009)
8. Kyasanur, P., Vaidya, N.: Routing and Interface assignment in multi channel multi-interface wireless networks. In: Proceedings of IEEE Wireless Communications and Networking Conference 2005, vol. 4, pp. 2051–2056. IEEE (2005)
9. Larjomaa, T., Popa, A.: The min-max edge q-coloring problem. CoRR abs/1302.3404 (2013)
10. Lipton, R., Tarjan, R.: Applications of a planar separator theorem. SIAM J. Comput. **9**(3), 615–627 (1980)
11. Muir, A., Garcia-Luma-Aceves, J.: A channel access protocol for multihop wireless networks with multiple channels. In: ICC 1998, vol. 3, pp. 1617–1621, June 1998
12. Raniwala, A., Chiueh, T.: Architecture and algorithms for an IEEE 802.11-based multi-channel wireless mesh network. In: INFOCOM, pp. 2223–2234. IEEE (2005)
13. Raniwala, A., Gopalan, K., Chiueh, T.: Centralized channel assignment and routing algorithms for multi-channel wireless mesh networks. Mob. Comput. Commun. Rev. **8**(2), 50–65 (2004)
14. Schaefer, T.: The complexity of satisfiability problems. In: STOC 1978, pp. 216–226. ACM, New York (1978)
15. So, J., Vaidya, N.: Multi-channel mac for ad hoc networks: handling multi-channel hidden terminals using a single transceiver. In: MobiHoc 2004, pp. 222–233. ACM (2004)

Speeding up Graph Algorithms
via Switching Classes

Nathan Lindzey[✉]

Computer Science Department, Colorado State University,
Fort Collins, CO 80523-1873, USA
lindzey@cs.colostate.edu

Abstract. Given a graph G, a *vertex switch* of $v \in V(G)$ results in
a new graph where neighbors of v become nonneighbors and vice versa.
This operation gives rise to an equivalence relation over the set of labeled
digraphs on n vertices. The equivalence class of G with respect to the
switching operation is commonly referred to as G's *switching class*. The
algebraic and combinatorial properties of switching classes have been
studied in depth; however, they have not been studied as thoroughly
from an algorithmic point of view. The intent of this work is to further
investigate the algorithmic properties of switching classes. In particular, we show that switching classes can be used to asymptotically speed
up several super-linear unweighted graph algorithms. The current techniques for speeding up graph algorithms are all somewhat involved insofar that they employ sophisticated pre-processing, data-structures, or use
"word tricks" on the RAM model to achieve at most a $O(\log(n))$ speed
up for sufficiently dense graphs. Our methods are much simpler and can
result in super-polylogarithmic speedups. In particular, we achieve better bounds for diameter, transitive closure, bipartite maximum matching,
and general maximum matching.

1   Introduction

The runtime of an algorithm is intimately related to how an instance is represented. Recall that the runtimes of the first generation of graph algorithms
were expressed solely in terms of n, the number of vertices. This analysis was
natural since at this time graphs were represented in $\Theta(n^2)$ space via their adjacency matrix. It was soon noticed that if $m = o(n^2)$, then a variety of graph
algorithms could be sped up by first computing the adjacency list from the adjacency matrix, then running the algorithm on the more efficient adjacency list
representation. This motivated the introduction of m to the runtime of graph
algorithms and it is now customary in algorithm design to assume that a graph
instance is given in the form of its adjacency list.

   We introduce $\widetilde{m}$ as a measure of complexity and show many classical graph
algorithms can be analyzed in terms of $\widetilde{m}$. This is a significant measure of com-
plexity since $\widetilde{m} = O(m)$ but $\widetilde{m} \neq \Theta(m)$. In particular, if $\widetilde{m} = o(m)$, then several
graph algorithms can be asymptotically sped up by computing the so-called

© Springer International Publishing Switzerland 2015
J. Kratochvíl et al. (Eds.): IWOCA 2014, LNCS 8986, pp. 238–249, 2015.
DOI: 10.1007/978-3-319-19315-1_21

partially complemented adjacency list (pc-list) from an adjacency list, then running the algorithm on the more efficient partially complemented adjacency list representation.

The pc-list [4] is a natural generalization of the adjacency list that involves an additional $O(n)$ bits of storage to represent *vertex switches*. When a vertex is switched, its neighbors become nonneighbors and nonneighbors become neighbors. A (di)graph afforded such a switching operation is commonly referred to as a *switching class* [17]. Figure 1 demonstrates how a pc-list can represent a switching class which can in turn be used to obtain a more compact representation of a graph. Algebraic and combinatorial properties of switching classes have been studied in depth [2,17]; however, they have not been studied as thoroughly from an algorithmic point of view. The intent of this work is extend [4] by further investigating algorithmic properties of switching classes.

In [4] canonical $\Theta(n + m)$ unweighted graph algorithms were developed for switching classes; however, due to the linear-time solvability of these problems, the pc-list provided no asymptotic speed up in runtime. We extend this work by developing switching class algorithms for classical unweighted graph problems for which no linear-time algorithm is known. We show that for sufficiently dense graphs, the pc-list can provide super-polylogarithmic speed ups in runtime.

This is notable since the current techniques for speeding up algorithms over dense instances are all somewhat involved and achieve at most a $O(\log(n))$ speed up. A data-structure in [11] is given that allows one to work on the complement of a graph without constructing it; however, the algorithms they consider are linear and do not improve any of the results established in [4]. The techniques in [3] are notable in that they achieve a $O(\log(n))$ speed-up for several canonical graph problems over arbitrary dense graphs (assuming the RAM model). Clever

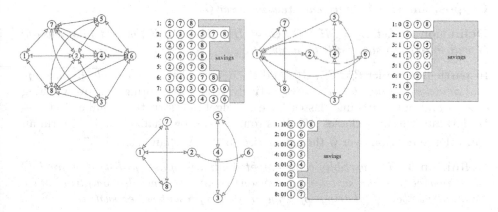

Fig. 1. An illustration of the partially complemented adjacency list. Setting G as the graph given in the leftmost picture, it is not hard to see that the graph obtained by out switching (the rightmost picture) and Gale-Berlekamp switching (the lower picture) are minimum representatives of $\mathcal{C}_G^+$ and $\mathcal{C}_G^{\pm}$ respectively. This demonstrates that $\overline{G}$ is not necessarily a minimum representative of G.

but complicated preprocessing in [6] allows for an asymptotic speedup that is logarithmic in the density of the graph that is at most $O(\log(n))$.

Our approach is much simpler insofar that it involves only basic preprocessing of the graph and slight modifications to existing algorithms.

2   Preliminaries

All graphs are assumed to be finite, labeled, directed, unweighted, and simple unless stated otherwise, and let $\mathcal{G}$ denote the class of all such graphs on n vertices. Let $V(G)$ and $E(G)$ denote vertex set and edge set of G respectively. Let $E(A, B)$ denote the set of edges that have exactly one endpoint in $A \subseteq V$ and exactly one endpoint in $B \subseteq V$. Let $G[X]$ denote the subgraph induced by the vertex set $X \subseteq V$. An *out-switch* (*in-switch*) of a vertex changes out-neighbors (in-neighbors) to non out-neighbors (in-neighbors) and vice versa. A *Seidel-switch* of a vertex in an undirected graph changes neighbors to nonneighbors and nonneighbors to neighbors. Performing an out-switch and in-switch on the same vertex of an undirected graph is equivalent to Seidel-switching that vertex. Let $\neg_v^+(G)$, $\neg_v^-(G)$, and $\neg_v(G)$ be the graphs obtained by out, in, and Seidel switching on a vertex $v \in V(G)$ respectively. It is easy to see that the order in which vertices are switched does not matter, so let $\neg_U^+(G)$, $\neg_U^-(G)$, and $\neg_U(G)$ be the graph obtained by out, in, and Seidel switching on a set $U \subseteq V(G)$. If a mixed sequence of in and out switches are permitted, then let $\neg_{I,O}^\pm(G)$ be the graph obtained by *Gale-Berlekamp switching* where $I, O \subseteq V(G)$ are the subsets of vertices that have been in-switched and out-switched respectively.

Definition 1. *Let $G, H \in \mathcal{G}$. Then $G \sim^* H$ iff $\exists U \subseteq V$ such that $\neg_U^*(G) \cong H$ with respect to some switching operation $*$.*

Proposition 1. $\sim^x$ *equivalence relation over $\mathcal{G}$.*

Definition 2. *Let $\mathcal{C}_G^* = \{H \in \mathcal{G} : G \sim^x H\}$. Then $\mathcal{C}_G^*$ is the switching class of G with respect to some switching operation $*$.*

In particular, we let $\mathcal{C}_G^+$, $\mathcal{C}_G^-$, $\mathcal{C}_G^\pm$, and $\mathcal{C}_G$ denote the *in, out, Gale-Berlekamp, and Seidel switching class* of G respectively. It is worth noting that in-switching classes and out-switching classes have an algebraic structure similar to Gale-Berlekamp switching classes [16]. It is routine to show that $\sim^+$ and $\sim^-$ form an equivalence relation over $\mathcal{G}$ that gives rise to the Abelian group $\mathbb{Z}_2^{n^2-2n}$.

Definition 3. *The partially complemented adjacency list (pc-list) of a graph G (with respect to some switching operation) is an adjacency list outfitted with a constant number of bitstrings of length n that represent vertex switches.*

If a vertex v is switched, then we let $\widetilde{N}(v)$ denote its doubly-linked neighborlist in the pc-list of $\widetilde{G}$. If v is unswitched, then we let $N(v)$ denote the doubly-linked neighbor list of v in the pc-list of $\widetilde{G}$. For any switched vertex v, we still refer to $\widetilde{N}(v)$ as the neighborlist of v even though its elements are actually non-neighbors in the original graph.

Proposition 2. $\overline{G}$ *is the graph obtained by out-switching (in-switching) all of the vertices.*

The proposition above is useful due to the fact that some graph classes can be recognized by considering properties of their complements [14]. Unfortunately, constructing the complement graph $\overline{G}$ is an $\Omega(n^2)$ operation which precludes any linear-time bound. The pc-list has proved useful in this context since it represents $\overline{G}$ implicitly which obviates the $\Omega(n^2)$ cost of constructing $\overline{G}$ [4,14].

The pc-list was motivated by McConnell's *complement-equivalence classes* [13]. It is straightforward to see that the symmetric complement-equivalence classes of [4] coincide with Seidel switching classes and in-out complement-equivalence classes [4] coincide with Gale-Berlekamp switching classes. Due to this correspondence, it seems natural to couch the pc-list in terms of the existing theory of switching classes.

It is obvious that we should seek out small members of switching classes to obtain a more succinct representation of a given graph. Ideally, we should seek a member of a switching class with the fewest edges. We definte a *minimum representative* of a switching class $\mathcal{C}_G^*$ as a not necessarily unique graph $\widetilde{G} \in \mathcal{C}_G^*$ having minimum edge cardinality $\widetilde{m}$. If we limit ourselves to strictly out-switches or strictly in-switches, the following lemma shows that we can easily construct a minimum representative using a greedy algorithm.

Lemma 1. *[4] A minimum representative $\widetilde{G} \in \mathcal{C}_G^+$ ($\widetilde{G} \in \mathcal{C}_G^-$) can be constructed in $O(n+m)$ time.*

Proof. Visit each vertex and if switching it reduces the edge count, do so. For out switching, if there are more than $n/2$ elements in v's neighbor list, switch v and replace the neighbor list with non-neighbors of v. The work for creating the list of non-neighbors can be charged to visiting the neighbors of v. For in switching, if v appears more than $n/2$ times in the adjacency list of G, then switch v. The work is clearly $O(m)$.

Observe that if both in-switches and out-switches are allowed, then the algorithm in the proof of Lemma 1 no longer guarantees that the representative is a minimum. This is because edges can reappear while constructing the representative. It is known that computing minimum representatives for Gale-Berlekamp and Seidel switching classes is NP-hard and is even hard to approximate within a constant factor of the optimum [9,16]. There do however exist randomized linear-time $(1+\epsilon)$-approximation schemes for computing a minimum representative $\widetilde{G} \in \mathcal{C}_G^\pm$ [12]. This allows one to obtain a representative $G' \in \mathcal{C}_G^\pm$ such that $|G'| = \Theta(|\widetilde{G}|)$ in $O(n \log(n) + m \log(n))$ time with high probability.

3   Basic Algorithms for Switching Classes

3.1   Traversal

Traversal algorithms for out-switching classes first appeared in [4] where the existence of $O(n + \widetilde{m})$ algorithms for traversal on Seidel switching classes was left open. We show that these algorithms can obtained in a straightforward manner through a slight modification of the pc-list data-structure and the traversal algorithms of [4,10]. We refer the reader to [4,10] for a more thorough treatment. We begin with an intuitive explanation as to why the pc-list is able to provide asymptotic savings in runtime for graph traversal.

Let $\mathcal{A}$ be a graph traversal algorithm and assume there exists an oracle $\mathcal{O}$ such that for any current vertex v, it returns in $O(1)$ time either an undiscovered neighbor v or reports that all of v's neighbors have been discovered. If $\mathcal{A}$ considers a vertex that has already been discovered, then we shall call this a *bad query*. It is clear that the runtime of $\mathcal{A}$ with oracle $\mathcal{O}$ is $\Theta(n)$ since $\mathcal{A}$ can make no bad queries. This is no longer the case if we run $\mathcal{A}$ without $\mathcal{O}$ since we might have $\omega(n)$ bad queries for arbitrary graphs. In this case, the runtime of $\mathcal{A}$ is dominated by bad queries since we could have as many as $O(m)$. However, if the size of G's pc-list is asymptotically smaller than its adjacency list, then we can obtain a tighter upper bound on the number of bad queries that can occur during an execution of algorithm $\mathcal{A}$ without use of an oracle. This is due to the fact that every bad query of BFS or DFS can be charged to an element of the pc-list data-structure [4,10].

Theorem 1. *[4] Given* $\tilde{G} \in \mathcal{C}_G^+$, *BFS on G can be done* $O(n + \tilde{m})$ *time.*

Theorem 2. *[4, 10] Given* $\tilde{G} \in \mathcal{C}_G^+$, *DFS on G can be done in* $O(n + \tilde{m})$ *time.*

Proposition 3. *Let* $S \subseteq V$ *be a set of Seidel switched vertices and let* $H = G[S] \cup G[S - V]$. *Then* $\neg_S(G)$ *is isomorphic to the graph* $H' = (V(H), E(H) \cup \overline{E}(S, V-S))$ *where* $\overline{E}(S, V-S)$ *is the complement of the cut induced by* $(S, V-S)$.

Given an adjacency list representation of G and a set of Seidel switched vertices $S \subseteq V$ such that $|E(\neg_S(G))| < |E(G)|$, a pc-list data-structure that represents $\neg_S(G)$ can be constructed in $O(n + m)$ time as follows. Let v be an arbitrary vertex. If v is switched, set its bit to 1, add all of its neighbors in S and its nonneighbors in $V - S$ into its neighborlist. If v is unswitched, set its bit to 0, add all of its neighbors in $V - S$ and its nonneighbors in S into its neighborlist. Relabel the vertices so that the members of $V - S$ have a smaller label than members of S, then radix-sort the pc-list with respect to this new labeling. The sort has the effect of making all nonneighbors of v appear consecutively in v's neighborlist. Finally, insert a dummy vertex between the two elements u and w of v's neighborlist such that u is switched and w is unswitched.

The same algorithms given in [4,10] can be used on this pc-list representation with the following modification; once v's dummy vertex is visited during a scan of its neighbor list, flip v's bit in the pc-list. If v is switched, then the flip has the effect of treating elements after the dummy vertex as actual neighbors of v. If v is unswitched, then the flip has the effect of treating elements after the dummy vertex as nonneighbors of v. Accounting for this modification in the traversal algorithms of [4,10] is trivial. The foregoing gives the following result.

Theorem 3. *Given* $\tilde{G} \in \mathcal{C}_G$, *BFS and DFS on G can be done in* $O(n+\tilde{m})$ *time.*

3.2   Contraction

Henceforth we shall assume that all switching operations are out-switches for ease of exposition. Let $\widetilde{G} \in \mathcal{C}_G^+$ be a minimum representative, $n(B)$ be the number of vertices of a subset $B \subseteq V$, and $\widetilde{m}(B)$ be the number of edges incident to vertices of B in the graph $\widetilde{G}$. Without loss of generality, we assume that the pc-list of $\widetilde{G}$ is sorted by vertex label.

Lemma 2. *Given* $\widetilde{G} \in \mathcal{C}_G^+$, *a set* $B \subseteq V$ *can be contracted to vertex* $\hat{b}$ *in* $O(n(B) + \widetilde{m}(B))$ *time.*

Proof. To build the neighbor list of $\hat{b}$, we first build two doubly-linked neighbor lists X, Y that correspond to the contraction of all the switched vertices $S \subseteq B$ and unswitched vertices of $B - S$ respectively.

Initialize X to be $\widetilde{N}(s)$ some $s \in S$ and let $b, c \in S$. Then contracting b and c together corresponds to taking the intersection $\widetilde{N}(b) \cap \widetilde{N}(c)$. Since the neighborlists are sorted, taking the intersection $\bigcap_{b \in S} \widetilde{N}(b)$ can be computed in $O(n(B) + \widetilde{m}(B))$ using a routine similar to the merge routine of merge-sort as follows. Let $L[i]$ denote the ith element of a doubly-linked list L.

If $X[i] = \widetilde{N}(b)[j]$, then the comparison can be charged to the jth edge of b's neighbor list. If $X[i] > \widetilde{N}(b)[j]$, then the comparison can be charged to the jth edge of b's neighbor list. If $X[i] < \widetilde{N}(b)[j]$, then $X[i]$ is removed from the doubly-linked list X. Removing a vertex from X happens at most $\widetilde{m}(B)$ times since each deletion can be charged to an element of $\widetilde{N}(s)$.

Initialize Y to be $N(v)$ for some $v \in B - S$ and let $b, c \in B - S$. Suppose that $b, c \in B - S$ are unswitched, then contracting b and c together corresponds taking the union $N(b) \cup N(c)$. The union $Y = \bigcup_{b \in B - S} N(b)$ can clearly be computed in $O(n(B) + \widetilde{m}(B))$.

To combine X and Y it suffices to scan X and remove all elements from X that also exist in Y. This can be done using a routine similar to the merge routine. At the end of the routine, define $\widetilde{N}(\hat{b})$ to be X. Since the sizes of X and Y are each $O(n(B) + \widetilde{m}(B))$, it follows that $\widetilde{N}(\hat{b})$ can be constructed in $O(n(B) + \widetilde{m}(B))$ time.

Lemma 3. *Let* β *be a collection of vertex-disjoint subsets. Then the contracted graph* $\widetilde{G}/\beta$ *can be computed in* $O(n + \widetilde{m})$.

Proof. By Lemma 2 we can perform all of the contractions in time $\sum_{B \in \beta} n(B) + \widetilde{m}(B) = O(n + \widetilde{m})$. After performing the contractions, the pc-list must be cleaned up so that vertices subsumed by contractions are no longer referenced. This can be done by radix-sorting and removing duplicates in $O(n + \widetilde{m})$ time.

4   Super-Linear Graph Algorithms

In this section, we show that given a graph $G = (V, E)$, spending $O(n + m)$ time to compute an $O(n + \widetilde{m})$ space pc-list representation of G gives rise to better bounds for several canonical unweighted graph algorithms.

4.1 Diameter and Transitive Closure

At present, the most efficient combinatorial algorithm (ignoring log factors) for computing the diameter and transitive closure of a graph to our knowledge is the naive $O(n^2 + nm)$ algorithm, that is, calling BFS from each vertex. The following results are straightforward.

Theorem 4. *The diameter of a graph G can be computed in $O(n^2 + n\widetilde{m})$ time.*

Theorem 5. *The transitive closure of a graph G can be computed in $O(n^2 + n\widetilde{m})$ time.*

Proof. Compute $\widetilde{G}$ in $O(n + m)$ time, then run the naive algorithm from each vertex using the BFS algorithm of [4]. The runtime of this algorithm is $O((n + m) + n(n + \widetilde{m})) = O(n^2 + n\widetilde{m})$.

Since $n^2 + n\widetilde{m}$ is never worse than $n^2 + nm$ and is sometimes better, this is indeed a better bound for both diameter and transitive closure.

At this point it is natural to ask when a graph *benefits* from its pc-list representation, that is, when its pc-list representation is asymptotically smaller than its adjacency-list representation. It is easy to see that there are at least twice as many graphs that benefit from pc-lists as there are graphs that benefit from adjacency-lists. This is because the pc-list is a generalization of the adjacency-list and the complement of any graph whose adjacency-list representation is asymptotically smaller than its adjacency-matrix representation must have a minimum representative $\widetilde{G} \in \mathcal{C}_G^+$ such that $\widetilde{m} = o(m)$. For instance, the class of graphs whose complement is sparse benefits from its pc-list representation simply by out-switching all of its vertices.

It would be interesting to give precise conditions for when a graph benefits from its pc-list representation, but for now, our intuition tells us that very dense graphs and "unbalanced" graphs (graphs with vanishingly few vertices of average valency) appear to benefit most from the pc-list representation. On the other hand, since almost all graphs are roughly $n/2$-regular, the pc-list does not provide an asymptotically smaller representation for most graphs. The techniques of [3,6] are notable since they give logarithmic speedups in runtime for almost all graphs.

4.2 Hopcroft-Karp Bipartite Maximum Matching

Notice that switching in general does not preserve the bipartite property. Given a bipartite graph G, define the *bipartite switching class* of G (with respect to some switching operator), such that neighbors of v become nonneighbors and nonneighbors of v that lie outside of v's partition class become neighbors. In this section, all switches are assumed to be bipartite out-switches. Familiarity with the bipartite maximum matching problem is assumed.

The Hopcroft-Karp algorithm consists of *phases*, each of which strictly increases the size of a current matching. In [8] it is shown that a phase consists of a call to a modified BFS routine followed by a call to a modified DFS and that only $O(\sqrt{n})$ phases are needed to compute a maximum matching. These routines can be implemented to run in $O(n+m)$ time which gives rise to a $O(\sqrt{n}(n+m))$ bound for computing a maximum matching of a bipartite graph.

The purpose of running the modified BFS is to discover a directed acyclic level graph L such that any path connecting two unmatched vertices in L corresponds to a shortest augmenting path in G. A modified DFS is then conducted on L in order to find a maximal set of vertex-disjoint augmenting paths $\mathcal{P}$ in G. These steps are repeated until a phase is encountered such that $\mathcal{P} = \emptyset$, in which case the matching M must be maximum by a theorem of Berge.

It suffices to show that given $\widetilde{G}$, a phase of Hopcroft-Karp can be implemented to run in $O(n + \widetilde{m})$ time. Let BFS* and DFS* be pc-list implementations of the aforementioned modified BFS and DFS routines. The BFS* routine is essentially the same as the BFS algorithm of [4] except for the following modifications. Let $level(v)$ denote the BFS level of a vertex v.

1. The undiscov ered vertices are divided into two doubly-linked lists U_A and U_B. If the current vertex $v \in A$, then BFS* only considers vertices in U_B (similarly for $v \in B$).
2. The discovered vertices are kept in an array of doubly-linked lists $\mathcal{L}$ that represents a partition of the discovered vertices by BFS level. In particular, a discovered vertex v resides in the doubly-linked list at index $level(v)$ of $\mathcal{L}$. This array represents the levels in the DAG L.
3. Let k be the level of the first unmatched vertex in B that has been dequeued. Once all vertices at level k have been dequeued, the routine returns $\mathcal{L}$.

It is clear that the aforementioned modifications to the BFS routine of [4] can be accomplished in $O(n + \widetilde{m})$ time

The vertices of $\mathcal{L}$ form the initial set of undiscovered vertices for the modified DFS routine. Notice that explicitly constructing the level DAG L from $\mathcal{L}$ would exceed the $O(n+\widetilde{m})$ time bound; however, this is unnecessary since $\mathcal{L}$ provides an implicit representation of L. A precondition to the DFS algorithm of [10] is that doubly-linked list of undiscovered vertices is ordered according to the ordering of the vertices of the pc-list. For each $i \in \{1 \cdots k\}$, the ordering of $\mathcal{L}[i]$ might not respect the ordering of the vertices in the sorted pc-list; however, this can be corrected by performing a radix-sort on $\mathcal{L}$ in $O(n)$ time. The DFS* routine is the same as the DFS algorithm in [10] except for the following modifications.

1. If a vertex $v \in \mathcal{L}[level(v)]$ is current, then U points to the doubly-linked list $\mathcal{L}[level(v)+1]$ so that v only considers undiscovered neighbors at $level(v)+1$.
2. When a path P from the designated source vertex s to an unmatched vertex in B has been found, the routine adds $P - s$ to the set of vertex-disjoint augmenting paths and restarts DFS* from s.

The vertices of $P - s$ cannot be considered in subsequent DFS* searches since they are removed from $\mathcal{L}$ once they are discovered. Since the union of vertex-disjoint paths $\mathcal{P}$ has size $O(n)$, it is clear that DFS* can be implemented to run in $O(n + \widetilde{m})$ time.

Finally, since a set of matched edges M has size $O(n)$, it is clear that the symmetric difference between M and the edges of the paths in $\mathcal{P}$ can be computed in $O(n)$ time. This completes the description of a phase. Because BFS* and DFS* are $O(n + \tilde{m})$ and updating a matching takes $O(n)$ time, we obtain the following result.

Lemma 4. *A phase of Hopcroft-Karp can be implemented in* $O(n + \tilde{m})$ *time.*

Theorem 6. *A maximum matching of a bipartite graph G can be computed in* $O(n^{1.5} + \sqrt{n}\tilde{m} + m)$ *time.*

In [1] it is shown that finding maximum matchings of complement biclique graphs can be used as a heuristic to enumerate large cliques in graphs. It is possible that the pc-list could make this heuristic more efficient in practice by circumventing the construction of the complement.

4.3   Gabow-Tarjan Maximum Matching

A full discussion of the Gabow-Tarjan maximum cardinality matching algorithm [7] is beyond the scope of this paper. The following is only a rough sketch of the algorithm. Familiarity with the maximum matching problem is assumed.

It is well known that the maximum matching problem for general graphs is complicated by the existence of alternating odd cycles (blossoms) [5]. The Gabow-Tarjan maximum matching algorithm consists of $\lceil\sqrt{n}\rceil$ iterations of so-called *phase 1* (not to be confused with Hopcroft-Karp's phases) followed by $O(\sqrt{n})$ calls to *find_ap_set* [7]. Phase 1 consists of running *find_ap_set* on a contracted graph G/β and expanding unweighted blossoms afterwards. The *find_ap_set* routine is a depth-first based search responsible for discovering augmenting paths and blossoms of the input graph. A pre-condition to *find_ap_set* is that the input graph G/β is contracted with respect to a set of blossoms β. Once *find_ap_set* halts, the routine returns a maximal set of vertex-disjoint augment paths in the contracted graph. These paths in the contracted graph are translated into vertex-disjoint augmenting paths in the original graph in $O(n)$ time and the current matching is updated with respect to those paths. In addition to finding these paths, an execution discovers a set of new blossoms as well as a set of previously discovered blossoms to be expanded. This gives rise to a new set of blossoms to be contracted before the next call to *find_ap_set*. A blossom is expanded if its dual variable z becomes non-positive, which establishes an invariant that contracted blossoms $\hat{b}$ always have positive weight after the first execution of *find_ap_set*. Unweighted blossoms are expanded because they might be "hiding augmenting paths" [5]. Expanding these blossoms gives *find_ap_set* a chance to find these hidden augmenting paths in the next iteration. The maximum cardinality case is treated as a special case of their minimum weight matching algorithm and is shown to run in time $O(\sqrt{n}(n + m))$ since the algorithm performs $O(\sqrt{n})$ iterations of *find_ap_set* which runs in time $O(n+m)$. In [7] it is shown that blossom, augmenting path, and dual variable maintenance all take $O(n)$ time and space during an iteration of phase 1.

Lemma 3 shows that we can build the contracted graph in time proportional to $\widetilde{G}$, so it remains to show that $find\_ap\_set$ can be conducted in $O(n + \widetilde{m})$ time. We shall assume that we have obtained a minimum member $\widetilde{G}$ of G's out-switching class. Recall that a vertex of $\widetilde{G}$ is switched if and only if it has $n/2$ or more neighbors in the original graph. It follows that we can build an adjacency lookup vector $\mathbf{v}[]$ for each switched vertex v in $O(n + m)$ time. This lookup vector will allow us to answer if some vertex u is adjacent to a switched vertex v in the original graph in $O(1)$ time.

Lemma 5. *The find_ap_set routine can be implemented in $O(n + \widetilde{m})$ time.*

It suffices to show that the subroutine $find\_ap$ can be implemented in $O(n + \widetilde{m})$ time [7]. Let $b(v)$ denote the blossom that v currently belongs to. It is important to note that $find\_ap$ only performs grow steps and blossom steps. We show how to perform each of these steps when the current vertex is switched. The following invariants will be needed to simplify the analysis of our modification to $find\_ap$.

1. Let x be the current vertex. If an edge xy is scanned and $b(y)$ became outer after $b(x)$, then a blossom step is performed and every vertex in $b(y)$ has been completely scanned [7].
2. The order in which vertices become outer is given by the ordering of a doubly-linked list OUT.
3. If the current vertex performs a blossom step, then no more grow steps are possible from the current vertex.

The second and third invariants are not part of the specification of their algorithm; however, it can be implemented to maintain these invariants without affecting the correctness or resource bounds of the algorithm.

It is straightforward to modify the DFS algorithm of [10] so that when a current outer vertex u is considering an inner undiscovered neighbor v, the next outer current vertex becomes v' where $vv' \in M$ is a matched edge. The details of the blossom step are slightly more involved. It helps to view the blossom steps as DFS where the "undiscovered vertices" are those vertices that have an outer label that is greater than the current vertex's outer label. Once a blossom step is performed, the inner vertices along that blossom become current (and therefore outer) in order of their DFS discovery time and the routine proceeds recursively.

Assume that a current vertex x has no more grow steps, that is, there are no more undiscovered vertices reachable from x. If x is not switched, then proceed as usual in $find\_ap$; otherwise, assume that x is switched. The invariants guarantee that the blossom steps from x can be conducted by performing DFS where the "undiscovered vertices" are those vertices to the right of x in OUT. Let us refer to this ordered set of vertices as $OUT_{>x}$. The main obstacle is that we cannot use the DFS routine of [10] immediately to solve this problem since the ordering of the pc-list does not respect the ordering of $OUT_{>x}$. For this reason we must use $\mathbf{x}[]$ to determine adjacency in $O(1)$ time.

Let $y \in OUT_{>x}$. If $\mathbf{x}[y] = 1$, then a blossom step is performed. When a blossom step is performed, then $b(y)$ is removed from OUT since invariant 1 implies that these vertices have already been completely scanned and cannot be further used to discover an augmenting path. Assuming the blossom data-structures in [7], it follows that blossom steps take $O(n + \widetilde{m})$ time.

If $\mathbf{x}[y] = 0$, then no blossom step is performed, but we can charge the lookup to y's entry in $\widetilde{N}(x)$. But as in the routines of [4, 10], we must guarantee that x considers y only $O(1)$ times throughout the entire DFS execution on $OUT_{>x}$. To ensure this, each time we encounter a y such that $\mathbf{x}[y] = 0$, we add it to the end of an auxiliary doubly-linked neighborlist for $\widetilde{N}'(x)$. This has the effect of building an ordered neighborlist for x that respects the ordering of OUT. Using this neighborlist, we can follow the restart step of the DFS algorithm in [10] to correctly and efficiently resume DFS on $OUT_{>x}$ when x returns from a recursive call. As in the bipartite case, keeping track of augmenting paths and updating the current matching can be accomplished in $O(n)$ time. Since $find\_ap\_set$ can be implemented in $O(n + \widetilde{m})$ time, we obtain the following result.

Theorem 7. *A maximum matching of a graph G can be computed in $O(n^{1.5} + \sqrt{n\widetilde{m}} + m)$ time.*

5   Conclusions

We have demonstrated that switching classes can be used to obtain asymptotically better bounds for several graph algorithms through use of the pc-list data-structure. These improvements on algorithm resource bounds suggest that the pc-list is a more efficient data-structure than the adjacency list for several unweighted graph problems. But like any graph representation, it has its trade-offs. For instance, finding an arbitrary neighbor of a switched vertex v in the original graph takes $\Theta(|\widetilde{N}(v)|)$ time whereas finding an arbitrary neighbor of an unswitched vertex takes $\Theta(1)$ time.

It may be tempting to believe that any unweighted graph algorithm can be implemented to work with a pc-list representation; however, it seems unlikely that the maximum matching algorithm of [15] is amenable to pc-lists. This is due to the fact that their approach requires a number of edges (so-called bridges) to be queued and processed at a later point in the execution of the algorithm. When a bridge is processed, it does not always progress the algorithm, that is, it may not lead to a grow step, produce an augmenting path, or discover a blossom. In light of this, processing a bridge cannot always be charged to a vertex or edge of $\widetilde{G}$. We were unable to obtain a bound tighter than $O(m)$ for the number of bridges queued throughout the algorithm; however, it might be possible to modify the algorithm so that only bridges that lead to an augmenting path or blossom are considered.

An obvious line of future work would be towards a more formal characterization of graphs that benefit from pc-list representations as well as the development of more pc-list algorithms. We conclude by thanking Ross M. McConnell for his insightful comments.

References

1. Balas, E., Niehaus, W.: Finding large cliques in arbitrary graphs by bipartite matching. In: Johnson, D.S., Trick, M.A. (eds.) Cliques, Colouring, and Satisfiability, Second DIMACS Implementations Challenge. DIMACS Series in Discrete Mathematics and Theoretical Computer Science, vol. 26, pp. 29–52. American Mathematical Society, Providence (1996)
2. Cheng, Y., Wells, A.L.: Switching classes of directed graphs. J. Comb. Theory, Ser. B **40**(2), 169–186 (1986). http://www.sciencedirect.com/science/article/pii/0095895686900754
3. Cheriyan, J., Mehlhorn, K.: Algorithms for dense graphs and networks on the random access computer. Algorithmica **15**(6), 521–549 (1996)
4. Dahlhaus, E., Gustedt, J., McConnell, R.M.: Partially complemented representations of digraphs. Discrete Math. Theor. Comput. Sci. **5**(1), 147–168 (2002)
5. Edmonds, J.: Paths, trees, and flowers. Can. J. Math. **17**, 449–467 (1965). http://dx.doi.org/10.4153/CJM-1965-045-4
6. Feder, T., Motwani, R.: Clique partitions, graph compression and speeding-up algorithms. J. Comput. Syst. Sci. **51**(2), 261–272 (1995)
7. Gabow, H.N., Tarjan, R.E.: Faster scaling algorithms for general graph-matching problems. J. ACM **38**(4), 815–853 (1991)
8. Hopcroft, J.E., Karp, R.M.: An $n^{5/2}$ algorithm for maximum matchings in bipartite graphs. SIAM J. Comput. **2**(4), 225–231 (1973)
9. Jelínková, E., Suchý, O., Hliněný, P., Kratochvíl, J.: Parameterized problems related to seidel's switching. Discrete Math. Theor. Comput. Sci. **13**(2), 19–44 (2011)
10. Joeris, B., Lindzey, N., McConnell, R.M., Osheim, N.: Simple DFS on the complement of a graph and on partially complemented digraphs. Inf. Process. Lett., arxiv.org (2013, submitted)
11. Kao, M.Y., Occhiogrosso, N., Teng, S.H.: Simple and efficient graph compression schemes for dense and complement graphs. J. Comb. Optim. **2**(4), 351–359 (1998)
12. Karpinski, M., Schudy, W.: Linear time approximation schemes for the gale-berlekamp game and related minimization problems. In: STOC, pp. 313–322 (2009)
13. McConnell, R.M.: Complement-equivalence classes on graphs. In: Mycielski, J., Rozenberg, G., Salomaa, A. (eds.) Structures in Logic and Computer Science. LNCS, vol. 1261, pp. 174–191. Springer, Heidelberg (1997)
14. McConnell, R.M., Spinrad, J.: Modular decomposition and transitive orientation. Discrete Math. **201**(1–3), 189–241 (1999)
15. Micali, S., Vazirani, V.V.: An o(sqrt(n) m) algorithm for finding maximum matching in general graphs. In: FOCS, pp. 17–27 (1980)
16. Roth, R.M., Viswanathan, K.: On the hardness of decoding the gale-berlekamp code. IEEE Trans. Inf. Theory **54**(3), 1050–1060 (2008)
17. Seidel, J.J.: A survey of two-graphs. In: Colloquio Internazionale sulle Teorie Combinatorie, pp. 481–511 (1976)

Study of $\kappa(D)$ for $D = \{2, 3, x, y\}$

Daniel Collister and Daphne Der-Fen Liu$^{(\boxtimes)}$

California State University Los Angeles, Los Angeles, CA 90032, USA
collister.d@gmail.com, dliu@calstatela.edu

Abstract. Let D be a set of positive integers. The *kappa value* of D, denoted by $\kappa(D)$, is the parameter involved in the so called "lonely runner conjecture." Let x, y be positive integers, we investigate the kappa values for the family of sets $D = \{2, 3, x, y\}$. For a fixed positive integer $x > 3$, the exact values of $\kappa(D)$ are determined for $y = x + i$, $1 \le i \le 6$. These results lead to some asymptotic behavior of $\kappa(D)$ for $D = \{2, 3, x, y\}$.

1 Introduction

Let D be a set of positive integers. For any real number x, let $||x||$ denote the minimum distance from x to an integer, that is, $||x|| = \min\{\lceil x \rceil - x, x - \lfloor x \rfloor\}$. For any real t, denote $||tD||$ the smallest value of $||td||$ among all $d \in D$. The *kappa value of D*, denoted by $\kappa(D)$, is the supremum of $||tD||$ among all real t. That is,

$$\kappa(D) := \sup\{\alpha : ||tD|| \ge \alpha \text{ for some } t \in \Re\}.$$

Wills [20] conjectured that $\kappa(D) \ge 1/(|D| + 1)$ is true for all finite sets D. This conjecture is also known as the *lonely runner conjecture* by Bienia et al. [2]. Suppose m runners run laps on a circular track of unit circumference. Each runner maintains a constant speed, and the speeds of all the runners are distinct. A runner is called *lonely* if the distance on the circular track between him or her and every other runner is at least $1/m$. Equivalently, the conjecture asserts that for each runner, there is some time t when he or she becomes lonely. The conjecture has been proved true for $|D| \le 6$ (cf. [1,3,6,7]), and remains open for $|D| \ge 7$.

For a set D of positive integers, the parameter $\kappa(D)$ is closely related to another parameter of D called the "density of integral sequences with missing differences." For a set D of positive integers, a sequence S of non-negative integers is called a *D-sequence* if $|x - y| \notin D$ for any $x, y \in S$. Denote $S(n)$ as $|S \cap \{0, 1, 2, \cdots, n-1\}|$. The upper density $\overline{\delta}(S)$ and the lower density $\underline{\delta}(S)$ of S are defined, respectively, by $\overline{\delta}(S) = \overline{\lim}_{n \to \infty} S(n)/n$ and $\underline{\delta}(S) = \underline{\lim}_{n \to \infty} S(n)/n$. We say S has density $\delta(S)$ if $\overline{\delta}(S) = \underline{\delta}(S) = \delta(S)$. The parameter of interest is the *density of D*, $\mu(D)$, defined by

$$\mu(D) := \sup \{ \delta(S) : S \text{ is a } D\text{-sequence}\}.$$

Supported in part by the National Science Foundation under grant DMS-1247679.

J. Kratochvíl et al. (Eds.): IWOCA 2014, LNCS 8986, pp. 250–261, 2015.
DOI: 10.1007/978-3-319-19315-1_22

It is known that for any set D (cf. [4,14]):

$$\mu(D) \geq \kappa(D). \tag{1}$$

For two-element sets $D = \{a, b\}$, Cantor and Gordon [4] proved that $\kappa(D) = \mu(D) = \frac{\lfloor \frac{a+b}{2} \rfloor}{a+b}$. For 3-element sets D, if $D = \{a, b, a + b\}$ it was proved that $\kappa(D) = \mu(D)$ and the exact values were determined (see Theorem 1 below). For the general case $D = \{i, j, k\}$, various lower bounds of $\kappa(D)$ were given by Gupta [11], in which the values of $\mu(D)$ were also studied. In addition, among other results it was shown in [11] that if D is an arithmetic sequence then $\kappa(D) = \mu(D)$ and the value was determined.

The parameters $\kappa(D)$ and $\mu(D)$ are closely related to coloring parameters of distance graphs. Let D be a set of positive integers. The *distance graph generated by D*, denoted as $G(\mathbb{Z}, D)$, has all integers $\mathbb{Z}$ as the vertex set. Two vertices are adjacent whenever their absolute value difference falls in D. The *chromatic number* (minimum number of colors in a proper vertex-coloring) of the distance graph generated by D is denoted by $\chi(D)$. It is known that $\chi(D) \leq \lceil 1/\kappa(D) \rceil$ for any set D (cf. [21]).

The *fractional chromatic number* of a graph G, denoted by $\chi_f(G)$, is the minimum ratio m/n $(m, n \in \mathbb{Z}^+)$ of an (m/n)-coloring, where an (m/n)-coloring is a function on $V(G)$ to n-element subsets of $[m] = \{1, 2, \cdots, m\}$ such that if $uv \in E(G)$ then $f(u) \cap f(v) = \emptyset$. It is known that for any graph G, $\chi_f(G) \leq \chi(G)$ (cf. [14,21]).

Denote the fractional chromatic number of $G(\mathbb{Z}, D)$ by $\chi_f(D)$. Chang et al. [5] proved that for any set of positive integers D, it holds that $\chi_f(D) = 1/\mu(D)$. Together with (1) we have

$$\frac{1}{\mu(D)} = \chi_f(D) \leq \chi(D) \leq \lceil \frac{1}{\kappa(D)} \rceil. \tag{2}$$

The chromatic number of distance graphs $G(\mathbb{Z}, D)$ with $D = \{2, 3, x, y\}$ was studied by several authors (cf. [8,9]). For prime numbers x and y, the values of $\chi(D)$ for this family were first studied by Eggleton, Erdős and Skilton [10] and later on completely solved by Voigt and Walther [18]. For general values of x and y, Kemnitz and Kolberg [13] and Voigt and Walther [19] determined $\chi(D)$ for some values of x and y. This problem was completely solved for all values of x and y by Liu and Setudja [15], in which $\kappa(D)$ was utilized as one of the main tools. In particular, it was proved in [15] that $\kappa(D) \geq 1/3$ for many sets in the form $D = \{2, 3, x, y\}$. Hence, by (2), for those sets it holds that $\chi(D) = 3$.

In this article we further investigate those previously established lower bounds of $\kappa(D)$ for the family of sets $D = \{2, 3, x, y\}$. In particular, we determine the exact values of $\kappa(D)$ for $D = \{2, 3, x, y\}$ with $|x - y| \leq 6$. Furthermore, for some cases we prove $\kappa(D) = \mu(D)$. Our results also lead to some asymptotic behavior of $\kappa(D)$.

2   Preliminaries

We introduce terminologies and known results that will be used to determine the exact values of $\kappa(D)$. It is easy to see that if the elements of D have a common factor r, then $\kappa(D) = \kappa(D')$ and $\mu(D) = \mu(D')$, where $D' = D/r = \{d/r : d \in D\}$. Thus, throughout the article we assume that $\gcd(D) = 1$, unless it is otherwise indicated.

The following proposition is derived directly from definitions.

Proposition 1. *If $D \subseteq D'$ then $\kappa(D) \geq \kappa(D')$ and $\mu(D) \geq \mu(D')$.*

The next result was established by Liu and Zhu [16], after confirming a conjecture of Rabinowitz and Proulx [17].

Theorem 1. [16] *Suppose $M = \{a, b, a + b\}$ for some positive integers a and b with $\gcd(a, b) = 1$. Then*

$$\mu(M) = \kappa(M) = \max\left\{ \frac{\lfloor \frac{2b+a}{3} \rfloor}{2b+a}, \frac{\lfloor \frac{2a+b}{3} \rfloor}{2a+b} \right\}.$$

By Proposition 1, if $\{a, b, a + b\} \subseteq D$ for some a and b, then Theorem 1 gives an upper bound for $\kappa(D)$.

For a D-sequence S, denote $S[n] = |\{0, 1, 2, \ldots, n\} \cap S|$. The next result was proved by Haralambis [12].

Lemma 1. [12] *Let D be a set of positive integers, and let $\alpha \in (0, 1]$. If for every D-sequence S with $0 \in S$ there exists a positive integer n such that $\frac{S[n]}{n+1} \leq \alpha$, then $\mu(D) \leq \alpha$.*

For a given D-sequence S, we shall write elements of S in an increasing order, $S = \{s_0, s_1, s_2, \ldots\}$ with $s_0 < s_1 < s_2 < \ldots$, and denote its *difference sequence* by

$$\Delta(S) = \{\delta_0, \delta_1, \delta_2, \ldots\} \text{ where } \delta_i = s_{i+1} - s_i.$$

We say a subsequence of consecutive terms in $\Delta(S)$, $\delta_a, \delta_{a+1}, \ldots, \delta_{a+b-1}$, generates a *periodic interval* of k copies, $k \geq 1$, if $\delta_{j(a+b)+i} = \delta_{a+i}$ for all $0 \leq i \leq b - 1$, $1 \leq j \leq k - 1$. We denote such a periodic subsequence of $\Delta(S)$ by $(\delta_a, \delta_{a+1}, \ldots, \delta_{a+b-1})^k$. If the periodic interval repeats infinitely, we simply denote it by $(\delta_a, \delta_{a+1}, \ldots, \delta_{a+b-1})$. If $\Delta(S)$ is infinite periodic, except the first finite number of terms, with the periodic interval $(t_1, t_2, \ldots, t_k)$, then the density of S is $(\sum_{i=1}^{k} t_i)/k$.

Proposition 2. *A sequence of non-negative integers S is a D-sequence if and only if $\sum_{i=a}^{b} \delta_i \notin D$ for every $a \leq b$.*

Proposition 3. *Assume* $2,3 \in D$. *If S is a D-sequence, then $\delta_i + \delta_{i+1} \geq 5$ for all i. Moreover, the equality holds only when $\{\delta_i, \delta_{i+1}\} = \{1,4\}$. Consequently, $\mu(D) \leq 2/5$.*

Lemma 2. *Let $D = \{2,3\} \cup A$. Then $\kappa(D) = 2/5$ if and only if $A \subseteq \{x : x \equiv 2,3 \pmod{5}\}$. Furthermore, if $\kappa(D) = 2/5$, then $\mu(D) = 2/5$.*

Proof. Let $D = \{2,3\} \cup A$. Assume $A \subseteq \{x : x \equiv 2,3 \pmod 5\}$. Let $t = 1/5$. Then $\|td\| \geq 2/5$ for all $d \in D$. Hence $\kappa(D) \geq 2/5$. On the other hand, the density of the infinite periodic D-sequence S with $\Delta(S) = (1,4)$ is $2/5$. By Proposition 3, this is an optimal D-sequence. Hence, $\mu(D) = 2/5$, implying $\kappa(D) = 2/5$.

Conversely, assume $\kappa(D) = 2/5$. Then $\mu(D) \geq 2/5$. By Proposition 3, $\mu(D) = 2/5$. By Proposition 2, this implies that if $d \in D$, then $d \not\equiv 0,1,4 \pmod 5$. Thus the result follows.   □

Note, in $D = \{2,3,x,y\}$, if $x = 1$, then it is known [16] and easy to see that $\mu(D) = \kappa(D) = 1/4$ if y is not a multiple of 4 (with $\Delta(S) = (4)$); otherwise $y = 4k$ and $\mu(D) = \kappa(D) = k/(4k+1)$ (with $\Delta(S) = ((4)^{k-1}5)$). Hence throughout the article we assume $x > 3$.

Another method we will utilize is an alternative definition of $\kappa(D)$. In this definition, for a projected lower bound α of $\kappa(D)$, for each element z in D the valid *time* t for z to achieve α is expressed as a union of disjoint intervals. Let $\alpha \in (0, \frac{1}{2})$. For positive integer i, define $I_i(\alpha) = \{t \in (0,1) : \| ti \| \geq \alpha\}$. Equivalently,

$$I_i(\alpha) = \{t : n + \alpha \leq ti \leq n + 1 - \alpha, 0 \leq n \leq i - 1\}.$$

That is, I_i consists of intervals of reals with length $(1 - 2\alpha)/i$ and centered at $(2n + 1)/(2i)$, $n = 0, 1, \ldots, i - 1$. By definition, $\kappa(D) > \alpha$ if and only if $\bigcap_{i \in D} I_i(\alpha) \neq \emptyset$. Thus,

$$\kappa(D) = \sup \left\{ \alpha \in (0, \frac{1}{2}) : \bigcap_{i \in D} I_i(\alpha) \neq \emptyset \right\}.$$

Observe that if $\bigcap_{i \in D} I_i(\alpha)$ consists of only isolated points, then $\kappa(D) \leq \alpha$. Hence, we have the following:

Proposition 4. *For a set D, $\kappa(D) = d/c$ if $\bigcap_{i \in D} I_i$ is a set of isolated points,* where

$$I_i = \bigcup_{n=0}^{i-1} \left[\frac{d + cn}{i}, \frac{c - d + cn}{i} \right].$$

3 $D = \{2,3,x,y\}$ for $y = x+1, x+2, x+3$

Theorem 2. *Let $D = \{2,3,x,x+1\}$, $x \geq 4$. Then*

$$\kappa(D) = \mu(D) = \begin{cases} \dfrac{2\lfloor \frac{x+3}{5} \rfloor + 1}{x+3} & \text{if } x \equiv 1 \pmod 5; \\[2mm] \dfrac{2\lfloor \frac{x+3}{5} \rfloor}{x+3} & \text{otherwise.} \end{cases}$$

Proof. We prove the following cases.

Case 1. $x = 5k + 2$. The result follows by Lemma 2.

Case 2. $x = 5k + 3$. Let $t = (k+1)/(5k+6)$. Then $||dt|| \geq (2k+2)/(5k+6)$ for every $d \in D$. Hence $\kappa(D) \geq (2k+2)/(5k+6)$.

By (1) it remains to show that $\mu(D) \leq (2k+2)/(5k+6)$. Assume to the contrary that $\mu(D) > (2k+2)/(5k+6)$. By Lemma 1, there exists a D-sequence S with $S[n]/(n+1) > (2k+2)/(5k+6)$ for all $n \geq 0$. This implies, for instance, $S[0] \geq 1$, so $s_0 = 0$; $S[2] \geq 2$, so $s_1 = 1$ (as $2, 3 \in D$); $S[5] \geq 3$, so $s_3 = 5$. Moreover, $S[5k+5] \geq 2k+3$. By Proposition 3, it must be $(\delta_0, \delta_1, \delta_2, \ldots, \delta_{2k+1}) = (1, 4, 1, 4, \ldots, 1, 4)$. This implies $5k + 5 \in S$, which is impossible since $1 \in S$ and $5k + 4 \in D$. Therefore, $\mu(D) = \kappa(D) = (2k+2)/(5k+6)$.

Case 3. $x = 5k + 4$. Let $t = (k+1)/(5k+7)$. Then $||dt|| \geq (2k+2)/(5k+7)$ for all $d \in D$. Hence $\kappa(D) \geq (2k+2)/(5k+7)$.

By (1) it remains to show that $\mu(D) \leq (2k+2)/(5k+7)$. Assume to the contrary that $\mu(D) > (2k+2)/(5k+7)$. By Lemma 1, there exists a D-sequence S with $S[n]/(n+1) > (2k+2)/(5k+7)$ for all $n \geq 0$. This implies, for instance, $S[0] \geq 1$, so $s_0 = 0$; $S[3] \geq 2$, so $s_1 = 1$ (as $2, 3 \in D$); and $S[5k + 6] \geq 2k + 3$. By Proposition 3, either $5k + 5$ or $5k + 6$ is an element in S. This is impossible since $0, 1 \in S$ and $5k + 4, 5k + 5 \in D$. Thus $\mu(D) = \kappa(D) = (2k+2)/(5k+7)$.

Case 4. $x = 5k + 5$. Let $t = (k+1)/(5k+8)$. Then $||dt|| \geq (2k+2)/(5k+8)$ for all $d \in D$. Hence $\kappa(D) \geq (2k+2)/(5k+8)$.

It remains to show that $\mu(D) \leq (2k+2)/(5k+8)$. Assume to the contrary that $\mu(D) > (2k+2)/(5k+8)$. By Lemma 1, there exists a D-sequence S with $S[n]/(n+1) > (2k+2)/(5k+8)$ for all $n \geq 0$. Similar to the above, one has $0, 1 \in S$ and $S[5k+7] \geq 2k+3$. This implies that one of $5k+5, 5k+6$, or $5k+7$ is an element in S, which is again impossible. Therefore, $\mu(D) = \kappa(D) = (2k+2)/(5k+8)$.

Case 5. $x = 5k + 1$. Let $t = (k+1)/(5k+4)$. Then $||dt|| \geq (2k+1)/(5k+4)$ for all $d \in D$. Hence $\kappa(D) \geq (2k+1)/(5k+4)$.

Now we show $\mu(D) \leq (2k+1)/(5k+4)$. Assume to the contrary that $\mu(D) > (2k+1)/(5k+4)$. By Lemma 1, $(s_0, s_1) = (0, 1)$, and $S[5k+3] \geq 2k+2$. Because $S[5k] \leq 2k+1$, so $S \cap \{5k+1, 5k+2, 5k+3\} \neq \emptyset$, which is impossible. Therefore, $\mu(D) = \kappa(D) = (2k+1)/(5k+4)$. □

By the above proof, one can the sets D in Theorem 2 to the following:

Corollary 1. Let $D^* = \{2, 3, x, x+1\} \cup D'$, where $D' \subseteq \{y : y \equiv \pm2, \pm3 \pmod{(x+3)}\}$. Then $\mu(D^*) = \kappa(D^*) = \mu(\{2, 3, x, x+1\})$.

Corollary 2. Let $D = \{2, 3, x, x+1\}$. Then $\lim_{x \to \infty} \kappa(D) = \frac{2}{5}$.

Theorem 3. Let $D = \{2, 3, x, x+2\}$, $x \geq 4$. Assume $x + 4 = 6\beta + r$ with $0 \leq r \leq 5$. Then

$$\kappa(D) = \begin{cases} \frac{\lfloor \frac{x+4}{3} \rfloor}{x+4} & \text{if } 0 \leq r \leq 2; \\ \\ \frac{\lfloor \frac{2x+1}{3} \rfloor}{2x+2} & \text{if } 3 \leq r \leq 5. \end{cases}$$

Furthermore, $\kappa(D) = \mu(D)$ if $r \neq 3$.

Proof. We prove the following cases.

Case 1. $x = 6k + 2$. Then $r = 0$. Let $t = 1/6$. Then $\|dt\| \geq 1/3$ for all $d \in D$. Hence $\kappa(D) \geq 1/3$.

Now we prove $\mu(D) \leq 1/3$. Let $M' = \{2, x, x + 2\} = \{2, 6k + 2, 6k + 4\}$. By Theorem 1 with $M = \{1, 3k + 1, 3k + 2\}$, we obtain $\mu(M') = \mu(M) = 1/3$. Because $M' \subseteq D$, so $\mu(D) \leq \mu(M') = 1/3$.

Case 2. $x = 6k + 3$. Then $r = 1$. Let $t = (k + 1)/(6k + 7)$. Then $\|dt\| \geq (2k + 2)/(6k + 7)$ for all $d \in D$. Hence $\kappa(D) \geq (2k + 2)/(6k + 7)$.

By Theorem 1 with $M = \{2, x, x + 2\} = \{2, 6k + 3, 6k + 5\}$, we get $\mu(M) = (2k+2)/(6k+7)$. Because $M \subseteq D$, so $\mu(D) \leq (2k+2)/(6k+7)$. Thus, the result follows.

Case 3. $x = 6k + 4$. Then $r = 2$. Let $t = (k + 1)/(6k + 8)$. Then $\|dt\| \geq (2k + 2)/(6k + 8)$ for all $d \in D$. Hence $\kappa(D) \geq (2k + 2)/(6k + 8)$.

By Theorem 1 with $M = \{2, x, x + 2\} = \{2, 6k + 4, 6k + 6\}$ which can be reduced to $M' = \{1, 3k + 2, 3k + 3\}$, we obtain $\mu(M) = (k + 1)/(3k + 4)$. Therefore, $\mu(D) \leq \mu(M) = (2k + 2)/(6k + 8)$. So the result follows.

Case 4. $x = 6k + 5$. Then $r = 3$. Let $t = (2k + 3)/(12k + 12)$. Then $\|dt\| \geq (4k + 3)/(12k + 12)$ for all $d \in D$. Hence $\kappa(D) \geq (4k + 3)/(12k + 12)$.

By Proposition 4, it remains to show that $\bigcap\limits_{i=2,3,x,x+2} I_i$ is a set of isolated points, where

$$I_i = \bigcup_{n=0}^{i-1} \left[\frac{4k + 3 + n(12k + 12)}{i}, \frac{8k + 9 + n(12k + 12)}{i} \right].$$

Let $I = \bigcap\limits_{i=2,3,x,x+2} I_i$. By symmetry it is enough to consider the interval $I \cap [0, (12k + 12)/2]$. In the following we claim $I \cap [0, 6k + 6] = \{2k + 3\}$. (Indeed, this single point is the numerator of the t value at the beginning of the proof.) Note that $I_2 \cap I_3 \cap [0, 6k+6] = [(4k+3)/2, (8k+9)/3]$. Denote this interval by

$$I_{2,3} = \left[\frac{4k + 3}{2}, \frac{8k + 9}{3} \right].$$

We then begin to investigate possible values of n for I_x and I_{x+2}, respectively, that will fall within $I_{2,3}$. First, we compare the I_x intervals with $I_{2,3}$. Recall

$$I_x = \left[\frac{3 + 4k + n(12 + 12k)}{6k + 5}, \frac{8k + 9 + n(12 + 12k)}{6k + 5} \right], \quad 0 \leq n \leq 6k + 4.$$

By calculation, the intervals of I_x that intersect with $I_{2,3}$ are those with $n \geq k$. Similarly, we compare I_{x+2} intervals with $I_{2,3}$. Recall

$$I_{x+2} = \left[\frac{3 + 4k + n(12 + 12k)}{6k + 7}, \frac{8k + 9 + n(12 + 12k)}{6k + 7} \right], \quad 0 \leq n \leq 6k + 6.$$

By calculation, the intervals of I_{x+2} that intersect with $I_{2,3}$ are those with $n \geq k+1$.

Next, we consider the intersection between intervals of I_x and I_{x+2}. Let $n = k + a$ for some $a \geq 0$ for the I_x interval, and let $n = k + a'$ for some $a' \geq 1$ for the I_{x+2} interval. By taking the common denominator of the I_x and I_{x+2} intervals we obtain the following numerators of those intervals:

$$\text{for } I_x : [21 + 84a + 130k + 156ak + 180k^2 + 72ak^2 + 72k^3,$$

$$63 + 84a + 194k + 156ak + 204k^2 + 72ak^2 + 72k^3];$$

$$\text{for } I_{x+2} : [15 + 60a' + 98k + 132a'k + 156k^2 + 72a'k^2 + 72k^3,$$

$$45 + 60a' + 154k + 132a'k + 180k^2 + 72a'k^2 + 72k^3].$$

Using $a = a' = 1$, we get

$$\text{for } I_x : [105 + 286k + 252k^2 + 72k^3, 147 + 350k + 276k^2 + 72k^3]$$

$$\text{for } I_{x+2} : [75 + 230k + 228k^2 + 72k^3, 105 + 286k + 252k^2 + 72k^3].$$

Thus, there is a single point intersection for I_x and I_{x+2} when $a = a' = 1$, which is $\{2k+3\}$. This single point intersection is also within the $I_{2,3}$ interval. Hence, $\{2k+3\} \in I \cap [0, 6k+6]$.

In addition, through inspection it is clear that making $n = k$ (i.e. $a = 0$) for the I_x interval and $n \geq k+1$ ($a' \geq 1$) for the I_{x+2} interval removes I_x and I_{x+2} from intersecting one another. For all other cases, ($a = 1$ and $a' \geq 2$), ($a \geq 2$ and $a' = 1$), or ($a, a' \geq 2$), there will never be an intersection of intervals for all elements in D, either because the $I_{2,3}$ interval is too small or because the I_{x+2} elements become too big. Thus, $I \cap [0, 6k+6] = \{2k+3\}$.

Case 5. $x = 6k + 6$. Then $r = 4$. Let $t = (2k+3)/(12k+14)$. Then $\|dt\| \geq (4k+4)/(12k+14)$ for all $d \in D$. Hence $\kappa(D) \geq (4k+4)/(12k+14)$.

By Theorem 1 with $M = \{2, x, x+2\} = \{2, 6k+6, 6k+8\}$ which can be reduced to $M' = \{1, 3k+3, 3k+4\}$, we get $\mu(M) = \kappa(M) = (2k+2)/(6k+7)$. Hence, $\mu(D) \leq \mu(M) = (2k+2)/(6k+7)$.

Case 6. $x = 6k + 7$. Then $r = 5$. Let $t = (2k+3)/(12k+16)$. Then $\|dt\| \geq (4k+5)/(12k+16)$ for all $d \in D$. Hence $\kappa(D) \geq (4k+5)/(12k+16)$.

By Theorem 1 with $M = \{2, x, x+2\} = \{2, 6k+7, 6k+9\}$, we obtain $\mu(M) = \kappa(M) = (4k+5)/(12k+16)$. Therefore, $\mu(D) \leq (4k+5)/(12k+16)$.   □

Theorem 4. *Let* $D = \{2, 3, x, x+3\}$, $x \geq 4$. *Assume* $(2x+3) = 9\beta + r$ *with* $0 \leq r \leq 8$. *Then*

$$\kappa(D) = \begin{cases} \frac{4}{15} & \text{if } x = 10; \\ \frac{3\lfloor \frac{2x+3}{9} \rfloor}{2x+3} & \text{if } 0 \leq r \leq 5 \text{ and } x \neq 10; \\ \frac{\lfloor \frac{x+5}{3} \rfloor}{x+6} & \text{if } 6 \leq r \leq 8. \end{cases}$$

Furthermore, if $r = 0, 1, 3, 6, 8$ *then* $\kappa(D) = \mu(D)$.

Proof. We prove the following cases:

Case 1. $x = 9k + 3$. Then $r = 0$. Let $t = 2/9$. Then $||dt|| \geq 1/3$ for all $d \in D$. Hence $\kappa(D) \geq (6k + 3)/(18k + 9) = 1/3$.

By Theorem 1 with $M = \{3, x, x + 3\} = \{3, 9k + 3, 9k + 6\}$, which can be reduced to $M' = \{1, 3k + 1, 3k + 2\}$, resulting in $\mu(M) = \kappa(M) = 1/3$. Because $M \subseteq D$, so $\mu(D) = \mu(M) = 1/3$.

Case 2. $x = 9k + 8$. Then $r = 1$. Let $t = (4k + 4)/(18k + 19)$. Then $||dt|| \geq (6k + 6)/(18k + 19)$ for all $d \in D$. Hence $\kappa(D) \geq (6k + 6)/(18k + 19)$.

By Theorem 1 with $M = \{3, x, x + 3\}$, we get $\kappa(M) = (6k + 6)/(18k + 19)$. Hence, $\mu(D) \leq \kappa(M) = (6k + 6)/(18k + 19)$.

Case 3. $x = 9k + 4$. Then $r = 2$. Let $t = (4k + 2)/(18k + 11)$. Then $||dt|| \geq (6k + 3)/(18k + 11)$ for all $d \in D$. Thus, $\kappa(D) \geq (6k + 3)/(18k + 11)$.

The proof for the other direction is similar to the proof of Case 4 in Theorem 3. Let $I = \bigcap_{i=2,3,x,x+3} I_i$. By calculation we have $I \cap [0, 9k + (11/2)] = \{4k + 2\}$. This single point of intersection occurs when $n = 2k$ in the I_x interval, and $n = 2k + 1$ in the I_{x+3} interval.

Case 4. $x = 9k$. Then $r = 3$. Let $t = 4k/(18k + 3)$. Then $||dt|| \geq (6k)/(18k + 3)$ for all $d \in D$. Thus $\kappa(D) \geq (2k)/(6k + 1)$.

By Theorem 1 with $M = \{3, x, x + 3\} = \{3, 9k, 9k + 3\}$, $\mu(M) = \kappa(M) = (2k)/(6k + 1)$. Hence, the result follows.

Case 5. $x = 9k + 5$. Then $r = 4$. Let $t = (4k + 2)/(18k + 13)$. Then $||dt|| \geq (6k + 3)/(18k + 13)$ for all $d \in D$. Thus $\kappa(D) \geq (6k + 3)/(18k + 13)$.

The proof for the other direction is similar to the proof of Case 4 in Theorem 3. Let $I = \bigcap_{i=2,3,x,x+3} I_i$. By calculation we have $I \cap [0, 9k + (13/2)] = \{4k + 2\}$. This single point of intersection occurs when $n = 2k$ in the I_x interval, and $n = 2k + 1$ in the I_{x+3} interval.

Case 6. $x = 9k + 1$. Then $r = 5$. If $k = 1$, then $x = 10$. Let $t = 2/15$. Then $||dt|| \geq 4/15$ for all $d \in D$. Hence $\kappa(D) \geq 4/15$.

Assume $k \geq 2$. Let $t = (4k)/(18k + 5)$. Then $||dt|| \geq (6k)/(18k + 5)$ for all $d \in D$. Thus $\kappa(D) \geq (6k)/(18k + 5)$.

The proof for $\kappa(D) \leq (6k)/(18k + 5)$ is similar to the proof of Case 4 in Theorem 3. Let $I = \bigcap_{i=2,3,x,x+3} I_i$. By calculation we have $I \cap [0, 9k + (5/2)] = \{4k\}$. This single point of intersection occurs when $n = 2k - 1$ in the I_x interval, and $n = 2k$ in the I_{x+3} interval.

Case 7. $x = 9k + 6$. Then $r = 6$. Let $t = (2k + 3)/(9k + 12)$. Then $||dt|| \geq (3k + 3)/(9k + 12)$ for all $d \in D$. Thus $\kappa(D) \geq (k + 1)/(3k + 4)$.

By Theorem 1 with $M = \{3, x, x + 3\}$ with $M = \{3, x, x + 3\} = \{3, 9t + 6, 9t + 9\}$, which can be reduced to $M' = \{1, 3t + 2, 3t + 3\}$, we get $\mu(M) = \kappa(M) = (k + 1)/(3k + 4)$. Because $M \subseteq D$, so $\kappa(D) \leq \mu(D) \leq \mu(M) \leq \kappa(M) = (k + 1)/(3k + 4)$.

Case 8. $x = 9k + 11$. Then $r = 7$. Let $t = (2k + 4)/(9k + 17)$. Then $\|dt\| \geq (3k + 5)/(9k + 17)$ for all $d \in D$. Thus $\kappa(D) \geq (3k + 5)/(9k + 17)$.

The proof for the other direction is similar to the proof of Case 4 in Theorem 3. Let $I = \bigcap_{i=2,3,x,x+3} I_i$. By calculation we have $I \cap [0, 9k+(17/2)] = \{2k+4\}$. This single point of intersection occurs when $n = 2k + 2$ in the I_x interval, and $n = 2k + 3$ in the I_{x+3} interval.

Case 9. $x = 9k + 7$. Then $r = 8$. Let $t = (2k + 3)/(9k + 13)$. Then $\|dt\| \geq (3k + 4)/(9k + 13)$ for all $d \in D$. Thus $\kappa(D) \geq (3k + 4)/(9k + 13)$.

By Theorem 1 with $M = \{3, x, x + 3\} = \{3, 9t + 7, 9t + 10\}$, we get $\kappa(M) = (3k + 4)/(9k + 13)$. Because $M \subseteq D$, so $\kappa(D) \leq \mu(D) \leq \mu(M) = \kappa(M) = (3k + 4)/(9k + 13)$. $\square$

Corollary 3. *Let $D = \{2, 3, x, y\}$ where $y \in \{x+2, x+3\}$. Then $\lim\limits_{x \to \infty} \kappa(D) = \frac{1}{3}$.*

4 $D = \{2, 3, x, y\}$ for $y = x + 4, x + 5, x + 6$

By similar proofs to the previous section, we obtain the following results.

Theorem 5. *Let $D = \{2, 3, x, x + 4\}$, $x \geq 4$. Assume $(x + 4) = 5\beta + r$ with $0 \leq r \leq 4$. Then*

$$
\kappa(D) = \begin{cases} \frac{2\beta+r}{x+7} & \text{if } 0 \leq r \leq 1; \\[2mm] \mu(D) = \frac{2}{5} & \text{if } r = 2; \\[2mm] \frac{2\beta}{x+2} & \text{if } 3 \leq r \leq 4. \end{cases}
$$

Proof. The case for $r = 2$ is from Lemma 2. The following table gives the corresponding t, $\kappa(D)$, and the n values of I_x and I_{x+4} where the single intersection point occurs.

x	r	t	n in I_x	n in I_{x+4}	$\kappa(D)$
$5k + 4$	3	$(k + 1)/(5k + 6)$	k	$k + 1$	$(2k + 2)/(5k + 6)$
$5k + 5$	4	$(k + 1)/(5k + 7)$	k	$k + 1$	$(2k + 2)/(5k + 7)$
$5k + 6$	0	$(k + 3)/(5k + 13)$	$k + 1$	$k + 2$	$(2k + 4)/(5k + 13)$
$5k + 7$	1	$(k + 3)/(5k + 14)$	$k + 1$	$k + 2$	$(2k + 5)/(5k + 14)$
$5k + 8$	2	$1/5$			$2/5$

$\square$

Theorem 6. *Let $D = \{2, 3, x, x + 5\}$, $x \geq 4$. Assume $(x + 3) = 5\beta + r$ with $0 \leq r \leq 4$. Then*

$$
\kappa(D) = \begin{cases} \mu(D) = \frac{2}{5} & \text{if } 0 \leq r \leq 1; \\[2mm] \frac{2\beta}{x+2} & \text{if } 2 \leq r \leq 3; \\[2mm] \frac{2\beta+1}{x+3} & \text{if } r = 4. \end{cases}
$$

Proof. The cases for $r = 0, 1$ are by Lemma 2. The following table gives the corresponding t, $\kappa(D)$, and the n values of I_x and I_{x+5} where the single intersection point occurs.

x	r	t	n in I_x	n in I_{x+5}	$\kappa(D)$
$5k+4$	2	$(k+1)/(5k+6)$	$k+1$	$k+1$	$(2k+2)/(5k+6)$
$5k+5$	3	$(k+1)/(5k+7)$	$k+1$	$k+1$	$(2k+2)/(5k+7)$
$5k+6$	4	$(k+2)/(5k+9)$	$k+1$	$k+2$	$(2k+3)/(5k+9)$
$5k+7$	0	$1/5$			$2/5$
$5k+8$	1	$1/5$			$2/5$

$\square$

Theorem 7. *Let* $D = \{2,3,x,x+6\}$, $x \geq 4$. *Assume* $(x+8) = 5\beta + r$ *with* $0 \leq r \leq 4$. *Then*

$$\kappa(D) = \begin{cases} \mu(D) = \frac{2}{7} & \text{if } x = 5; \\[2mm] \mu(D) = \frac{2}{5} & \text{if } r = 0; \\[2mm] \frac{2\beta}{x+8} & \text{if } 1 \leq r \leq 3 \text{ and } x \neq 5; \\[2mm] \frac{2\beta+1}{x+3} & \text{if } r = 4. \end{cases}$$

Proof. Assume $x = 5$. That is $D = \{2,3,5,11\}$. Letting $t = 1/7$ we get $\|td\| \geq 2/7$ for every $d \in D$. Hence, $\kappa(D) \geq 2/7$. On the other hand, by Theorem 1, $\mu(\{2,3,5\}) = 2/7$. Therefore, by (2), we have $\kappa(D) \leq \mu(D) \leq 2/7$.

The case for $r = 0$ is from Lemma 2. The following table gives the corresponding t, $\kappa(D)$, and the n values of I_x and I_{x+6} where the single intersection point occurs.

x	r	t	n in I_x	n in I_{x+6}	$\kappa(D)$
$5k+4$	2	$(k+2)/(5k+12)$	k	$k+1$	$(2k+4)/(5k+12)$
$5k+5$	3	$(k+2)/(5k+13)$	k	$k+1$	$(2k+4)/(5k+13)$
$5k+6$	4	$(k+2)/(5k+9)$	$k+1$	$k+2$	$(2k+3)/(5k+9)$
$5k+7$	0	$1/5$			$2/5$
$5k+8$	1	$(k+3)/(5k+16)$	$k+1$	$k+2$	$(2k+6)/(5k+16)$

$\square$

Corollary 4. *Let* $D = \{2,3,x,y\}$, $y \in \{x+4, x+5, x+6\}$. *Then* $\lim\limits_{x \to \infty} \kappa(D) = \frac{2}{5}$.

Concluding Remark and Future Study. Similar to Corollary 1, one can obtain sets D^* that are extensions of the sets D studied in this article, $D \subset D^*$, such that $\kappa(D) = \kappa(D^*)$. Furthermore, the methods used in this article can be applied to other sets $D = \{2, 3, x, x + c\}$ with $c \geq 7$. For a fixed c, preliminary results we obtained indicate that the values of $\kappa(D)$ might be inconsistent only for the first finite terms; after a certain threshold of the values of x, they become consistent (that is, they can be described by a single formula). It is interesting to investigate the correlation between c and the asymptotic values of $\kappa(D)$. For instance, would the conclusions of Corollaries 2, 3, or 4 hold for other values of c? Are there other asymptotic values of $\kappa(D)$ besides $1/3$ and $2/5$ as indicated in Corollaries 2, 3 and 4? In particular, whether $c = 2, 3$ are the only values of c such that the conclusion of Corollary 3 is true?

References

1. Barajas, J., Serra, O.: The lonely runner with seven runners. Electron. J. Combin. **15**, #R48 (2008)
2. Bienia, W., Goddyn, L., Gvozdjak, P., Sebő, A., Tarsi, M.: Flows, view obstructions, and the lonely runner. J. Combin. Theory Ser. B **72**, 1–9 (1998)
3. Bohman, T., Holzman, R., Kleitman, D.: Six lonely runners. Electron. J. Comb. **8**, 49 (2001). Research Paper 3
4. Cantor, D., Gordon, B.: Sequences of integers with missing differences. J. Comb. Theory Ser. A **14**, 281–287 (1973)
5. Chang, G., Liu, D., Zhu, X.: Distance graphs and T-coloring. J. Comb. Theory Ser. B **75**, 159–169 (1999)
6. Cusick, T.: View-obstruction problems in n-dimensional geometry. J. Comb. Theory Ser. A **16**, 1–11 (1974)
7. Cusick, T., Pomerance, C.: View-obstruction problems, III. J. Number Theory **19**, 131–139 (1984)
8. Eggleton, R., Erdős, P., Skilton, D.: Colouring the real line. J. Comb. Theory Ser. A **39**, 86–100 (1985)
9. Eggleton, R., Erdős, P., Skilton, D.: Research problem 77. Discrete Math. **58**, 323 (1986)
10. Eggleton, R., Erdős, P., Skilton, D.: Colouring prime distance graphs. Graphs Comb. **6**, 17–32 (1990)
11. Gupta, S.: Sets of integers with missing differences. J. Comb. Theory Ser. A **89**, 55–69 (2000)
12. Haralambis, N.: Sets of integers with missing differences. J. Comb. Theory Ser. A **23**, 22–33 (1997)
13. Kemnitz, A., Kolberg, H.: Coloring of integer distance graphs. Discrete Math. **191**, 113–123 (1998)
14. Liu, D.: From rainbow to the lonely runner: a survey on coloring parameters of distance graphs. Taiwanese J. Math. **12**, 851–871 (2008)
15. Liu, D., Sutedja, A.: Chromatic number of distance graphs generated by the sets $\{2, 3, x, y\}$. J. Comb. Optim. **25**, 680–693 (2013)
16. Liu, D., Zhu, X.: Fractional chromatic number and circular chromatic number for distance graphs with large clique size. J. Graph Theory **47**, 129–146 (2004)
17. Rabinowitz, J., Proulx, V.: An asymptotic approach to the channel assignment problem. SIAM J. Alg. Discrete Methods **6**, 507–518 (1985)
18. Voigt, M., Walther, H.: Chromatic number of prime distance graphs. Discrete Appl. Math. **51**, 197–209 (1994)

19. Voigt, M., Walther, H.: On the chromatic number of special distance graphs. Discrete Math. **97**, 395–397 (1991)
20. Wills, J.: Zwei Sätze über inhomogene diophantische appromixation von irrationlzahlen. Monatsch. Math. **71**, 263–269 (1967)
21. Zhu, X.: Circular chromatic number: a survey. Discrete Math. **229**, 371–410 (2001)

Some Hamiltonian Properties
of One-Conflict Graphs

Christian Laforest and Benjamin Momège[(⊠)]

LIMOS, CNRS UMR 6158 – Université Blaise Pascal, Clermont-Ferrand,
Campus des Cézeaux, 24 Avenue des Landais, 63173 Aubiére Cedex, France
{laforest,momege}@isima.fr

Abstract. Dirac's and Ore's conditions (1952 and 1960) are well known
and classical sufficient conditions for a graph to contain a Hamiltonian
cycle and they are generalized in 1976 by the Bondy-Chvátal Theorem.
In this paper, we add constraints, called *conflicts*. A conflict in a graph G
is a pair of distinct edges of G. We denote by $(G, Conf)$ a graph G with
a set of conflicts $Conf$. A *path without conflict* P in $(G, Conf)$ is a path
P in G such that for any edges e, e' of P, $\{e, e'\} \notin Conf$. In this paper
we consider graph with conflicts such that each vertex is not incident
to the edges of more than one conflict. We call such graphs *one-conflict
graphs*. We present sufficient conditions for one-conflict graphs to have a
Hamiltonian path or cycle without conflict.

Keywords: Graph · Conflict · Hamiltonian · Path · Cycle

1 Introduction

These last decades many works have been done to prove sufficient conditions for
a graph to have a Hamiltonian cycle. The most classical ones in a graph with n
vertices are the following: Dirac's condition [3] (the degree of each vertex is at
least $\frac{n}{2}$), Ore's condition [8] (for any non adjacent vertices u and v, degree of u
plus degree of v is at least n), Bondy-Chvátal's condition (based on closure of
graphs). For more results in this area, see the recent survey [7]. More recently,
several papers were devoted to the introduction of *conflicts* into graphs. A conflict is a pair of edges of the graph and, as they are in conflict, they cannot
be both in a structure as a path or a tree. Conflicts are useful to model situations in which it is forbidden to use two incompatible objects (because of their
nature, functions, etc.) in the same structure. In [5,9] the authors investigated
the problem to find a path without conflict between two vertices in a graph with
a given set of restrictive conflicts (the two edges of a conflict are adjacent). For
these graphs, the problem of finding two-factors[1] was considered in [4] and a
dichotomy between tractable and intractable instances was also given. In [6],
the authors investigated the problem of finding a spanning tree without conflict.

B. Momège has a PhD grant from CNRS and région Auvergne.

[1] A subgraph such that for any vertex its in-degree and its out-degree is exactly one.

© Springer International Publishing Switzerland 2015
J. Kratochvíl et al. (Eds.): IWOCA 2014, LNCS 8986, pp. 262–273, 2015.
DOI: 10.1007/978-3-319-19315-1_23

All the known results show that adding conflicts to a graph considerably increases the complexity of problems. In this paper we try to extend the classical Dirac's, Ore's and Bondy-Chvátal's results to graphs in which each vertex is part of at most one conflict (called *one-conflict graphs*). We show that it is not possible in all cases. We then propose sufficient conditions (inspired by Dirac's and Ore's ones) for a one-conflict graph to contain a Hamiltonian path or cycle without conflict and Bondy-Chvátal-type conditions.

2   Preliminaries, Notations and Definitions

In this paper, we only consider undirected, unweighted and simple graphs. We refer to [2] for definitions and undefined notations. The vertex set of a graph G is denoted by V_G and its edge set by E_G. If $|V_G| = n$, the graph G is called an n-vertex graph. An edge between u and v in a graph G is denoted by uv. The two endpoints of an edge are said to be adjacent to each other. The *complete graph* K_n is a graph with n vertices in which every vertex is adjacent to every other. A path in G consists of a sequence of vertices with each two consecutive vertices in the sequence adjacent to each other in the graph. A simple path is a path with no repetitions of vertices. A cycle in G consists of a sequence of vertices starting and ending at the same vertex, with each two consecutive vertices in the sequence adjacent to each other in the graph. We only consider simple cycle *i.e.* no repetitions of vertices allowed, other than the repetition of the starting and ending vertex. A path (or a cycle) of G is *Hamiltonian* if it contains all the vertices of G exactly once (all paths and cycles are elementary here). The *length* of a path or a cycle is the number of edges it contains. For example, in an n-vertex graph the length of a Hamiltonian path is $n-1$ and the length of a Hamiltonian cycle is n. An edge and a vertex on that edge are called *incident*. The *degree* of a vertex v of G is the number of edges incident to v. It is denoted by $deg_G(v)$. The *minimum degree* of G is denoted by $\delta(G)$ (*i.e.* $\delta(G) = \min_{v \in V_G} deg_G(v)$). A *matching* in a graph is a subgraph where each vertex has degree 1 (*i.e.* a set of edges without common vertices). If G is a graph, a *conflict* in G is a pair $\{e_1, e_2\}$ of distinct edges of G. We denote by $(G, Conf)$ a graph G with a set of conflicts $Conf$. A *path without conflict* P in $(G, Conf)$ is a path P in G such that for any e, e' of P, $\{e, e'\} \notin Conf$ (similarly for a simple path without conflict, Hamiltonian path without conflict, cycle without conflict and Hamiltonian cycle without conflict). In this paper we only consider graph with conflicts such that each vertex is not incident to the edges of more than one conflict (each vertex is not involved in more than one conflict). We call such graphs *one-conflict graphs*. From now, $(G, Conf)$ is an n-vertex one-conflict graph.

3   Dirac-Type Conditions in One-Conflict Graphs

We recall the result of Dirac (1952) for "classical" graphs (*i.e.* without conflict):

Theorem 1 (Dirac). *[3] An n-vertex graph ($n \geq 3$) G s.t. $\delta(G) \geq \frac{n}{2}$ contains a Hamiltonian cycle.*

Firstly, one can deduce from Theorem 1 the following result:

Corollary 2. *An n-vertex one-conflict graph $(G, Conf)$ s.t. $\delta(G) \geq \frac{n}{2} + 1$ contains a Hamiltonian cycle without conflict.*

Proof. Simply remove one edge of each conflict and note that the resulting graph satisfies the conditions of Theorem 1 and therefore admits a Hamiltonian cycle that is a Hamiltonian cycle without conflict in $(G, Conf)$. □

We want to show the following result:

Theorem 3. *An n-vertex one-conflict graph $(G, Conf)$ s.t. $\delta(G) \geq \frac{n}{2}$ contains a Hamiltonian path without conflict.*

Our proof is constructive. For $n = 1, 2, 3$ or 4, it is easy to see that the theorem is true. From now, we suppose that n is greater than or equal to 5.

Lemma 4. *An n-vertex one-conflict graph $(G, Conf)$ s.t. $\delta(G) \geq \frac{n}{2}$ contains a simple path without conflict of length at least $\frac{n}{2} - 1$ that can be constructed using the method described in the proof.*

Proof. We start from any vertex of $(G, Conf)$ and gradually construct a simple path without conflict, as long as possible, by adding an adjacent vertex of one of its endpoints. As long as the length of the path is less than $\frac{n}{2} - 1$ each of its endpoints is adjacent to at least two vertices that are exterior to it and it can be extended while remaining without conflict. By doing this as much as possible we are sure to get a simple path without conflict of length at least $\frac{n}{2} - 1$. □

Lemma 5. *Let $P = v_1, \ldots, v_{k+1}$ be a simple path without conflict of length k in an n-vertex one-conflict graph $(G, Conf)$ s.t. $\delta(G) \geq \frac{n}{2}$, that cannot be extended using the method described in the proof of Lemma 4. If $k \leq n - 3$, $(G, Conf)$ contains a cycle without conflict of length $k + 1$ that can be constructed using the method described in the proof.*

Proof. Begin by noting that as $n \geq 5$ we have $k \geq 2$. We remove one edge outside P from each conflict. We denote by G' the graph obtained. A Hamiltonian path in G' is then a Hamiltonian path without conflict in $(G, Conf)$. Each vertex of G' is of degree greater than or equal to $\frac{n}{2} - 1$. In the new graph v_1 and v_{k+1} are each adjacent to at least $\frac{n}{2} - 1$ vertices of P (otherwise we could extend P using the method described in the proof of Lemma 4). Moreover, there is $1 \leq i \leq k$ such that v_1 is adjacent to v_{i+1} and v_{k+1} is adjacent to v_i. Indeed, if for any neighbor v_{i+1} of v_1, v_i is not adjacent to v_{k+1}, at least $\frac{n}{2} - 1$ vertices of P are not adjacent to v_{k+1}. Thus, v_{k+1} would have at most $k - (\frac{n}{2} - 1)$ neighbors in P, and as $k - (\frac{n}{2} - 1) = k - \frac{n}{2} + 1 \leq n - 3 - \frac{n}{2} + 1 = \frac{n}{2} - 2 < \frac{n}{2} - 1$, this is impossible. Finally, if $i = 1$ or $i = k$, $v_1, \ldots, v_{k+1}, v_1$ is a cycle without conflict of length $k + 1$ and if $1 < i < k - 1$, $v_1, \ldots, v_i, v_{k+1}, \ldots v_{i+1}, v_1$ is a cycle without conflict of length $k + 1$. □

Lemma 6. *Let $C = v_1, \ldots, v_{k+1}, v_1$ be a cycle without conflict of length $k+1$ in an n-vertex one-conflict graph $(G, Conf)$ s.t. $\delta(G) \geq \frac{n}{2}$, with $\frac{n}{2} - 1 \leq k \leq n - 3$. We can construct in $(G, Conf)$ a simple path without conflict of length $k + 1$ from C by following the method described in the proof.*

Proof. Assume first $k + 1 = \frac{n}{2}$. Take any vertex v_i of the cycle C. As the degree of v_i is greater than or equal to $\frac{n}{2}$ it is adjacent to a vertex v outside the cycle. If there is no edge of C in conflict with vv_i then we add vv_i to C and remove one edge of C containing v_i to obtain a simple path without conflict of length $k + 1$. Otherwise, we denote by ab the edge of the cycle in conflict with vv_i. We then have $a \neq v_i$ or $b \neq v_i$. By symmetry we may assume $a \neq v_i$. Since a cannot be in more than one conflict, we can we can choose a initially and perform the above operations to get the desired path.

If $k + 1 > \frac{n}{2}$, take a vertex v outside the cycle. As the degree of v is greater than or equal to $\frac{n}{2}$ it is adjacent to at least two vertices v_i and v_j of C. Since v is not in more than one conflict, one of the edges vv_i or vv_j is not involved in a conflict. For example if it is vv_i, we add vv_i to C and remove one edge of C containing v_i to obtain a simple path without conflict of length $k + 1$.   □

In an n-vertex one-conflict graph $(G, Conf)$ s.t. $\delta(G) \geq \frac{n}{2}$, any simple path without conflict of length $k \leq n - 3$ can give rise to a simple path without conflict of length $k + 1$ in the following way:

- If possible, apply the techniques presented in Lemma 4,
- Otherwise, construct a cycle without conflict of length $k + 1$ using the techniques presented in Lemma 5 and then construct a simple path without conflict of length $k + 1$ from this cycle using the techniques presented in Lemma 6.

So using those operations we get:

Lemma 7. *Starting from any vertex of an n-vertex one-conflict graph $(G, Conf)$ s.t. $\delta(G) \geq \frac{n}{2}$, one can construct a simple path without conflict of length at least $n - 2$.*

If the simple path without conflict obtained with Lemma 7 is of length $n-1$, then it is a Hamiltonian path without conflict in $(G, Conf)$. Otherwise, the following result is used to construct a Hamiltonian path without conflict in $(G, Conf)$.

Lemma 8. *Any simple path without conflict $P = v_1, \ldots, v_{n-1}$ of length $n - 2$ in an n-vertex one-conflict graph $(G, Conf)$ s.t. $\delta(G) \geq \frac{n}{2}$ can give rise to a Hamiltonian path without conflict in $(G, Conf)$ by following the method described in the proof.*

Proof. We denote by v the vertex outside P. We remove one edge outside P from each conflict. We denote by G' the graph obtained. Each vertex of G' is of degree greater than or equal to $\frac{n}{2} - 1$. A Hamiltonian path in G' is then a Hamiltonian path without conflict in $(G, Conf)$. If $vv_1 \in E_{G'}$ or $vv_{n-1} \in E_{G'}$, then there exists a Hamiltonian path without conflict in $(G, Conf)$. So assume that $vv_1 \notin E_{G'}$ and $vv_{n-1} \notin E_{G'}$. If $v_1v_{n-1} \in E_{G'}$, then for every vertex v_i of P such that $vv_i \in E_{G'}$, $v, v_i, \ldots, v_{n-1}, v_1, \ldots v_{i-1}$ is a Hamiltonian path without conflict in $(G, Conf)$. Suppose then that $v_1v_{n-1} \notin E_{G'}$.

If there is $k \in \{2, \ldots, n-3\}$ such that v is adjacent to v_k and v_{k+1} then $v_1, \ldots, v_k, v, v_{k+1}, \ldots v_{n-1}$ is a Hamiltonian path without conflict in $(G, Conf)$.

Otherwise, n is even and v is adjacent to the $\frac{n}{2} - 1$ vertices of

$$A := \{v_2, v_4, \ldots, v_{n-2}\}.$$

In this case, if the vertex v_1 (resp. v_{n-1}) is adjacent to a vertex $v_i \notin A$, then $v_i, v_1, \ldots, v_{i-1}, v, v_{i+1} \ldots, v_{n-1}$ (resp. $v_1, \ldots, v_{i-1}, v, v_{i+1} \ldots, v_{n-1}, v_i$) is a Hamiltonian path without conflict in $(G, Conf)$.

Otherwise, v_1 and v_{n-1} are adjacent to the $\frac{n}{2} - 1$ vertices of A. In this case, if we have deleted an edge incident to v to obtain G', we add it to G' and we remove the edge with which it is in conflict to obtain a new graph G''. We show that G'' contains a Hamiltonian path, which will be a Hamiltonian path without conflict in $(G, Conf)$. If $vv_1 \in E_{G''}$ or $vv_{n-1} \in E_{G''}$, then there exists a Hamiltonian path without conflict in $(G, Conf)$.

Otherwise, there is $k \in \{2, \ldots, n-3\}$ such that v is adjacent to v_k and v_{k+1} in G', but it can miss an edge of P. If all edges of P are in G'' then $v_1, \ldots, v_k, v, v_{k+1}, \ldots v_{n-1}$ is a Hamiltonian path without conflict in $(G, Conf)$.

Otherwise, let v_jv_{j+1} be the edge of P that is not in G''. If $j = k$ then $v_1, \ldots, v_k, v, v_{k+1}, \ldots v_{n-1}$ is a Hamiltonian path without conflict in $(G, Conf)$.

If $j < k$ and $j \in A$ then $v_{j+1}, \ldots, v_k, v, v_{k+1}, \ldots v_{n-1}, v_j, \ldots, v_1$ is a Hamiltonian path without conflict in $(G, Conf)$.

If $j < k$ and $j \notin A$ then $v_j, \ldots, v_1, v_{j+1}, \ldots, v_k, v, v_{k+1}, \ldots v_{n-1}$ is a Hamiltonian path without conflict in $(G, Conf)$.

If $j > k$ and $j \in A$ then $v_1, \ldots, v_k, v, v_{k+1}, \ldots, v_j, v_{n-1} \ldots, v_{j+1}$ is a Hamiltonian path without conflict in $(G, Conf)$.

If $j > k$ and $j \notin A$ then $v_{n-1}, \ldots, v_{j+1}, v_1, \ldots, v_k, v, v_{k+1}, \ldots, v_j$ is a Hamiltonian path without conflict in $(G, Conf)$. □

We can now prove the Theorem 3:

Proof (of Theorem 3). Lemma 7 shows that one can construct a simple path without conflict of length at least $n - 2$. If necessary, Lemma 8 shows how to transform it into a Hamiltonian path without conflict. □

Theorem 3 proves that under Dirac's condition, a one-conflict graph contains a Hamiltonian *path* without conflict. We prove now that, in general, it is impossible to get a Hamiltonian *cycle* without conflict. For example, if $n = 4$, the cycle on 4 vertices C_4 with a conflict satisfies the conditions of Theorem 3 but doesn't contain a Hamiltonian cycle without conflict. For $n = 6$, consider a graph G_1 with 2 vertices and no edge, and a graph G_2 with 4 vertices and only 2 disjoint edges. If G is the graph obtained from the disjoint union of G_1 and G_2 by adding all possible edges between G_1 and G_2, any Hamiltonian cycle in G necessarily contains the two edges of G_2. If $Conf$ contains the conflict composed of these two edges, $(G, Conf)$ satisfies the conditions of Theorem 3 but contains no Hamiltonian cycle without conflict. We will see in Corollary 21, that when the set of conflicts $Conf$ satisfies $|Conf| \leq \frac{n}{4} - 1$, the graph $(G, Conf)$ always contains a Hamiltonian cycle without conflict.

4   Ore and Bondy-Chvátal-type Conditions in One-Conflict Graphs

We recall the result of Ore (1960) for "classical" graphs (*i.e.* without conflict):

Theorem 9 (Ore) [8]. *If for every pair of non-adjacent vertices of an n-vertex graph $(n \geq 3)$ the sum of their degrees is at least n then it contains a Hamiltonian cycle.*

This result becomes false when we consider conflicts. Indeed, let G be the graph obtained from the disjoint union of K_1 and K_{n-1} by adding two edges between K_1 and K_{n-1}. A Hamiltonian cycle in G necessarily contains these two edges. If $Conf$ contains the conflict composed of these two edges, $(G, Conf)$ satisfies the conditions of Theorem 9 but contains no Hamiltonian cycle without conflict.

Definition 1. *A circular representation C of an n-vertex one-conflict graph $(G, Conf)$ with $n \geq 3$ is obtained by ordering the vertices of G in some cyclic order $v_1, v_2, \ldots, v_n, v_1$ and by retaining only the edges between two consecutive vertices and only the conflicts in $Conf$ containing two of these edges. We often refer to a circular representation C by a natural sequence of its vertices $C = v_1, v_2, \ldots, v_n, v_1$. The two natural sequences $v_1, v_2, \ldots, v_n, v_1$ and $v_1, v_n, \ldots, v_2, v_1$ denote the same circular representation. Still, if often helps to fix one of this two orderings of V_G notationally: we may then speak of things like the successor of a vertex, etc. There is a hole $\overline{ab}$ between two consecutive vertices a and b if there is no edge connecting them in E_G. In a graph G (without conflict), a circular representation is a Hamiltonian cycle with missing edges: the holes, and a Hamiltonian cycle is a circular representation without hole.*

Lemma 10. *Let $(G, Conf)$ be an n-vertex one-conflict graph with $n \geq 3$. If C is a circular representation of $(G, Conf)$ with a hole $\overline{ab}$ such that:*

– *both a and b are not incident to any edge of any conflict in $Conf$ and,*
– *$deg_G(a) + deg_G(b) \geq n$.*

Then there exists a circular representation of $(G, Conf)$ whose holes are included in the holes of C, with at least one hole less and no more conflicts than C.

Proof. If $n = 3$, a circular representation satisfying the conditions of Lemma 10 cannot have a hole. So, in this case we have $G = K_3$.

   For $n \geq 4$, we denote C by

$$C = a, b, x_1, \ldots, x_{n-2}, a$$

and we show first there are two consecutive vertices x_i and x_{i+1} with $1 \leq i \leq n-3$ such that x_i is adjacent to a and x_{i+1} is adjacent to b. Indeed, if this was not the case, b cannot be adjacent to the successors of the neighbors of a in G (because $\overline{ab}$ is a hole) and we would have:

$$deg_G(b) \leq n - 1 - deg_G(a)$$

that is to say:

$$deg_G(a) + deg_G(b) \leq n - 1$$

which contradicts the assumptions of the lemma.

Now

$$C' := a, x_i, \ldots, x_1, b, x_{i+1}, \ldots, x_{n-2}, a$$

(C to which we add edges au and bv and remove edge uv) is a circular representation of $(G, \mathcal{C}onf)$ whose holes are included in the holes of C and with a hole less than C (if uv is a hole of C, C' has two holes less). As a and b are not in any conflict, the conflicts of C' are included in the conflicts of C. As C' does not contain the edge uv, if it is in conflict with another edge of C then C' contains a conflict less than C.   □

Definition 2. *If the two edges of a conflict don't share a common vertex, the conflict is called* parallel.

We have the following result:

Theorem 11. *If $(G, \mathcal{C}onf)$ is an n-vertex one-conflict graph ($n \geq 3$) s.t. all conflicts in $\mathcal{C}onf$ are parallel and $\forall u, v \in V_G$ ($u \neq v$) : $uv \notin E_G \Rightarrow deg_G(u) + deg_G(v) \geq n + 1$ then $(G, \mathcal{C}onf)$ contains a Hamiltonian cycle without conflict.*

Proof. By Theorem 9, G contains a Hamiltonian cycle but it may have conflicts. Consider a Hamiltonian cycle C in G with the minimum number of conflicts. If it has no conflict, the result is true. Otherwise, take a conflict $\{ab, cd\}$ (Note: a, b, c and d are four distinct vertices because conflicts are parallel and we have $n \geq 4$) such that we have:

$$C = a, b, x_1, \ldots, x_i, c, d, \ldots, a$$

Consider the circular representation

$$C' = a, c, x_i, \ldots, x_1, b, d, \ldots, a$$

We note that, as $\{ab, cd\}$ is a conflict and as each vertex is in at most one conflict, C' has a conflict less than C. C' is a circular representation with at most two holes $\overline{ac}$ and $\overline{bd}$. If C' has no holes, it is a Hamiltonian cycle in G with a conflict less than C. This contradicts the minimality of the number of conflicts of C. So C' has at least one hole. We can assume without loss of generality that $\overline{ac}$ is a hole. C' is also a circular representation of $(G' := G - ab, \mathcal{C}onf \smallsetminus \{ab, cd\})$ with:

- both a and c are not incident to any edge of any conflict in $Conf \smallsetminus \{ab, cd\}$ and,
- $deg_{G'}(a) + deg_{G'}(c) \geq n$.

According to Lemma 10, there is a circular representation C'' of G' whose holes are included in the holes of C', with at least one hole less and not more conflicts than C'. Then C'' has at most one hole. Using the Lemma 10 as above if necessary we obtain a Hamiltonian cycle in G' and hence in G with no more conflicts than C'. This contradicts the minimality of the number of conflicts of C. Thus, C has no conflict.   □

We now turn our attention to Bondy-Chvátal's conditions that uses the notion of closure that we generalize here.

Theorem 12. *For any integer $k \geq 0$ and for any n-vertex graph $G = (V, E_G)$, Algorithm 1 constructs a unique graph $\overline{G}^k$ (whatever the order of addition of the edges). This graph is called the k-closure of G.*

Proof. We show that the k-closure of an n-vertex graph $G = (V, E_G)$ is unique, whatever the order of addition of the edges, because it is equal to the intersection $\mathcal{H}$ of all the graphs $H = (V, E_H)$ satisfying the following properties (P):

- $E_G \subsetneq E_H$ and,
- $uv \notin E_H \Rightarrow deg_H(u) + deg_H(v) < n + k$.

Indeed, consider $\overline{G}^k$ a result of Algorithm 1 on G with integer $k \geq 0$. As clearly $H := \overline{G}^k$ satisfies the properties (P) we have $E_{\mathcal{H}} \subseteq E_{\overline{G}^k}$. Suppose that $E_{\mathcal{H}} \neq E_{\overline{G}^k}$. Let xy be the first edge of $E_{\overline{G}^k} \smallsetminus E_{\mathcal{H}}$ added by Algorithm 1. We have $deg_{\mathcal{H}}(x) + deg_{\mathcal{H}}(y) \geq deg_G(x) + deg_G(y) \geq n + k$. As $xy \notin E_{\mathcal{H}}$, there is H satisfying properties (P) s.t. $xy \notin E_H$. This implies $deg_H(x) + deg_H(y) < n + k$ and as $\mathcal{H} \subseteq H$ we have $deg_{\mathcal{H}}(x) + deg_{\mathcal{H}}(y) < n + k$ which contradicts the previous inequality. Finally we have $E_{\mathcal{H}} = E_{\overline{G}^k}$.   □

Algorithm 1. Construction of the k-closure $\overline{G}^k$ of G ($k \geq 0$).

Input: An n-vertex graph G
Output: The closure $\overline{G}^k$ of G
begin

> $\overline{G}^k \leftarrow G$
> **while** *There are two vertices u and v of $\overline{G}^k$ such that $uv \notin E_{\overline{G}^k}$ and* $deg_{\overline{G}^k}(u) + deg_{\overline{G}^k}(v) \geq n + k$ **do**
>> $\overline{G}^k \leftarrow \overline{G}^k + uv$
>
> **end**
> **return** $\overline{G}^k$;

end

Remark 13. $G \neq \overline{G}^k \Rightarrow n \geq 4 + k.$

Proof. As $G \neq \overline{G}^k$ there are two vertices u and v of G such that $uv \notin E_G$ and $deg_G(u) + deg_G(v) \geq n + k$. As $deg_G(u) \leq n - 2$ and $deg_G(v) \leq n - 2$ then $deg_G(u) + deg_G(v) \leq 2n - 4$ hence $n + k \leq 2n - 4$ and therefore $n \geq 4 + k$. □

We recall the result of Bondy and Chvátal (1976) for "classical" graphs (*i.e.* without conflict):

Theorem 14 (Bondy-Chvátal) *[1]. A graph (without conflict) contains a Hamiltonian cycle if and only if its 0-closure contains a Hamiltonian cycle.*

In one-conflict graphs, we have the following result:

Theorem 15. *An n-vertex one-conflict graph $(G, Conf)$ contains a Hamiltonian path without conflict if and only if $(\overline{G}^1, Conf)$ too.*

Proof. If $n < 5$ the result is true by Remark 13. We can assume that $n \geq 5$. It is clear that if $(G, Conf)$ contains a Hamiltonian path without conflict then $(\overline{G}^1, Conf)$ too. Conversely, suppose that $(\overline{G}^1, Conf)$ contains a Hamiltonian path without conflict and not $(G, Conf)$. Consider any sequence of graphs starting with $(G, Conf)$ and ending with $(\overline{G}^1, Conf)$ such that we move from one graph to the next by adding one edge:

$$(G_0 := G, Conf), \ldots, (G_i, Conf), (G_{i+1}, Conf), \ldots, (\overline{G}^1, Conf)$$

There is a smallest integer i such that $(G_{i+1}, Conf)$ contains a Hamiltonian path without conflict and not $(G_i, Conf)$. We denote by ab the edge added to go from G_i to G_{i+1} and by P a Hamiltonian path without conflict in $(G_{i+1}, Conf)$. We therefore have:

$$deg_{G_i}(a) + deg_{G_i}(b) \geq n + 1.$$

Consider the graph G_i. If a is incident to an edge of a conflict, it admits (at least) one edge outside P. We remove it from E_{G_i}. If b is incident to an edge of a conflict, we remove it from $Conf$. We denote by $(G_i', Conf')$ the one-conflict graph obtained. In the graph G_i' we have:

$$deg_{G_i'}(a) + deg_{G_i'}(b) \geq n. \tag{1}$$

In $(G_i', Conf')$, the circular representation naturally associated to $P - ab$ satisfies the conditions of Lemma 10 and $(G_i', Conf')$ contains a Hamiltonian cycle without conflict. Now, by removing one edge if b is incident to an edge of a conflict in $Conf$, with its two edges in P we obtain a Hamiltonian path without conflict in $(G_i, Conf)$. This contradicts the existence of an integer i such that $(G_{i+1}, Conf)$ contains a Hamiltonian path without conflict and not $(G_i, Conf)$. Thus $(G, Conf)$ contains a Hamiltonian path without conflict. □

From Theorem 15, we deduce an Ore-type result for Hamiltonian paths without conflict in one-conflict graphs.

Corollary 16. *If $(G, Conf)$ is an n-vertex one-conflict graph s.t. $\forall u, v \in V_G :$ $uv \; (u \neq v) \notin E_G \Rightarrow deg_G(u) + deg_G(v) \geq n + 1$ then $(G, Conf)$ contains a Hamiltonian path without conflict.*

Proof. $(\overline{G}^1, Conf)$ is a complete one-conflict graph and therefore it admits a Hamiltonian path without conflict by Theorem 3 and $(G, Conf)$ contains a Hamiltonian path without conflict by Theorem 15.

Theorem 17. *An n-vertex one-conflict graph $(G, Conf)$ contains a Hamiltonian cycle without conflict if and only if $\overline{G}^2$ too.*

Proof. If $n < 6$ the result is true by Remark 13. We assume $n \geq 6$. It is clear that if $(G, Conf)$ contains a Hamiltonian cycle without conflict then $\overline{G}^2$ too. Conversely, suppose that $\overline{G}^2$ contains a Hamiltonian cycle without conflict and not $(G, Conf)$. Consider any sequence of graphs starting with $(G, Conf)$ and ending with $(\overline{G}^2, Conf)$ such that we move from one graph to the next by adding an edge:

$$(G_0 := G, Conf), \ldots, (G_i, Conf), (G_{i+1}, Conf), \ldots, (\overline{G}^2, Conf)$$

There is a smallest integer i such that $(G_{i+1}, Conf)$ contains a Hamiltonian cycle without conflict and not $(G_i, Conf)$. We denote by ab the edge added to go from G_i to G_{i+1} and by C a Hamiltonian cycle without conflict in $(G_{i+1}, Conf)$. We therefore have:

$$deg_{G_i}(a) + deg_{G_i}(b) \geq n + 2.$$

Consider the graph G_i. If a is incident to an edge of a conflict, it admits (at least) one edge outside C. We remove it from E_{G_i}. The same is done for b. We denote by $(G_i', Conf')$ the one-conflict graph obtained. In the graph G_i' we have:

$$deg_{G_i'}(a) + deg_{G_i'}(b) \geq n.$$

In $(G_i', Conf')$, the circular representation $C - ab$ satisfies the conditions of Lemma 10 and $(G_i', Conf')$ contains a Hamiltonian cycle without conflict and therefore G_i too. This contradicts the existence of an integer i such that $(G_{i+1}, Conf)$ contains a Hamiltonian cycle without conflict and not $(G_i, Conf)$. Finally $(G, Conf)$ contains a Hamiltonian cycle without conflict. $\qquad\square$

From Theorem 17, we deduce an Ore-type result for Hamiltonian cycles without conflict in one-conflict graphs.

Corollary 18. *If $(G, Conf)$ is n-vertex one-conflict graph s.t. $n \geq 4$ and $\forall u, v \in V_G \; (u \neq v) : uv \notin E_G \Rightarrow deg_G(u) + deg_G(v) \geq n + 2$ then $(G, Conf)$ contains a Hamiltonian cycle without conflict.*

Proof. $(\overline{G}^2, Conf)$ is a complete one-conflict graph on n vertices. From $(\overline{G}^2, Conf)$ we construct a new graph H (without conflict) by removing one edge of each conflict. $\forall u \in V_H$ we have $deg_H(u) = n - 2$ and when $n \geq 4$ we have $deg_H(u) \geq \frac{n}{2}$. By Theorem 1, H contains a Hamiltonian cycle and therefore $(\overline{G}^2, \mathcal{C}onf)$ contains a Hamiltonian cycle without conflict. Finally, by Theorem 17, $(G, \mathcal{C}onf)$ contains a Hamiltonian cycle without conflict.   □

Now we will give a result for "classical" graphs (*i.e.* without conflict) from which we can draw corollaries in one-conflict graphs.

Lemma 19. *Let H be an n-vertex graph and X a matching in H. Define the integer $k := |V_X|$ (V_X is the set of vertices of X). If $\forall u, v \in V_H$ ($u \neq v$) : $uv \notin E_H \Rightarrow deg_H(u) + deg_H(v) \geq n$ and if we can order the vertices of X as $(x_1, x_2, \ldots, x_k)$ such that for each $1 \leq i \leq k$*

$$deg_H(x_i) \geq i + 2,$$

then the graph H deprived of the edges of X contains a Hamiltonian cycle.

Proof. We denote by H' the graph H deprived of the edges of X. We will show that $\overline{H'}^0$ is a complete graph and use Theorem 14 to deduce that H' contains a Hamiltonian cycle.

As in H the sum of the degrees of any two non-adjacent vertices is at least n, $\overline{H'}^0$ contains all edges between two vertices which are not in V_X and these vertices are thus of degree at least $n - k - 1$ in $\overline{H'}^0$.

We now show that $\overline{H'}^0$ contains all edges between vertices of V_X and vertices that are not in V_X. Suppose the contrary. Consider the largest integer i such that x_i is not connected to all the vertices of H that are not in V_X. $\overline{H'}^0$ therefore contains all edges between vertices of H that are not in V_X and vertices of V_X of index strictly greater than i. Thus, the vertices of H not in V_X have a degree greater than or equal to $n - k - 1 + k - i = n - i - 1$. But as the degree of x_i in H' is greater than or equal to $i + 1$ (because we removed a edge), it must be connected to all vertices of H that are not in V_X. This contradicts the initial assumption and therefore $\overline{H'}^0$ contains all the edges between vertices of V_X and vertices that are not in V_X.

It remains to show that $\overline{H'}^0$ contains all the edges of X. To do this, we will show that in $\overline{H'}^0$ each vertex x_j is connected to all the vertices of index greater than j. Suppose the contrary. Consider then the largest integer j such the vertex x_j is not connected to all the vertices of index greater than j. The vertices of index strictly greater than j of V_X are of degree greater than or equal to $n - k + k - j - 1 = n - j - 1$. But as the degree of x_j in H' is greater than or equal to $j + 1$ (because we removed an edge), it must be connected to all vertices of index greater than j. This contradicts the initial hypothesis and therefore x_j is connected to all the vertices of index strictly greater than j. Thus $\overline{H'}^0$ contains all edges of X.

Finally $\overline{H'}^0$ is a complete graph. So it contains a Hamiltonian cycle and H also by Theorem 14.   □

Theorem 20. *If* $(G, \mathcal{C}onf)$ *is an* n-*vertex one-conflict graph s.t.* $\forall u, v \in V_G$ $(u \neq v)$: $uv \notin E_G \Rightarrow deg_G(u) + deg_G(v) \geq n$ *and if there is a matching* X *in* G *containing exactly one edge of each conflict of* $\mathcal{C}onf$ *such that we can order the vertices of* X *as* $(x_1, x_2, \ldots, x_k)$ *with for each* $1 \leq i \leq k$:

$$deg_G(x_i) \geq i + 2,$$

then $(G, \mathcal{C}onf)$ *contains a Hamiltonian cycle without conflict.*

Proof. By Lemma 19, the graph G (without conflict) deprived of the edges of X contains a Hamiltonian cycle and such a cycle is a Hamiltonian cycle without conflict in $(G, \mathcal{C}onf)$. □

Corollary 21. *If* $(G, \mathcal{C}onf)$ *is* n-*vertex one-conflict graph s.t. each vertex has degree greater than or equal to* $\frac{n}{2}$ *and* $|\mathcal{C}onf| \leq \frac{n}{4} - 1$ *then* $(G, \mathcal{C}onf)$ *contains a Hamiltonian cycle without conflict.*

Proof. Let X be a matching in G having exactly one edge in each conflict of $\mathcal{C}onf$. As $|E_X| = |\mathcal{C}onf| \leq \frac{n}{4} - 1$, we have $|V_X| \leq \frac{n}{2} - 2$ and $\frac{n}{2} \geq |V_X| + 2$. Now, take any ordering $(x_1, x_2, \ldots, x_k)$ of the vertices of X. We obtain $deg_H(x_i) \geq \frac{n}{2} \geq |V_X| + 2 \geq i + 2$ and therefore by Theorem 20 $(G, \mathcal{C}onf)$ contains a Hamiltonian cycle without conflict. □

Acknowledgements. We thank Mamadou M. Kanté and anonymous referees for reading and helping to improve a first version of this work.

References

1. Bondy, J.A., Chvátal, V.: A method in graph theory. Discrete Math. **15**, 111–135 (1976)
2. Bondy, J.A., Murty, U.S.R.: Graph Theory. Springer, London Ltd (2010)
3. Dirac, G.A.: Some theorems on abstract graphs. Proc. London Math. Soc. **2**, 69–81 (1952)
4. Dvořák, Z.: Two-factors in orientated graphs with forbidden transitions. Discrete Math. **309**(1), 104–112 (2009)
5. Kanté, M.M., Laforest, C., Momège, B.: An exact algorithm to check the existence of (elementary) paths and a generalisation of the cut problem in graphs with forbidden transitions. In: van Emde Boas, P., Groen, F.C.A., Italiano, G.F., Nawrocki, J., Sack, H. (eds.) SOFSEM 2013. LNCS, vol. 7741, pp. 257–267. Springer, Heidelberg (2013)
6. Kanté, M.M., Laforest, C., Momège, B.: Trees in graphs with conflict edges or forbidden transitions. In: Chan, T.-H.H., Lau, L.C., Trevisan, L. (eds.) TAMC 2013. LNCS, vol. 7876, pp. 343–354. Springer, Heidelberg (2013)
7. Li, H.: Generalizations of Dirac's theorem in hamiltonian graph theory - a survey. Discrete Math. **313**(19), 2034–2053 (2013)
8. Ore, Ø.: Note on Hamiltonian circuits. Am. Math. Mon. **67**, 55 (1960)
9. Szeider, S.: Finding paths in graphs avoiding forbidden transitions. Discrete Appl. Math. **126**(2–3), 261–273 (2003)

Sequence Covering Arrays and Linear Extensions

Patrick C. Murray and Charles J. Colbourn[✉]

School of Computing, Informatics, and Decision Systems Engineering,
Arizona State University, P.O. Box 878809, Tempe, AZ 85287, USA
colbourn@asu.edu

Abstract. Covering subsequences by sets of permutations arises in numerous applications. Given a set of permutations that cover a specific set of subsequences, it is of interest not just to know how few permutations can be used, but also to find a set of size equal to or close to the minimum. These permutation construction problems have proved to be computationally challenging; few explicit constructions have been found for small sets of permutations of intermediate length, mostly arising from greedy algorithms. A different strategy is developed here. Starting with a set that covers the specific subsequences required, we determine local changes that can be made in the permutations without losing the required coverage. By selecting these local changes (using linear extensions) so as to make one or more permutations less 'important' for coverage, the method attempts to make a permutation redundant so that it can be removed and the set size reduced. A post-optimization method to do this is developed, and preliminary results on sequence covering arrays show that it is surprisingly effective.

1 Introduction

In order to motivate our study, consider the following question from [2]: Given an n-vertex m-edge graph G, what is the smallest number k of dimensions so that m axis-parallel k-dimensional boxes in $\mathbb{R}^k$ can be found whose intersection graph is the line graph of G? Remarkably, they recast this as a question about permutations: What is the smallest number of permutations so that for every two vertex-disjoint edges $\{w, x\}$ and $\{y, z\}$ of G, in at least one of the permutations w and x both precede y and z, or w and x both follow y and z? Similar problems abound. In [17], in a problem in event sequence testing, one asks for the fewest permutations so that for each of the $t!$ orders of each subset of t elements, some permutation contains the specified elements in the specified order.

In numerous problems of this type, strong asymptotic bounds on the minimum number of permutations as a function of the length of the permutations have been established. Our concern here is quite different; for any practical application we must explicitly construct a set of permutations, and asymptotic results are often not well suited to addressing construction problems for moderate lengths. Probabilistic arguments typically establish that choosing a certain number of permutations uniformly at random can yield a non-zero probability of success; yet for practical purposes this is not satisfactory, because the number

© Springer International Publishing Switzerland 2015
J. Kratochvíl et al. (Eds.): IWOCA 2014, LNCS 8986, pp. 274–285, 2015.
DOI: 10.1007/978-3-319-19315-1_24

of permutations required to have a reasonable chance of success is often much more than the minimum. To treat the construction of such sets of permutations, one avenue is to find mechanisms to translate knowledge about related combinatorial structures to underlie a direct construction, and another is to make sets of permutations of greater length recursively from those with smaller. General mechanisms to do this are not known, although in specific cases these can provide fast construction of sets of permutations far smaller than those from random selections. But when such direct or recursive constructions are not available, one resorts to computation.

Not surprisingly, exact methods such as backtracking or integer programming can be applied effectively only for small lengths and simple requirements. Heuristic methods extend the size and complexity of problems for which useful sets of permutations can be found. As lengths increase and requirements become more complex, best known results often arise from greedy algorithms that select one permutation at a time. In the problems being discussed, even deciding what is the best permutation to include next can be challenging, and hence greedy methods often make selections that are sub-optimal even locally.

This is not a good state of affairs. If we want to test sequences of 75 events so that every four of them appears in each of the 24 orders, we need an explicit set of permutations, and the best method in the literature to find them is a greedy method that selects permutations sub-optimally (but efficiently). In cases for which this greedy strategy has been implemented [5], it provides the smallest known sets of permutations in broad ranges of interest, despite the myopic nature of greedy methods and the sub-optimal selection of permutations. How can we do better? Rather than trying to improve the greedy methods further, we propose a different local optimization approach that we call *post-optimization*. Post-optimization repeatedly modifies the set of permutations with the general goal of making one of the permutations "less useful" in meeting the requirements. Ultimately, if it can be made redundant, it is deleted and the set size is reduced.

Our strategy determines what each permutation contributes, expresses this information as a partial order, and chooses a linear extension (which is ensured to contribute at least as much). This strategy can be effectively implemented, but the real surprise is that it reduces the number of permutations in best known solutions for event sequencing, sometimes dramatically.

The remainder of the paper is organized as follows. In Sect. 2 we give a precise formulation of a general set of problems and then discuss a special case, the sequence covering arrays that arise in event sequence testing. Then in Sect. 3 we focus on sequence covering arrays; we describe the current state of computational methods, and motivate the idea of post-optimization. Section 4 develops a framework and some details for the post-optimization method using linear extensions, and Sect. 5 describes preliminary computational results that are quite encouraging.

2 Background

Let $\Sigma = \{0, \ldots, v - 1\}$ be a set of *symbols* or *elements*. A *t-subsequence* of Σ is a *t*-tuple $(x_1, \ldots, x_t)$ with $x_i \in \Sigma$ for $1 \le i \le t$, and $x_i \ne x_j$ when $i \ne j$.

A *genus* of t-subsequences for the t-subset $\{x_1, \ldots, x_t\}$ is a non-empty subset of $\{(y_1, \ldots, y_t) : \{y_1, \ldots, y_t\} = \{x_1, \ldots, x_t\}\}$, A permutation π of Σ *covers* the t-subsequence $(x_1, \ldots, x_t)$ if $\pi^{-1}(x_i) < \pi^{-1}(x_j)$ whenever $i < j$. A permutation *covers* a genus when it covers one of the t-subsequences in the genus.

By way of example, suppose that $v = 8$ and $t = 4$. Then $(2, 4, 3, 5)$ is a 4-subsequence, and one genus of subsequences for $\{2, 3, 4, 5\}$ is $\{(y_1, y_2, y_3, y_4) : \{\{y_1, y_2\}, \{y_3, y_4\}\} = \{\{2, 4\}, \{3, 5\}\}\}$. This genus contains eight 4-subsequences $((2,4,3,5), (2,4,5,3), (4,2,3,5), (4,2,5,3), (3,5,2,4), (3,5,4,2), (5,3,2,4),$ and $(5,3,4,2))$. The permutation $\pi = 40251376$ has $\pi^{-1}(2) = 2$, $\pi^{-1}(3) = 5$, $\pi^{-1}(4) = 0$, and $\pi^{-1}(5) = 3$. It therefore covers the 4-subsequence $(4, 2, 5, 3)$, and hence covers the genus given. However, the same permutation fails to cover the genus $\{(y_1, y_2, y_3, y_4) : \{\{y_1, y_2\}, \{y_3, y_4\}\} = \{\{2, 3\}, \{4, 5\}\}\}$.

Let $\mathcal{G}$ be a set of genera of t-subsequences on symbols Σ. A set $\Pi = \{\pi_1, \ldots, \pi_N\}$ of N permutations is a $\mathcal{G}$-*permutation covering* if, for every genus $G \in \mathcal{G}$, there exists a permutation $\pi_j \in \Pi$ for which π_j covers genus G. To continue the example, the box genus, $box(\{x_1, x_2\}, \{x_3, x_4\})$, is $\{(y_1, y_2, y_3, y_4) : \{\{y_1, y_2\}, \{y_3, y_4\}\} = \{\{x_1, x_2\}, \{x_3, x_4\}\}\}$. Let

$$\mathcal{B}_\Sigma = \{box(\{x_1, x_2\}, \{x_3, x_4\}) : \{x_1, x_2, x_3, x_4\} \text{ is a 4-subset of } \Sigma\}.$$

A $\mathcal{B}_{\{0,\ldots,7\}}$-permutation covering is:
$$
\begin{array}{l}
4\,3\,2\,0\,6\,1\,5\,7 \\
3\,5\,0\,6\,2\,1\,7\,4 \\
7\,3\,0\,1\,2\,6\,4\,5 \\
4\,0\,2\,5\,1\,3\,7\,6 \\
2\,7\,5\,4\,0\,6\,3\,1
\end{array}
$$

Genus $box(\{2, 4\}, \{3, 5\})$ is covered by 40251376. Now $box(\{2, 3\}, \{4, 5\})$ is covered by 73012645 and $box(\{2, 5\}, \{3, 4\})$ is covered by 27540631. In this example, there are 1680 4-subsequences forming 210 box genera, so the remaining verification is best left to a machine.

No matter what genera are to be covered, some number of permutations suffices to cover them, because each t-subsequence has a linear extension to a permutation that necessarily covers the subsequence. Our interest is to determine the minimum number of permutations needed to cover a set $\mathcal{G}$ of genera.

Many permutation covering problems have been explored; we mention a few. Dushnik [8] examined the existence of sets of permutations in which for every subset S of k elements and every element σ not in this subset, one permutation has all elements in S preceding σ. In other words, he examined $\mathcal{G}_k$-permutation coverings with $\mathcal{G}_k = \{(x_1, \ldots, x_k, x_{k+1}) : \{x_1, \ldots, x_k\} = S\}$ for $S \cup \{x_{k+1}\}$ a $(k+1)$-subset of $\Sigma\}$. Spencer [25] established bounds on the number of permutations needed. In defining the "dimension of hypergraphs," Fishburn and Trotter [10] examined coverage of fewer genera in which the set S corresponds to hyperedges of an input hypergraph.

Füredi [11] explored *3-mixing sets*, which are $\mathcal{M}$-permutation coverings with $\mathcal{M} = \{(x_1, x_3, x_2), (x_2, x_3, x_1) : \{x_1, x_2, x_3\} \text{ is a 3-subset of } \Sigma\}$. In other words, for every pair of elements $\{a, b\}$ and third element c, there must be a permutation in which c is between a and b; see also [23] and [6].

Basaravaju *et al.* [2] discuss the relevance of these permutation coverings to geometric representations of graphs and hypergraphs. In particular, they define the *separation dimension* of a graph in a manner equivalent to the following. Let $G = (V, E)$ be a graph. Let $\mathcal{B}_G$ be the set of all box genera $box(\{x_1, x_2\}, \{x_3, x_4\})$ for which $\{x_1, x_2\}$ and $\{x_3, x_4\}$ are vertex-disjoint edges of G. They establish that the smallest number of permutations in a $\mathcal{B}_G$-permutation covering, which they term the separation dimension of G, is precisely the same as the "boxicity" of the line graph of G.

Recently, applications of permutation coverings in event sequence testing have attracted attention as well. In this case, the genera each contain a single t-subsequence, so the terminology need not refer to genera at all. A *sequence covering array* of order v and *strength* t, or $\mathsf{SeqCA}(N; t, v)$, is a set $\Pi = \{\pi_1, \ldots, \pi_N\}$ where π_i is a permutation of Σ, and every t-subsequence of Σ is covered by at least one of the permutations $\{\pi_1, \ldots, \pi_N\}$. Often the permutations are written as an $N \times v$ array. (Every permutation of every t of the v letters appears in the specified order in at least one of the N permutations.) Kuhn *et al.* [17, 18] describe the application to testing. Suppose that a process involves a sequence of v tasks or events. The operator may perform the tasks in an incorrect order, resulting in system faults. Often errors result from the improper ordering of a small number of events. When each permutation of a sequence covering array is used as a test order for the events, if faults result from the improper ordering of t or fewer tasks and are not masked, the presence of a fault will be detected. To reduce testing cost, we examine $\mathsf{SeqCAN}(t, v)$, the smallest N for which a $\mathsf{SeqCA}(N; t, v)$ exists.

Spencer [25] examined equivalent sets of permutations, *completely t-scrambling permutations*, obtained by interchanging the roles of symbols and columns in a sequence covering array [5]. For subsequent research, see Füredi [11], Ishigami [14, 15], Radhakrishnan [24], and Tarui [26]. Chee *et al.* [5] explore the relationship to so-called "directed t-coverings" as well.

From this point onwards, we restrict to cases when, for some strength t, every t-subsequence appears in one of the genera to be covered. Even then, for each of the permutation covering problems mentioned thus far, few exact results for the fewest permutations needed are known. More precisely, when each genus (of strength t) contains all $t!$ orderings, a single permutation suffices to cover all genera. When the genera can be named as $G_1, \ldots G_g, H_1, \ldots H_g$ so that for $1 \leq i \leq g$, $G_i \cup H_i$ contains all $t!$ orderings of t symbols, and both G_i and H_i contain each t-subsequence of these t symbols or its reversal, two permutations suffice: Simply take any permutation and its reversal. In these "trivial" situations, there is no dependence on the size of Σ. When the strength t is at most two, only these trivial cases arise. When Σ is "small," exact values are also sometimes known. For example, $\mathsf{SeqCAN}(t, t + 1) = t!$ [19].

Unfortunately, except in these situations, current knowledge of sizes of minimum permutation coverings has focussed on asymptotic results, determining the relationship between the size of the covering and the number of symbols, as the latter goes to infinity. While informative, these methods typically do not provide explicit solutions for small numbers of symbols. Yet in the testing application, the construction of sequence covering arrays is essential. In [26], an elegant direct construction for $\mathsf{SeqCA}(N; 3, v)$ is given that, while typically the smallest known, is known not to realize the minimum when v is small [5]. In [5], a direct construction produces a $\mathsf{SeqCA}(N + M; 3, vw)$ from a $\mathsf{SeqCA}(N; 3, v)$ and a $\mathsf{SeqCA}(M; 3, w)$; occasionally this produces the smallest sequence covering array that is known, but it does not do so in general. For strength $t \geq 4$, no such direct or recursive methods are known. Hence we turn to computational methods. Although we focus on sequence covering arrays to make the presentation more self-contained, most of the method to be described operates *mutatis mutandis* for permutation coverings with more complicated genera.

3   Computational Constructions

In [18], a simple greedy method is used to compute upper bounds on $\mathsf{SeqCAN}(t, v)$ for $t \in \{3, 4\}$ and small values of v. A more sophisticated greedy method was developed by Erdem *et al.* [9]. A conditional expectation greedy algorithm that derandomizes a randomized method establishes:

Theorem 1. [5] *For fixed t and input v, there is an algorithm to construct a* $\mathsf{SeqCA}(N; t, v)$ *having at most* $N \leq 2(\log(\frac{v!}{(v-t)!}))/(\log(\frac{t!}{t!-2}))$ *permutations in time that is polynomial in v.*

Chee *et al.* [5] observe that the bound in Theorem 1 is quite pessimistic. Implementation of the conditional expectation methods yields substantially better results for $t \in \{3, 4, 5\}$ than guaranteed by the theorem. One might expect that a greedy method can fare well, but it appears unlikely that it will yield optimum coverings. Indeed, using answer set programming, improvements on the greedy methods have been developed when $t \in \{3, 4\}$ [1,3,9]. A cooperative search strategy (the "bees algorithm") is explored in [12]. These methods are compared in [5], and we summarize the conclusions here.

For strength $t = 3$, the answer set programming methods outperform all of the greedy methods. Nevertheless when $v \geq 30$ they do not fare as well as Tarui's direct construction or the recursive construction. Tarui's direct construction is not optimum, however; for small values of v, answer set programming wins, and for certain values of v (such as $v = 128$), the recursive method wins.

Proceeding to strengths four and five, no direct or recursive method is available. Surprisingly, the answer set programming methods do not report the best known results except when v is very small; the conditional expectation greedy method yields the best known result. The explanation is almost certainly that the time and storage required for the answer set programming methods and the cooperative search methods are prohibitive.

15|2 3 0 4 1 5   15|5 4 0 1 3 2
15|3 1 4 0 2 5   15|1 0 5 2 3 4
15|4 3 5 2 1 0   15|2 0 5 1 4 3
15|4 1 2 5 3 0   15|1 3 5 0 4 2
15|5 0 2 4 3 1   15|0 3 2 1 5 4
15|3 2 4 5 0 1   14|1 2 4 0 3 5
15|5 3 1 2 0 4   14|0 3 4 5 1 2
13|0 4 2 1 3 5   13|5 2 1 3 4 0
12|4 2 0 5 3 1   12|0 1 4 5 3 2
12|2 3 5 4 1 0   11|4 3 0 1 5 2
10|2 1 5 0 3 4   9|1 5 4 2 3 0
10|4 5 1 0 2 3   9|5 3 4 0 2 1
7|3 0 5 2 1 4   7|0 1 3 2 4 5
5|2 1 0 4 5 3   5|4 1 3 2 0 5
4|0 2 5 4 1 3   4|5 2 0 3 1 4
3|3 1 4 5 2 0   3|0 5 1 4 2 3
2|1 0 3 2 5 4   1|2 5 4 3 0 1

The conditional expectation algorithm is greedy, and hence it is reasonable to expect that the later permutations chosen are less useful in the coverage of t-subsequences. Let us examine this more carefully. Consider the SeqCA(34;4,6) shown at left; the permutations are shown in the last six columns. Although the permutation covering need not order the permutations, in the sequence covering array they are ordered, and the greedy method added these permutations in this order. Therefore we can count, for each permutation, the 4-subsequences covered by this permutation that are covered by no earlier one (this is precisely the quantity that the greedy algorithm attempts to maximize). These counts are shown in the first column on the left.

No permutation in this example can cover more than $\binom{6}{4} = 15$ 4-subsequences. Mathon and Tran Van Trung [21] give a SeqCA(24;4,6) in which every permutation necessarily covers 15 4-subsequences for the first time. However, the myopic nature of the greedy method has resulted in permutations that cover fewer and fewer 4-subsequences for the first time, so that the last only covers a single 4-subsequence.

When the permutations are listed in this order, the last appears to be less useful. Can we avoid using some of these later permutations? The last permutation covers only the 4-subsequence $(2, 4, 3, 0)$ for the first time, and hence any of 30 different permutations would serve as well. A similar problem arises in the construction of related combinatorial objects known as covering arrays. In that setting, Nayeri et al. [22] devised a post-optimization method, which repeatedly reorders the array, attempting to reduce the amount of coverage required from the last row. If the last row can be made to provide no coverage not provided by an earlier row, it can be deleted to yield a solution with fewer rows. Surprisingly, this works well! For a variety of covering arrays from different constructions, such post-optimization eliminates many rows, sometimes more than 10 % of the rows in an initial (best known) solution. Arguably this is because the best known solutions can often be far from optimal due to deficiencies in the constructions that we know; nevertheless in that context such post-optimization has proved useful. Indeed covering arrays arise as a type of "t-restriction" problem, and post-optimization is effective more generally for such problems [7].

4   Post-Optimization and Linear Extensions

The main contribution of this paper is to develop a post-optimization technique for sequence covering arrays. Define the *effective coverage* of a permutation in an ordered list of permutations to be the number of t-subsequences covered by this permutation but by no earlier one. The basic algorithm follows:

repeat until "termination condition"
  choose an ordering of the permutations $\pi_1, \ldots, \pi_N$.
  compute the effective coverage for each permutation.
  select a permutation with least effective coverage and place it last in the order.
  repeat until "iteration condition"
    if any permutation has effective coverage zero, delete it.
    for each permutation π, replace π by any permutation that covers all of the
      t-subsequences covered by π for the first time.
    reorder the first $N - 1$ permutations
    recompute the effective coverage for each permutation.
    if any permutation has effective coverage lower than the last one,
      interchange it with the last one.

As an iteration condition, we terminate the inner loop when at least one permutation is removed or when an iteration limit is exceeded. The termination condition enforces a limit on the number of times a complete reordering of the array is undertaken. The determination of specific iteration and termination conditions dictate the number of times that an improvement is attempted, and can be set based on experimentation (that we do not describe here). We adopt random reordering. While it is reasonable to instead reorder so that permutations with larger effective coverage appear earlier in the ordering, we found that the method appeared more likely to be trapped in a local optimum and fail to make improvement.

The key aspect of the algorithm is to determine when a permutation can be replaced by another without loss of (effective) coverage. Patterned on the approach for covering arrays [7], call an entry σ in a permutation π_j *necessary* if there is some t-subsequence containing σ that is covered by π_j for the first time, and *flexible* otherwise. By iterating through all t-subsequences, finding their first occurrence in the array, and marking the t corresponding symbols as necessary, any symbol left unmarked is flexible. Any permutation π that contains the necessary elements in π_j in the same order can be used to replace π_j without reducing the effective coverage. Hence a basic post-optimization method can locate all flexible symbols in permutations, and move them (perhaps randomly) within the permutation; the result remains a sequence covering array.

Despite the fact that the SeqCA(34;4,6) is far from optimal, it has only two flexible symbols (1 and 5), both in the last permutation. Nevertheless, it has much more flexibility; we pursue this next. For each permutation π_j define a partial order $\prec_j$ on Σ so that whenever $(x_1, \ldots, x_t)$ is covered for the first time by π_j, we have $x_i \prec_j x_{i+1}$ for $1 \leq i < t$. Evidently π_j is a linear extension of $\prec_j$, but any linear extension of $\prec_j$ serves to provide at least the same effective coverage. In our SeqCA(34;4,6) example, the last permutation has partial order $2 \prec 4 \prec 3 \prec 0$ incomparable to 1 and 5. (This just restates the earlier observation about flexible symbols.) The second last permutation has effective coverage 2 (covering (1,0,3,2) and (0,3,5,4)) and partial order $1 \prec 0 \prec 3 \prec 5 \prec 4$ and $3 \prec 2$. While there are no flexible symbols, the partial order has three linear extensions. The third last permutation, with effective coverage of 3 (covering

(0,1,2,3), (0,5,1,4), and (5,1,2,3)), has partial order $0 \prec 5 \prec 1 \prec 2 \prec 3$ and $1 \prec 4$, again with three linear extensions. The fourth last permutation, with effective coverage of 3 (covering (3,1,2,0), (3,1,5,0), and (3,4,5,2)), has partial order $3 \prec 1 \prec 5 \prec 2 \prec 0$ and $3 \prec 4 \prec 5$, with two linear extensions.

The partial orders $\prec_j$ are computed at the same time as the effective coverage. Replacement of permutations is carried out by choosing a linear extension of $\prec_j$ to replace π_j. A strategy for choosing the 'best' linear extension for each of the partial orders is not clear; deterministic rules appear to stall the method in local optima. Hence one might prefer random linear extensions. Counting linear extensions is #P-complete [4], but there is a polynomial expected time algorithm for generating a random linear extension [13,16]. We do not need such sophisticated machinery, because we have no need for extensions to be selected uniformly at random. For our purposes, any method that has non-zero probability of obtaining each linear extension suffices. Such a method is easy: A simple greedy algorithm that repeatedly selects a minimum element and removes it exhibits this behaviour.

One improvement is worth mentioning. Rather than computing all of the partial orders $\prec_j$ and then forming a linear extension of each, once each partial order $\prec_j$ is determined, we immediately replace it by a linear extension. This can sometimes cover an additional t-subsequence previously only covered by a later permutation, reducing its effective coverage. Indeed, if in considering partial order $\prec_j$ some t-subsequence T covered only in the last permutation is consistent with $\prec_j$, we add the comparability relations from T to $\prec_j$ before choosing a linear extension; in this way, we guarantee that the last permutation has less effective coverage than before (making it a better candidate for removal). Further effort could be made to search for t-subsequences covered only after the jth permutation that are consistent with $\prec_j$ to make some permutation other than the last have lower effective coverage, but this can require checking a large number of t-subsequences for consistency with $\prec_j$. We have concentrated instead on reducing the effective coverage of the last permutation.

With these implementation decisions, the algorithm requires only $O(vN)$ storage for the array and $O(v^2)$ storage for the partial orders. The time is dominated by the time to determine the partial orders, which involves examining up to N permutations for $O(v^t)$ different t-subsequences. While this appears to be quite large, a single iteration of the post-optimization involves essentially the same effort as checking that the array is indeed a sequence covering array. In practice, the execution time depends not only on the time per iteration of the inner loop, but the termination and iteration conditions that determine the number of iterations. Our interest is not in theoretical efficiency, although our decisions have been guided by ensuring that a single iteration is not too computationally intensive; rather our concern is with whether the post-optimization method can be used in a practical sense to improve our knowledge about sequence covering arrays. For this, we turn to some computational results.

5   Some Computations

The objective is to find best explicit constructions for small sequence covering arrays, so the 'acid test' for post-optimization is whether it can improve upon the current best sequence covering numbers, and to what extent. We have implemented the method, and performed a number of preliminary experiments for strengths 3, 4, and 5 in the ranges of lengths reported in [5]. Results are reported in Table 1.

Table 1. Post-optimization for strengths 3, 4, 5, and 6

$t = 3$				$t = 4$				$t = 5$				$t = 6$		
v	Best	In	Out	v	Best	In	Out	v	Best	In	Out	v	In	Out
4	6	8	6	5	24	26	24	6	120	148	122	7	991	836
5	7	8	7	6	24	34	24	7	198	198	175	8	1342	1179
6	8	10	8	7	38	41	37	8	242	242	218	9	1662	1535
7	8	12	8	8	44	47	42	9	282	284	261	10	1970	1873
8	9	12	9	9	50	52	46	10	318	322	294			
9	9	12	9	10	55	57	53	11	354	354	330			
10	9	14	10	15	78	78	67	12	384	386	360			
15	10	15	12	20	92	92	80	13	416	419	390			
20	12	16	13	25	104	104	90	14	446	446	418			
25	14	18	14	30	113	113	98	15	470	475	448			
30	14	19	15	40	128	128	112	16	496	501	474			
40	16	21	16	50	141	141	123	17	518	518	496			
50	16	23	17	60	151	151	133	18	540	547	520			
60	16	26	18	70	160	160	142	19	560	570	541			
70	16	28	19	80	168	168	150	20	582	590	568			
80	17	30	20	90	176	180	162	25	674	674	656			
90	18	30	21					30	748	748	725			

For each strength, different lengths examined. For each, Best reports the best result known from [5], In reports the size of the array – usually produced by the conditional expectation greedy algorithm – to which post-optimization is applied, and Out reports the size when post-optimization was terminated.

First consider the results for strength $t = 3$. In this case, more sophisticated computational methods have earlier been applied, and useful direct and recursive constructions are known. Perhaps then it is no surprise that post-optimization has not improved upon any of the best known sizes. The improvements upon the sizes produced by the greedy method are nevertheless substantial. When $v = 80$, for example, the conditional expectation method gives a SeqCA(30;3,80);

ensuring that reversals are present yields a SeqCA(26;3,80) [5]. The answer set programming techniques give a SeqCA(24;3,80) [1] and a SeqCA(23;3,80) [3]. Post-optimization improves the SeqCA(30;3,80) by removing ten permutations, yielding a SeqCA(20;3,80). While Tarui's method [26] yields a SeqCA(18;3,80) and the recursive construction a SeqCA(17;3,80) [5], it remains striking that post-optimization fares so well.

Turning to strengths 4 and 5 shows the potential of post-optimization. In each of the cases examined, post-optimization matches or improves upon the best known result. In light of the comparison with the direct and recursive constructions for strength 3, it is unlikely that post-optimization has produced optimal arrays except when v is quite small. Nevertheless, in the absence of such constructions for strength greater than 3, it does yield the best known sizes, giving a non-trivial improvement on other methods applied.

Even when powerful direct or recursive constructions (as for strength three) are known, post-optimization may nevertheless prove useful. If only some of the t-subsequences are to be covered ("partial coverage"), the direct and recursive constructions do not exploit this, whereas post-optimization can eliminate permutations while ensuring that every t-subsequence that is covered initially remains covered. In principle, there is no obstacle to incorporating constraints as well. A *constraint* is an ℓ-subsequence that is not permitted to appear in any permutation. Sets of constraints may make it impossible to find an array covering some specified t-subsequences consistent with the constraints [5,20]. When some array with appropriate coverage meeting all constraints does exist, however, post-optimization can be applied by ensuring that every linear extension chosen does not violate any constraint. We have conducted limited experiments with partial coverage and with constraints; the extensions are natural. We have also conducted a more thorough set of experiments with covering various genera, such as the box genus. Our computational results, not discussed here, demonstrate that post-optimization is practical and effective in these problems. We argue that the ability to cope with such variants is a positive feature of post-optimization.

Finally we remark on execution times. Every iteration of post-optimization on a SeqCA(180;4,90) examines 61,324,560 4-subsequences, determining for each the first permutation in which it is covered and forming 180 partial orders on 90 elements each. Linear extensions of each order are then selected and a random reordering of the permutations is done before going on to the next iteration. Thus each iteration can be substantial. The results shown reflect computations after between 1000 and 10000 iterations for the most part, yielding times that are comparable to the initial construction cost by the conditional expectation method. Because our concern until this point has been with the extent of improvement possible, we have not optimized execution times. We plan a more detailed examination of the time to conduct one iteration, and the numbers of iterations employed to see different reductions.

As a proof of concept for post-optimization, we believe that the results shown succeed in demonstrating its potential.

6   Conclusions

In describing and implementing post-optimization, we have concentrated on sequence covering arrays. The extensions to requiring specified partial coverage, and to incorporating constraints on the ordering of pairs, are immediate. The extension to permutation coverings with more complicated genera follows similar lines. Both will be reported fully elsewhere.

At the outset, we asked a question: How ought one, in practice, construct a 'small' set of permutations of specified length with specified coverage properties? When the length and the strength of coverage are small enough, exhaustive methods will do. For somewhat larger strength and length, clever metaheuristic methods apply. For large enough length, randomized methods can be used. But sadly there typically remains a substantial intermediate range of lengths for which none of these methods applies. In these cases, randomized methods yield far too many permutations. Random selection gives a SeqCA(361;4,90), but greedy methods produce a SeqCA(176;4,90). In any real application the reduction from 361 to 176 is important. Our post-optimization provides a mechanism to obtain even smaller solutions, in this case a SeqCA(162;4,90).

Naturally one prefers powerful explicit constructions in the intermediate range of interest, as has been done in part for sequence covering arrays of strength 3. However, for most of the permutation coverage problems mentioned here, such powerful explicit constructions remain elusive; this is particularly the case when considering partial coverage or constraints. Our argument is that a sensible and practical strategy in these situations is to first apply a greedy method to get a 'reasonably sized' initial array, and then to post-optimize it.

Acknowledgments. Thanks to Sunil Chandran, Marty Golumbic, Rogers Mathew, and Deepak Rajendraprasad for interesting discussions about permutation coverings and geometric representations of graphs and hypergraphs.

References

1. Banbara, M., Tamura, N., Inoue, K.: Generating event-sequence test cases by answer set programming with the incidence matrix. In: Technical Communications of the 28th International Conference on Logic Programming (ICLP 2012), pp. 86–97 (2012)
2. Basavaraju, M., Chandran, L.S., Golumbic, M.C., Mathew, R., Rajendraprasad, D.: Boxicity and separation dimension. In: Kratsch, D., Todinca, I. (eds.) WG 2014. LNCS, vol. 8747, pp. 81–92. Springer, Heidelberg (2014)
3. Brain, M., Erdem, E., Inoue, K., Oetsch, J., Pührer, J., Tompits, H., Yilmaz, C.: Event-sequence testing using answer-set programming. Int. J. Adv. Softw. 5(3–4), 237–251 (2012)
4. Brightwell, G., Winkler, P.: Counting linear extensions. Order 8(3), 225–242 (1991)
5. Chee, Y.M., Colbourn, C.J., Horsley, D., Zhou, J.: Sequence covering arrays. SIAM J. Discrete Math. 27(4), 1844–1861 (2013)
6. Chor, B., Sudan, M.: A geometric approach to betweenness. SIAM J. Discrete Math. 11(4), 511–523 (1998)

7. Colbourn, C.J., Nayeri, P.: Randomized Post-optimization for t-Restrictions. In: Aydinian, H., Cicalese, F., Deppe, C. (eds.) Ahlswede Festschrift. LNCS, vol. 7777, pp. 597–608. Springer, Heidelberg (2013)
8. Dushnik, B.: Concerning a certain set of arrangements. Proc. Amer. Math. Soc. **1**, 788–796 (1950)
9. Erdem, E., Inoue, K., Oetsch, J., Pührer, J., Tompits, H., Yilmaz, C.: Answer-set programming as a new approach to event-sequence testing. In: Proceedings of the Second International Conference on Advances in System Testing and Validation Lifecycle, pp. 25–34. Xpert Publishing Services (2011)
10. Fishburn, P.C., Trotter, W.T.: Dimensions of hypergraphs. J. Combin. Theory Ser. B **56**(2), 278–295 (1992)
11. Füredi, Z.: Scrambling permutations and entropy of hypergraphs. Random Struct. Alg. **8**(2), 97 104 (1996)
12. Hazli, M.M.Z., Zamli, K.Z., Othman, R.R.: Sequence-based interaction testing implementation using bees algorithm. In: 2012 IEEE Symposium on Computers and Informatics, pp. 81–85. IEEE (2012)
13. Huber, M.: Fast perfect sampling from linear extensions. Discrete Math. **306**(4), 420–428 (2006)
14. Ishigami, Y.: Containment problems in high-dimensional spaces. Graphs Combin. **11**(4), 327–335 (1995)
15. Ishigami, Y.: An extremal problem of d permutations containing every permutation of every t elements. Discrete Math. **159**(1–3), 279–283 (1996)
16. Karzanov, A., Khachiyan, L.: On the conductance of order Markov chains. Order **8**(1), 7–15 (1991)
17. Kuhn, D.R., Higdon, J.M., Lawrence, J.F., Kacker, R.N., Lei, Y.: Combinatorial methods for event sequence testing. CrossTalk: J. Defense Software Eng. **25**(4), 15–18 (2012)
18. Kuhn, D.R., Higdon, J.M., Lawrence, J.F., Kacker, R.N., Lei, Y.: Combinatorial methods for event sequence testing. In: IEEE Fifth International Conference on Software Testing, Verification and Validation (ICST), pp. 601–609 (2012)
19. Levenshtein, V.I.: Perfect codes in the metric of deletions and insertions. Diskret. Mat. **3**(1), 3–20 (1991)
20. Margalit, O.: Better bounds for event sequence testing. In: The 2nd International Workshop on Combinatorial Testing (IWCT 2013), pp. 281–284 (2013)
21. Mathon, R.: Tran Van Trung: Directed t-packings and directed t-Steiner systems. Des. Codes Cryptogr. **18**(1–3), 187–198 (1999)
22. Nayeri, P., Colbourn, C.J., Konjevod, G.: Randomized postoptimization of covering arrays. Eur. J. Comb. **34**, 91–103 (2013)
23. Opatrný, J.: Total ordering problem. SIAM J. Comput. **8**(1), 111–114 (1979)
24. Radhakrishnan, J.: A note on scrambling permutations. Random Struct. Alg. **22**(4), 435–439 (2003)
25. Spencer, J.: Minimal scrambling sets of simple orders. Acta Math. Acad. Sci. Hungar. 22, 349–353 (1971/72)
26. Tarui, J.: On the minimum number of completely 3-scrambling permutations. Discrete Math. **308**(8), 1350–1354 (2008)

Minimum r-Star Cover of Class-3 Orthogonal Polygons

Leonidas Palios$^{(\boxtimes)}$ and Petros Tzimas

Department of Computer Science and Engineering,
University of Ioannina, 45110 Ioannina, Greece
{palios,ptzimas}@cs.uoi.gr

Abstract. We are interested in the problem of covering simple orthogonal polygons with the minimum number of r-stars; an orthogonal polygon is an r-star if it is star-shaped. The problem has been considered by Worman and Keil [13] who described an algorithm running in $O(n^{17}\text{poly-log }n)$ time where n is the size of the input polygon.

In this paper, we consider the above problem on simple class-3 orthogonal polygons, i.e., orthogonal polygons that have dents along at most 3 different orientations. By taking advantage of geometric properties of these polygons, we give an output-sensitive $O(n+k\log k)$-time algorithm where k is the size of a minimum r-star cover; this is the first purely geometric algorithm for this problem. Ideas in this algorithm may be generalized to yield faster algorithms for the problem on general simple orthogonal polygons.

Keywords: Orthogonal polygon · Cover · Decomposition · r-star · Visibility · Output-sensitive

1 Introduction

A polygon is *orthogonal* if its edges are either horizontal or vertical; an edge e of such a polygon is a N-edge (S-edge, E-edge, and W-edge, resp). if the outward-pointing normal vector to e is directed towards the North (South, East, and West, resp.). Of particular importance are the *dents* (i.e., edges whose endpoints are reflex vertices of the polygon) and the *extremities* (i.e., edges whose endpoints are convex vertices); see Fig. 1(left). Orthogonal polygons can be classified in terms of the types of dents that they contain [2]: a *class-k* orthogonal polygon ($0 \le k \le 4$) is defined to have dents along at most k different orientations. Class-2 polygons can be further classified into class-2a where the 2 dent orientations are parallel (i.e., N and S, or E and W), and class-2b where the 2 dent orientations are perpendicular to each other; class-2a orthogonal polygons are monotone.

We are interested in minimum covers of simple orthogonal polygons by r-stars. A *cover* of a polygon by a set S of pieces (or subpolygons) requires that the union of the pieces in S be equal to the polygon. If additionally the pieces are required to be mutually disjoint (except along boundaries), then we have

© Springer International Publishing Switzerland 2015
J. Kratochvíl et al. (Eds.): IWOCA 2014, LNCS 8986, pp. 286–297, 2015.
DOI: 10.1007/978-3-319-19315-1_25

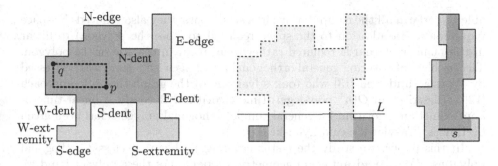

Fig. 1. (left) Illustration of the main definitions (the r-visibility polygon of p is shown dark); (middle) two trousers with their covered parts shown dark; (right) the histogram of the line segment s is shown dark.

a *partition*. Clearly, a minimum-size cover of a polygon involves at most as many pieces as a minimum-size partition of the polygon into the same type of pieces. Yet, covering problems prove to be harder than their corresponding partition problems and there are cases where the former are NP-hard whereas the latter admit polynomial solutions (e.g., finding a minimum-size Steiner-free partition of a simple polygon into star-shaped polygons can be computed in polynomial time [6], whereas the corresponding covering problem is NP-complete [1]). Covers and partitions are very important as they can be used for decomposition into simpler pieces. Recent applications of rectangulations include planar self-assembly with local information [8] and DNA self-assembly (M.Y. Kao and A. Sterling).

An orthogonal polygon is an r-*star* if it is star-shaped; clearly, an r-star is orthogonally convex. The term r-star comes from its formal definition with respect to the r-*visibility*: in an orthogonal polygon P, two points p, q of P are r-*visible* from one another if and only if the axis-parallel rectangle with p, q at opposite corners lies within P (Fig. 1(left) shows two such points p and q); then, a polygon P is an r-*star* if there exists a point p of P such that every point $q \in P$ is r-visible from p. (For completeness, we mention that in orthogonal polygons another notion of visibility, the s-*visibility*, is also defined; see [12].) Clearly, the problem of determining a minimum cover of a simple orthogonal polygon by r-stars is equivalent to determining a minimum set of r-visibility guards to watch the polygon. We note that computing the minimum number of guards to watch a general simple polygon is NP-complete [1], and that orthogonal polygons require fewer guards (in terms of the size of the polygon) [5].

Covering by r-stars has been investigated early enough. Keil [7] described an $O(n^2)$-time algorithm to cover a class-2a orthogonal polygon by r-stars. Culberson and Reckhow [2] showed that Keil's algorithm is worst-case optimal if the r-stars need to be explicitly reported and presented an $O(n)$-time algorithm to count the minimum number of r-stars needed; they also gave $O(n^2)$-time algorithms for minimally covering class-2a and class-2b orthogonal polygons. Gewali, Keil, and Ntafos [4] gave a 2-pass $O(n)$-time algorithm to report the locations of a minimum-cardinality set of r-visibility guards for class-2a orthogonal polygons. Their algorithm was improved by Lingas et al. [9,10] who were able to perform all the computations in a single pass; they also reduced the space requirement (in addition to the space required to store the polygon) to linear in the number of guards required rather than linear in the size of the polygon. The problem of covering general orthogonal polygons with r-stars was addressed by Worman and Keil [13] who took advantage of the graph-theoretic approach [12] to describe an $O(n^{17}\text{poly-}\log n)$-time algorithm. Recently, a linear-time 3-approximation algorithm for general simple orthogonal polygons has been given by Lingas, Wasylewicz, and Żyliński [11].

In this paper, we study the r-star covering problem on class-3 orthogonal polygons. We take advantage of geometric properties of these polygons and we present an 1-pass output-sensitive $O(n + k \log k)$-time algorithm to report the locations of a minimum-cardinality set of r-visibility guards to watch the entire polygon, where k is the (minimum) number of such guards. This is the first purely geometric algorithm for this problem. Ideas in this algorithm may be generalized to yield faster algorithms for the problem on general simple orthogonal polygons.

2   Theoretical Framework

We consider simple orthogonal polygons; so, in the following, we omit the adjective "simple." In a cartesian coordinate system, consider an orthogonal polygon P that has no N-dents. The intersection of P with a horizontal line L may consist of several line segments. Since P has no N-dents, these line segments correspond to *disjoint* parts of the polygon P *below the line* L; for convenience, we call each such part of P a *trouser* (in Fig. 1(middle) the sweep-line at its current position defines two trousers). It is important to note that, due to their definition, the trousers can be ordered along the x-axis.

We extend the notions of "grid segment" and "level" used in [4]: a *grid segment* of P or of a trouser T is a maximal (closed) horizontal line segment in P or T; the *level* of a point or a horizontal line segment (which may be a grid segment or a horizontal edge) is its y-coordinate. We also use the notion of orthogonal projection in an orthogonal polygon P given in [10]: the *orthogonal projection* $o(s)$ of a horizontal line segment s at level ℓ in P onto the grid segment s' at level $\ell' \geq \ell$ is the maximal subsegment of s' such that for each point a of $o(s)$ there exists a vertical line segment in P that goes through a and intersects s. Finally, for a horizontal line segment s (edge or grid segment) we define its *x-range* to be the set of x-coordinates of the points of s. (Although a polygon is considered a closed set, we consider edges to be open sets (i.e., they do not include their endpoints) and thus their x-ranges are open sets as well).

The following lemma provides three important properties of class-3 orthogonal polygons.

Lemma 1. *Let P be a class-3 orthogonal polygon without N-dents. Then:*

(i) The polygon P has a single topmost edge.
(ii) Consider sweeping the polygon P from bottom to top. Each edge encountered other than a S-extremity is incident with the swept part of P's boundary.

(iii) Let s_1 and s_2 be grid segments of P at levels ℓ_1 and ℓ_2, respectively, where $\ell_2 > \ell_1$, such that there exists a vertical line segment in P intersecting both s_1 and s_2. Then, the orthogonal projection of s_1 onto a level $\ell \geq \ell_2$ is a subset of the orthogonal projection of s_2 onto ℓ.

For a horizontal line segment s in an orthogonal polygon P, the *histogram* of s is the set of points p of P such that the vertical line segment pq, where $q \in s$, lies entirely in P (in Fig. 1(right), the histogram of line segment s is shown dark). It is important to note that the lack of N-dents implies that the histogram of any line segment is horizontally convex (note that by definition it is also vertically convex); see Fig. 1(right). The topmost edge of the histogram of s is a part of an edge of the polygon P and is the *top ceiling* of s [4].

Lemma 2. *Let P be a class-3 orthogonal polygon without N-dents, s a horizontal line segment, and let $s' \subseteq s$ be the projection of the top ceiling of s vertically onto s. Then, any point of P at a level $\geq level(s)$ that is seen by any point of s is also seen by any point of s'.*

Let us consider a polygon P and a trouser T of P defined by a horizontal sweep-line L. Then, T is partitioned into the set of points $p \in T$ such that the vertical line segment pq, where $q \in L$, contains points outside P (these points form the *covered* part of T) and the set of the remaining points (these points form the *non-covered* part of T); in Fig. 1(middle) the covered parts of the two trousers are shown dark. Then it is not difficult to see the following:

Lemma 3. *Let P be a class-3 orthogonal polygon without N-dents swept by a horizontal sweep-line and let T be a trouser of P.*

(i) The non-covered part of T is an orthogonally convex region bounded from above by the intersection of the sweep-line with T and from below by an x-monotone orthogonal chain whose horizontal edges are (entire or parts of) S-edges of P.

(ii) Consider points p, q in the non-covered part of T such that the line segment pq is vertical and p is higher than q. Then any point in P not belonging to the covered part of T seen by q is also seen by p.

Finally, by using induction in the number of dents and since a guard can see at most 1 extremity in each direction, we show the following:

Lemma 4. *Let P be an orthogonal polygon, k be the minimum number of r-visibility guards needed to watch P, and I and J be the number of extremities and dents of P, respectively. Then: (i) $I = J + 4$; (ii) $I \leq 4k$ (and hence $J \leq 4k - 4$).*

3   The Algorithm

Our algorithm applies plane-sweeping; we assume that the given class-3 polygon P does not have N-dents and we sweep it from bottom to top stopping at each horizontal edge (thus we can take advantage of Lemmas 1, 2, and 3). The invariant that we maintain is that:

Invariant: At any time, every point of the currently covered part of P is seen by at least one guard in the current set of guards.

At each S-edge we do some preparatory work but do not place guards as such edges can be watched by guards located at a higher level. N-edges may "cover" parts of the polygon from guards located higher. We check this and only if a guard is needed, it is located at the level of the N-edge (see Lemma 3(ii) and the Invariant); the x-coordinate of the guard's location is determined later as discussed below. In the end, the algorithm reports the locations of a minimum-cardinality set of r-visibility guards that watch the entire input polygon.

3.1   Determining When a Guard is Needed and Where to be Placed

Consider any S-edge e of the given polygon P. If the x-range of a N-edge d intersects e's x-range (Fig. 2(left)), then a guard must be located at a level *between (and including) the levels of e and d* since no other guard can watch the darker shaded area. Additionally, if such a guard is to be placed at level ℓ, it has to be placed at any point of *the orthogonal projection of the grid segment containing e onto level ℓ*, in order to watch e.

Therefore, in order to enforce the above observations, each encountered S-edge e submits a *guard-request* with which we maintain:

- a *forcing-range*, or *f-range* for short, which is the x-range of the edge e (we need to make sure that there is a guard watching e if the x-range of a N-edge above e and e's f-range have non-empty intersection);
- a *placement-range*, or *p-range* for short, which is the range of x-coordinates of the grid segment containing e (this is the initial range of x-coordinates of the location of a guard watching e).

Each of these ranges is a *single interval* of x-coordinates (the f-range is *open*, the p-range is closed), and it always holds that the f-range of a S-edge is a subset of its p-range. Figure 2(right) shows the initial values of the f-range (shown dotted) and p-range (shown dashed) for the S-edge e.

Here is how the f- and p-range of a guard-request r submitted by an edge e are used: During the sweeping, as long as we encounter N-edges whose x-ranges do not intersect either range, no change occurs. After we have encountered and

Fig. 2. (left) A guard is needed no higher than the N-edge d to watch the darker shaded area; (right) the initial values of the f-range (shown dotted) and the p-range (shown dashed) of the S-edge e.

processed a N-edge d whose x-range intersects the p-range of r, we clip r's p-range about d so that it always is the x-range of the orthogonal projection of the initial p-range onto a level slightly above the level of d; thus, it is the range of x-coordinates of the location of a guard at that level watching e). However, if we encounter a N-edge d whose x-range intersects the f-range of r, then a guard is needed immediately (in this way, our Invariant will keep holding); any guard located at a point with level between (and including) the levels of e and d, and with x-coordinate in the p-range of r will do. If the x-range of the N-edge d is a superset of the f-range of r, then r is deleted; if the x-range intersects the f-range of r but it is not a superset of the f-range (the N-edge partially covers e) then we keep r but we update its f-range to the difference of the f-range minus tho x rango of d. Thus, we have guard-requests for all the S-edges (or parts of them) that bound the non-covered part of the trousers of P from below.

3.2   Maintaining and Processing Guards

In order to be able to manage the guards, with each guard g we maintain:

- its *level* (i.e., the y-coordinate of g's location)   and
- its *location-range*, or *loc-range* for short, which is the range of x-coordinates of the points at which the guard g can currently be placed.

For a guard g to be placed at a grid segment s_ℓ at level ℓ in a trouser T, initially its loc-range coincides with the x-range of s_ℓ. As the sweep-line moves upward, the loc-range gets clipped by N-edges whose x-ranges intersect it. Thus, since there are no N-dents, the loc-range is a *single closed interval* of x-coordinates. If y is chosen to fulfill a guard-request r then the loc-range of y is set equal to r's p-range (in such a case, g's loc-range must intersect r's p-range; in fact, g's loc-range is a superset of r's p-range due to Lemma 1(iii) and the fact that both ranges are x-ranges of orthogonal projections of grid segments with the grid segment corresponding to g's loc-range being higher than that of r's p-range).

Finally, when a N-edge d is encountered such that the (possibly clipped) loc-range of g is a subset of d's x-range, then we say that g is *positioned* and the x-coordinate of g's location is set equal to the left bound value of g's loc-range right before d was encountered (in accordance with the convention followed in [4,9]). (Note that g cannot see any points in the polygon P above the level of d). In this way, if g was used to satisfy a set A of guard-requests and s is the (non-empty) intersection of the p-ranges of the requests in A, it is ensured that

P1: g is placed at the leftmost point of the vertical projection of the top-ceiling of s onto the level of g.

In addition to the histograms of the S-edges whose requests are satisfied by a guard g that got positioned when processing N-edge d, g may (entirely or partially) watch histograms of other edges (e.g., in Fig. 3(left), when processing N-edge d_1, guard g_1 is used to satisfy the request of edge e_1; later, when processing d_2, g_1 is used to also satisfy the requests of e_2 and e_3, and g_1 gets positioned; g_1 also watches parts of the histograms of edges e_4, e_5, e_6, e_7). We concentrate

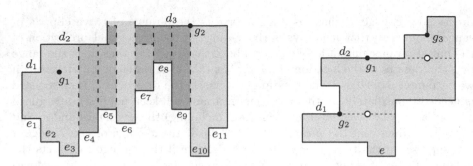

Fig. 3. (left) The histogram of e_7 is entirely watched by positioned guards g_1 and g_2; (right) not selecting the lowermost candidate guard may lead to a non-minimum number of guards.

only on the *histograms whose points below the part seen by g (if any) are seen by other positioned guards* (these are the histograms of e_4, e_5, e_7 for g_1 and of e_5, e_7, e_8, e_9 for g_2 in Fig. 3(left)). Any other histogram H_e (of an edge e) partly watched by g has a set Q of points below the part seen by g that are not seen by any positioned guard. For Q to be watched, (i) either another positioned guard g' will see part of Q but since g' will be positioned after g (when processing a N-edge d' at a level higher than d's), it will watch the entire part of the histogram H_e seen by g (no matter whether the level of g' is above or below that of g), (ii) or another guard g'' will be used to satisfy the guard-request submitted by e, and thus g'' will watch the entire histogram H_e.

In order to find the histograms we are interested in, for each trouser T we need to compute the top segments of the (seen by positioned guards) parts of the histograms in the non-covered part of T that are *right-* and *left-maxima*.

3.3 Selecting a Guard to Watch a S-Edge

Many guards at different levels in the polygon may be able to watch a S-edge e' when the f-range in the guard-request submitted by e' is intersected by the x-range of a N-edge. In order to make a good choice among them, we apply the following policy:

> **P2:** Whenever a guard-request needs to be fulfilled, among all guards that can fulfill it, the lowermost one is chosen.

The correctness of this approach follows from Lemma 3(ii); in this way, among the guards fulfilling the guard-request, we choose a guard g whose visibility polygon in the non-covered part of the trouser T to which it belongs is as small as possible, saving guards with larger visibility polygons to help watch portions of the non-covered part of T that g cannot see.

In fact, there are cases where by choosing a guard other than the lowermost available we get an incorrect result; see Fig. 3(right). When encountering the N-edges d_1 and d_2, we realize that guards are needed at these levels. If when assigning a guard to watch the S-edge e, we select a guard at the level of d_2

(see guard g_1 in Fig. 3(right)), then a third guard g_3 will also be needed; yet, two guards (located at the white circles) suffice to watch the entire polygon.

3.4 Description and Complexity of the Algorithm

As mentioned, we sweep the given class-3 orthogonal polygon P from bottom to top maintaining information on the current trousers (for simplicity, we assume that no two edges of the polygon are collinear). With each trouser T, we maintain T's guard ranges, guard-request ranges, left- and right-maxima, and the set of guards in T that have not yet been positioned. Additionally, we maintain a set *Positioned* storing all the positioned guards.

For the sweeping, we apply the idea of Hertel and Mehlhorn [3]: we have a global sweep-line that stops at each S-extremity (a new trouser is defined), S-dent (two neighboring trousers get merged), and the topmost edge (the last trouser gets deleted) in increasing y-coordinate, and each trouser T has its own sweep-line that may lag behind the level of the global sweep-line; the sweep-line of T gets advanced to the level of the global sweep-line (located at an edge e) only if we want to find the relative position of e with respect to T.

While advancing the sweep-line of a trouser T, we process the horizontal edges that are adjacent to T's endpoints in increasing y-coordinate. If the currently processed horizontal edge e is a S-edge, we set up and insert a corresponding guard-request and update the trouser information. If e is a N-edge, we process the guard-requests whose f-ranges are intersected by e's x-range, position and process the guards whose loc-ranges are subsets of e's x-range, update the maxima information, and clip the guard-requests' p-ranges and the guards' loc-ranges. After all the edges have been processed, the resulting guard set *Positioned* gives us the locations of a minimum-cardinality set of r-visibility guards.

The correctness of the algorithm follows from the discussion in Sects. 3.1–3.3 and the following:

1. the guards in the resulting set *Positioned* watch the histograms of all S-edges and thus watch the entire input polygon,
2. we use a new guard only when a portion of the polygon (i) is not watched by the currently used guards and (ii)cannot be watched by any guard above the current position of the sweep-line,
3. Lemmas 1–4 and approach P2, and
4. any guard g used to handle a number of guard-requests is positioned in the vertical projection of the top-ceiling of the intersection of the p-ranges of these requests onto the level of g (see P1), and thus g's visibility polygon contains the visibility polygon of any guard handling the same requests.

Time and Space Complexity. Let n be the number of vertices of the given class-3 polygon P and $k = O(n)$ be the minimum number of r-visibility guards needed to watch P. Then, at any given time, the number of guard-requests is $O(n)$ (we have at most 1 guard-request for each of the S-edges) whereas the number of trousers is $O(k)$ since each trouser contains at least one S-extremity and due to Lemma 4(ii).

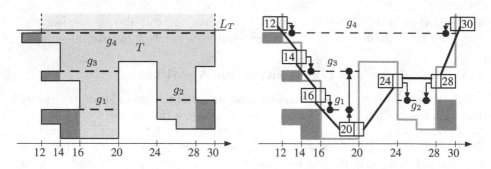

Fig. 4. A trouser T (sweep-line L_T) and the data structure for the guards' loc-ranges (rectangles indicate main doubly-linked list nodes, black disks indicate sublist nodes).

Data Structures. Let us now discuss the data structures used. The guard set *Positioned* is stored in an $O(k)$-size array. For the trousers, since we need to be able to insert, to delete, and to search the current trousers to locate the one incident with an edge (see Lemma 1(ii)), we maintain them in an $O(k)$-size balanced binary search tree D_t storing them in order from left to right; then every insertion, deletion, and search operation takes $O(\log k)$ time.

The f-ranges of the guard-requests associated with a trouser T are stored with T in a doubly-linked list with pointers at both ends so that insertion and deletion at either end, and list concatenation can be done in $O(1)$ time. The overall size of the lists in all the trousers is $O(n)$.

The clipping of the guards' loc-ranges is done in an implicit way; thus, the loc-ranges are stored in a special doubly-linked list as shown in Fig. 4. Each node corresponds to a vertical edge (which either defined the endpoint of a loc-range or clipped a previously defined loc-range) and stores the x-coordinate of that edge and a y-ordered (from bottom to top) sublist (with pointers at both its start and its end nodes) of loc-ranges ending at that vertical edge; moreover, the sublist nodes corresponding to the endpoints of the same loc-range are linked with double pointers to each other and to the corresponding guard's record. Note that for any two consecutive nodes in the list corresponding both to E-edges or both to W-edges, the y-coordinates of the loc-ranges in the sublist of the one node are all larger than the y-coordinates of all the loc-ranges in the sublist of the other node; thus, by simply concatenating the sublists of two such nodes we get a y-ordered list of all the loc-ranges. If the clipping affects only the first or last node in the list, then we simply update the x-coordinate stored in the node in $O(1)$ time. If the clipping affects more nodes, then these nodes are merged into a single one with updated x-coordinate and sublist that is the concatenation of the merged nodes' sublists (in y-order); the $O(1)$-time concatenation of the sublists of two consecutive nodes t_1, t_2 is *charged* to the vertical edge corresponding to the topmost node between t_1 and t_2, which will not be encountered, and thus will not be charged, again. For example, in Fig. 4, if a N-edge whose right endpoint has x-coordinate equal to 15 is encountered above the top left corner of the

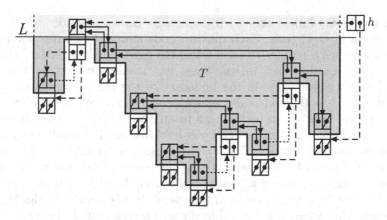

Fig. 5. The data structure for the right-maxima of trouser T: non-dashed pointers link maxima list-nodes; dashed pointers from a node p (other than the head-node h) point to the first and last nodes of the (lower-level) list of maxima of T's non-covered part between the edges corresponding to p and to the next maxima node.

trouser shown, the two leftmost nodes of the list will be merged into one with associated x-coordinate equal to 15, their sublists will be concatenated, and the cost will be charged to the vertical edge with x-coordinate equal to 12.

The guard-requests' p-ranges are stored together with the guards' loc-ranges in the same data structure (a bit distinguishes the sublist nodes storing loc-ranges from those storing p-ranges). Thus, the clipping of p-ranges is done as above. Getting the lowermost guard to fulfill a guard-request r is done as follows: Consider a N-edge d (incident with the right endpoint of a trouser T (see Lemma 1(ii))) that covers the right end of r's f-range (the case of T's left endpoint is treated symmetrically). Then the p-range of r extends to the right endpoint of T since the f-range of a request is a subset of its p-range. We traverse the sublist (from bottom to top) of the rightmost node of the ranges data structure looking for a loc-range of a guard. If such a guard g is found, the sublist node corresponding to the left endpoint of the loc-range of g is marked for deletion while its info is copied on a copy of the sublist node for the left endpoint of the lowermost p-range in the sublist, thus setting g's loc-range equal to the smallest p-range; if no loc-range is found, we place a new guard g and update its loc-range as previously described. It is important to observe that in either case this guard g fulfills all the requests whose p-ranges were traversed until g was found or placed. Thus, we mark the p-ranges traversed, unlink their right endpoint sublist nodes, and place them in a p-range super-node so that we spend $O(1)$ time traversing all of them in subsequent steps. Therefore, finding the lowermost guard or placing a new one as well as updating its loc-range can be done in $O(1 + t)$ time where t is the number of p-ranges bundled together. In total, these add up to $O(n)$ time.

The data structure for the right-maxima of a trouser T is illustrated in Fig. 5 (the one for the left-maxima is symmetric). The structure is hierarchical and is based on doubly-linked lists with pointers at both end nodes which allows concatenation in $O(1)$ time. Each maxima node stores the current level of the maximum and has a pointer to (and from) the corresponding S-edge, a pointer to the next maximum, and, in case these two maxima do not correspond to consecutive horizontal edges in the boundary of the polygon, pointers (shown dashed in Fig. 5) to the end nodes of a *lower-level maxima list* (i.e., a similar structure for the maxima in between), a pointer to the previous maximum, and in case this is the first maximum of a lower-level maxima list pointed by a node t, a pointer to t (shown dotted in Fig. 5). It is important to note that we can walk from either end towards the other in $O(1)$ time per step and that we can update this structure in $O(1)$ time if a S-edge is added at the upper left or upper right corner of T or in $O(1)$ time plus $O(1)$ time per deleted edge if a N-edge is added. We also note that whenever a positioned guard g sees parts of histograms of S-edges other than those whose requests g was used to satisfy (e.g., edges e_4, e_5, e_7 for g_1 in Fig. 3(left)) then the maxima corresponding to these edges along with the structure of the maxima in between them are stored in a super-node so that the level of all of them can be updated in $O(1)$ time; from then on, these maxima are treated all together except when they are deleted. Here again, the total time to process these structures is $O(n)$.

Complexity Analysis. Finding the S-extremities, S-dents, and unique topmost edge (Lemma 1(i)) takes $O(n)$ time while sorting them takes $O(k \log k)$ time (Lemma 4(ii)). Then, for each S-extremity or S-dent e in order, we need to locate e with respect to the existing trousers in the $O(k)$-size D_t; this takes $O(\log k)$ time for each such edge for a total of $O(k \log k)$ time. The remaining processing for each such edge takes $O(1)$ time plus the time to process the edges encountered while advancing the sweep-line of the neighboring trouser(s). Processing a S-edge implies a constant-size change in the data structures and can be done in $O(1)$ time. Let us now consider the processing of each N-edge e. The requests whose f-ranges are intersected by the x-range of e are located at one of the two ends of the f-range linked list and thus we can find them in $O(1)$ time each. Determining whether a guard is needed and getting and updating one also takes $O(1)$ time per request plus time proportional to the number of bundled p-ranges. Processing a guard that gets positioned due to edge e takes $O(1)$ time plus time proportional to the number of maxima collected (which, as mentioned, are inserted in a single super-node). Thus the total time to collect maxima over all processed edges is $O(n)$. Moreover, updating the maxima data structure takes $O(n)$ total time. Clipping also takes $O(1)$ time per N-edge plus $O(1)$ time per charged vertical edge, for $O(n)$ time in total. In summary, processing the entire polygon takes $O(n + k \log k)$ time. The space complexity is $O(n)$.

Since reporting the guards takes $O(k)$ time, we have:

Theorem 1. *Let P be a simple class-3 orthogonal polygon with n vertices that can be covered with no fewer than k r-stars. Then, a minimum-cardinality set of r-visibility guards watching the entire P can be computed in $O(n + k \log k)$ time and $O(n)$ space.*

4   Concluding Remarks

We presented an output-sensitive $O(n + k \log k)$-time algorithm for computing a minimum r-star cover of a class-3 orthogonal polygon on n vertices where k is the size of the minimum r-star cover. We leave as open problems the following on r-star covers: obtaining faster algorithms for general simple orthogonal polygons, investigating the complexity of the problem for orthogonal polygons with holes, and studying extensions to 3 dimensions. It would also be interesting to obtain algorithms that run in less than $\Theta(n^8)$ time (see [12]) for the s-star covering problem on general simple orthogonal polygons.

Acknowledgments. This research has been co-financed by the European Union (European Social Fund - ESF) and Greek national funds through the Operational Program "Education and Lifelong Learning" of the National Strategic Reference Framework (NSRF) - Research Funding Program: THALIS UOA (MIS 375891) - Investing in knowledge society through the European Social Fund.

References

1. Aggarwal, A.: The art gallery theorem: its variations, applications, and algorithmic aspects. Ph.D. thesis, Department of Electrical Engineering and Computer Science, Johns Hopkins University (1984)
2. Culberson, J., Reckhow, R.A.: Orthogonally convex coverings of orthogonal polygons without holes. J. Comput. Syst. Sci. **39**(2), 166–204 (1989)
3. Hertel, S., Mehlhorn, K.: Fast triangulation of simple polygons. In: FCT 1983: Proceedings of the 4th International Conference on Fundamentals of Computation Theory, pp. 207–218 (1983)
4. Gewali, L., Keil, M., Ntafos, S.C.: On covering orthogonal polygons with star-shaped polygons. Inf. Sci. **65**, 45–63 (1992)
5. Kahn, J., Klawe, M., Kleitman, D.: Traditional galleries require fewer watchmen. SIAM J. Algebraic Discrete Methods **4**(2), 194–206 (1983)
6. Keil, J.M.: Decomposing a polygon into simpler components. SIAM J. Comput. **14**, 799–817 (1985)
7. Keil, J.M.: Minimally covering a horizontally convex orthogonal polygon. In: SoCG 1986: Proceedings of the 2nd Annual ACM Symposium Computational Geometry, pp. 43–51 (1986)
8. Li, G., Zhang, H.: A rectangular partition algorithm for planar self-assembly. In: IROS 2005: Proceedings of the IEEE/RSJ International Conference on Intelligent Robots and Systems, pp. 3213–3218 (2005)
9. Lingas, A., Palios, L., Wasylewicz, A., Żyliński, P.: Corrigendum: note on covering orthogonal polygons. Inf. Process. Lett. **114**, 646–654 (2014)
10. Lingas, A., Wasylewicz, A., Żyliński, P.: Note on covering orthogonal polygons with star-shaped polygons. Inf. Process. Lett. **104**(6), 220–227 (2007)
11. Lingas, A., Wasylewicz, A., Żyliński, P.: Linear-time 3-approximation algorithm for the r-star covering problem. Int. J. Comput. Geom. Appl. **22**(2), 103–141 (2012)
12. Motwani, R., Raghunathan, A., Saran, H.: Covering orthogonal polygons with star polygons: the perfect graph approach. J. Comput. Syst. Sci. **40**, 19–48 (1990)
13. Worman, C., Keil, J.M.: Polygon decomposition and the orthogonal art gallery problem. Int. J. Comput. Geom. Appl. **17**(2), 105–138 (2007)

Embedding Circulant Networks into Butterfly and Benes Networks

R. Sundara Rajan[1]($\boxtimes$), Indra Rajasingh[1], Paul Manuel[2], T.M. Rajalaxmi[3], and N. Parthiban[4]

[1] School of Advanced Sciences, VIT University, Chennai 600 127, India
vprsundar@gmail.com
[2] Department of Information Science, Kuwait University,
13060 Safat, Kuwait
[3] Department of Mathematics, SSN College of Engineering, Chennai 603 110, India
[4] School of Computing Sciences and Engineering,
VIT University, Chennai 600 127, India

Abstract. The dilation of an embedding is defined as the maximum distance between pairs of vertices of host graph that are images of adjacent vertices of guest graph. An embedding with a long dilation faces many problems, such as long communication delay, coupling problems and the existence of different types of uncontrolled noise. In this paper, we compute the minimum dilation of embedding circulant networks into butterfly and benes networks.

Keywords: Embedding · Dilation · Circulant network · Butterfly and benes networks

1 Introduction

Graph embedding is an important technique that maps a guest graph into a host graph, usually an interconnection network. Many applications can be modeled as graph embedding. In architecture simulation, graph embedding has been known as a powerful tool for implementation of parallel algorithms or simulation of different interconnection networks [1].

The quality of an embedding can be measured by certain cost criteria. One of these criteria which is considered very often is the *dilation*. The dilation of an embedding is defined as the maximum distance between pairs of vertices of host graph that are images of adjacent vertices of the guest graph. It is a measure for the communication time needed when simulating one network on another [2]. Dilation of an embedding has been well studied for a number of networks [2–7].

An interconnection network is a scheme that connects the units of a multiprocessing system. It plays a central role in determining the overall performance of a multicomputer system. A suitable interconnection network is an important part

I. Rajasingh—This work is supported by Project No. SR/S4/MS: 846/13, Department of Science and Technology, SERB, Government of India.

© Springer International Publishing Switzerland 2015
J. Kratochvíl et al. (Eds.): IWOCA 2014, LNCS 8986, pp. 298–306, 2015.
DOI: 10.1007/978-3-319-19315-1_26

for the design of a multicomputer or multiprocessor system and is usually modeled by a symmetric graph, where the nodes represent the processing elements and the edges represent the communication channels. Desirable properties of an interconnection network include symmetry, embedding capabilities, relatively small degree, small diameter, scalability, robustness, and efficient routing [8].

The circulant network is a natural generalization of the double loop network [9]. It has been used for decades in the design of computer and telecommunication networks due to its optimal fault-tolerance and routing capabilities [10]. It is also used in VLSI design and distributed computation [11–13]. Every circulant graph is a vertex transitive graph and a Cayley graph [14]. Most of the earlier research concentrated on using the circulant graphs to build interconnection networks for distributed and parallel systems [10,11].

Butterfly network is an important and well known topological structure of interconnection networks. It is a bounded-degree derivative of the hypercube which aims at overcoming some drawbacks of hypercube. It is used to perform fast Fourier transform, which is intensively used in the field of signal processing. The benes network consists of back-to-back butterflies [15]. In this paper, we embed circulant networks into butterfly and benes networks with minimum dilation.

2   Basic Concepts

In this section we give the basic definitions and preliminaries required for our subsequent work.

Definition 1. *[3] Let G and H be finite graphs. An embedding $\phi = (f, P_f)$ of G into H is defined as follows:*

1. *f is a one-to-one map from $V(G) \to V(H)$*
2. *P_f is a one-to-one map from $E(G)$ to $\{P_f(u,v) : P_f(u,v)$ is a path in H between $f(u)$ and $f(v)$ for $(u,v) \in E(G)\}$.*

For brevity, we denote the pair (f, P_f) as f. The *expansion* [3] of an embedding f is the ratio of the number of vertices of H to the number of vertices of G. In this paper, we consider embeddings with expansion one.

Definition 2. *[3] Let f be an embedding of G into H. If $e = (u,v) \in E(G)$, then the length of $P_f(u,v)$ in H is called the dilation of the edge e. The maximum dilation over all edges of G is called the dilation of the embedding f. The dilation of embedding G into H is the minimum dilation taken over all embeddings of G into H and is denoted by $dil(G, H)$.*

Definition 3. *[4, 11] The undirected circulant graph $G(n; \pm S)$, $S \subseteq \{1, 2, \ldots, j\}$, $1 \leq j \leq \lfloor n/2 \rfloor$ is a graph with vertex set $V = \{0, 1, \ldots, n-1\}$ and the edge set $E = \{(i, k) : |k - i| \equiv s (mod\ n), s \in S\}$.*

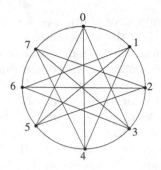

Fig. 1. Circulant graph $G(8; \pm\{1,3,4\})$

The circulant graph shown in Fig. 1 is $G(8; \pm\{1,3,4\})$. It is clear that $G(n; \pm1)$ is the undirected cycle C_n and $G(n; \pm\{1,2,\ldots,\lfloor n/2 \rfloor\})$ is the complete graph K_n. The cycle $G(n; \pm1) \simeq C_n$ contained in $G(n; \pm\{1,2,\ldots,j\})$, $1 \leq j \leq \lfloor n/2 \rfloor$ is sometimes referred to as the outer cycle C of G.

Definition 4. *[15] The r-dimensional butterfly BF_r has $n = 2^r$ $(r+1)$ nodes arranged in $r+1$ levels of 2^r nodes each. Each node has a distinct label $\langle w, i \rangle$ where i is the level of the node $(0 \leq i \leq r)$ and w is a r-bit binary number that denotes the column of the node. All nodes of the form $\langle w, i \rangle$, $0 \leq i \leq r$, are said to belong to column w. Similarly, the ith level L_i consists of all of the nodes $\langle w, i \rangle$, where w ranges over all r-bit binary numbers. Two nodes $\langle w, i \rangle$ and $\langle w', i' \rangle$ are linked by an edge if $i' = i+1$ and either w and w' are identical or w and w' differ only in the bit in position i'.*

Definition 5. *[15] An r-dimensional Benes network B_r has $2r+1$ levels, each level with 2^r nodes. The level 0 to level r nodes in the network form an r dimensional butterfly. The middle level of the Benes network is shared by these butterflies.*

Remark 1. The diameter of both $BF(r)$ and $B(r)$ is $2r$, $r \geq 1$.

Figure 2 illustrate the butterfly network $BF(3)$ and the benes network $B(3)$.

3   Main Results

For an embedding f of G into H, the sum of the dilations in H of edges in G is called the wirelength of f. The minimum taken over all the embeddings f is called the wirelength of embedding G into H. The dilation problem and the wirelength problem are different in the sense that an embedding that gives the minimum dilation need not give the minimum wirelength and vice-versa. Even though there are numerous results and discussions on the dilation problem, there is no efficient method to compute the exact dilation of graph embeddings [4–6].

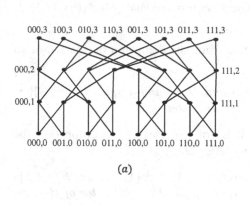

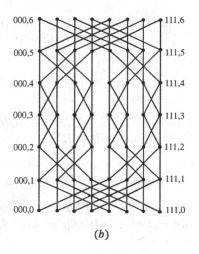

Fig. 2. (a) Butterfly network $BF(3)$ (b) Benes network $B(3)$

In 2012, Manuel et al. obtained a lower bound for dilation of an embedding using exact wirelength and formulated the result as IPS Lemma [7]. Since wirelength problem itself is NP-complete [16,17], the question arises whether it is possible to obtain a lower bound for dilation without computing wirelength of an embedding. In this direction, Manuel et al. obtained a lower bound for dilation of an embedding without using exact wirelength [18], which we quote as Generalized Dilation Lemma in this paper.

In this section, we prove that the lower bounds obtained for embedding circulant networks into butterfly and benes networks are sharp.

Lemma 1. *(Generalized Dilation Lemma) [18] Let G be an undirected circulant graph $G(n; \pm\{1, 2, \ldots, j\})$, $1 \leq j < \lfloor n/2 \rfloor$. Let H be a graph on n vertices with diameter δ such that for $u_1, u_2, \ldots, u_l \in V(H)$, $D_\delta(u_i) \neq \phi$, $1 \leq i \leq l$, where $D_\alpha(u_i)$ denotes the set of all vertices in H which are at distance α from u_i. Let k_i be the least integer such that $|D_\delta(u_i)| + |D_{\delta-1}(u_i)| + \cdots + |D_{\delta-k_i}(u_i)| > n - 2j - l$ and let $k = \min\limits_i k_i$. Then the dilation of embedding G into H is at least $\delta - k$.*

Remark 2. The Dilation Lemma [19] is a particular case of Generalized Dilation Lemma, when $l = 1$.

Lemma 2. *Let u be a vertex in $BF(r)$ with $D_{\delta_r}(u) \neq \phi$, where δ_r is the diameter of $BF(r)$. Then $|D_{\delta_r}(u)| = 2^{r-1}$ and $|D_{\delta_r-1}(u)| = 2^{r-1}$, $r \geq 1$.*

Proof. We prove the result by induction on r. When $r = 1$, there is exactly one vertex at distance $\delta_1 = 2$ from u, since $BF(1) \simeq C_4$. Thus the result is true for $r = 1$. Let us assume that the result is true for $r = k$. That is, $|D_{\delta_k}(u)| = 2^{k-1}$ in $BF(k)$. Now consider $r = k+1$. $BF(k+1)$ contains two vertex disjoint subgraphs of $BF(k)$, say $BF^1(k)$ and $BF^2(k)$, and diametrically opposite vertices in $BF(r)$

belong to the 0th level of $BF(r-1)$, for all r. Therefore, $|D_{\delta_{k+1}}(u)| = 2 \cdot |D_{\delta_k}(u)| = 2^k$. Hence the result. Similarly, we can prove that $|D_{\delta_r-1}(u)| = 2^{r-1}$, $r \geq 1$.

The following result is an easy observation from the definition of dilation of an embedding.

Theorem 1. *Let G be a complete graph and H be any graph with $|V(G)| = |V(H)|$. Then $dil(G,H) = \delta$, where δ is the diameter of H.*

The following theorem illustrates that it is not necessary for a graph G to be complete to get δ as the dilation of an embedding.

Theorem 2. *Let G be the circulant graph $G((r+1)2^r; \pm\{1,2,\ldots,r\cdot 2^{r-1}+1\})$ and H be the butterfly network $BF(r)$. Then $dil(G,H) = $ diameter of $H = 2r$, where $r \geq 1$.*

Proof. Let $u_i \in V(G)$ be such that $D_\delta(u_i) \neq \phi$, $1 \leq i \leq 2^{r-1}$. By the definition of circulant graph, there are exactly $n-2j-2^{r-1} = (r+1)2^r - 2[r\cdot 2^{r-1}+1]-2^{r-1} = 2^{r-1}-2$ vertices which are not adjacent to the vertices $u_1, u_2, \ldots, u_{2^{r-1}}$ in G. By Lemma 2, $|D_\delta(f(u_i))| = 2^{r-1} > 2^{r-1}-2$. Hence by Lemma 1, $dil(G,H) \geq \delta$, the diameter of H. For any embedding of G into H, $dil(G,H) \leq \delta$. Hence $dil(G,H) = \delta = 2r$.

The problem of finding lower bounds for dilation of an embedding is challenging. In what follows, we obtain the same for embedding a class of circulant graphs into butterfly and benes networks and prove that the bound is sharp.

Theorem 3. *Let G be the circulant graph $G((r+1)2^r; \pm\{1,2,\ldots,(2r-1)2^{r-2}+1\})$ and H be the butterfly network $BF(r)$. Then $dil(G,H) \geq 2r-1$, where $r \geq 1$.*

Proof. Let $u_i \in V(G)$ be such that $D_\delta(f(u_i)) \neq \phi$, $1 \leq i \leq 2^{r-1}$. By the definition of circulant graph, there are exactly $n - 2j - 2^{r-1} = (r+1)2^r - 2[(2r-1)2^{r-2}+1] - 2^{r-1} = 2^r - 1$ vertices which are not adjacent to the vertices $u_1, u_2, \ldots, u_{2^{r-1}}$ in G. Since $|D_\delta(f(u_i))| + |D_{\delta-1}(f(u_i))| = 2^r > 2^r - 1$, by Lemma 1, $dil(G,H) \geq \delta - 1 = 2r - 1$.

Now we embed the circulant graph $G((r+1)2^r; \pm\{1,2,\ldots,(2r-1)2^{r-2}+1\})$ into $BF(r)$ with dilation $2r-1$, which is one less than the diameter of H.

Dilation Algorithm A

Input: The circulant graph $G((r+1)2^r; \pm\{1,2,\ldots,(2r-1)2^{r-2}+1\})$ and butterfly network $BF(r)$, $r \geq 1$.

Algorithm: Label the consecutive vertices of the outer cycle C in $G((r+1)2^r; \pm\{1,2,\ldots,(2r-1)2^{r-2}+1\})$ as $0, 1, \ldots, (r+1)2^r - 1$ in the clockwise sense. Label the vertices in the 0th level of $BF(r)$ as $0, 1, \ldots, 2^{r-1}-1, (r+1)2^{r-1}, (r+1)2^{r-1}+1, \ldots, (r+2)2^{r-1}-1$ and label the remaining vertices in $BF(r)$ arbitrarily. See Fig. 3.

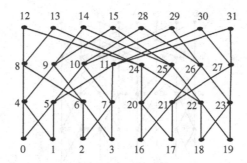

Fig. 3. Labeling of $BF(3)$ using Dilation Algorithm A

Output: An embedding f of $G((r+1)2^r; \pm\{1, 2, \ldots, (2r-1)2^{r-2}+1\})$ into $BF(r)$ given by $f(x) = x$ with dilation $2r - 1$.

Theorem 4. *Let G be the circulant graph $G((r+1)2^r; \pm\{1, 2, \ldots, (2r-1)2^{r-2}+1\})$ and H be the butterfly network $BF(r)$. Then $dil(G, H) = 2r - 1$, $r \geq 1$.*

Proof. Label the vertices of G and H using Dilation Algorithm A. We assume that the labels represent the vertices to which they are assigned. $BF(r)$ contains two vertex disjoint subgraphs isomorphic to $BF(r-1)$, say $BF^1(r-1)$ and $BF^2(r-1)$. This labeling implies that there is no edge $c = (u, v)$ of G with $f(u)$ mapped to a vertex in the 0th level of $BF^1(r-1)$ and $f(v)$ mapped to a vertex in the 0th level of $BF^2(r-1)$. Thus $dil(G, H) \leq \delta - 1$, where $\delta = 2r$. By Theorem 3, $dil(G, H) \geq 2r - 1$. Thus $dil(G, H) = 2r - 1$.

Now we embed the circulant graph $G((2r+1)2^r; \pm\{1, 2, \ldots, (4r-7)2^{r-2}+1\})$ into benes network $B(r)$ with dilation $2r-1$, which is one less than the diameter of H.

Theorem 5. *Let G be the circulant graph $G((2r+1)2^r; \pm\{1, 2, \ldots, (4r-7)2^{r-2}+1\})$ and H be the benes network $B(r)$. Then $dil(G, H) \geq 2r - 1$, where $r \geq 1$.*

Proof. Let $u_i \in V(G)$ be such that $D_\delta(f(u_i)) \neq \phi$, $1 \leq i \leq 3 \cdot 2^{r-1}$. By the definition of circulant graph, there are exactly $n - 2j - 3 \cdot 2^{r-1} = (r+1)2^r - 2[(2r-1)2^{r-2}+1] - 2^{r-1} = 2^r - 1$ vertices which are not adjacent to the vertices $u_1, u_2, \ldots, u_{3 \cdot 2^{r-1}}$ in G. Since $|D_\delta(f(u_i))| + |D_{\delta-1}(f(u_i))| = 2^r > 2^r - 1$, by Lemma 1, $dil(G, H) \geq \delta - 1 = 2r - 1$.

Dilation Algorithm B

Input: The circulant graph $G((2r+1)2^r; \pm\{1, 2, \ldots, (4r-7)2^{r-2}+1\})$ and benes network $B(r)$, $r \geq 1$.

Algorithm: Label the consecutive vertices of the outer cycle C in $G((2r+1)2^r; \pm\{1, 2, \ldots, (4r-7)2^{r-2}+1\})$ as $0, 1, \ldots, (4r-7)2^{r-2}$ in the clockwise sense. Label the vertices of $B(r)$ as follows:

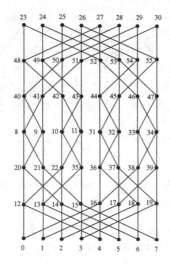

Fig. 4. Labeling of $B(3)$ using Dilation Algorithm B

1. Label the vertices in the 0th level as $0, 1, \ldots, 2^r - 1$ beginning with $(\underbrace{00 \cdots 00}_{r \text{ times}}, 0)$.

2. Label the vertices in the rth level as $2^r, 2^r + 1, \ldots, 3 \cdot 2^{r-1} - 1, 2^{r-2}(4r+3) + 1, 2^{r-2}(4r+3) + 2, \ldots, 2^{r-2}(4r+5)$ beginning with $(\underbrace{00 \cdots 00}_{r \text{ times}}, r)$.

3. Label the vertices in the $2r$-th level as $2^{r-2}(4r-1) + 1, 2^{r-2}(4r-1) + 2, \ldots, 2^{r-2}(4r+3)$ beginning with $(\underbrace{00 \cdots 00}_{r \text{ times}}, 2r)$.

4. Label the remaining vertices in $B(r)$ arbitrarily. See Fig. 4.

Output: An embedding f of $G((2r+1)2^r; \pm\{1, 2, \ldots, (4r-7)2^{r-2} + 1\})$ into $B(r)$ given by $f(x) = x$ with dilation $2r - 1$.

Theorem 6. *Let G be the circulant graph $G((2r+1)2^r; \pm\{1, 2, \ldots, (4r-7)2^{r-2} + 1\})$ and H be the benes network $B(r)$. Then $dil(G, H) = 2r - 1$, $r \geq 1$.*

Proof. Label the vertices of G and H using Dilation Algorithm B. We assume that the labels represent the vertices to which they are assigned. This labeling implies that there is no edge $e = (u, v)$ of G with $f(u)$ mapped to a vertex in the 0th level of $B(r)$ and $f(v)$ mapped to a vertex in the $2r$-th level of $B(r)$. Further, there is no edge $e = (u, v)$ of G with $f(u)$ mapped to a vertex in the set $\{(\underbrace{00 \cdots 00}_{r \text{ times}}, r), (\underbrace{00 \cdots 001}_{r-1 \text{ times}}, r), \ldots, (\underbrace{011 \cdots 11}_{r-1 \text{ times}}, r)\}$ of $B(r)$ and $f(v)$ mapped to a vertex in the set $\{(\underbrace{100 \cdots 00}_{r-1 \text{ times}}, r), (\underbrace{100 \cdots 001}_{r-2 \text{ times}}, r), \ldots, (\underbrace{11 \cdots 11}_{r \text{ times}}, r)\}$ of $B(r)$. Thus $dil(G, H) \leq \delta - 1$, where $\delta = 2r$. By Theorem 5, $dil(G, H) \geq 2r - 1$. Thus $dil(G, H) = 2r - 1$.

4   Concluding Remarks

In this paper, we compute the minimum dilation of embedding certain classes of circulant networks into butterfly and benes networks. Finding the wirelengths of embedding circulant networks into butterfly and benes networks is under investigation.

Acknowledgement. The authors would like to thank the anonymous referees for their comments and suggestions. These comments and suggestions were very helpful for improving the quality of this paper.

References

1. Chaudhary, V., Aggarwal, J.K.: Generalized mapping of parallel algorithms onto parallel architectures. In: Proceeding of International Conference on Parallel Processing, pp. 137–141 (1990)
2. Dvořák, T.: Dense sets and embedding binary trees into hypercubes. Discrete Appl. Math. **155**(4), 506–514 (2007)
3. Bezrukov, S.L., Chavez, J.D., Harper, L.H., Röttger, M., Schroeder, U.P.: Embedding of hypercubes into grids. In: Mortar Fire Control System, pp.693–701 (1998)
4. Rajasingh, I., Rajan, B., Rajan, R.S.: Embedding of special classes of circulant networks, hypercubes and generalized Petersen graphs. Int. J. Comput. Math. **89**(15), 1970–1978 (2012)
5. Gupta, A.K., Nelson, D., Wang, H.: Efficient embeddings of ternary trees into hypercubes. J. Parallel Distrib. Comput. **63**(6), 619–629 (2003)
6. Bezrukov, S.L.: Embedding complete trees into the hypercube. Discrete Appl. Math. **110**(2–3), 101–119 (2001)
7. Manuel, P., Rajasingh, I., Rajan, R.S.: Embedding variants of hypercubes with dilation 2. J. Interconnect. Netw. **13**(1–2), 1–16 (2012)
8. Ramanathan, P., Shin, K.G.: Reliable broadcast in hypercube multicomputers. IEEE Trans. Comput. **37**(12), 1654–1657 (1988)
9. Wong, G.K., Coppersmith, D.A.: A combinatorial problem related to multimodule memory organization. J. Assoc. Comput. Mach. **21**(3), 392–401 (1994)
10. Boesch, F.T., Wang, J.: Reliable circulant networks with minimum transmission delay. IEEE Trans. Circuit Syst. **32**(12), 1286–1291 (1985)
11. Bermond, J.C., Comellas, F., Hsu, D.F.: Distributed loop computer networks: a survey. Surv. J. Parallel Distrib. Comput. **24**(1), 2–10 (1995)
12. Beivide, R., Herrada, E., Balcazar, J.L., Arruabarrena, A.: Optimal distance networks of low degree for parallel computers. IEEE Trans. Comput. **40**(10), 1109–1124 (1991)
13. Wilkov, R.S.: Analysis and design of reliable computer networks. IEEE Trans. Commun. **20**(3), 660–678 (1972)
14. Xu, J.M.: Topological Structure and Analysis of Interconnection Networks. Kluwer Academic Publishers, Dordrecht (2001)
15. Manuel, P., Abd-El-Barra, M.I., Rajasingh, I., Rajan, B.: An efficient representation of benes networks and its applications. J. Discrete Algorithms **6**(1), 11–19 (2008)

16. Garey, M.R., Johnson, D.S.: Computers and Intractability: A Guide to the Theory of NP-Completeness. Freeman, San Francisco (1979)
17. Harper, L.H.: Global Methods for Combinatorial Isoperimetric Problems. Cambridge University Press, Cambridge (2004)
18. Rajan, R.S., Miller, M., Rajasingh, I., Manuel, P.: Embedding circulant networks into certain trees. J. Comb. Optim. (submitted)
19. Rajan, R.S., Manuel, P., Rajasingh, I., Parthiban, N., Miller, M.: A lower bound for dilation of an embedding. Comput. J. (2015). http://comjnl.oxfordjournals.org/content/early/2015/04/01/comjnl.bxv021

Kinetic Reverse k-Nearest Neighbor Problem

Zahed Rahmati[1,2]([✉]), Valerie King[1], and Sue Whitesides[1]

[1] Department of Computer Science, University of Victoria, Victoria, Canada
{rahmati,val,sue}@uvic.ca
[2] Cheriton School of Computer Science, University of Waterloo, Waterloo, Canada
zrahmati@uwaterloo.ca

Abstract. This paper provides the first solution to the kinetic reverse k-nearest neighbor (RkNN) problem in $\mathbb{R}^d$, which is defined as follows: Given a set P of n moving points in arbitrary but fixed dimension d, an integer k, and a query point $q \notin P$ at any time t, report all the points $p \in P$ for which q is one of the k-nearest neighbors of p.

Keywords: Reverse k-nearest neighbor query · Moving points · k-nearest neighbors · Kinetic data structure · Continuous monitoring · Continuous queries

1 Introduction

The *reverse k-nearest neighbor* (RkNN) problem is a popular variant of the k-nearest neighbor (kNN) problem and asks for the influence of a query point on a point set. Unlike the kNN problem, the exact number of reverse k-nearest neighbors of a query point is not known in advancem, but as we prove in this paper the number is upper-bounded by $O(k)$. The RkNN problem is formally defined as follows: Given a set P of n points in $\mathbb{R}^d$, an integer k, $1 \leq k \leq n-1$, and a query point $q \notin P$, find the set RkNN(q) of all p in P for which q is one of k-nearest neighbors of p. Thus RkNN$(q) = \{p \in P : |pq| \leq |pp_k|\}$, where $|.|$ denotes Euclidean distance, and p_k is the kth nearest neighbor of p among the points in P. The *kinetic RkNN* problem is to answer RkNN queries on a set P of moving points, where the trajectory of each point $p \in P$ is a function of time. Here, we assume the trajectories are polynomial functions of maximum degree bounded by some constant s.

Related Work. The reverse k-nearest neighbor problem was first posed by Korn and Muthukrishnan [13] in the database community, and then considered extensively in this community due to its many applications, *e.g.*, decision support systems, profile-based marketing, traffic networks, business location planning, clustering and outlier detection, and molecular biology. The reverse k-nearest neighbor queries for a set of continuously moving objects has also attracted the

This work was partially supported by a British Columbia Graduate Student Fellowship and by NSERC discovery grants.

© Springer International Publishing Switzerland 2015
J. Kratochvíl et al. (Eds.): IWOCA 2014, LNCS 8986, pp. 307–317, 2015.
DOI: 10.1007/978-3-319-19315-1_27

attention of the database community; see [8] and references therein. Examples of moving objects include players in multi-player game environments, soldiers in a battlefield, tourists in dangerous environments, and mobile devices in wireless ad-hoc networks.

To our knowledge, in computational geometry, there exist two data structures [9,14] that give solutions to the RkNN problem. Both of these solutions answer RkNN queries for a set P of stationary points and both only work for $k = 1$. Maheshwari *et al.* (2002) [14] gave a data structure to solve the R1NN problem in $\mathbb{R}^2$. Their data structure creates an arrangement of largest empty circles centered at the points of P and answers R1NN queries by point location in the arrangement. Their data structure uses $O(n)$ space and $O(n \log n)$ preprocessing time, and an R1NN query can be answered in time $O(\log n)$. Cheong *et al.* (2011) [9] considered the R1NN problem in $\mathbb{R}^d$, where $d = O(1)$. Their method, which uses a compressed quadtree, partitions space into cells such that each cell contains a small number of candidate points. To answer an R1NN query, their solution finds a cell that contains the query point and then checks all the points in the cell. Their approach uses $O(n)$ space and $O(n \log n)$ preprocessing time, and can answer an R1NN query in $O(\log n)$ time. It seems that the approach by Cheong *et al.* can be extended to answer RkNN queries with preprocessing time $O(kn \log n)$, space $O(kn)$, and query time $O(\log n + k)$.

For a set P of n stationary points, one can report all the 1-nearest neighbors in time $O(n \log n)$ [18], and all the k-nearest neighbors, for any $k \geq 1$, in time $O(kn \log n)$ [12], where the neighbors are reported in order of increasing distance from each point; reporting the unordered set takes time $O(n \log n + kn)$ [5,10,12].

For a set of moving points, there are three kinetic data structures (KDS's) [2, 16,17] to maintain all the k-nearest neighbors, but they only work for $k = 1$.

Our Contribution. For a set P of n continuously moving points in $\mathbb{R}^d$, where the trajectory of each point is a polynomial function of at most constant degree s, we provide a simple kinetic approach to answer RkNN queries on the moving points. In fact, we provide the *first* solution to the kinetic RkNN problem for *any* $k \geq 1$ in *any* fixed dimension d. To answer an RkNN query for a query point $q \notin P$ at any time t, we partition the d-dimensional space into a constant number of cones around q, and then among the points of P in each cone, we examine the k points having shortest projections on the cone axis. We obtain $O(k)$ candidate points for q such that q might be one of their k-nearest neighbors at time t. To check which if any of these candidate points is a reverse k-nearest neighbor of q, we maintain the kth nearest neighbor p_k of each point $p \in P$ over time. By checking whether $|pq| \leq |pp_k|$ we can easily check whether a candidate point p is one of the reverse k-nearest neighbors of q at time t.

In the preprocessing step, we introduce a method for reporting all the k-nearest neighbors for all the points $p \in P$ in order of increasing distance from p. For $k = \Omega(\log^{d-1} n)$, both our method and the method of Dickerson and Eppstein [12] give the same complexity, but in our view, our method is simpler in practice.

In order to answer RkNN queries, our kinetic approach maintains all the k-nearest neighbors over time. This is the *first* KDS for maintenance of all the k-nearest neighbors in $\mathbb{R}^d$, for any $k \geq 1$. Our KDS uses $O(n \log^{d+1} n + kn)$ space and $O(n \log^{d+1} n + kn \log n)$ preprocessing time, and processes $O(\phi(s, n) * n^2)$ events, each in amortized time $O(\log n)$. Here, $\phi(s, n)$ is the complexity of the k-level of a set of n partially-defined polynomial functions, such that each pair of them intersects at most s times. The current bounds on $\phi(s, n)$ are as follows [6, 7].

$$\phi(s, n) = \begin{cases} O(n^{3/2} \log n), & \text{for } s = 2; \\ O(n^{5/3} \text{poly} \log n), & \text{for } s = 3; \\ O(n^{31/18} \text{poly} \log n), & \text{for } s = 4; \\ O(n^{161/90 - \delta}), & \text{for } s = 5, \text{ for some constant } \delta > 0; \\ O(n^{2 - 1/2s - \delta_s}), & \text{for odd } s, \text{ for some constant } \delta_s > 0; \\ O(n^{2 - 1/2(s-1) - \delta_s}), & \text{for even } s, \text{ for some constant } \delta_s > 0. \end{cases}$$

At any time t, an RkNN query can be answered in time $O(\log^d n + k)$. Note that if an event occurs at the same time t, we first spend amortized time $O(\log n)$ to update all the k-nearest neighbors, and then we answer the query.

Outline. Section 2 provides two key lemmas, and in fact introduces a new supergraph, namely the k-*Semi-Yao graph*, of the k-nearest neighbor graph. In Sect. 3, we show how to report all the k-nearest neighbors. Section 4 gives a (kinetic) data structure for answering RkNN queries on moving points, where the trajectory of each point is a bounded-degree polynomial. Section 5 concludes.

2 Key Lemmas

Partition the plane around the origin o into six wedges, $W_0, ..., W_5$, each of angle $\pi/3$ (see Fig. 1(a)). Denote by $W_l(p)$ the translation of wedge W_l, $0 \leq l \leq 5$, such that its apex moves from o to point p (see Fig. 1(b)). Denote by x_l (resp. $x_l(p)$) the vector along the bisector of W_l (resp. $W_l(p)$) directed outward from the apex at o (resp. p). Denote the reflection of $W_l(p)$ through p by $W_{l'}(p)$. Note that $l' = (l + 3)$ mod 6; see Fig. 1(b). Consider the ith nearest neighbor p_i of p. Denote by $L(P \cap W_l(p_i))$ the list of the points in $P \cap W_l(p_i)$, sorted by increasing order of their x_l-coordinates (projections). The following lemma provides a key insight. The short proof is omitted (see the full version of the paper in Chap. 6 of the first author's PhD dissertation [15]).

Lemma 1. *Let p_i be the ith nearest neighbor of p among a set P of points in $\mathbb{R}^2$, and let $W_l(p_i)$ be the wedge of p_i that contains p. Then point p is among the first i points in $L(P \cap W_l(p_i))$.*

The k-*nearest neighbor graph* (k-NNG) of a point set P is constructed by connecting each point in P to all its k-nearest neighbors. If we connect each point

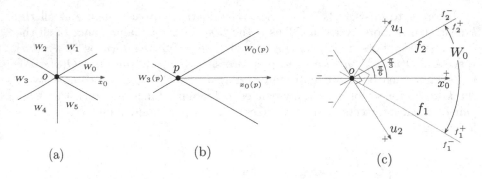

Fig. 1. (a) A Partition of the plane into six wedges with common apex at o. (b) A translation of W_0 that moves apex to p. The wedge $W_0(p)$ is the reflection through p of $W_3(p)$ and vise-versa. (c) The wedge W_0 in $\mathbb{R}^2$ is bounded by f_1 and f_2. The coordinate axes u_1 and u_2 are orthogonal to f_1 and f_2.

$p \in P$ to the first k points in the sorted list $L(P \cap W_l(p))$, for $l = 0, ..., 5$, we obtain what we call the k-*Semi-Yao graph* (k-SYG). Lemma 1 gives a necessary condition for p_i to be the ith nearest neighbor of p: the point p is among the first i points in $L(P \cap W_l(p_i))$, where l is such that $p \in W_l(p_i)$. Therefore, the edge set of the k-SYG covers the edges of the k-NNG. In summary, we have the following.

Lemma 2. *The k-NNG of a set P of points in $\mathbb{R}^2$ is a subgraph of the k-SYG of the set P.*

3   Reporting All k-Nearest Neighbors

Here we give a simple method for reporting all the k-nearest neighbors via a construction of the k-SYG.

Let C be a *right circular cone* in $\mathbb{R}^d$ with opening angle θ with respect to some given unit vector v. Thus C is the set of points $x \in \mathbb{R}^d$ such that the angle between $\vec{ox}$ and $\vec{v}$ is at most $\theta/2$. The angle between any two rays inside C emanating from the apex o is at most θ. From now on, we assume $\theta \leq \pi/3$.

Now consider a *polyhedral cone* inscribed in the right circular cone C where the polyhedral cone is formed by the intersection of d distinct half-spaces, bounded by $f_1, ..., f_d$, passing through the apex of C. Assuming d is arbitrary but fixed, the d-dimensional space around the origin o can be tiled by a constant number of polyhedral cones $W_0, ..., W_{c-1}$ [1,2]. Denote by C_l the associated right circular cone of the polyhedral cone W_l. Let x_l be the vector in the direction of the symmetry of C_l. Denote by $W_l(p)$ the translation of the wedge (polyhedral cone) W_l where o moves to p.

A similar approach and analysis as that in Sect. 2 can be easily used to state (key) Lemmas 1 and 2 for a set of points in $\mathbb{R}^d$.

To construct the k-SYG efficiently, we need a data structure to perform the following operation efficiently: For each $p \in P$ and any of its wedges $W_l(p)$, $0 \le l \le c - 1$, find the first k points in $L(P \cap W_l(p))$. Such an operation can be performed by using *range tree* data structures. For each wedge W_l with apex at origin o, we construct an associated d-dimensional range tree T_l as follows.

Consider a particular wedge W_l with apex at o. The wedge W_l is the intersection of d half-spaces $f_1^+, ..., f_d^+$ bounded by $f_1, ..., f_d$ (see Fig. 1(c)). Let $\hat{u}_j$ denote the normal to f_j pointing to f_j^+. We define d coordinate axes u_j, $j = 1, ..., d$, through $\hat{u}_j$, where $\hat{u}_j$ gives the respective directions of increasing u_j-coordinate values.

The range tree T_l is a regular d-dimensional range tree based on the u_j-coordinates, $j - 1, ..., d$. The points at level j are sorted at the leaves according to their u_j-coordinates (for more details about range trees, see Chap. 5 of [4]). Any d-dimensional range tree, e.g., T_l, uses $O(n \log^{d-1} n)$ space and can be constructed in time $O(n \log^{d-1} n)$; for any point $r \in \mathbb{R}^d$, the points of P inside the query wedge $W_l(r)$ whose sides are parallel to f_j, $j = 1, ..., d$, can be reported in time $O(\log^{d-1} n + z)$, where z is the cardinality of the set $P \cap W_l(r)$ [4].

Now we add a new level to T_l, based on the coordinate x_l. Let $C_l(p)$ be the set of the first k points in $L(P \cap W_l(p))$. To find $C_l(p)$ in an efficient time, we use the level $d+1$ of T_l, which is constructed as follows: For each internal node v at level d of T_l, we create a list $L(P(v))$ sorted by increasing order of x_l-coordinates of the points in $P(v)$. For the set P of n points in $\mathbb{R}^d$, the range tree T_l, which now is a $(d+1)$-dimensional range tree, uses $O(n \log^d n)$ space and can be constructed in time $O(n \log^d n)$.

The following lemma establishes the processing time for obtaining a $C_l(p)$. The short proof is omitted (see the full version of the paper).

Lemma 3. *Given T_l, the set $C_l(p)$ can be found in time $O(\log^d n + k)$.*

By Lemma 3, we can efficiently find all the $C_l(p)$, for all the points $p \in P$. This gives the following lemma.

Lemma 4. *Using a data structure of size $O(n \log^d n)$, the edges of the k-syg of a set of n points in fixed dimension d can be reported in time $O(n \log^d n + kn)$.*

Next, suppose we are given the k-SYG and we want to report all the k-nearest neighbors. Let E_p be the set of edges incident to the point p in the k-SYG. By sorting these edges in non-decreasing order according to their Euclidean lengths, which can be done in time $O(|E_p| \log |E_p|)$, we can find the k-nearest neighbors of p ordered by increasing distance from p. Since the number of edges in the k-SYG is $O(kn)$ and each edge pp' belongs to exactly two sets E_p and $E_{p'}$, the time to find all the k-nearest neighbors, for all the points $p \in P$, is $\sum_p O(|E_p| \log |E_p|) = O(kn \log n)$.

From the above discussion and Lemmas 2 and 4, the following results.

Theorem 1. *For a set of n points in fixed dimension d, our data structure can report all the k-nearest neighbors, in order of increasing distance from each point, in time $O(n \log^d n + kn \log n)$. The data structure uses $O(n \log^d n + kn)$ space.*

4   RkNN Queries on Moving Points

We are given a set P of n continuously moving points, where the trajectory of each point in P is a polynomial function of bounded degree s. To answer RkNN queries on the moving points, we must keep a valid range tree and track all the k-nearest neighbors during the motion. This section first shows how to maintain a (ranked-based) range tree, and then provides a KDS for maintenance of the k-SYG, which in fact gives a supergraph of the k-NNG over time. Using the kinetic k-SYG, we can easily maintain all the k-nearest neighbors over time. Finally we show how to answer RkNN queries on the moving points.

Kinetic RBRT. Let u_j, $1 \leq j \leq d$, be the coordinate axis orthogonal to the half-space f_j of the wedge W_l, $0 \leq l \leq c - 1$ (see Fig. 1(c)). Abam and de Berg [1] introduced a variant of the range tree, namely the *ranked-based range tree* (RBRT), which has the following properties. Denote by T_l the RBRT corresponding to the wedge W_l.

- T_l can be described as a set of pairs $\Psi_l = \{(B_1, R_1), ..., (B_m, R_m)\}$ such that:
  - For any two points p and q in P where $q \in W_l(p)$, there is a unique pair $(B_i, R_i) \in \Psi_l$ such that $p \in B_i$ and $q \in R_i$.
  - For any pair $(B_i, R_i) \in \Psi_l$, if $p \in B_i$ and $q \in R_i$, then $q \in W_l(p)$ and $p \in W_{l'}(q)$; here $W_{l'}(q)$ is the reflection of $W_l(q)$ through q.
  The Ψ_l is called a *cone separated pair decomposition* (CSPD) for P with respect to W_l. Each pair (B_i, R_i) is generated from an internal node v at level d of the RBRT T_l.
- Each point $p \in P$ is in $O(\log^d n)$ pairs of (B_i, R_i), which means that the number of elements of all the pairs (R_i, B_i) is $O(n \log^d n)$.
- For any point $p \in P$, all the sets B_i (resp. R_i) where $p \in B_i$ (resp. $p \in R_i$) can be found in time $O(\log^d n)$.
- The set $P \cap W_l(p)$ is the union of $O(\log^d n)$ sets R_i, where $p \in B_i$.
- When the points are moving, T_l remains unchanged as long as the order of the points along axes u_j, $1 \leq j \leq d$, remains unchanged.
- When a u-swap event occurs, meaning that two points exchange their u_j-order, the RBRT T_l can be updated in worst-case time $O(\log^d n)$ without rebalancing operations.

4.1   Kinetic k-SYG

Here we give a KDS for the k-SYG, for any $k \geq 1$, extending [16].

To maintain the k-SYG, we must track the set $C_l(p)$ for each point $p \in P$. So, for each $1 \leq i \leq m$, we need to maintain a sorted list $L(R_i)$ of the points in R_i in ascending order according to their x_l-coordinates over time. Note that each set R_i is some $P(v)$, the set of points at the leaves of the subtree rooted at some internal node v at level d of T_l. To maintain these sorted lists $L(R_i)$, we add a new level to the RBRT T_l; the points at the new level are sorted at the leaves in ascending order according to their x_l-coordinates. Therefore, in the modified RBRT T_l, in addition to the u-swap events, we handle new events, called *x-swap events*, when two points exchange their x_l-order. The modified RBRT $\mathcal{T}_l$ behaves like a $(d + 1)$-dimensional RBRT. From the last property of an RBRT above, when a u-swap event or an x-swap event occurs, the RBRT $\mathcal{T}_l$ can be updated in worst-case time $O(\log^{d+1} n)$.

Denote by $\ddot{p}_{l,k}$ the kth point in $L(P \cap W_l(p))$. To track the sets $\mathcal{C}_l(p)$, for all the points $p \in P$, we need to maintain the following over time.

- A set of $d + 1$ *kinetic sorted lists* $L_j(P)$, $j = 1, ..., d$, and the $L_l(P)$ of the point set P. We use these kinetic sorted lists to track the order of the points in the coordinates u_j and x_l, respectively.
- For each B_i, a sorted list $L(B_i')$ of the points in B_i', where $B_i' = \{(p, \ddot{p}_{l,k}) | \ p \in B_i\}$. The order of the points in $L(B_i')$ is according to a *label* of the second points $\ddot{p}_{l,k}$. This sorted list $L(B_i')$ is used to answer the following query efficiently: Given a query point q and a B_i, find all the points $p \in B_i$ such that $\ddot{p}_{l,k} = q$.
- The kth point $r_{i,k}$ in the sorted list $L(R_i)$. We track the values $r_{i,k}$ in order to make necessary changes to the k-SYG when an x-swap event occurs.

Handling u-swap Events. W.l.o.g., let $q \in W_l(p)$ before the event. When a u-swap event between p and q occurs, the point q moves outside the wedge $W_l(p)$; after the event, $q \notin W_l(p)$. Note that the changes that occur in the k-SYG are the deletions and insertions of the edges incident to p inside the wedge $W_l(p)$.

Whenever two points p and q exchange their u_j-order, we do the following updates.

- We update the kinetic sorted list $L_j(P)$. Each swap event in a kinetic sorted list can be handled in time $O(\log n)$.
- We update the RBRT $\mathcal{T}_l$ and if a point is deleted or inserted into a B_i, we update the sorted list $L(B_i')$. Since each insertion/deletion to $L(B_i')$ takes $O(\log n)$ time, and since each point is in $O(\log^d n)$ sets B_i, this takes $O(\log^{d+1} n)$ time.
- We update the values of $r_{i,k}$. After updating the RBRT $\mathcal{T}_l$, point q might be inserted or deleted from some R_i and change the values of $r_{i,k}$. So, for all R_i where $q \in R_i$, before and after the event, we do the following. We check whether the x_l-coordinate of q is less than or equal to the x_l-coordinate of $r_{i,k}$; if so, we take the successor or predecessor point of $r_{i,k}$ in $L(R_i)$ as the new value for $r_{i,k}$. This takes $O(\log^{d+1} n)$ time.
- We query to find $\mathcal{C}(p)$. By Lemma 3, this takes $O(\log^d n + k)$ time.
- If we get a new value for $\ddot{p}_{l,k}$, we update all the sorted lists $L(B_i')$ such that $p \in B_i$. This takes $O(\log^{d+1} n)$ time.

Considering the complexity of each step above, and assuming the trajectory of each point is a bounded degree polynomial, the following results.

Lemma 5. *Our KDS for maintenance of the k-SYG handles $O(n^2)$ u-swap events, each in worst-case time $O(\log^{d+1} n + k)$.*

Handling x-swap Events. When an x-swap event between two consecutive points p and q with p preceding q occurs, it does not change the elements of the pairs (B_i, R_i) of the CSPD Ψ_l. Such an event changes the k-SYG if both p and q are in the same $W_l(w)$, for some $w \in P$, and $w_{l,k} = p$.

We apply the following updates to our KDS when two points p and q exchange their x_l-order.

1. We update the kinetic sorted list $L_l(P)$; this takes $O(\log n)$ time.
2. We update the RBRT $\mathcal{T}_l$, which takes $O(\log^{d+1} n)$ time.
3. We find all the sets R_i where both p and q belong to R_i and such that $r_{i,k} = p$. Also, we find all the sets R_i where $r_{i,k} = q$. This takes $O(\log^d n)$ time.
4. For each R_i, we extract all the pairs $(w, \ddot{w}_{l,k})$ from the sorted lists $L(B_i')$ such that $\ddot{w}_{l,k} = p$. Note that each change to the pair $(w, \ddot{w}_{l,k})$ is a change to the k-SYG.
5. For each w, we update all the sorted lists $L(B_i')$ where $(w, \ddot{w}_{l,k}) \in B_i'$: we replace the previous value of $\ddot{w}_{l,k}$, which is p, by the new value q.

Denote by χ_k the number of exact changes to the k-SYG of a set of moving points over time. For each found R_i, the fourth step takes $O(\log n + \xi_i)$ time, where ξ_i is the number of pairs $(w, \ddot{w}_{l,k})$ such that $\ddot{w}_{l,k} = p$. For all these $O(\log^d n)$ sets R_i, this step takes $O(\log^{d+1} n + \sum_i \xi_i)$ time, where $\sum_i \xi_i$ is the number of exact changes to the k-SYG when an x-swap event occurs. Therefore, for all the $O(n^2)$ x-swap events, the total processing time for this step is $O(n^2 \log^{d+1} n + \chi_k)$.

The processing time for the fifth step is a function of χ_k. For each change to the k-SYG, this step spends $O(\log^{d+1} n)$ time to update the sorted lists $L(B_i')$. Therefore, the total processing time for all the x-swap events in this step is $O(\chi_k * \log^{d+1} n)$.

From the above discussion and an upper bound for χ_k in Lemmas 6, 7 results. The proof of Lemma 6 is omitted (see the full version of the paper).

Lemma 6. *The number of changes to the k-SYG of a set of n moving points, where the trajectory of each point is a polynomial function of at most constant degree s, is $\chi_k = O(\phi(s, n) * n)$.*

Lemma 7. *Our KDS for maintenance of the k-SYG handles $O(n^2)$ x-swap events with a total cost of $O(\phi(s, n) * n \log^{d+1} n)$.*

From Lemmas 5 and 7, the following theorem results.

Theorem 2. *For a set of n moving points in $\mathbb{R}^d$, where the trajectory of each point is a polynomial function of at most constant degree s, our k-SYG KDS uses $O(n \log^{d+1} n)$ space and handles $O(n^2)$ events with a total cost of $O(kn^2 + \phi(s, n) * n \log^{d+1} n)$.*

4.2   Kinetic All k-Nearest Neighbors

Given a KDS for maintenance of the k-SYG (from Theorem 2), a supergraph of the k-NNG, this section shows how to maintain all the k-nearest neighbors over time. For maintenance of the k-nearest neighbors of each point $p \in P$, we only need to track the order of the edges incident to p in the k-SYG according to their Euclidean lengths. This can easily be done by using a kinetic sorted list. The following theorem summarizes the complexity of our kinetic approach. The proof is omitted (see the full version of the paper).

Theorem 3. *For a set of n moving points in $\mathbb{R}^d$, where the trajectory of each point is a polynomial of at most constant degree s, our KDS for maintenance of all the k-nearest neighbors, ordered by distance from each point, uses $O(n \log^{d+1} n + kn)$ space and $O(n \log^{d+1} n + kn \log n)$ preprocessing time. Our KDS handles $O(\phi(s, n) * n^2)$ events, each in $O(\log n)$ amortized time.*

4.3 RkNN Queries

Suppose we are given a query point $q \notin P$ at some time t. To find the reverse k-nearest neighbors of q, we seek the points in $P \cap W_l(q)$ and find $C_l(q)$, the set of the first k points in $L(P \cap W_l(q))$. The set $\cup_l C_l(q)$ contains $O(k)$ candidate points for q such that q might be one of their k-nearest neighbors. In time $O(\log^d n)$ we can find a set of R_i where $P \cap W_l(q) = \sum_i R_i$. From Lemma 3, and since we have sorted lists $L(R_i)$ at level $d + 1$ of T_l, the $O(k)$ candidate points for the query point q can be found in worst-case time $O(\log^d n + k)$. Now we check whether these candidate points are the reverse k-nearest neighbors of the query point q at time t or not; this can be easily done by application of Theorem 3, which in fact maintain the kth nearest neighbor p_k of each $p \in P$. Therefore, checking a candidate point can be done in $O(1)$ time by comparing distance $|pq|$ to distance $|pp_k|$. This implies that checking which elements of $C_l(q)$, for $l = 0, ..., c - 1$, are reverse k-nearest neighbors of the query point q takes time $O(k)$.

If a query arrives at a time t that is simultaneous with the time when one of the $O(\phi(s, n) * n^2)$ events occurs, our KDS first spends amortized time $O(\log n)$ to handle the event, and then spends time $O(\log^d n + k)$ to answer the query. Thus we have the following.

Theorem 4. *Consider a set P of n moving points in $\mathbb{R}^d$, where the trajectory of each one is a bounded-degree polynomial. The number of reverse k-nearest neighbors for a query point $q \notin P$ is $O(k)$. Our KDS uses $O(n \log^{d+1} n + kn)$ space, $O(n \log^{d+1} n + kn \log n)$ preprocessing time, and handles $O(\phi(s, n) * n^2)$ events. At any time t, an RkNN query can be answered in time $O(\log^d n + k)$. If an event occurs at time t, the KDS spends amortized time $O(\log n)$ on updating itself.*

5 Discussion

In the kinetic setting, where the trajectories of the points are polynomials of bounded degree, to answer the RkNN queries over time we have provided a KDS for maintenance of all the k-nearest neighbors. Our KDS is the first KDS for maintenance of all the k-nearest neighbors in $\mathbb{R}^d$, for any $k \geq 1$. It processes $O(\phi(s, n) * n^2)$ events, each in amortized time $O(\log n)$. An open problem is to design a KDS for all k-nearest neighbors that processes less than $O(\phi(s, n) * n^2)$ events.

Arya *et al.* [3] have a kd-tree implementation to approximate the nearest neighbors of a query point that is in use by practitioners [11] who have found challenging to implement the theoretical algorithms [5,10,12,18]. Since to report all the k-nearest neighbors ordered by distance from each point our method uses multidimensional range trees, which can be easily implemented, we believe our method may be useful in practice.

Acknowledgments. We thank Timothy M. Chan for his helpful comments and suggestions.

References

1. Abam, M.A., de Berg, M.: Kinetic spanners in $\mathbb{R}^d$. Discrete Comput. Geom. **45**(4), 723–736 (2011)
2. Agarwal, P.K., Kaplan, H., Sharir, M.: Kinetic and dynamic data structures for closest pair and all nearest neighbors. ACM Trans. Algorithms **5**(4), 1–37 (2008)
3. Arya, S., Mount, D.M., Netanyahu, N.S., Silverman, R., Wu, A.Y.: An optimal algorithm for approximate nearest neighbor searching in fixed dimensions. J. ACM **45**(6), 891–923 (1998)
4. de Berg, M., Cheong, O., van Kreveld, M., Overmars, M.: Computational Geometry: Algorithms and Applications, 3rd edn. Springer-Verlag TELOS, Santa Clara (2008)
5. Callahan, P.B., Kosaraju, S.R.: A decomposition of multidimensional point sets with applications to k-nearest-neighbors and n-body potential fields. J. ACM **42**(1), 67–90 (1995)
6. Chan, T.M.: On levels in arrangements of curves, ii: A simple inequality and its consequences. Discrete Comput. Geom. **34**(1), 11–24 (2005)
7. Chan, T.M.: On levels in arrangements of curves, iii: further improvements. In: Proceedings of the 24th annual Symposium on Computational Geometry (SoCG 2008), pp. 85–93. ACM, New York (2008)
8. Cheema, M.A., Zhang, W., Lin, X., Zhang, Y., Li, X.: Continuous reverse k nearest neighbors queries in euclidean space and in spatial networks. VLDB J. **21**(1), 69–95 (2012)
9. Cheong, O., Vigneron, A., Yon, J.: Reverse nearest neighbor queries in fixed dimension. Int. J. Comput. Geom. Appl. **21**(02), 179–188 (2011)
10. Clarkson, K.L.: Fast algorithms for the all nearest neighbors problem. In: Proceedings of the 24th Annual Symposium on Foundations of Computer Science (FOCS 1983), pp. 226–232. IEEE Computer Society, Washington, DC (1983)
11. Connor, M., Kumar, P.: Fast construction of k-nearest neighbor graphs for point clouds. IEEE Trans. Vis. Comput. Graph. **16**(4), 599–608 (2010)
12. Dickerson, M.T., Eppstein, D.: Algorithms for proximity problems in higher dimensions. Int. J. Comput. Geom. Appl. **5**(5), 277–291 (1996)
13. Korn, F., Muthukrishnan, S.: Influence sets based on reverse nearest neighbor queries. In: Proceedings of the 2000 ACM SIGMOD International Conference on Management of Data (SIGMOD 2000), pp. 201–212. ACM, New York (2000)

14. Maheshwari, A., Vahrenhold, J., Zeh, N.: On reverse nearest neighbor queries. In: Proceedings of the 14th Canadian Conference on Computational Geometry (CCCG 2002), pp. 128–132 (2002)
15. Rahmati, Z.: Simple, faster kinetic data structures. Ph.D. thesis, University of Victoria (2014). http://zahedrahmati.com
16. Rahmati, Z., Abam, M.A., King, V., Whitesides, S.: Kinetic data structures for the Semi-Yao graph and all nearest neighbors in $\mathbb{R}^d$. In: Proceedings of the 26th Canadian Conference on Computational Geometry (CCCG 2014) (2014)
17. Rahmati, Z., Abam, M.A., King, V., Whitesides, S., Zarei, A.: A simple, faster method for kinetic proximity problems. Comput. Geom. **48**(4), 342–359 (2015)
18. Vaidya, P.M.: An $O(n \log n)$ algorithm for the all-nearest-neighbors problem. Discrete Comput. Geom. **4**(2), 101–115 (1989)

Efficiently Listing Bounded Length st-Paths

Romeo Rizzi[1], Gustavo Sacomoto[2,3]([✉]), and Marie-France Sagot[2,3,4]

[1] Dipartimento di Informatica, Università di Verona, Verona, Italy
[2] Université de Lyon, 69000 Lyon, France
sacomoto@gmail.com
[3] CNRS, UMR5558, Laboratoire de Biométrie et Biologie Évolutive,
Université Lyon 1, 69622 Villeurbanne, France
[4] INRIA Grenoble Rhône-Alpes, Montbonnot-saint-martin, France

Abstract. The problem of listing the K shortest simple (loopless) st-paths in a graph has been studied since the early 1960s. For a non-negatively weighted graph with n vertices and m edges, the most efficient solution is an $O(K(mn + n^2 \log n))$ algorithm for directed graphs by Yen and Lawler [Management Science, 1971 and 1972], and an $O(K(m + n \log n))$ algorithm for the undirected version by Katoh et al. [Networks, 1982], both using $O(Kn + m)$ space. In this work, we consider a different parameterization for this problem: instead of bounding the number of st-paths output, we bound their length. For the bounded length parameterization, we propose new non-trivial algorithms matching the time complexity of the classic algorithms but using only $O(m+n)$ space. Moreover, we provide a unified framework such that the solutions to both parameterizations – the classic K-shortest and the new length-bounded paths – can be seen as two different traversals of a same tree, a Dijkstra-like and a DFS-like traversal, respectively.

1 Introduction

The K-shortest simple paths problem has been studied for more than 50 years (see the references in [6]). The first efficient algorithm for this problem in directed graphs with non-negative weights only appeared 10 years later independently by Yen [18] and Lawler [12]. Given a non-negatively weighted directed graph $G = (V, E)$ with $n = |V|$ vertices and $m = |E|$ edges, using modern data structures [1], their algorithm lists the K distinct shortest *simple* st-paths by non-decreasing order of the their lengths in $O(K(mn + n^2 \log n))$ time. For undirected graphs, Katoh et al. [11] gave an improved $O(K(m+n \log n))$ algorithm. Both algorithms use $O(Kn + m)$ memory.

The best known algorithm for directed *unweighted* graphs is an $\widetilde{O}(Km\sqrt{n})$ randomized algorithm [16], where $\widetilde{O}(f(n))$ is a shorthand for $O(f(n) \log^k n)$. In a different direction, Roditty [15] noticed that the K-shortest simple paths can

GS and MFS were partially supported by the ERC programme FP7/2007-2013/ERC grant agreement no. [247073]10, and the French project ANR-12-BS02-0008 (Colib'read).

J. Kratochvíl et al. (Eds.): IWOCA 2014, LNCS 8986, pp. 318–329, 2015.
DOI: 10.1007/978-3-319-19315-1_28

be efficiently approximated. Building upon his work, Bernstein [2] presented an $\widetilde{O}(Km/\epsilon)$ time algorithm for a $(1 + \epsilon)$-approximation. Moreover, Eppstein [7] showed that if the paths are allowed to repeat vertices, *i.e.* they are not *simple*, then the problem can be solved in $O(K + m + n\log n)$ time. However, when the paths are *simple* and to be computed exactly, no improvement has been made on Yen and Lawler's for directed graphs or Katoh's algorithm for undirected graphs. The main bottleneck of these algorithms is their memory consumption.

Here, we consider the problem of listing all st-paths with length at most α. This is a different parameterization of the K-shortest path problem, where we impose an upper-bound on the length of the output paths instead of their number. This is a natural variant of the K-shortest path problem. There are situations where it is necessary to consider all paths that are a given percentage of the optimal (*e.g.* [4]). Moreover, the bounded length problem is *almost* a particular case of the K-shortest path problem. Given any solution to the K-shortest path problem, such that the st-paths are generated one at a time in non-decreasing length order, we can use the following simple approach to solve the α-bounded length variant: choose a sufficiently large K and halt the enumeration when the length of the paths is larger than α. The main disadvantage of this algorithm is its space complexity which is proportional to the number of paths output hence, in the worst case, exponential in the size of the graph.

Our first and main contribution are new polynomial delay algorithms to list st-paths with length at most α matching the time complexity (per path) of Yen and Lawler's algorithm for directed graphs (Sect. 3) and Katoh's for undirected graphs (Sect. 4), but using only $O(n + m)$ internal memory. This represents an exponential improvement in memory consumption.

The main differences between the classic solutions to the K-shortest paths problem and our solutions to the α-bounded paths problem are the order in which the solutions are output and the memory complexity of the algorithms.

Our second contribution is thus a unified framework where both problems can be represented in such a way that those differences arise in a natural manner (Sect. 3). Intuitively, we show that both families of algorithmic solutions correspond to two different traversals of a *same* rooted tree: a Dijkstra-like traversal for the K-shortest and a DFS-like traversal for the α-bounded paths.

2 Preliminaries

Given a directed graph $G = (V, E)$ with $n = |V|$ vertices and $m = |E|$ arcs, the in and out-neighborhoods of $v \in V$ are denoted by $N^-(v)$ and $N^+(v)$, respectively. Given a (directed or undirected) graph G with weights $w : E \mapsto \mathbb{Q}$, the weight, or *length*, of a path π is $\sum_{(u,v) \in \pi} w(u, v)$ and is denoted by $w(\pi)$. We say that a path π is α-*bounded* if its *length* satisfies $w(p) \leq \alpha$ and $\alpha \in \mathbb{Q}$; in the particular case of unit weights (*i.e.* of unweighted graphs), we say that p is k-*bounded* if $w(p) \leq k$ with $k \in \mathbb{Z}_{\geq 0}$. A listing algorithm is *polynomial delay* if it generates the solutions, one after the other in some order, and the time elapsed until the first is output, and thereafter the time elapsed (delay) between any two consecutive solutions, is bounded by a polynomial in the input size [9]. The general problem which we are concerned in this work is listing α-bounded st-paths in G.

Problem 1 (Listing α-bounded st-paths). Given a weighted directed graph $G = (V, E)$, two vertices $s, t \in V$, and an upper bound $\alpha \in \mathbb{Q}$, output all α-bounded st-paths.

Clearly, any solution to the K-shortest path problem is also a solution to Problem 1, with the same (total/delay) time and space complexities. Thus Problem 1 is no harder than the classic K-shortest path problem.

   We assume all directed graphs are weakly connected and all undirected graphs are connected, hence $m \geq n - 1$. Moreover, we assume hereafter the weights are non-negative. We remark however that a weaker assumption suffices to the applicability of our algorithms. Indeed, it is a well known fact that, when the graph G and the weights $w : E \mapsto \mathbb{Q}$ are such that no cycle is negative, then, using Johnson's reweighting strategy [10], we can compute non-negative weights w' such that, for some constant C, we have that $w'(\pi) = w(\pi) + C$ for any st-path π. This reweighting can be done in $O(mn)$ preprocessing steps.

3   An $O(mn + n^2 \log n)$-Delay Algorithm

In this section, we present an $O(mn + n^2 \log n)$-delay algorithm to list all st-paths with length at most α in a weighted directed graph G. Thus matching the time complexity (per path) of Yen and Lawler's algorithm, while using only space linear in the *input* size.

   The new algorithm, inspired by the binary partition method [3,14], recursively partitions the solution space at every call until the considered subspace is a singleton (contains only one solution) and in that case outputs the corresponding solution. In order to have an efficient algorithm is important to explore only non-empty partitions. Moreover, it should be stressed that the order in which the solutions are output is fixed, but arbitrary.

   Let us describe the partition scheme. Let $\mathcal{P}_\alpha(s, t, G)$ be the set of all α-bounded paths from s to t in G, and $(x, s) \cdot \mathcal{P}_\alpha(s, t, G)$ denote the concatenation of (x, s) to each path of $\mathcal{P}_\alpha(s, t, G)$. Assuming $s \neq t$, we have that

$$\mathcal{P}_\alpha(s, t, G) = \bigcup_{v \in N^+(s)} (s, v) \cdot \mathcal{P}_{\alpha'}(v, t, G - s), \tag{1}$$

where $\alpha' = \alpha - w(s, v)$. In words, the set of paths from s to t can be *partitioned* into the disjoint union of $(s, v) \cdot \mathcal{P}_{\alpha'}(v, t, G - s)$, the sets of paths beginning with an arc (s, v), for each $v \in N^+(s)$. Indeed, since $s \neq t$, every path in $\mathcal{P}_\alpha(s, t, G)$ necessarily begins with an arc (s, v), where $v \in N^+(s)$.

   Algorithm 1 implements this recursive partition strategy. The solutions are only output in the leaves of the recursion tree (line 2), where the partition is always a singleton. Moreover, in order to guarantee that every leaf in the recursion tree outputs one solution, we have to test if $\mathcal{P}_{\alpha'}(v, t, G - u)$, where

$\alpha' = \alpha - w(u,v)$, is not empty before the recursive call (line 7). This set is not empty if and only if the weight of the shortest path from v to t in $G - u$ is at most α', i.e. $d_{G-u}(v,t) \leq \alpha' = \alpha - w(u,v)$. Hence, to perform this test it is enough to compute all the distances from t in the graph $G^R - u$, where G^R is the graph G with all arcs reversed.

Consider a generic execution of Algorithm 1 for a graph G, vertices $s, t \in V$ and an upper bound α. We can represent this execution by a rooted tree $\mathcal{T}$, i.e. the recursion tree, where each node corresponds to a call with arguments $\langle u, t, \alpha, \pi_{su}, G' \rangle$. The children of a given node (call) in $\mathcal{T}$ are the recursive calls with arguments $\langle v, t, \alpha', \pi_{su}(u,v), G' - u \rangle$ of line 8. This tree plays an important role in the unified framework of Sect. 5.

Lemma 1. *The recursion tree $\mathcal{T}$ has the following properties:*

1. *The leaves of $\mathcal{T}$ are in one-to-one correspondence with the paths in $\mathcal{P}_\alpha(s, t, G)$.*
2. *The leaves in the subtree rooted on a node $\langle u, t, \alpha, \pi_{su}, G' \rangle$ correspond to the paths in $\pi_{su} \cdot \mathcal{P}_{\alpha'}(u, t, G')$.*
3. *The height of $\mathcal{T}$ is bounded by n.*

Algorithm 1. list_paths$(u, t, \alpha, \pi_{su}, G)$

---

1   **if** $u = t$ **then**
2   output(π_{su})
3   **return**
4   **end**
5   compute the distances from t in $G^R - u$
6   **for** $v \in N^+(u)$ **do**
7   **if** $d(v,t) \leq \alpha - w(u,v)$ **then**
8   list_paths$(v, t, \alpha - w(u,v), \pi_{su} \cdot (u,v), G - u)$
9   **end**
10 **end**

The correctness of Algorithm 1 follows directly from the relation given in Eq. 1 and the correctness of the tests of line 7.

Let us now analyze its running time. The cost of a node in $\mathcal{T}$ is the time spent by the operations inside the corresponding call, without including its recursive calls. This cost is dominated by the tests of line 7. They are performed in $O(1)$ time by pre-computing the distances from t to all vertices in the reverse graph $G^R - u$ (line 5). This takes $O(t(n, m))$ time, where $t(n, m)$ is the cost of a single source shortest path computation. By Lemma 1 the height of $\mathcal{T}$ is bounded by n, so the path between any two leaves (solutions) in the recursion tree has at most $2n$ nodes. Thus, the time elapsed between two solutions being output is $O(nt(n, m))$. Moreover, the algorithm uses $O(m)$ space, since each recursive call has to store only the difference with the its parent graph. Recall that each solution is immediately output (line 2), not stored by the algorithm.

Theorem 1. *Algorithm 1 has delay $O(nt(n,m))$, where $t(n,m)$ is the cost of a single source shortest path computation, and uses $O(m)$ space.*

For unweighted (directed and undirected) graphs, the single source shortest paths can be computed using breadth-first search (BFS) running in $O(m)$ time, so Theorem 1 guarantees an $O(km)$ delay to list all k-bounded st-paths, since the height of the recursion tree is bounded by k instead of n. More generally, the single source shortest paths can be computed using Dijkstra's algorithm in $O(m + n \log n)$ time (we are assuming non-negative weights), resulting in an $O(nm + n^2 \log n)$ delay.

4   An Improved Algorithm for Undirected Graphs

The total time complexity of Algorithm 1 is equal to the delay times the number of solutions, *i.e.* $O(nt(n,m)\gamma)$, where $\gamma = |\mathcal{P}_\alpha(s,t,G)|$ is the number of α-bounded st-paths. We now improve its total time complexity from $O(nt(n,m)\gamma)$ to $O((m + t(n,m))\gamma)$ in the case of weighted *undirected* graphs. On average the algorithm spends $O(m + t(n,m))$ per solution (amortized delay), thus matching the time complexity (per path) of Katoh's algorithm. The (worst-case) delay, however, remains the same as Algorithm 1.

The main idea to improve the complexity of Algorithm 1 is to explore the structure of the set of paths $\mathcal{P}_\alpha(s,t,G)$ to reduce the number of nodes in the recursion tree. We avoid redundant partition steps by guaranteeing that every node in the recursion tree has at least two children. More precisely, at every call, we identify the longest common prefix of $\mathcal{P}_\alpha(s,t,G)$, *i.e.* the longest (considering the number of edges) path $\pi_{ss'}$ such that $\mathcal{P}_\alpha(s,t,G) = \pi_{ss'} \cdot \mathcal{P}_\alpha(s',t,G)$, and append it to the current path prefix being considered in the recursive call. The intuition here is that by doing so we identify and "merge" all the consecutive single-child nodes in the recursion tree, thus guaranteeing that the remaining nodes have at least two children.

The pseudocode for this algorithm is very similar to Algorithm 1 and, for the sake of completeness, is given in Algorithm 2. We postpone the description of the $\mathtt{lcp}(u,t,\alpha,G)$ function to the next section, along with a discussion about the difficulties to extend it to directed graphs.

The correctness of Algorithm 2 follows directly from the correctness of Algorithm 1. The space used is the same of Algorithm 1, provided that $\mathtt{lcp}(u,t, \alpha, G)$ uses linear space, which, as we show in the next section, is indeed the case (Theorem 3).

Let us now analyze the total complexity of Algorithm 2 as a function of the input size and of γ, the number of α-bounded st-paths. Let R be the recursion tree of Algorithm 2 and $T(r)$ the cost of a given node $r \in R$. The total cost of the algorithm can be split in two parts, which we later bound individually, in the following way:

$$\sum_{r \in R} T(r) = \sum_{r:internal} T(r) + \sum_{r:leaf} T(r). \qquad (2)$$

Algorithm 2. list_paths$(u, t, \alpha, \pi_{su}, G)$

---

1 $\pi_{uu'} = $ lcp(u, t, α, G)
2 if $u' = t$ **then**
3 $\quad$ output$(\pi_{su}\pi_{uu'})$
4 $\quad$ **return**
5 else
6 $\quad$ compute a shortest path tree T_t' from t in $G^R - \pi_{uu'}$
7 $\quad$ **for** $v \in N(u')$ **do**
8 $\quad\quad$ **if** $d(v, t) + w(u, v) \leq \alpha$ **then**
9 $\quad\quad\quad$ list_paths$(v, t, \alpha - w(\pi_{uu'}) - w(u', v), \pi_{su} \cdot \pi_{uu'} \cdot (u', v), G - \pi_{uu'})$
10 $\quad\quad$ **end**
11 $\quad$ **end**
12 end

We have that $\sum_{r:leaf} T(r) = O((m + t(m, n))\gamma)$, since leaves and solutions are in one-to-one correspondence and the cost for each leaf is dominated by the cost of lcp(u, t, α, G), that is $O(m + t(m, n))$ (Theorem 3). Now, we have that every internal node of the recursion has at least two children, otherwise $\pi_{uu'}$ would not be the longest common prefix of $\mathcal{P}_\alpha(u, t, G)$. Thus, $\sum_{r:internal} T(r) = O((m+t(m, n))\gamma)$ since in any tree the number of branching nodes is at most the number of leaves, and the cost of each internal node is dominated by the $O(m + t(m, n))$ cost of the longest prefix computation. Therefore, the total complexity of Algorithm 2 is $O((m + t(n, m))\gamma)$. This completes the proof of Theorem 2.

Theorem 2. *Algorithm 2 outputs all α-bounded st-paths in $O((m + t(n, m))\gamma)$ time using $O(m)$ space.*

This means that for unweighted graphs, it is possible to list all k-bounded st-paths in $O(m)$ time per path. In addition, for weighted graphs, it is possible to list all α-bounded st-paths in $O(m + n \log n)$ time per path.

4.1 Computing the Longest Common Prefix of $\mathcal{P}_\alpha(s, t, G)$

The problem of computing the longest common prefix of $\mathcal{P}_\alpha(s, t, G)$ can be seen as a special case of the *replacement paths problem* [8]. Let π be a shortest st-path in G. In this problem we want to compute, for each edge e on π, the shortest st-path that avoids e. Given a solution to the replacement path problem we can compute the longest common prefix of $\mathcal{P}_\alpha(s, t, G)$ using the following procedure. For each edge e along the path π, check whether the shortest st-path avoiding e is shorter than α. There is an $O(m+n \log n)$ algorithm to compute the replacement path in undirected graphs [13], but for directed graphs the best solutions is a trivial $O(nm + n^2 \log n)$ algorithm.

In this section, we present an alternative, arguably simpler, algorithm to compute the longest common prefix of the set of α-paths from s to t, completing the description of Algorithm 2. The naive algorithm for this problem runs in

$O(nt(n, m))$ time, so that using it in Algorithm 2 would not improve the total complexity compared to Algorithm 1. Basically, the naive algorithm computes a shortest path π_{st} and then for each prefix in increasing order of length tests if there are at least two distinct extensions each with total weight less than α. In order to test the extensions, for each prefix π_{su}, we recompute the distances from t in the graph $G - \pi_{su}$, thus performing n shortest path tree computations (k computations in the unweighted case) in the worst case.

Algorithm 3 improves the naive algorithm by avoiding those recomputations. However, before entering the description of Algorithm 3, we need a better characterization of the structure of the longest common prefix of $\mathcal{P}_\alpha(s, t, G)$. Lemma 2 gives this. It does so by considering a shortest path tree rooted at s, denoted by T_s. Recall that T_s is a subgraph of G and induces a partition of the edges of G into tree edges and non-tree edges. In this tree, the longest common prefix of $\mathcal{P}_\alpha(s, t, G)$ is a prefix of the tree path from the root s to t. Additionally, any st-path in G, excluding the tree path, necessarily passes through at least one non-tree edge. The lemma characterizes the longest common prefix in terms of the non-tree edges from the subtrees rooted at siblings of the vertices in the tree path from s to t. For instance, in Fig. 1(b) the common prefix π_{su} can be extended to $\pi_{su} \cdot (u, v)$ only if there is no α-bounded path that passes through the subtree T_w and a non-tree edge (x, z), where v belongs to tree path from s to t and w is one of its siblings.

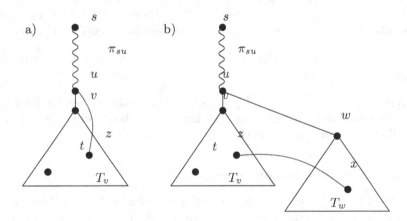

Fig. 1. The common prefix π_{su} of $\mathcal{P}_\alpha(s, t, G)$ can always be extended into an st-path using the tree path of T_s from u to t. The path π_{su} is the longest common prefix if and only if it can also be extended with a path containing a non-tree edge (x, z) such that $z \in T_v$ and (a) $x = u$ or (b) $x \in T_w$ and w is sibling of v; and $d_{G'}(s, x) + w(x, z) + d_{G'}(z, t) \leq \alpha$, where $G' = G - (u, v)$.

Lemma 2. *Let* $\pi_{su} = (s = v_0, v_1), \ldots, (v_{l-1}, v_l = u)$ *be a common prefix of all paths in* $\mathcal{P}_\alpha(s, t, G) \neq \emptyset$ *and* T_s *a shortest path tree rooted at* s. *Then,*

1. *the path $\pi_{su}(u,v)$ is a common prefix of $\mathcal{P}_\alpha(s,t,G)$, if there is no edge (x,z) such that $d_{G'}(s,x) + w(x,z) + d_{G'}(z,t) \leq \alpha$, where $G' = G - (u,v)$, $z \in T_v$, and (a) $x = u$ or (b) $x \in T_w$ with w a sibling of v (see Fig. 1);*
2. π_{su} *is the longest common prefix of $\mathcal{P}_\alpha(s,t,G)$, otherwise.*

In order to use the characterization of Lemma 2 for the longest prefix of $\mathcal{P}_\alpha(s,t,G)$, we need to efficiently test the weight condition given in item 1, namely $d_{G'}(s,x) + w(x,z) + d_{G'}(z,t) \leq \alpha$, where $G' = G - (u,v)$ and (u,v) belongs to the tree path from s to t. We have that $d_{G'}(s,x) = d_G(s,x)$, since x does not belong to the subtree of v in the shortest path tree T_s. Indeed, only the distances of vertices in the subtree T_v can possibly change after the removal of the tree edge (u,v). However, in principle we have no guarantee that $d_{G'}(z,t)$ also remains unchanged: recall that to maintain the distances from t we need a tree rooted at t not at s. Clearly, we cannot compute the shortest path tree from t for each G'; in the worst case, this would imply the computation of n shortest path trees. For this reason, we need Lemma 3. It states that, in the specific case of the vertices z we need to compute the distance to t in G', we have that $d_{G'}(z,t) = d_G(z,t)$.

Lemma 3. *Let T_s be a shortest path tree rooted at s and t a vertex of G. Then, for any edge (u,v), with v closer to t, in the shortest path π_{st} in the tree T_s, we have that $d_G(z,t) = d_{G'}(z,t)$, where $z \in T_v$ and $G' = G - (u,v)$.*

It is not hard to verify that Lemma 2 is also valid for directed graphs. However, the non-negative hypothesis for the weights is necessary; more specifically, we need the monotonicity property for path weights which states that for any path the weight of any subpath is not greater than the weight of the full path. Now, in Lemma 3 both the path monotonicity property and the fact that the graph is undirected are necessary. Since these two lemmas are the basis for the efficiency of Algorithm 3, it seems difficult to extend it to directed graphs.

Algorithm 3 implements the strategy suggested by Lemma 2. Given a shortest path tree T_s of G rooted at s, the algorithm traverses each vertex v_i in the tree path $s = v_0 \ldots v_n = t$ from the root s to t, and at every step finds all non-tree edges (x,z) entering the subtree rooted at v_{i+1} from a sibling subtree, *i.e.* a subtree rooted at $w \in N^+(v_i) \backslash \{v_{i+1}\}$. For each non-tree (x,z) linking the sibling subtrees found, it checks if it satisfies the weight condition $d_{G'}(s,x) + w(x,z) + d_{G'}(z,t) \leq \alpha$, where $G' = G - (v_i, v_{i+1})$. Item 2 of the same lemma implies that the first time an edge (x,z) satisfies the weight condition, the tree path traversed so far is the longest common prefix of $\mathcal{P}_\alpha(s,t,G)$. In order to test the weight conditions, as stated previously, we have that $d_{G'}(s,x) = d_G(s,x)$, since x does not belong to the subtree of v in T_s. In addition, Lemma 3 guarantees that $d_{G'}(z,t) = d_G(z,t)$. Thus, it is sufficient for the algorithm to compute only the shortest path trees from t and from s in G.

Theorem 3. *Algorithm 3 finds the longest common prefix of $\mathcal{P}_\alpha(s,t,G)$ in $O(m + t(n,m))$ time using $O(m)$ space.*

Algorithm 3. $\mathtt{lcp}(s, t, \alpha, G)$

---

1 compute T_s, a shortest path tree from s in G
2 compute T_t, a shortest path tree from t in G
3 let $\pi_{st} = (s = v_0, v_1) \ldots (v_{n-1}, v_n = t)$ be the shortest path in T_s
4 **for** $v_i \in \{v_1, \ldots, v_n\}$ **do**
5   **for** $w \in N^+(v_i) \setminus \{v_{i+1}\}$ **do**
6   let T_w be the subtree of T_s rooted at w
7   **for** $(x, z) \in G$ s.t. $x \in T_w$ or $x = v_i$ **do**
8   **if** $z \in T_{v_{i+1}}$ **and** $d_G(s, x) + w(x, z) + d_G(z, t) \le \alpha$ **then**
9 | break
10   **end**
11   **end**
12   **end**
13 **end**
14 **return** $\pi_{sv_{i-1}}$

Proof. The cost of the algorithm can be divided in two parts: the cost to compute the shortest path trees T_s and T_t, and the cost of the loop in line 4. The first part is bounded by $O(t(n, m))$. Let us now prove that the second part is bounded by $O(m + n)$. The cost of each execution of line 8 is $O(1)$, since we only need distances from s and t and the shortest path trees from s and t are already computed, and we pre-process the tree to decide in $O(1)$ if a vertex belongs to a subtree. Hence, the cost of the loop is bounded by the number of times line 8 is executed. The neighborhood of each vertex $x \in T_w$ is visited exactly once, since for each $w \in N^+(v_i) \setminus \{v_{i+1}\}$ and $w' \in N^+(v_j) \setminus \{v_{j+1}\}$ the subtrees T_w and $T_{w'}$ are disjoint, where v_i and v_j belong to the tree path from s to t. □

5 K-Shortest and α-Bounded Paths: A Unified View

The two main differences between the solutions to the K-shortest and α-bounded paths problems are: (i) the order in which the paths are output and (ii) the space complexity of the algorithms. In this section, we show that both problems can be placed in a unified framework such that those differences arise in a natural way. More precisely, we show that their solutions correspond to two different traversals of the *same* rooted tree: a Dijkstra-like traversal for the K-shortest and a DFS-like traversal for the α-bounded paths. This tree is a weighted version of the recursion tree of Algorithm 1, so the height is bounded by n and each leaf corresponds to an α-bounded st-path (see Lemma 1).

The space complexity of the algorithms then follows from the fact that, in addition to the memory to store the tree, Dijkstra's algorithm uses memory proportional to the number of nodes, whereas the DFS uses memory proportional to the height of the tree. In addition, the order in which the solutions are output is precisely the order in which the leaves of the tree are visited, a Dijkstra-like traversal visits the leaves in increasing order of their distance from the root, whereas a DFS-like traversal visits them in an arbitrary but fixed order.

We first modify Algorithm 1 to obtain an iterative *generic* variant. The pseudocode is shown in Algorithm 4. Observe that each node in the recursion tree of Algorithm 1 corresponds to some tuple $\langle u, t, \pi_{ut}, G' \rangle$ in line 3 of Algorithm 4. By generic we mean that the container Q is not specified in the pseudocode, the only requirement is the support for two operations: *push*, to insert a new element in Q; and *pop*, to remove and return an element of Q. It should be clear now that depending on the container, the algorithm will perform a different traversal in the underlying recursion tree of Algorithm 1.

Algorithm 4. list_paths_iterative($u, t, \alpha, \pi_{su}, G$)

---

1 push $\langle s, t, \emptyset, G \rangle$ in Q
2 **while** Q *is not empty* **do**
3      $\langle u, t, \pi_{su}, G' \rangle = Q.pop()$
4   **if** $u = t$ **then**
5 | output(π_{su})
6   **else**
7   compute a shortest path tree T_t from t in $G^R - u$
8   **for** $v \in N^+(u)$ **do**
9   **if** $d(v, t) \leq \alpha - w(u, v)$ **then**
10 | push $\langle v, t, \alpha - w(u, v), \pi_{su} \cdot (u, v), G' - u \rangle$ in Q
11   **end**
12   **end**
13 **end**
14 **end**

Algorithm 4 uses the same strategy to partition the solution space (Eq. 1). Of course, the order in which the partitions are explored depends on the type of container used for Q. We show that if Q is a stack, then the solutions are output in the reverse order of Algorithm 1 and the maximum size of the stack is linear in the size of the input. If on the other hand, Q is a priority queue, using a suitable key, the solutions are output in increasing order of their lengths, but in this case the maximum size of the priority queue is linear in the number of solutions, which is not polynomial in the size of the input.

Let T be the recursion tree of Algorithm 1 (see Lemma 1). In Algorithm 4, each element $\langle u, t, \pi_{su}, G' \rangle$ corresponds to the arguments of a call of Algorithm 1, *i.e.* a node of T. For any container Q supporting push and pop operations, Algorithm 4 visits each node of T exactly once, since at every iteration a node from Q is deleted and its children are inserted in Q, and T is a tree. In particular, this guarantees that every leaf of T is visited exactly once, thus proving the following lemma.

Lemma 4. *Algorithm 4 outputs all α-bounded st-paths.*

Let us consider the case where Q is a stack. It is not hard to prove that Algorithm 1 is a DFS traversal of T starting from the root, while Algorithm 4 is an *iterative* DFS [17] traversal of T also starting from the root. Basically, an iterative DFS keeps the vertices of the fringe of the non-visited subgraph in a stack, at each iteration the next vertex to be explored is popped from the stack, and recursive calls are replaced by pushing vertices in the stack. Now, for a fixed permutation of the children of each node in $\mathcal{T}$, the nodes visited in an iterative DFS traversal are in the reverse order of the nodes visited in a recursive DFS traversal, thus proving Lemma 5.

Lemma 5. *If Q is a stack, then Algorithm 4 outputs the α-bounded st-path in the reverse order of Algorithm 1.*

For any rooted tree, at any moment during an iterative DFS traversal, the number of nodes in the stack is bounded by the sum of the degrees of the root-to-leaf path currently being explored. Recall that every leaf in $\mathcal{T}$ corresponds to a path in $\mathcal{P}_\alpha(s, t, G)$. Actually, there is a one-to-one correspondence between the nodes of a root-to-leaf path P in $\mathcal{T}$ and the vertices of the α-bounded st-path π associated to that leaf. Hence, the sum of the degrees of the nodes of P in $\mathcal{T}$ is equal to the sum of the degrees of the vertices π in G, which is bounded by m, thus proving Lemma 6.

Lemma 6. *The maximum number of elements in the stack of Algorithm 4 over all iterations is bounded by m.*

Let us consider now the case where Q is a priority queue. There is a one-to-many correspondence between arcs in G and arcs in $\mathcal{T}$, *i.e.* if $\mathcal{P}_{\alpha''}(v, t, G'')$ is a child of $\mathcal{P}_{\alpha'}(u, t, G')$ in $\mathcal{T}$ then (u, v) is an arc of G. For every arc of $\mathcal{T}$, we give the weight of the corresponding arc in G. Now, Algorithm 4 using a priority queue with $w(\pi_{su}) + d_G(u, t)$ as keys performs a Dijkstra-like traversal in this weighted version of $\mathcal{T}$ starting from the root. Indeed, for a node $\langle u, t, \pi_{su}, G \rangle$ the distance from the root is $w(\pi_{su})$, and $d_G(u, t)$ is a (precise) estimation of the distance from $\langle u, t, \pi_{su}, G \rangle$ to the closest leaf of $\mathcal{T}$. In other words, it is an A^* traversal [5] in the weighted rooted tree $\mathcal{T}$, using the (optimal) heuristic $d_G(u, t)$. As such, Algorithm 4 explores first the nodes of $\mathcal{T}$ leading to the cheapest non-visited leaf. This is formally stated in Lemma 7.

Lemma 7. *If Q is a priority queue with $w(\pi_{su}) + d_{G'}(u, t)$ as the priority key of $\langle u, t, \pi_{su}, G' \rangle$, then Algorithm 4 outputs the α-bounded st-paths in increasing order of their lengths.*

For any choice of the container Q, each node of $\mathcal{T}$ is visited exactly once, that is, each node of $\mathcal{T}$ is pushed at most once in Q. This proves Lemma 8.

Lemma 8. *The maximum number of elements in a priority queue of Algorithm 4 over all iterations is bounded by γ.*

Algorithm 4 uses $O(m\gamma)$ space since for every node inserted in the priority queue, we also have to store the corresponding graph. Moreover, using a binary heap as a priority queue, the push and pop operations can be performed in $O(\log \gamma)$ each, where γ is the maximum size of the heap. Therefore, combining this with Lemma 7, we obtain the following theorem.

Theorem 4. *Algorithm 4 using a binary heap outputs all α-bounded st-paths in increasing order of their lengths in $O((nt(n, m) + \log \gamma)\gamma)$ total time, using $O(m\gamma)$ space.*

References

1. Ahuja, R.K., Mehlhorn, K., Orlin, J.B., Tarjan, R.E.: Faster algorithms for the shortest path problem. J. ACM **37**, 213–223 (1990)
2. Bernstein, A.: A nearly optimal algorithm for approximating replacement paths and k shortest simple paths in general graphs. In: Proceedings of the 20th ACM-SIAM Symposium on Discrete Algorithms (SODA), pp. 742–755 (2010)
3. Birmelé, E., Ferreira, R., Grossi, R., Marino, A., Pisanti, N., Rizzi, R., Sacomoto, G.: Optimal listing of cycles and st-paths in undirected graphs. In: Proceedings of the 24th Symposium on Discrete Algorithms (SODA), pp. 1884–1896 (2013)
4. Böhmová, K., Mihalák, M., Pröger, T., Srámek, R., Widmayer, P.: Robust routing in urban public transportation: how to find reliable journeys based on past observations. In: 13th Workshop on Algorithmic Approaches for Transportation Modelling, Optimization, and Systems (ATMOS), pp. 27–41 (2013)
5. Dechter, R., Pearl, J.: Generalized best-first search strategies and the optimality of A*. J. ACM **32**(3), 505–536 (1985)
6. Dreyfus, S.E.: An appraisal of some shortest-path algorithms. Oper. Res. **17**(3), 395–412 (1969)
7. Eppstein, D.: Finding the k shortest paths. SIAM J. Comput. **28**(2), 652–673 (1999)
8. Hershberger, J., Suri, S.: Vickrey prices and shortest paths: what is an edge worth? In: Proceedings of the 42nd Symposium on Foundations of Computer Science (FOCS), pp. 252–259. IEEE Computer Society (2001)
9. Johnson, D.S., Papadimitriou, C.H., Yannakakis, M.: On generating all maximal independent sets. Inf. Process. Lett. **27**(3), 119–123 (1988)
10. Johnson, D.B.: Efficient algorithms for shortest paths in sparse networks. J. ACM **24**(1), 1–13 (1977)
11. Katoh, N., Ibaraki, T., Mine, H.: An efficient algorithm for K shortest simple paths. Networks **12**(4), 411–427 (1982)
12. Lawler, E.L.: A procedure for computing the K best solutions to discrete optimization problems and its application to the shortest path problem. Manage. Sci. **18**, 401–405 (1972)
13. Malik, K., Mittal, A.K., Gupta, S.K.: The k most vital arcs in the shortest path problem. Oper. Res. Lett. **8**(4), 223–227 (1989)
14. Ferreira, R., Grossi, R., Rizzi, R., Sacomoto, G., Sagot, M.-F.: Amortized $\tilde{O}(|V|)$-delay algorithm for listing chordless cycles in undirected graphs. In: Schulz, A.S., Wagner, D. (eds.) ESA 2014. LNCS, vol. 8737, pp. 418–429. Springer, Heidelberg (2014)
15. Roditty, L.: On the k-simple shortest paths problem in weighted directed graphs. In: Proceedings of the 18th ACM-SIAM Symposium on Discrete Algorithms (SODA). SIAM (2007)
16. Roditty, L., Zwick, U.: Replacement paths and k simple shortest paths in unweighted directed graphs. In: Caires, L., Italiano, G.F., Monteiro, L., Palamidessi, C., Yung, M. (eds.) ICALP 2005. LNCS, vol. 3580, pp. 249–260. Springer, Heidelberg (2005)
17. Sedgewick, R.: Algorithms in C, Part 5: Graph Algorithms, 3rd edn. Addison-Wesley Professional, Reading (2001)
18. Yen, J.Y.: Finding the K shortest loopless paths in a network. Manage. Sci. **17**, 712–716 (1971)

Metric Dimension for Amalgamations of Graphs

Rinovia Simanjuntak[✉], Saladin Uttunggadewa, and Suhadi Wido Saputro

Combinatorial Mathematics Research Group, Faculty of Mathematics
and Natural Sciences, Institut Teknologi Bandung, Bandung 40132, Indonesia
{rino,s_uttunggadewa,suhadi}@math.itb.ac.id

Abstract. A set of vertices S resolves a graph G if every vertex is uniquely determined by its vector of distances to the vertices in S. The metric dimension of G is the minimum cardinality of a resolving set of G.

Let $\{G_1, G_2, \ldots, G_n\}$ be a finite collection of graphs and each G_i has a fixed vertex v_{0_i} or a fixed edge e_{0_i} called a terminal vertex or edge, respectively. The *vertex-amalgamation* of $G_1, G_2, \ldots, G_n$, denoted by $Vertex - Amal\{G_i; v_{0_i}\}$, is formed by taking all the G_i's and identifying their terminal vertices. Similarly, the *edge-amalgamation* of $G_1, G_2, \ldots, G_n$, denoted by $Edge - Amal\{G_i; e_{0_i}\}$, is formed by taking all the G_i's and identifying their terminal edges.

Here we study the metric dimensions of vertex-amalgamation and edge-amalgamation for finite collection of arbitrary graphs. We give lower and upper bounds for the dimensions, show that the bounds are tight, and construct infinitely many graphs for each possible value between the bounds.

1 Introduction

In this paper we consider finite, simple, and connected graphs. The vertex and edge sets of a graph G are denoted by $V(G)$ and $E(G)$, respectively.

The *distance* $d(u, v)$ between two vertices u and v in a connected graph G is the length of a shortest $u - v$ path in G. For an ordered set $W = \{w_1, w_2, \ldots, w_k\} \subseteq V(G)$, we refer to the k-vector $r(v|W) = (d(v, w_1), d(v, w_2), \ldots, d(v, w_k))$ as the *(metric) representation* of v with respect to W. The set W is called a *resolving set* for G if $r(u|W) = r(v|W)$ implies that $u = v$ for all $u, v \in G$. In a graph G, a resolving set with minimum cardinality is called a *basis* for G. The *metric dimension*, $dim(G)$, is the number of vertices in a basis for G.

The metric dimension problem was first introduced in 1975 by Slater [24], and independently by Harary and Melter [11] in 1976; however the problem for hypercube was studied (and solved asymptotically) much earlier in 1963 by Erdős and Rényi [7]. In general, it is difficult to obtain a basis and metric dimension for arbitrary graph. Garey and Johnson [10], and also Khuller *et al.* [18], showed that determining the metric dimension of an arbitrary graph is an NP-complete

This research is partially supported by Penelitian Unggulan Perguruan Tinggi (Desentralisasi) 2013.

© Springer International Publishing Switzerland 2015
J. Kratochvíl et al. (Eds.): IWOCA 2014, LNCS 8986, pp. 330–337, 2015.
DOI: 10.1007/978-3-319-19315-1_29

problem. The problem is still NP-complete even if we consider some specific families of graphs, such as bipartite graphs [19] or planar graphs [6].

Until today, only graphs of order n with metric dimension 1 (the paths), $n - 3$, $n - 2$, and $n - 1$ (the complete graphs) have been characterized [5,12,16]. On the other hand, researchers have determined metric dimensions for many particular classes of graphs, such as trees, cycles, grids, complete multipartite graphs, hypercube, wheels, fans, unicyclic graphs, honeycombs, and circulant graphs. Recently, Bailey and Cameron [1] established relationship between the base size of automorphism group of a graph and its metric dimension. This result then motivated researchers to study metric dimensions of distance regular graphs. Bollobas, Mitsche, and Pralat have also studied the metric dimension of some random graphs [2].

There are also some research on metric dimensions of graphs resulting from graph operations; for instance: Cartesian product graphs [4], joint product graphs [3], corona product graphs [14,26], lexicographic product graphs [21], hierarchical product graphs [8], and line graphs [9,17]. Note that if we could determine whether a particular graph is constructed from certain graph operations, then the aforementioned results could be utilized to approximate metric dimensions of large graphs. We would only have to recognize the subgraphs and graph operations used to produce the large graph under consideration; and then utilize the dimensions of the subgraphs to determine the dimension of the large graph.

In this paper, we study metric dimension of graphs resulting from another type of graph operations, i.e., vertex-amalgamation and edge-amalgamation of a finite collection of arbitrary graphs. Compare with other graph operations, it is easier to recognize which graphs are constructed from vertex-amalgamation or edge-amalgamation; and thus our results would be useful in approximating the dimensions such graphs. Previous study has been done for vertex-amalgamation of two arbitrary graphs [20] and vertex-amalgamation and edge-amalgamation of particular families of graphs, which include cycles, complete graphs, and prisms [13,22,23]. We present these known results in the next section and then provide more general results in the last section: give lower and upper bounds for the dimensions, show that the bounds are tight, and construct infinitely many graphs for each possible value between the bounds.

2   Previous Results

Let $\{G_1, G_2, \ldots, G_n\}$ be a finite collection of graphs and each *block* G_i has a fixed vertex v_{0_i} or a fixed edge e_{0_i} called a *terminal vertex* or *edge*, respectively. The *vertex-amalgamation* of $G_1, G_2, \ldots, G_n$, denoted by $Vertex - Amal\{G_i; v_{0_i}\}$, is formed by taking all the G_i's and identifying their terminal vertices. Similarly, the *edge-amalgamation* of $G_1, G_2, \ldots, G_n$, denoted by $Edge - Amal\{G_i; e_{0_i}\}$, is formed by taking all the G_i's and identifying their terminal edges.

In [20], Poisson and Zhang studied vertex-amalgamation of two nontrivial connected graphs G_1, G_2 and provide a lower bound as follow.

Theorem 1 [20]. *Let G be the vertex-amalgamation of nontrivial connected graphs G_1 and G_2 with terminal vertices v_{0_1} and v_{0_2}. Then*

$$dim(G) \geq dim(G_1) + dim(G_2) - 2.$$

Other known results are vertex-amalgamation and edge-amalgamation of particular families of graphs, as presented in the following theorems. We denote by C_n the cycle of order n, by K_n the complete graph of order n, and by Pr_n the prism of order $2n$.

Theorem 2 [13,22]. *Let $\{C_{c_1}, C_{c_2}, \ldots, C_{c_n}\}$ be a collection of n cycles with n_e cycles of even order. Suppose that G is the vertex-amalgamation of $C_{c_1}, C_{c_2}, \ldots, C_{c_n}$ and H is the edge-amalgamation of $C_{c_1}, C_{c_2}, \ldots, C_{c_n}$. Then*

$$dim(G) = \begin{cases} n & , n_e = 0, \\ n + n_e - 1 & , n_e \geq 1 \end{cases}$$

and

$$n - 2 \leq dim(H) \leq n.$$

Theorem 3 [23]. *Let $\{K_{k_1}, K_{k_2}, \ldots, K_{k_n}\}$ be a collection of n complete graphs with n_2 complete graphs of order 2 and n_3 complete graphs of order 3. Suppose that G is the vertex-amalgamation of $K_{k_1}, K_{k_2}, \ldots, K_{k_n}$ and H is the edge-amalgamation of $K_{k_1}, K_{k_2}, \ldots, K_{k_n}$. Then*

$$dim(G) = \begin{cases} \sum_{i=1}^{n}(k_i - 2) + n_2 - 1, & n_2 \geq 2, \\ \sum_{i=1}^{n}(k_i - 2), & otherwise \end{cases}$$

and

$$dim(H) = \begin{cases} \sum_{i=1}^{n}(k_i - 3) + 1, & n_3 = 0, \\ \sum_{i=1}^{n}(k_i - 3) + 2, & n_3 = 1 \text{ and } n = 2, \\ \sum_{i=1}^{n}(k_i - 3) + n_3, & otherwise. \end{cases}$$

Theorem 4 [23]. *Let $\{Pr_{p_1}, Pr_{p_2}, \ldots, Pr_{p_n}\}$ be a collection of n prisms with n_o prisms of p_is. Suppose that G is the vertex-amalgamation of $Pr_{p_1}, Pr_{p_2}, \ldots, Pr_{p_n}$ and H is the edge-amalgamation of $Pr_{p_1}, Pr_{p_2}, \ldots, Pr_{p_n}$. Then*

$$dim(G) = \begin{cases} 2n & , n_o = 0, \\ 2n - n_0 - 1 & , n_o \geq 1 \end{cases}$$

and

$$dim(H) = 2n - n_0 - 1.$$

3   Main Results

The next theorem provide the sharp lower and upper bounds for the metric dimension of vertex-amalgamation of finite collection of arbitrary graphs, as well as a construction showing that all values between the bounds are attainable.

Theorem 5. *Let $\{G_1, G_2, \ldots, G_n\}$ be a finite collection of graphs and v_{0_i} is a terminal vertex of G_i, $i = 1, 2, \ldots, n$. If G is the vertex-amalgamation of $G_1, G_2, \ldots, G_n$, $Vertex - Amal\{G_i; v_{0_i}\}$, then*

$$\sum_{i=1}^{n} dim(G_i) - n \leq dim(G) \leq \sum_{i=1}^{n} dim(G_i) + n - 1.$$

Moreover, the bounds are sharp and there are infinitely many graphs with dimension equal to all values within the range of the bounds.

Proof. For the lower bound, consider a vertex set W with cardinality less than $\sum_{i=1}^{n} dim(G_i) - n$. Consequently, there exists a G_i which the cardinality of its intersection with W is less than $dim(G_i) - 1$. Therefore W could not be a resolving set of G and so

$$dim(G) \geq \sum_{i=1}^{n} dim(G_i) - n.$$

For the upper bound, consider two arbitrary G_i and G_j of G and their basis R_i and R_j. Clearly, at most two vertices in $V(G_i) \cup V(G_j)$, say x and y, could have the same representation with respect to $R_i \cup R_j$, since otherwise there exist two vertices in either G_i or G_j, say G_i, having the same representation with respect to R_i, a contradiction with R_i being a resolving set. Thus, to guarantee all vertices in $V(G_i) \cup V(G_j)$ have different representation, we have to add either x or y to $R_i \cup R_j$. If we consider each pair of arbitrary G_i and G_j in G, we obtain

$$dim(G) \leq \sum_{i=1}^{n} dim(G_i) + n - 1.$$

Now let us start our construction by considering $\{G_1, G_2, \ldots, G_n\}$ as a finite collection of complete graphs of order at least 3, where $G_i = K_{k_i}, k_i \geq 3, i = 1, \ldots, n$. By Theorem 3, $dim(G) = \sum_{i=1}^{n}(k_i - 2)$, which can be rewritten as $dim(G) = \sum_{i=1}^{n} dim(G_i) - n$, which achieve the lower bound. We then replace G_1 with a path consisting non-leaf terminal vertex. Let B be the union of all the G_is' basis. Since the path has dimension 1 and its basis vertex is a leaf vertex, then the two vertices of the path adjacent to the terminal vertex will have the same representation with respect to B. Thus we have to add one vertex, i.e. one of the two vertices of the path adjacent to the terminal vertex, to B in order to obtain a basis for G. This results in $dim(G) = \sum_{i=1}^{n} dim(G_i) - n + 1$, which increases the lower bound by one. We continue this process by replacing the G_is one at a time until all complete graphs are replaced with paths (see Fig. 1). The resulting graph is a subdivided star, whose dimension achieves the upper bound. $\qquad\square$

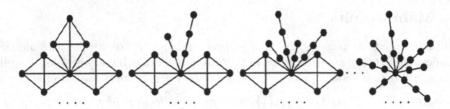

Fig. 1. Vertex-amalgamations of graphs whose dimensions attaining all values between the lower and upper bounds.

Note that the lower bound in the previous theorem generalizes the result of Poisson and Zhang in Theorem 1. From Theorems 2 and 4, we can see that there exist amalgamations of particular cycles and prisms whose dimensions attaining the lower bounds. These graphs could be used in the construction of the proof of Theorem 5.

To prove the result for edge-amalgamation of a finite collection of graphs, we need to know the dimensions of two special graphs. The first graph is complete bipartite graphs $K_{m,n}$. It is known that $dim(K_{m,n}) = m + n - 2$ and the basis consists of all vertices in $K_{m,n}$ except for one vertex from each partite set. The second graph is a variation of a cycle of order n, C_n. For $n \geq 6$, suppose that $V(C_n) = \{x_1, x_2, \ldots, x_n\}$ and $E(C_n) = \{x_1 x_n, x_i x_{i+1}, i = 1, 2, \ldots, n-1\}$. We add two vertices y_2, y_5 and six edges $y_2 x_i, i = 1, 2, 3, y_5 x_i, i = 4, 5, 6$. We call the resulting graph a *double-hats cycle*, denoted by DHC_n (see Fig. 2). It is easy to see that a resolving set of DHC_n must consist two vertices: either x_2 or x_5 and either y_2 or y_5. On the other hand, the set $\{x_2, y_5\}$ is a resolving set of DHC_n, and so $dim(DHC_n) = 2$.

Fig. 2. The double-hats cycle, DHC_n.

Theorem 6. *Let $\{G_1, G_2, \ldots, G_n\}$ be a finite collection of graphs and e_{0_i} is a terminal edge of G_i, $i = 1, 2, \ldots, n$. If H is the edge-amalgamation of $G_1, G_2, \ldots, G_n$, $Edge - Amal\{G_i; e_{0_i}\}$, then*

$$\sum_{i=1}^{n} dim(G_i) - 2n \leq dim(H) \leq \sum_{i=1}^{n} dim(G_i) + n - 1.$$

Moreover, the bounds are sharp and there are infinitely many graphs with dimension equal to all values within the range of the bounds.

Proof. For the lower bound, consider a vertex set W with cardinality less than $\sum_{i=1}^{n} dim(G_i) - 2n$. Consequently, there exists a G_i which the cardinality of its intersection with W is less than $dim(G_i) - 2$. Therefore W could not be a resolving set of H and so

$$dim(G) \geq \sum_{i=1}^{n} dim(G_i) - 2n.$$

For the upper bound, consider two arbitrary G_i and G_j of G and their basis R_i and R_j. Clearly, at most two vertices in $V(G_i) \cup V(G_j)$, say x and y, could have the same representation with respect to $R_i \cup R_j$, since otherwise there exist two vertices in either G_i or G_j, say G_i, having the same representation with respect to R_i, a contradiction with R_i being a resolving set. Thus, to guarantee all vertices in $V(G_i) \cup V(G_j)$ have different representation, we have to add either x or y to $R_i \cup R_j$. If we consider each pair of arbitrary G_i and G_j in G, we obtain

$$dim(G) \leq \sum_{i=1}^{n} dim(G_i) + n - 1.$$

Similarly to the construction for vertex-amalgamation of graphs in the proof of Theorem 5, we start by considering $\{G_1, G_2, \ldots, G_n\}$ as a finite collection of symmetric complete bipartite graphs K_{m_i, m_i} with vertex-set partitioned into $\{x_1, x_2, \ldots, x_{m_i}\}$ and $\{y_1, y_2, \ldots, y_{m_i}\}$. Let $x_{m_i} y_{m_i}$ be the terminal edge in each K_{m_i, m_i}. By the first part of the theorem, we have $dim(H) \geq \sum_{i=1}^{n} dim(G_i) - 2n$. Now, consider the set $R = \bigcup_{i=1}^{n} \{x_1, x_2, \ldots, x_{m_i-2}\}$. It is easy to see that R is a resolving set for H, and thus $dim(H) = \sum_{i=1}^{n} dim(G_i) - 2n$. Therefore we have amalgamation of graphs attaining the lower bound.

We then replace G_1 with a double-hats cycle DHC_n with terminal edge $x_6 x_7$ (refer to the standard vertices notation of DHC_n). Let B be the union of all the G_is' basis. The two vertices of DHC_n adjacent to the terminal edge will have the same representation with respect to B. Thus we have to add one vertex, i.e. one of the two vertices of DHC_n adjacent to the terminal edge, to B in order to obtain a basis for H. This results in $dim(H) = \sum_{i=1}^{n} dim(G_i) - n + 1$, which increases the lower bound by one. We continue this process by replacing the G_is one at a time until all complete bipartite graphs are replaced with DHC_ns. The dimension of the resulting graph then achieves the upper bound. □

Notice that there exists edge-amalgamation of some complete graphs with dimension equal to the lower bound (see Theorem 3), and so these graphs could be used in the construction of the proof of the previous theorem. We could also see from Theorems 2, 3, and 4, that the dimensions of edge-amalgamations of cycles and prims are the middle values between the lower and upper bounds, while the dimensions of edge-amalgamation of complete graphs are values around the lower bound. Determining which collection of graphs whose vertex-amalgamation or edge-amalgamation have small dimensions, i.e. close to the lower bounds, might be seen as interesting problems. Other problems to be considered are recognizing which vertices or edges should be chosen as terminals in order to obtain amalgamated graphs attaining the lower bounds.

We would like to conclude by a remark that recognizing which graphs constructed from vertex-amalgamation and edge-amalgamation is relatively easier than from other graph operations. Thus our main results could be utilized to approximate the dimensions of such graphs. Moreover, if a graph contains a bridge, we could subdivide the bridge by inserting one vertex; and the resulting graph could be seen as a vertex-amalgamation of two graphs. Similarly, if a graph contains two bridges, we could insert one vertex in each of the bridges and connect those two vertices with an edge; and so the resulting graph could be seen as an edge-amalgamation of two graphs. In such cases, our main results enable us to approximate the dimension of (large) graphs containing either one or two bridges.

References

1. Bailey, R.F., Cameron, P.J.: Base size, metric dimension and other invariants of groups and graphs. Bull. Lond. Math. Soc. 43, 209–242 (2011)
2. Bollobas, B., Mitsche, D., Pralat, P.: Metric dimension for random graphs. Electron. J. Comb. 20, ♯P1 (2013)
3. Buczkowski, P.S., Chartrand, G., Poisson, C., Zhang, P.: On k-dimensional graphs and their bases. Period. Math. Hung. 46, 9–15 (2003)
4. Caceres, J., Hernando, C., Mora, M., Pelayo, I.M., Puertas, M.L., Seara, C., Wood, D.R.: On the metric dimension of Cartesian products of graphs. SIAM J. Discrete Math. 21, 423–441 (2007)
5. Chartrand, G., Eroh, L., Johnson, M.A., Oellermann, O.R.: Resolvability in graphs and the metric dimension of a graph. Discrete Appl. Math. 105, 99–113 (2000)
6. Díaz, J., Pottonen, O., Serna, M., van Leeuwen, E.J.: On the complexity of metric dimension. In: Epstein, L., Ferragina, P. (eds.) ESA 2012. LNCS, vol. 7501, pp. 419–430. Springer, Heidelberg (2012)
7. Erdős, P., Rényi, A.: On two problems of information theory. Magyar Tud. Akad. Mat. Kutat Int. Kzl. 8, 229–243 (1963)
8. Feng, M., Wang, K.: On the metric dimension and fractional metric dimension of the hierarchical product of graphs. Appl. Anal. Discrete Math. 7, 302–313 (2013)
9. Feng, M., Xu, M., Wang, K.: On the metric dimension of line graphs. Discrete Appl. Math. 161, 802–805 (2013)
10. Garey, M.R., Johnson, D.S.: Computers and Intractability: A Guide to the Theory of NP Completeness. W.H. Freeman and Company, San Francisco (1979)
11. Harary, F., Melter, R.A.: On the metric dimension of a graph. Ars Comb. 2, 191–195 (1976)
12. Hernando, C., Mora, M., Pelayo, I.M., Seara, C., Wood, D.R.: Extremal graph theory for metric dimension and diameter. Electron. J. Comb. 17 ♯R30 (2010)
13. Iswadi, H., Baskoro, E.T., Salman, A.N.M., Simanjuntak, R.: The metric dimension of amalgamation of cycles. Far East J. Math. Sci. 41, 19–31 (2010)
14. Iswadi, H., Baskoro, E.T., Simanjuntak, R.: On the metric dimension of corona product of graphs. Far East J. Math. Sci. 52, 155–170 (2011)

15. Jannesari, M., Omoomi, B.: The metric dimension of the lexicographic product of graphs. Discrete Math. **312**, 3349–3356 (2012)
16. Jannesari, M., Omoomi, B.: Characterization of n-vertex graphs with metric dimension $n - 3$. Math. Bohemica **139**, 1–23 (2014)
17. Klein, D.J., Yi, E.: A comparison on metric dimension of graphs, line graphs, and line graphs of the subdivision graphs. European J. Pure Appl. Math. **5**, 302–316 (2012)
18. Khuller, S., Raghavachari, B., Rosenfeld, A.: Landmarks in graphs. Discrete Appl. Math. **70**, 217–229 (1996)
19. Manuel, P.D., Abd-El-Barr, M.I., Rajasingh, I., Rajan, B.: An efficient representation of Benes networks and its applications. J. Discrete Algorithms **6**, 11–19 (2008)
20. Poisson, C., Zhang, P.: The metric dimension of unicyclic graphs. J. Comb. Math. Comb. Comput. **40**, 17–32 (2002)
21. Saputro, S.W., Simanjuntak, R., Uttunggadewa, S., Assiyatun, H., Baskoro, E.T., Salman, A.N.M., Baća, M.: The metric dimension of the lexicographic product of graphs. Discrete Math. **313**, 1045–1051 (2013)
22. Simanjuntak, R., Assiyatun, H., Baskoroputro, H., Iswadi, H., Setiawan, Y., Uttunggadewa, S.: Graphs with relatively constant metric dimensions (preprint)
23. Simanjuntak, R., Murdiansyah, D.: Metric dimension of amalgamation of some regular graphs (preprint)
24. Slater, P.J.: Leaves of trees. Congr. Numer. **14**, 549–559 (1975)
25. Tavakoli, M., Rahbarnia, F., Ashrafi, A.R.: Distribution of some graph invariants over hierarchical product of graphs. App. Math. Comput. **220**, 405–413 (2013)
26. Yero, I.G., Kuziak, D., Rodriguez-Velázquez, J.A.: On the metric dimension of corona product graphs. Comput. Math. Appl. **61**, 2793–2798 (2011)

A Suffix Tree Or Not a Suffix Tree?

Tatiana Starikovskaya[1] and Hjalte Wedel Vildhøj[2]([✉])

[1] National Research University Higher School of Economics (HSE), Moscow, Russia
[2] DTU Compute, Technical University of Denmark, Kongens Lyngby, Denmark
hwvi@dtu.dk

Abstract. In this paper we study the structure of suffix trees. Given an unlabeled tree τ on n nodes and suffix links of its internal nodes, we ask the question "Is τ a suffix tree?", i.e., is there a string S whose suffix tree has the same topological structure as τ? We place no restrictions on S, in particular we do not require that S ends with a unique symbol. This corresponds to considering the more general definition of *implicit* or *extended* suffix trees. Such general suffix trees have many applications and are for example needed to allow efficient updates when suffix trees are built online. We prove that τ is a suffix tree if and only if it is realized by a string S of length $n - 1$, and we give a linear-time algorithm for inferring S when the first letter on each edge is known. This generalizes the work of I et al. [Discrete Appl. Math. 163, 2014].

1 Introduction

The suffix tree was introduced by Peter Weiner in 1973 [21] and remains one of the most popular and widely used text indexing data structures (see [1] and references therein). In static applications it is commonly assumed that suffix trees are built only for strings with a unique end symbol (often denoted $\$$), thus ensuring the useful one-to-one correspondance between leaves and suffixes. In this paper we view such suffix trees as a special case and refer to them as $\$$-*suffix trees*. Our focus is on suffix trees of *arbitrary strings*, which we simply call *suffix trees* to emphasize that they are more general than $\$$-suffix trees[1]. Contrary to $\$$-suffix trees, the suffixes in a suffix tree can end in internal non-branching locations of the tree, called *implicit suffix nodes*.

Suffix trees for arbitrary strings are not only a nice generalization, but are required in many applications. For example in online algorithms that construct the suffix tree of a left-to-right streaming text (e.g., Ukkonen's algorithm [20]), it is necessary to maintain the implicit suffix nodes to allow efficient updates. Despite their essential role, the structure of suffix trees is still not well understood. For instance, it was only recently proved that each internal edge in a suffix tree can contain at most one implicit suffix node [4].

T. Starikovskaya—Partly supported by Dynasty Foundation.
[1] In the literature the standard terminology is *suffix trees* for $\$$-suffix trees and *extended/implicit suffix trees* [3,12] for suffix trees of strings not ending with $\$$.

J. Kratochvíl et al. (Eds.): IWOCA 2014, LNCS 8986, pp. 338–350, 2015.
DOI: 10.1007/978-3-319-19315-1_30

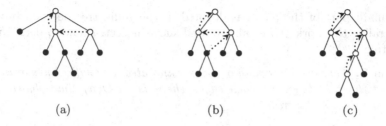

(a)   (b)   (c)

Fig. 1. Three potential suffix trees. (a) is a $-suffix tree, e.g. for **ababa$**. (b) is not a $-suffix tree, but it is a suffix tree, e.g. for **abaabab**. (c) is not a suffix tree.

In this paper we prove some new properties of suffix trees and show how to decide whether suffix trees can have a particular structure. Structural properties of suffix trees are not only of theoretical interest, but are essential for analyzing the complexity and correctness of algorithms using suffix trees.

Given an unlabeled ordered rooted tree τ and suffix links of its internal nodes, the *suffix tree decision problem* is to decide if there exists a string S such that the suffix tree of S is isomorphic to τ. If such a string exists, we say that τ *is a suffix tree* and that S *realizes* τ. If τ can be realized by a string S having a unique end symbol $, we additionally say that τ *is a $-suffix tree*. See Fig. 1 for examples of a $-suffix tree, a suffix tree, and a tree which is not a suffix tree. In all figures in this paper leaves are black and internal nodes are white.

I et al. [16] recently considered the suffix tree decision problem and showed how to decide if τ is a $-suffix tree in $O(n)$ time, assuming that the first letter on each edge of τ is also known. Concurrently with our work, another approach was developed in [5]. There the authors show how to decide if τ is a $-suffix tree without knowing the first letter on each edge, but also introduce the assumption that τ is an unordered tree.

Deciding if τ is a suffix tree is much more involved than deciding if it is a $-suffix tree, mainly because we can no longer infer the length of a string that realizes τ from the number of leaves. Without an upper bound on the length of such a string, it is not even clear how to solve the problem by an exhaustive search. In this paper, we give such an upper bound, show that it is tight, and give a linear time algorithm for deciding whether τ is a suffix tree when the first letter on each edge is known.

Our Results. In Sect. 2, we start by settling the question of the sufficient length of a string that realizes τ.

Theorem 1. *An unlabeled tree τ on n nodes is a suffix tree if and only if it is realized by a string of length $n-1$.*

As far as we are aware, there were no previous upper bounds on the length of a shortest string realizing τ. The bound implies an exhaustive search algorithm for solving the suffix tree decision problem, even when the suffix links are not provided. In terms of n, this upper bound is tight, since e.g. stars on n nodes are realized only by strings of length at least $n-1$.

The main part of the paper is devoted to the suffix tree decision problem. We generalize the work of I et al. [16] and show in Sect. 4 how to decide if τ is a suffix tree.

Theorem 2. *Let τ be a tree with n nodes, annotated with suffix links of internal nodes and the first letter on each edge. There is an $O(n)$ time algorithm for deciding if τ is a suffix tree.*

In case τ is a suffix tree, the algorithm also outputs a string S that realizes τ. To obtain the result, we show several new properties of suffix trees, which may be of independent interest.

For space reasons the proofs of Lemmas 6, 9 and 10 have been omitted. They can be found in the full version of this paper [19].

Related Work. The problem of revealing structural properties and exploiting them to recover a string realizing a data structure has received a lot of attention in the literature. Besides $-suffix trees, the problem has been considered for border arrays [8,18], parameterized border arrays [13–15], suffix arrays [2,10,17], KMP failure tables [9,11], prefix tables [6], cover arrays [7], directed acyclic word graphs [2], and directed acyclic subsequence graphs [2].

2   Suffix Trees

In this section we prove Theorem 1 and some new properties of suffix trees, which we will need to prove Theorem 2. We start by briefly recapitulating the most important definitions.

The *suffix tree* of a string S is a compacted trie on suffixes of S [12]. Branching nodes and leaves of the tree are called *explicit nodes*, and positions on edges are called *implicit nodes*. The *label* of a node v is the string on the path from the root to v, and the length of this label is called the *string depth* of v. The *suffix link* of an internal explicit node v labeled by $a_1 a_2 \ldots a_m$ is a pointer to the node u labeled by $a_2 a_3 \ldots a_m$. We use the notation $v \dashrightarrow u$ and extend the definition of suffix links to leaves and implicit nodes as well. We will refer to nodes that are labeled by suffixes of S as *suffix nodes*. All leaves of the suffix tree are suffix nodes, and unless S ends with a unique symbol $, some implicit nodes and internal explicit nodes can be suffix nodes as well. Suffix links for suffix nodes form a path starting at the leaf labeled by S and ending at the root. Following [4], we call this path the *suffix chain*.

Lemma 1 ([4]). *The suffix chain of the suffix tree can be partitioned into the following consecutive segments: (1) Leaves; (2) Implicit suffix nodes on leaf edges; (3) Implicit suffix nodes on internal edges; and (4) Suffix nodes that coincide with internal explicit nodes. (See Fig. 2a.)*

We define the parent $par(x)$ of a node x to be the deepest explicit node on the path from the root to x (excluding x). The distance between a node and one of its ancestors is defined to be the difference between the string depths of these nodes.

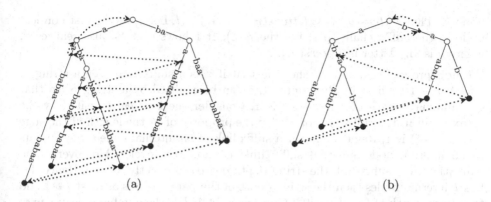

Fig. 2. (a) The suffix tree τ of a string $S = abaababaababaa$ with suffix nodes and the suffix chain. (b) The suffix tree of a prefix $S' = abaabab$ of S. Suffix links of internal nodes are not shown, but they are the same in both trees.

Lemma 2. *If $x_1 \dashrightarrow x_2$ is a suffix link, then the distance from x_1 to $par(x_1)$ cannot be less than the distance from x_2 to $par(x_2)$.*

Proof. If d is the distance between x_1 and $par(x_1)$, then the suffix link of $par(x_1)$ points to an explicit ancestor d characters above x_2.

Lemma 3. *Let x be an implicit suffix node. The distance between x and $par(x)$ is not bigger than the length of any leaf edge.*

Proof. It follows from Lemma 2 that as the suffix chain $y_0 \dashrightarrow y_1 \dashrightarrow \ldots \dashrightarrow y_l = root$ is traversed, the distance from each node to its parent is non-increasing. Since the leaves are visited first, the distance between any implicit suffix node and its parent cannot exceed the length of a leaf edge.   □

Lemma 4. *If τ is a suffix tree, then it can be realized by some string such that*

(1) The minimal length of a leaf edge of τ will be equal to one;
(2) Any edge of τ will contain at most one implicit suffix node at the distance one from its upper end.

Proof. Let S be a string realizing τ, and m be the minimal length of a leaf edge of τ. Consider a prefix S' of S obtained by deleting its last $(m - 1)$ letters. Its suffix tree is exactly τ trimmed at height $m - 1$. (See Fig. 2b.) The minimal length of a leaf edge of this tree is one. Applying Lemma 3, we obtain that the distance between any implicit suffix node x of this tree and $par(x)$ is one, and, consequently, any edge contains at most one implicit suffix node.   □

Lemma 5. *If τ is realized by a string of length l, then it is also realized by strings of length $l + 1, l + 2, l + 3$, and so on.*

Proof. Let $y_0 \dashrightarrow y_1 \dashrightarrow \ldots \dashrightarrow y_l = root$ be the suffix chain for a string S that realizes τ. Moreover let $letters(y_i)$ be the set of first letters immediately below node y_i. Then $letters(y_{i-1}) \subseteq letters(y_i)$, $i = 1, \ldots, l$. Let y_j be the first non-leaf node in the suffix chain (possibly the root). It follows that Sa also realizes τ, where a is any letter in $letters(y_j)$. □

We now prove Theorem 1 by showing that if τ is a suffix tree then a string of length $n - 1$ realizes it. By Lemma 4, τ can be realized by a string S' so that the minimal length of a leaf edge is 1. Consider the last leaf ℓ visited by the suffix chain in the suffix tree of S'. By the property of S' the length of the edge $(par(\ell) \to \ell)$ is 1. Remember that a suffix link of an internal node always points to an internal node and that suffix links cannot form cycles. Moreover, upon transition by a suffix link the string depth decreases exactly by one. Hence if τ has I internal nodes then the string depth of the parent of ℓ is at most $I - 1$ and the string depth of ℓ is at most I. Consequently, if L is the number of leaves in τ, the length of the suffix chain and thus the length of S' is at most $L + I - 1 = n - 1$, so by Lemma 5 there is a string of this length that realizes τ.

3   The Suffix Tour Graph

In their work [16] I et al. introduced a notion of *suffix tour graphs*. They showed that suffix tour graphs of \$-suffix trees must have a nice structure which ties together the suffix links of the internal explicit nodes, the first letters on edges, and the order of leaves of τ — i.e., which leaf corresponds to the longest suffix, which leaf corresponds to the second longest suffix, and so on. Knowing this order and the first letters on edges outgoing from the root, it is easy to infer a string realizing τ. We study the structure of suffix tour graphs of suffix trees. We show a connection between suffix tour graphs of suffix trees and \$-suffix trees and use it to solve the suffix tree decision problem.

Let us first formalize the input to the problem. Consider a tree $\tau = (V, E)$ annotated with a set of suffix links $\sigma : V \to V$ between internal explicit nodes, and the first letter on each edge, given by a labelling function $\lambda : E \to \Sigma$ for some alphabet Σ. For ease of description, we will always augment τ with an auxiliary node $\bot$, the parent of the root. We add the suffix link (root $\dashrightarrow \bot$) and label the edge $(\bot \to$ root) with a symbol "?", which matches any letter of the alphabet.

To construct the suffix tour graph of τ, we first compute values $\ell(x)$ and $d(x)$ for every explicit node x in τ. The value $\ell(x)$ is equal to the number of leaves y where $par(y) \dashrightarrow par(x)$ is a suffix link in σ, and $\lambda(par(y) \to y) = \lambda(par(x) \to x)$. Let L_x and V_x be the sets of leaves and nodes, respectively, of the subtree of τ rooted at a node x. Note that L_x is a subset of V_x. We define $d(x) = |L_x| - \sum_{y \in V_x} \ell(y)$. See Fig. 3 for an example.

Definition 1. The *suffix tour graph* of a tree $\tau = (V, E)$ is a directed graph $G = (V, E_G)$, where $E_G = \{(y \to x)^k \mid (y \to x) \in E, k = d(x)\} \cup \{(y \to x) \mid y \text{ is a leaf contributing to } \ell(x)\}$. Here $(y \to x)^k$ means the edge $y \to x$ with multiplicity k. If $k = d(x) < 0$, we define $(y \to x)^k$ to be $(x \to y)^{|k|}$. (See Fig. 3.)

Lemma 6 ([16], see [19] for proof). *The suffix tour graph G of a suffix tree τ is an Eulerian graph (possibly disconnected).*

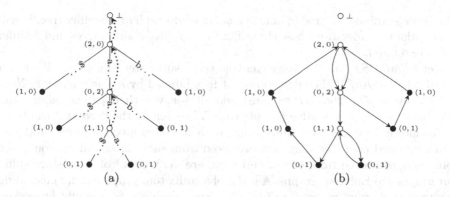

Fig. 3. (a) An input consisting of a tree, suffix links and the first letter on each edge. The input has been extended with the auxiliary node $\perp$, and each node is assigned values $(\ell(x), d(x))$. (b) The corresponding suffix tour graph. The input (a) is realized by the string abaaa$, which corresponds to an Euler tour of (b).

3.1 Suffix Tour Graph of a $-suffix Tree

The following proposition follows from the definition of a $-suffix tree.

Proposition 1 ([16]). *If τ is a $-suffix tree with a set of suffix links σ and first letters on edges defined by a labelling function λ, then*

(1) For every internal explicit node x in τ there exists a unique path $x = x_0 \cdots\!\!\to$ root such that $x_1 \cdots\!\!\to \ldots \cdots\!\!\to x_k = $ belongs to σ for all i;

(2) If y is the end of the suffix link for $par(x)$, there is a child z of y such that $\lambda(par(x) \to x) = \lambda(y \to z)$, and the end of the suffix link for x belongs to the subtree of τ rooted at y;

(3) For any node $x \in V$ the value $d(x) \geq 0$.

If all tree conditions hold, it can be shown that

Lemma 7 ([16]). *The tree τ is a $-suffix tree iff its suffix tour graph G contains a cycle C which goes through the root and all leaves of τ. Moreover, a string realizing τ can be inferred from C in linear time.*

In more detail, the authors proved that the order of leaves in the cycle C corresponds to the order of suffixes. That is, the ith leaf after the root corresponds to the ith longest suffix. Thus, the string can be reconstructed in linear time: its ith letter will be equal to the first letter on the edge in the path from the root to the ith leaf. Note that the cycle and hence the string is not necessarily unique. See Fig. 3 for an example.

3.2 Suffix Tour Graph of a Suffix Tree

The high level idea of our solution is to try to augment the input tree so that the augmented tree is a $-suffix tree. More precisely, we will try to augment the suffix tour graph of the tree to obtain a suffix tour graph of a $-suffix tree. It will be essential to understand how the suffix tour graphs of suffix trees and $-suffix trees are related.

Let ST and $ST_\$$ be the suffix tree and the $-suffix tree of a string. We call a leaf of $ST_\$$ a $-*leaf* if the edge ending at it is labeled by a single letter $. Note that to obtain $ST_\$$ from ST we must add all $-leaves, their parents, and suffix links between the consecutive parents to ST. We denote the deepest $-leaf by s.

An internal node x of a suffix tour graph has $d(x)$ incoming arcs produced from edges and $\ell(x)$ incoming arcs produced from suffix links. All arcs outgoing from x are produced from edges, and there are $d(x) + \ell(x)$ of them since suffix tour graphs are Eulerian graphs. A leaf x of a suffix tour graph has $d(x)$ incoming arcs produced from edges, $\ell(x)$ incoming arcs produced from suffix links, and one outgoing arc produced from a suffix link. Below we describe what happens to the values $d(x)$ and $\ell(x)$, and to the outgoing arcs produced from suffix links. These two things define the changes to the suffix tour graph.

Lemma 8. *For the deepest $-leaf s we have $\ell(s) = 0$ and $d(s) = 1$. The ℓ-values of other $-leaves are equal to one, and their d-values are equal to zero.*

Proof. Suppose that $\ell(s) = 1$. Then there is a leaf y such that $x_i \dashrightarrow x_{i+1}$ is a suffix link in σ, and the first letter on the edge from $par_\$(y)$ to y is $. That is, y is a $-leaf and its string depth is bigger than the string depth of s, which is a contradiction. Hence, $\ell(s) = 0$ and therefore $d(s) = 1$. The parent of any other $-leaf y will have an incoming suffix link from the parent of the previous $-leaf and hence $\ell(y) = 1$ and $d(y) = 0$.   □

The important consequence of Lemma 8 is that in the suffix tour graph of $ST_\$$ all the $-leaves are connected by a path starting in the deepest $-leaf and ending in the root.

Next, we consider nodes that are explicit in ST and $ST_\$$. If a node x is explicit in both trees, we denote its (explicit) parent in ST by $par(x)$ and in $ST_\$$ — by $par_\$(x)$. Below in this section we assume that each edge of ST contains at most one implicit suffix node at distance one from its parent.

Lemma 9 (see [19] for proof). *Consider a node x of ST. If a leaf y contributes to $\ell(x)$ either in ST or $ST_\$$, and $par_\$(y)$ and $par_\$(x)$ are either both explicit or both implicit in ST, then y contributes to $\ell(x)$ in both trees.*

Before we defined the deepest $-leaf s. If the parent of s is implicit in ST, the changes between ST and $ST_\$$ are more involved. To describe them, we first need to define the *twist node*. Let p be the deepest explicit parent of any $-leaf in ST. The node that precedes p in the suffix chain is thus an implicit node in ST, i.e., it has two children in $ST_\$$, one which is a $-leaf and another node y, which is either a leaf or an internal node. If y is a leaf, let t be the child of p such that y contributes to $\ell(t)$. We refer to t as the *twist* node.

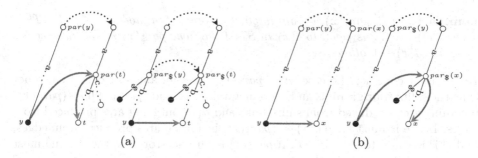

Fig. 4. Both figures show ST on the left and $ST_\$$ on the right. Edges of the suffix tour graphs that change because of the twist node t (Fig. 4a) and because of an implicit parent (Fig. 4b) are shown in grey.

Lemma 10. *Let x be a node of ST. Upon transition from ST to $ST_\$$, the ℓ-value of $x = t$ increases by one and the ℓ-value of its parent decreases by one. If $par_\$(x)$ is an implicit node of ST, then $\ell(x)$ decreases by $\ell(par_\$(x))$. Otherwise, $\ell(x)$ does not change.*

Proof (Sketch — see full version for all details [19]).
The value $\ell(x)$ can change when (1) A leaf y contributes to $\ell(x)$ in $ST_\$$, but not in ST; or (2) A leaf y contributes to $\ell(x)$ in ST, but not in $ST_\$$. In the first case $par_\$(y)$ is implicit in ST, and $par_\$(x)$ is explicit. Since $par_\$(x)$ is the first explicit suffix node and y is a leaf that contributes to $\ell(x)$, we have $x = t$, and $\ell(x) = \ell(t)$ in $ST_\$$ is bigger than $\ell(t)$ in ST by one (see Fig. 4).
Consider one of the leaves y satisfying (2). In this case exactly one of the nodes $par_\$(y)$ and $par_\$(x)$ must be implicit in ST. Hence, we have two subcases: (2a) $par_\$(y)$ is implicit in ST, and $par_\$(x)$ is explicit; (2b) $par_\$(y)$ is explicit in ST, and $par_\$(x)$ is implicit.
In the subcase (2a) the node x is the parent of the twist node t, and the value $\ell(x) = \ell(par_\$(t))$ is smaller by one in $ST_\$$ (see Fig. 4a). In the subcase (2b) the ℓ-value of x in ST is bigger than the ℓ-value of x in $ST_\$$ by $\ell(par_\$(x))$, as all leaves contributing to $par_\$(x)$ in $ST_\$$, e.g. y, switch to x in ST (see Fig. 4b). □

Lemma 11 *Let x be a node of ST. Upon transition from ST to $ST_\$$, the value $d(x)$ of a node x such that $par_\$(x)$ is implicit in ST increases by $\ell(par_\$(x))$. If x is the twist node t, its d-value decreases by one. Finally, the d-values of all ancestors of the deepest $\$-leaf s increase by one.*

Proof. Remember that $d(x) = |L_x| - \sum_{y \in V_x} \ell(y)$. If $par_\$(x)$ is implicit in ST, $\ell(x)$ decreases by $\ell(par_\$(x))$, i.e. $d(x)$ increases by $\ell(par_\$(x))$. Note that d-values of ancestors of x are not affected since for them the decrease of $\ell(x)$ is compensated by the presence of $par_\$(x)$. The value $\ell(t)$ increases by one and results in decrease of $d(t)$ by one, but for other ancestors of t increase of $\ell(t)$ will be compensated by decrease of $\ell(par_\$(t))$.
The value $\ell(s) = 0$ and the ℓ-values of other $\$-leaves are equal to one. Consequently, when we add the $\$-leaves to ST, d-values of ancestors of s increase by one, and d-values of ancestors of other $\$-leaves are not affected. □

Lemma 12. *Let $par_\$(x)$ be an implicit parent of a node $x \in ST$. Then $d(par_\$(x))$ in $ST_\$$ is equal to $d(x)$ in ST if the node $par_\$(x)$ is not an ancestor of s, and $d(x) + 1$ otherwise.*

Proof. First consider the case when $par_\$(x)$ is not an ancestor of s. Remember that the suffix tour graph is an Eulerian graph. The node $par_\$(x)$ has $\ell(par_\$(x))$ incoming arcs produced from suffix links and $d(x)$ outgoing arcs produced from edges. Hence it must have $d(x) - \ell(par_\$(x))$ incoming arcs produces from edges, and this is equal to $d(x)$ in ST. If $par_\$(x)$ is an ancestor of s, the d-value must be increased by one as in the previous lemma.   □

Speaking in terms of suffix tour graphs, we make local changes when the node is the twist node t or when the parent of a node is implicit in ST, and add a cycle from the root to s (increase of d-values of ancestors of s) and back via all \$-leaves.

4   A Suffix Tree Decision Algorithm

Given a tree $\tau = (V, E)$ annotated with a set of suffix links and a labelling function, we want to decide whether there is a string S such that τ is the suffix tree of S and it has all the properties described in Lemma 4.

We assume that τ satisfies Proposition 1(1) and Proposition 1(2), which can be verified in linear time. We will not violate this while augmenting τ. If τ is a suffix tree, the string depth of a node equals the length of the suffix link path starting at it. Consequently, string depths of all explicit internal nodes and lengths of all internal edges can be found in linear time.

We replace the original problem with the following one: Can τ be augmented to become a \$-suffix tree? The deepest \$-leaf s can either hang from a node of τ, or from an implicit suffix node $par_\$(s)$ on an edge of τ. In the latter case the distance from $par_\$(s)$ to the upper end of the edge is equal to one. That is, there are $O(n)$ possible locations of s. For each of the locations we consider a suffix link path starting at its parent. The suffix link paths form a tree which we refer to as the suffix link tree. The suffix link tree can be built in linear time: For explicit locations the paths already exist, and for implicit locations we can build the paths following the suffix link path from the upper end of the edge containing a location and exploiting the knowledge about lengths of internal edges. (Of course, if we see a node encountered before, we stop.)

If τ is a suffix tree, then it is possible to augment it so that its suffix tour graph will satisfy Proposition 1(3) and Lemma 7. We remind that Proposition 1(3) says that for any node x of the suffix tour graph $d(x) \geq 0$, and Lemma 7 says that the suffix tour graph contains a cycle going through the root and all leaves. We show that each of the conditions can be verified for all possible ways to augment τ by a linear time traverse of τ or the suffix link tree. We start with Proposition 1(3).

Lemma 13. *If τ can be augmented to become a \$-suffix tree, then $\forall x\ d(x) \geq -1$.*

Proof. The value $d(x)$ increases only when x is an ancestor of s or when $par_\$(x)$ is implicit in ST. In the first case it increases by one. Consider the second case. Remember that $d(par_\$(x))$ is equal to $d(x)$ or to $d(x)+1$ if it is an ancestor of s. Since in a \$-suffix tree all d-values are non-negative, we have $d(x) \geq -1$ for any node x.   □

Step 1. We first compute all d-values and all ℓ-values. If $d(x) \leq -2$ for some node x of τ, then τ cannot be augmented to become a \$-suffix tree and hence it is not a suffix tree. From now on we assume that τ does not contain such nodes. All nodes x with $d(x) = -1$, except for at most one, must be ancestors of s. If there is a node with a negative d-value that is not an ancestor of s, then it must be the lower end of the edge containing $par_\$(s)$, and the d-value must become non-negative after we augment τ.

We find the deepest node x with $d(x) = -1$ by a linear time traverse of τ. All nodes with negative d-values must be its ancestors, which can be verified in linear time. If this is not the case, τ is not a suffix tree. Otherwise, the possible locations for the parent of s are descendants of x and the implicit location on the edge to x if $d(x) + \ell(x)$, the d-value of x after augmentation, is at least zero. We cross out all other locations.

Step 2. For each of the remaining locations we consider the suffix link path starting at its parent. If the implicit node q preceding the first explicit node p in the path belongs to a leaf edge then the twist node t is present in τ and will be a child of p. We cannot tell which child though, since we do not know the first letter on the leaf edge outgoing from q. However, we know that $d(t)$ decreases by 1 after augmentation, and hence $d(t)$ must be at least 0. Moreover, if $d(t) = 0$ the twist node t must be an ancestor of s to compensate for the decrease of $d(t)$.

In other words, a possible location of s is crossed out if the twist node t is present but p has no child t that satisfies $d(t) > 0$ or $d(t) = 0$ and t is ancestor of s. For each of the locations of s we check if t exists, and if it does, we find p (i1). This can be done in linear time in total by a traverse of the suffix link tree. We also compute for every node if it has a child u such that $d(u) > 0$ (i2). Finally, we traverse τ in the depth-first order while testing the current location of s. During the traverse we remember, for any node on the path to s, its child which is an ancestor of s (i3). With the information (i1), (i2), and (i3), we can determine if we cross out a location of s in constant time, and hence the whole computation takes linear time.

Step 3. We assume that the suffix tour graph of τ is an Eulerian graph, otherwise τ is not a suffix tree by Lemma 7. This condition can be verified in linear time. When we augment τ, we add a cycle C from the root to the deepest \$-leaf s and back via \$-leaves. The resulting graph will be an Eulerian graph as well, and one of its connected components (cycles) must contain the root and all leaves of τ.

We divide C into three segments: the path from the root to the parent $par(x)$ of the deepest node x with $d(x) = -1$, the path from $par(x)$ to s, and the path from s to the root. We start by adding the first segment to the suffix tour graph.

This segment is present in the cycle C for any choice of s, and it might actually increase the number of connected components in the graph. (Remember that if C contains an edge $x \rightarrow y$ and the graph contains an edge $y \rightarrow x$, then the edges eliminate each other.)

The second segment cannot eliminate any edges of the graph, and if it touches a connected component then all its nodes are added to the component containing the root of τ. Since the third segment contains the $-leaves only, the second segment must go through all connected components that contain leaves of τ. We paint nodes of each of the components into some color. And then we perform a depth-first traverse of τ maintaining a counter for each color and the total number of distinct colors on the path from the root to the current node. When a color counter becomes equal to zero, we decrease the total number of colors by one, and when a color counter becomes positive, we increase the total number of colors by one. If a possible location of s has ancestors of all colors, we keep it.

Lemma 14. *The tree τ is a suffix tree iff there is a survived location of s.*

Proof. If there is such a location, then for any x in the suffix tour graph of the augmented tree we have $d(x) \geq 0$ and there is a cycle containing the root and all leaves. We are still to apply the local changes caused by implicit parents. Namely, for each node x with an implicit parent the edge from y to x is to be replaced by the path $y, par_\$(x), x$ (see Fig. 4b). The cycle can be re-routed to go via the new paths instead of the edges, and it will contain the root and the leaves of τ. Hence, the augmented tree is a $-suffix tree and τ is a suffix tree.

If τ is a suffix tree, then it can be augmented to become a $-suffix tree. The parent of s will survive the selection process. $\square$

Suppose that there is such a location. Then we can find the parent of the twist node if it exists. The parent must have a child t such that either $d(t) > 0$ or $d(t) = 0$ and t is an ancestor of s, and we choose t as the twist node. Let the first letter on the edge to the twist node be a. Then we put the first letter on all new leaf edges caused by the implicit nodes equal to a. The resulting graph will be the suffix tour graph of a $-suffix tree. We can use the solution of I et al. [16] to reconstruct a string $S\$$ realizing this $-suffix tree in linear time. The tree τ will be a suffix tree of the string S. This completes the proof of Theorem 2.

5   Conclusion and Open Problems

We have proved several new properties of suffix trees, including an upper bound of $n - 1$ on the length of a shortest string S realizing a suffix tree τ with n nodes. As noted this bound is tight in terms of n, since the number of leaves in τ, which can be $n - 1$, provides a trivial lower bound on the length of S.

Using these properties, we have shown how to decide if a tree τ with n nodes is a suffix tree in $O(n)$ time, provided that the suffix links of internal nodes and the first letter on each edge is specified. It remains an interesting open question whether the problem can be solved without first letters or, even, without suffix links (i.e., given only the tree structure).

Our results imply that the set of all $-suffix trees is a proper subset of the set all of suffix trees (e.g., the suffix tree of *abaabab* is not a $-suffix tree by Lemma 7), which in turn is a proper subset of the set of all trees (consider, e.g., Fig. 1c).

References

1. Apostolico, A., Crochemore, M., Farach-Colton, M., Galil, Z., Muthukrishnan, S.: Forty years of text indexing. In: Fischer, J., Sanders, P. (eds.) CPM 2013. LNCS, vol. 7922, pp. 1–10. Springer, Heidelberg (2013)
2. Bannai, H., Inenaga, S., Shinohara, A., Takeda, M.: Inferring strings from graphs and arrays. In: Rovan, B., Vojtáš, P. (eds.) MFCS 2003. LNCS, vol. 2747, pp. 208–217. Springer, Heidelberg (2003)
3. Breslauer, D., Hariharan, R.: Optimal parallel construction of minimal suffix and factor automata. Parallel Process. Lett. **06**(01), 35–44 (1996)
4. Breslauer, D., Italiano, G.F.: On suffix extensions in suffix trees. Theor. Comp. Sci. **457**, 27–34 (2012)
5. Cazaux, B., Rivals, E.: Reverse engineering of compact suffix trees and links: a novel algorithm. J. Discrete Algorithms **28**, 9–22 (2014)
6. Clément, J., Crochemore, M., Rindone, G.: Reverse engineering prefix tables. In: Proceedings of the 26th STACS, pp. 289–300 (2009)
7. Crochemore, M., Iliopoulos, C.S., Pissis, S.P., Tischler, G.: Cover array string reconstruction. In: Amir, A., Parida, L. (eds.) CPM 2010. LNCS, vol. 6129, pp. 251–259. Springer, Heidelberg (2010)
8. Duval, J.P., Lecroq, T., Lefebvre, A.: Border array on bounded alphabet. J. Autom. Lang. Comb. **10**(1), 51–60 (2005)
9. Duval, J.P., Lecroq, T., Lefebvre, A.: Efficient validation and construction of border arrays and validation of string matching automata. RAIRO Theor. Inform. Appl. **43**, 281–297 (2009)
10. Duval, J.P., Lefebvre, A.: Words over an ordered alphabet and suffix permutations. RAIRO Theor. Inform. Appl. **36**(3), 249–259 (2002)
11. Gawrychowski, P., Jeż, A., Jeż, Ł.: Validating the Knuth-Morris-Pratt failure function, fast and online. Theory Comput. Syst. **54**(2), 337–372 (2014)
12. Gusfield, D.: Algorithms on Strings, Trees and Sequences: Computer Science and Computational Biology. Cambridge University Press, New York (1997)
13. I, T., Inenaga, S., Bannai, H., Takeda, M.: Verifying and enumerating parameterized border arrays. Theor. Comput. Sci. **412**(50), 6959–6981 (2011)
14. I, T., Inenaga, S., Bannai, H., Takeda, M.: Counting parameterized border arrays for a binary alphabet. In: Dediu, A.H., Ionescu, A.M., Martín-Vide, C. (eds.) LATA 2009. LNCS, vol. 5457, pp. 422–433. Springer, Heidelberg (2009)
15. I., Tomohiro, Inenaga, Shunsuke, Bannai, Hideo, Takeda, Masayuki: Verifying a Parameterized Border Array in $O(n^{1.5})$ Time. In: Amir, Amihood, Parida, Laxmi (eds.) CPM 2010. LNCS, vol. 6129, pp. 238–250. Springer, Heidelberg (2010)
16. I, T., Inenaga, S., Bannai, H., Takeda, M.: Inferring strings from suffix trees and links on a binary alphabet. Discrete Appl. Math. **163**, 316–325 (2014)
17. Kucherov, G., Tóthmérész, L., Vialette, S.: On the combinatorics of suffix arrays. Inf. Process. Lett. **113**(22–24), 915–920 (2013)

18. Lu, W., Ryan, P.J., Smyth, W.F., Sun, Y., Yang, L.: Verifying a border array in linear time. J. Comb. Math. Comb. Comput. **42**, 223–236 (2002)
19. Starikovskaya, T., Vildhøj, H.W.: A suffix tree or not a suffix tree. arXiv (2014). http://arxiv.org/abs/1403.1364
20. Ukkonen, E.: On-line construction of suffix trees. Algorithmica **14**(3), 249–260 (1995)
21. Weiner, P.: Linear pattern matching algorithms. In: Proceedings of the 14th FOCS (SWAT), pp. 1–11 (1973)

Deterministic Algorithms for the Independent Feedback Vertex Set Problem

Yuma Tamura[1,2]([⊠]), Takehiro Ito[1], and Xiao Zhou[1]

[1] Graduate School of Information Sciences, Tohoku University,
Aoba-yama 6-6-05, Sendai 980-8579, Japan
{tamura,takehiro,zhou}@ecei.tohoku.ac.jp
[2] JST, ERATO, Kawarabayashi Large Graph Project,
c/o Global Research Center for Big Data Mathematics, NII,
2-1-2 Hitotsubashi, Chiyoda-ku, Tokyo 101-8430, Japan

Abstract. A feedback vertex set F of an undirected graph G is a vertex subset of G whose removal results in a forest. Such a set F is said to be independent if F forms an independent set of G. In this paper, we study the problem of finding an independent feedback vertex set of a given graph with the minimum number of vertices, from the viewpoint of graph classes. This problem is NP-hard even for planar bipartite graphs of maximum degree four. However, we show that the problem is solvable in linear time for graphs having tree-like structures, more specifically, for bounded treewidth graphs, chordal graphs and cographs. We then give a fixed-parameter algorithm for planar graphs when parameterized by the solution size. Such a fixed-parameter algorithm is already known for general graphs, but our algorithm is exponentially faster than the known one.

1 Introduction

The feedback vertex set problem for undirected graphs is one of the most classical NP-hard problems. A *feedback vertex set* F of an undirected graph $G = (V, E)$ is a vertex subset of G such that the subgraph of G induced by $V \setminus F$ is a forest. (See Fig. 1(b) as an example.) For a given undirected graph G, the *feedback vertex set problem* is to find a feedback vertex set of G with the minimum number of vertices. This problem has several applications such as in combinatorial circuit design [1], constraint satisfaction and Bayesian inference [2].

Recently, Misra et al. [14] introduced the "independence" variant of the feedback vertex set problem. Formally, a feedback vertex set F of a graph G is said to be *independent* if F forms an independent set of G. (See Fig. 1(c) as an example. Notice that the feedback vertex set in Fig. 1(b) is not independent.) Note that, in contrast to (original) feedback vertex sets, there are graphs which have no independent feedback vertex set; for example, consider the complete graph on four vertices. For a given undirected graph G, the *independent feedback vertex set problem* is to find an independent feedback vertex set of G with the minimum number of vertices if it exists.

© Springer International Publishing Switzerland 2015
J. Kratochvíl et al. (Eds.): IWOCA 2014, LNCS 8986, pp. 351–363, 2015.
DOI: 10.1007/978-3-319-19315-1_31

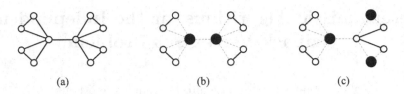

Fig. 1. (a) Graph G, (b) an optimal feedback vertex set of G, and (c) an optimal independent feedback vertex set of G, where each vertex in the feedback vertex sets is depicted by a large black circle.

For convenience, we sometimes call the feedback vertex set problem the *original problem*, and the independent feedback vertex set problem the *independence variant*.

1.1 Related Results and Known Results

The original problem has been intensively studied from various viewpoints, such as of approximation, fixed parameter tractability (FPT), and tractability on special graph classes [1,2,9,11,17]. The original problem is APX-complete [1], and remains NP-hard even for planar graphs of maximum degree four [17].

In contrast to the original problem, as far as we know, there are only two results for the independence variant. Misra et al. [14] gave an FPT algorithm for general graphs when parameterized by the solution size k, and gave a kernel of size $O(k^3)$. Their algorithm runs in time $O(5^k n^{O(1)})$, where n is the number of vertices in a graph. Recently, Song [16] improved this running time to $O(4^k n^2)$.

As Misra et al. pointed out in [14], the original problem can be reduced to the independence variant in polynomial time in an approximation-preserving manner. This implies that several hardness results for the original problem hold also for the independence variant: the independence variant is APX-hard for bipartite graphs, and remains NP-hard even for planar bipartite graphs of maximum degree four. However, such a relationship does not hold for tractable results: algorithms for the original problem do not work for the independence variant.

1.2 Our Contribution

In this paper, we embark on a systematic investigation of the computational status of the independent feedback vertex set problem from the viewpoint of graph classes. (Figure 2 summarizes known results and our results for the problem.)

We first show that the problem can be solved in linear time for graphs having tree-like structures, more specifically, for bounded treewidth graphs, chordal graphs and cographs. Note that these graph classes are incomparable with each other as shown in Fig. 2, and have "tree-like" structures in difference senses.

We then give a subexponential FPT algorithm for planar graphs when parameterized by the solution size k. Our algorithm runs in time $2^{O(\sqrt{k}\log k)}n$, where n is the number of vertices in a graph, and hence improves the best known running time $O(4^k n^2) = 2^{O(k)}n^2$ by Song [16] when restricted to planar graphs.

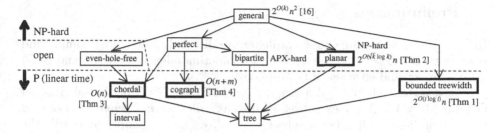

Fig. 2. Known results and our results, where n and m are the numbers of vertices and edges in a graph, respectively, and t is the treewidth of the graph. Each arrow represents the inclusion relationship between graph classes: $A \to B$ represents that B is properly included in A [6].

Note that the improvement is achieved in terms of both k and n; especially, the function depending on k is improved exponentially.

We emphasize that our algorithms are not developed independently: it is interesting that our FPT algorithm for planar graphs and linear-time algorithm for chordal graphs employ our linear-time algorithm for bounded treewidth graphs, although neither planar nor chordal graphs have bounded treewidth in general.

Due to the page limitation, we omit some proofs from this extended abstract.

1.3 Comparison with Known Techniques

It can be easily shown that the independence variant admits a linear-time algorithm for bounded treewidth graphs, because any optimization problem that can be expressed by Extended Monadic Second Order Logic (EMSOL) can be solved in linear time for bounded treewidth graphs [7]. However, the algorithm obtained by this method is very slow since the hidden constant factor of the running time depends on a tower of exponentials in treewidth [12]. On the other hand, our algorithm runs in time $2^{O(t \log t)}n$, where n is the number of vertices in a graph and t is the treewidth of the graph (defined in Sect. 3); thus, its constant factor is only $2^{O(t \log t)}$.

Recently, Bodlaender et al. [4] developed new techniques to obtain algorithms for "connectivity problems" which run in time $2^{O(t)}n^{O(1)}$. Both the original problem and the independence variant certainly belong to connectivity problems, and hence we can obtain $\left(2^{O(t)}n^{O(1)}\right)$-time algorithms for both problems. However, it seems difficult to obtain an algorithm for the independence variant whose running time is single exponential in t with keeping the dependence on n is linear, although the original problem can be solved in time $2^{O(t)}n$ [4]. To show this, the $\left(2^{O(t)}n\right)$-time algorithm uses the property that any super vertex-set of a feedback vertex set of a graph G forms a feedback vertex set of G, too. However, this property does not hold for the independence variant.

Therefore, our $\left(2^{O(t \log t)}n\right)$-time algorithm for bounded treewidth graphs should be interesting in its own right. Furthermore, we note that this running time is essential to obtain an FPT algorithm for planar graphs whose running time is subexponential in the parameter and is linear in n.

2   Preliminaries

In this paper, we assume that graphs are undirected, unweighted, simple and connected. Let $G = (V, E)$ be a graph; we sometimes denote by $V(G)$ and $E(G)$ the vertex set and edge set of G, respectively.

For a vertex subset V' of a graph $G = (V, E)$, let $G[V']$ be the subgraph of G induced by V'. A vertex subset V' of G is called an *independent set* of G if $G[V']$ contains no edge. For a subset $W \subseteq V$, we denote simply by $G \setminus W$ the induced subgraph $G[V \setminus W]$.

For a graph G, a vertex subset F of G is called a *feedback vertex set of G* if $G \setminus F$ is a forest. In particular, a feedback vertex set F of G is said to be *independent* if $G[F]$ forms an independent set of G. Let

$$\mathsf{OPT}(G) = \min\{|F| : F \text{ is an independent feedback vertex set of } G\};$$

we let $\mathsf{OPT}(G) = +\infty$ if G has no independent feedback vertex set. For a given graph G, the *independent feedback vertex set problem* is to find an independent feedback vertex set F of G such that $|F| = \mathsf{OPT}(G)$.

All algorithms given in this paper only compute the value $\mathsf{OPT}(G)$ for a given graph G. It is easy to modify them so that they actually find an independent feedback vertex set of G with the minimum number $\mathsf{OPT}(G)$ of vertices.

3   Our Results Based on Algorithm for Bounded Treewidth Graphs

As we have mentioned in Introduction, our results for planar graphs and chordal graphs employ a linear-time algorithm for bounded treewidth graphs. In this section, we explain how to obtain these results.

We first define the notion of treewidth. A *tree-decomposition* of a graph G is a pair $\langle \{X_i : i \in V_T\}, T \rangle$, where $T = (V_T, E_T)$ is a rooted tree, such that the following four conditions (1)–(4) hold [3]:

(1) each X_i is a subset of $V(G)$, and is called a *bag* for a node i;
(2) $\bigcup_{i \in V_T} X_i = V(G)$;
(3) for each edge $(u, v) \in E(G)$, there is at least one node $i \in V_T$ such that $u, v \in X_i$; and
(4) for each vertex $v \in V(G)$, the set $\{i \in V_T : v \in X_i\}$ induces a connected subgraph (subtree) of T.

For example, Fig. 3(b) illustrates a tree-decomposition of the graph G in Fig. 3(a). We will refer to a *node* in V_T in order to distinguish it from a vertex in $V(G)$. The *width* of a tree-decomposition $\langle \{X_i : i \in V_T\}, T \rangle$ is defined as $\max\{|X_i| - 1 : i \in V_T\}$, and the *treewidth* of G is the minimum t such that G has a tree-decomposition of width t. We denote by $\mathsf{tw}(G)$ the treewidth of G. Bodlaender et al. [5] gave the following useful algorithm.

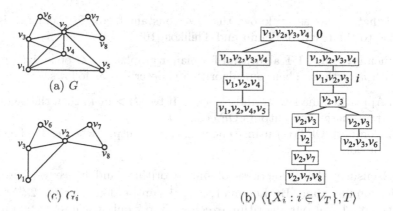

Fig. 3. (a) Graph G, (b) a nice tree-decomposition $\langle\{X_i : i \in V_T\}, T\rangle$ of G, and (c) the subgraph G_i of G for the node $i \in V_T$.

Lemma 1 ([5]). *For a graph G with n vertices and a positive integer t, there is a $\left(2^{O(t)}n\right)$-time algorithm which either outputs $\mathsf{tw}(G) > t$ or gives a tree-decomposition of G whose width is at most $5t + 4$.*

We now formally state our result for bounded treewidth graphs.

Theorem 1. *Let G be a graph whose treewidth is bounded by an integer t. Then, $\mathsf{OPT}(G)$ can be computed in time $2^{O(t \log t)}n$, where $n = |V(G)|$.*

Our proof for Theorem 1 will be given in Sect. 4. In the remainder of this section, we give two algorithms for planar graphs and chordal graphs, based on Theorem 1. We note again that neither planar graphs nor chordal graphs have bounded treewidth in general [6].

3.1   FPT Algorithm for Planar Graphs

The independent feedback vertex set problem is NP-hard even for planar bipartite graphs. In this subsection, we give a subexponential FPT algorithm for planar graphs when parameterized by the solution size.

Theorem 2. *Let k be a positive integer, and let G be a planar graph with n vertices. Then, one can determine whether $\mathsf{OPT}(G) \leq k$ or not in time $2^{O(\sqrt{k} \log k)}n$.*

Recall that our FPT algorithm is exponentially faster than the known one (for general graphs) which runs in time $O(4^k n^2) = 2^{O(k)}n^2$ [16].

We first give the following lemma, which enables us to apply the algorithm for bounded treewidth graphs to planar graphs.

Lemma 2. *Let G be a planar graph such that $\mathsf{OPT}(G) \leq k$. Then, $\mathsf{tw}(G) \leq c\sqrt{k}$, where c is a fixed constant.*

We note that, as far as we know, the best constant factor for Lemma 2 is $c = 9.546$, due to the result of Fomin and Thilikos [10].

We then give our FPT algorithm for planar graphs. Let G be a given planar graph with n vertices. Then, our algorithm is described as follows.

Step 1. Apply Lemma 1 to G and $t = c\sqrt{k}$. If $\mathsf{tw}(G) > c\sqrt{k}$, then the algorithm terminates and outputs $\mathsf{OPT}(G) > k$.

Step 2. Compute $\mathsf{OPT}(G)$ using Theorem 1, and output whether $\mathsf{OPT}(G) \leq k$ or not.

Lemma 2 ensures the correctness of our algorithm, and hence we here estimate the running time. By Lemma 1, Step 1 can be done in time $2^{O(\sqrt{k})}n$. It should be noted that our algorithm executes Step 2 only when we obtain a tree-decomposition of G with width at most $5c\sqrt{k} + 4$ in Step 1; this means that $\mathsf{tw}(G) \leq 5c\sqrt{k} + 4$. Therefore, by Theorem 1 we can compute $\mathsf{OPT}(G)$ in time $2^{O(\sqrt{k}\log k)}n$.

This completes the proof of Theorem 2. □

For a planar graph G with n vertices and a positive integer k, one can also obtain an FPT algorithm which determines whether $\mathsf{OPT}(G) \leq k$ or not in time $2^{O(\sqrt{k})}n^{O(1)}$, because there is a $\left(2^{O(t)}n'^{O(1)}\right)$-time algorithm which computes $\mathsf{OPT}(G')$ for a graph G' with n' vertices and $\mathsf{tw}(G') \leq t$ [4].

3.2   Linear-Time Algorithm for Chordal Graphs

We then consider the problem restricted to chordal graphs. A graph G is *chordal* if every cycle in G of length at least four has a chord, which is an edge joining non-consecutive vertices in the cycle [6].

Theorem 3. *The independent feedback vertex set problem can be solved in linear time for chordal graphs.*

We give such an algorithm as a proof of Theorem 3. Indeed, our algorithm for chordal graphs employs the same strategy as that for planar graphs. Then, together with Theorem 1, the following lemma establishes Theorem 3.

Lemma 3. *For any chordal graph G, $\mathsf{tw}(G) \leq 2$ if $\mathsf{OPT}(G) \neq +\infty$.*

Proof. For a graph G, we denote by $\omega(G)$ the size of a maximum clique in G. Then, we will prove the following two claims: for any chordal graph G,

(a) $\mathsf{tw}(G) \leq 2$ if $\omega(G) \leq 3$; and
(b) $\omega(G) \leq 3$ if $\mathsf{OPT}(G) \neq +\infty$.

Clearly, Lemma 3 follows from the claims (a) and (b) above.

The claim (a) holds, because Robertson and Seymour [15] proved that $\mathsf{tw}(G) = \omega(G) - 1$ for any chordal graph G.

We then prove the claim (b) above. Indeed, this claim holds for any graph which is not necessarily chordal. Suppose that $\mathsf{OPT}(G) \neq +\infty$ for a graph G, and hence G has an independent feedback vertex set F. Since $G \setminus F$ is a forest, we have $\omega(G \setminus F) \leq 2$. Furthermore, since $G[F]$ forms an independent set, we have $\omega(G[F]) = 1$. Therefore, $\omega(G) \leq \omega(G \setminus F) + \omega(G[F]) \leq 3$.

This completes the proof of lemma.   □

4   Algorithm for Bounded Treewidth Graphs

In this section, we give a constructive proof of Theorem 1.

4.1   Nice Tree-Decomposition

In Sect. 3, we have defined a tree-decomposition of a graph G, based on the conditions (1)–(4). In particular, a tree-decomposition $\langle \{X_i : i \in V_T\}, T \rangle$ of G is called a *nice tree-decomposition* if the following conditions (5)–(8) hold [8]:

(5) $|V_T| = O(n)$, where $n = |V(G)|$;
(6) every node in V_T has at most two children in T;
(7) if a node $i \in V_T$ has two children l and r, then $X_i = X_l = X_r$; and
(8) if a node $i \in V_T$ has only one child j, then
  • $|X_i| = |X_j| - 1$ and $X_i \subset X_j$ (such a node i is called a *forget* node); or
  • $|X_i| = |X_j| + 1$ and $X_i \supset X_j$ (such a node i is called an *introduce* node.)

Figure 3(b) illustrates a nice tree-decomposition $\langle \{X_i : i \in V_T\}, T \rangle$ of the graph G in Fig. 3(a) whose treewidth is three.

For a given tree-decomposition of G of width t', one can transform it into a nice one of width t' in time $O(t'^2 n)$ [8]. Therefore, together with Lemma 1, one can obtain the following proposition.

Proposition 1. *Let G be a graph with n vertices, and let t be a positive integer. Then, there exists a $\left(2^{O(t)} n\right)$-time algorithm which either outputs $\mathsf{tw}(G) > t$ or gives a nice tree-decomposition of G whose width is at most $5t + 4$.*

Let $\langle \{X_i : i \in V_T\}, T \rangle$ be a nice tree-decomposition of a graph G. Then, each node $i \in V_T$ corresponds to a subgraph G_i of G which is induced by the vertices that are contained in the bag X_i and the bags for all descendants of i in T. Therefore, if a node $i \in V_T$ has two children l and r in T, then G_i is the union of G_l and G_r which are the subgraphs corresponding to nodes l and r, respectively. Clearly, $G = G_0$ for the root 0 of T. For example, Fig. 3(c) illustrates the subgraph G_i of the graph G in Fig. 3(a) which corresponds to the node $i \in V_T$ in Fig. 3(b).

4.2   Idea and Definitions

Let G be a graph such that $\mathsf{tw}(G) \leq t$ for a positive integer t, and let $\langle \{X_i : i \in V_T\}, T \rangle$ be a nice tree-decomposition of G of width t'. By Proposition 1 we may assume that $t' \leq 5t + 4$.

For a node $i \in V_T$, let S be any subset of X_i and let $\pi : X_i \setminus S \to \{0, 1, \ldots, t'\}$ be any mapping; we call such a pair (S, π) a *pair on* X_i. Then, an independent feedback vertex set F of G_i is called an (S, π)-*set of* G_i if the following two conditions (i) and (ii) hold:

(i) $F \cap X_i = S$; and
(ii) two vertices $v, w \in X_i \setminus S$ are contained in the same connected component in $G_i \setminus F$ if and only if $\pi(v) = \pi(w)$.

Therefore, for an (S, π)-set F of G_i, the set S forms an independent set of $G[X_i]$, and $G_i \setminus F$ forms a forest. For a pair (S, π) on X_i, we define the value $f(G_i; S, \pi)$ as follows:

$$f(G_i; S, \pi) = \min\{|F| : F \text{ is an } (S, \pi)\text{-set of } G_i\}.$$

If G_i has no (S, π)-set for the pair (S, π), then we let $f(G_i; S, \pi) = +\infty$.

Our algorithm computes the values $f(G_i; S, \pi)$ for each node $i \in V_T$ and all pairs (S, π) on X_i, from the leaves of T to the root 0 of T, by means of dynamic programming. Since $G_0 = G$ for the root 0 of T, we can compute the optimal value $\mathsf{OPT}(G)$ from the values of $f(G_0; S, \pi)$, as follows:

$$\mathsf{OPT}(G) = \min\{f(G_0; S, \pi) : (S, \pi) \text{ is a pair on } X_0\}. \tag{1}$$

4.3 Algorithm

In this subsection, we explain how to compute the values $f(G_i; S, \pi)$.

(I) The node i is a leaf of T.

Note that $G_i = G[X_i]$ in this case. Thus, for any pair (S, π) on X_i, we can determine whether the vertex subset S forms an (S, π)-set of G_i. We thus have

$$f(G_i; S, \pi) = \begin{cases} |S| & \text{if } S \text{ is an } (S, \pi)\text{-set of } G_i; \\ +\infty & \text{otherwise.} \end{cases}$$

(II) The node i is an internal node of T.

Suppose that we have already computed the values $f(G_j; S_j, \pi_j)$ for all children j of an internal node i in T and their pairs (S_j, π_j). Then, we compute $f(G_i; S, \pi)$ for the node i and each pair (S, π) on X_i. Since $\langle \{X_i : i \in V_T\}, T \rangle$ is a nice tree-decomposition of G, there are three cases to consider, that is, i is a forget node, i is an introduce node, and i has two children. We here explain only one case; the remaining two cases are omitted from this extended abstract.

Case: The node $i \in V_T$ is a forget node. (See Fig. 4.)

In this case, the node i has exactly one child j in T such that $|X_i| = |X_j| - 1$ and $X_i \subset X_j$. Let v be the vertex in $X_j \setminus X_i$. Since $G_i = G_j$ in this case, any (S, π)-set F of G_i is an independent feedback vertex set of G_j and hence F forms an (S_j, π_j)-set of G_j for some pair (S_j, π_j) on X_j. However, since $v \notin X_i$, we do not know from the set S whether $v \in F$ or not (and hence $v \in S_j$ or not). We thus consider both the two sub-cases $v \in F$ and $v \notin F$, and define two values $f^a(G_i; S, \pi)$ and $f^b(G_i; S, \pi)$ for them; the value $f(G_i; S, \pi)$ can be computed by taking the better one.

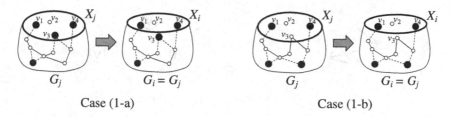

Case (1-a)   Case (1-b)

Fig. 4. Suppose $S = \{v_1, v_4\}$ and only v_3 is forgotten from X_j. Case (1-a) depicts an (S, π)-set F of G_i such that $v_3 \in F$, and Case (1-b) depicts an (S, π)-set F of G_i such that $v_3 \notin F$.

(1-a) $v \in F$.

Since $v \in X_j$, we have $S_j = S \cup \{v\}$ in this sub-case. Then, since $v \notin X_j \setminus S_j$, we have $\pi_j = \pi$. Therefore, F is an $(S \cup \{v\}, \pi)$-set of G_j, and hence we define $f^a(G_i; S, \pi)$ as follows:

$$f^a(G_i; S, \pi) = f(G_j; S \cup \{v\}, \pi).$$

(1-b) $v \notin F$.

We have $S_j = S$ in this sub-case. Then, since $v \in X_j \setminus S_j = X_j \setminus S$, F is an (S, π_j)-set of G_j for some mapping $\pi_j : X_j \setminus S \to \{0, 1, \ldots, t'\}$. Therefore, we define $f^b(G_i; S, \pi)$ as follows:

$$f^b(G_i; S, \pi) = \min f(G_j; S, \pi_j),$$

where the minimum above is taken over all mappings $\pi_j : X_j \setminus S \to \{0, 1, \ldots, t'\}$ such that $\pi_j(w) = \pi(w)$ for all vertices $w \in X_i \setminus S$.

Then, we can compute the value $f(G_i; S, \pi)$, as follows:

$$f(G_i; S, \pi) = \min\{f^a(G_i; S, \pi), \ f^b(G_i; S, \pi)\}.$$

4.4   Running Time

We first estimate the number of all pairs (S, π) on X_i for each node $i \in V_T$. For a nice tree-decomposition $\langle\{X_i : i \in V_T\}, T\rangle$ of width t', each bag X_i contains at most $t' + 1$ vertices of G. Since $S \subseteq X_i$ and $\pi : X_i \setminus S \to \{0, 1, \ldots, t'\}$, the number of all pairs (S, π) on X_i is bounded by

$$\sum_{k=0}^{t'+1} \binom{t'+1}{k} \cdot (t'+1)^{t'+1-k} \leq 2^{t'+1} \cdot (t'+1)^{t'+1} = 2^{O(t' \log t')}.$$

Recall that by Proposition 1 we have assumed $t' \leq 5t + 4$. Therefore, the number of all pairs (S, π) on X_i for each node $i \in V_T$ is bounded by $2^{O(t \log t)}$.

For each leaf i of T and every pair (S, π) on X_i, recall that we simply checked whether S is an (S, π)-set of G_i. This can be done in time $O(t^2)$ because $G_i =$

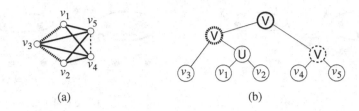

Fig. 5. (a) Cograph G and (b) its cotree, where each edge in G is depicted by the same (thick dotted, thin dotted, or thick) line as the corresponding join node of the cotree.

$G[X_i]$ and $|X_i| = O(t)$. Therefore, for each leaf i and all pairs (S, π) on X_i, the values $f(G_i; S, \pi)$ can be computed in time $O(t^2) \times 2^{O(t \log t)} = 2^{O(t \log t)}$. Since T has $O(n)$ leaves, the initialization can be done in time $2^{O(t \log t)}n$ in total.

Similar argument shows that, for each internal node i of T and all pairs (S, π) on X_i, the values $f(G_i; S, \pi)$ can be computed in time $2^{O(t \log t)}$ from the values $f(G_j; S_j, \pi_j)$ for its children j. Since T has $O(n)$ nodes, the values $f(G_0; S, \pi)$ can be computed in time $2^{O(t \log t)}n$ for all pairs (S, π) on the root 0 of T. By Eq. (1) the optimal value $\mathsf{OPT}(G)$ can be computed in time $2^{O(t \log t)}$ from the values $f(G_0; S, \pi)$.

This completes the proof of Theorem 1.   □

5   Our Result for Cographs

In this section, we consider the problem restricted to cographs. In contrast to chordal graphs, a cograph G may have a super-constant treewidth even if $\mathsf{OPT}(G) \neq +\infty$. For example, consider a complete bipartite graph $K_{p,q}$, which is a cograph. Then, $\mathsf{tw}(K_{p,q}) = \min\{p, q\}$ and hence it is not bounded by a constant, even though $\mathsf{OPT}(K_{p,q}) = \min\{p, q\} - 1 \neq +\infty$.

We first define the class of cographs (also known as P_4-free graphs) [6]. For two graphs $G_1 = (V_1, E_1)$ and $G_2 = (V_2, E_2)$, their *union* $G_1 \cup G_2$ is the graph such that $V(G_1 \cup G_2) = V_1 \cup V_2$ and $E(G_1 \cup G_2) = E_1 \cup E_2$, while their *join* $G_1 \vee G_2$ is the graph such that $V(G_1 \vee G_2) = V_1 \cup V_2$ and $E(G_1 \vee G_2) = E_1 \cup E_2 \cup \{(v, w) : v \in V_1, w \in V_2\}$. Then, a *cograph* can be recursively defined, as follows (see Fig. 5(a) as an example):

(1) a graph consisting of a single vertex is a cograph;
(2) if G_1 and G_2 are cographs, then the union $G_1 \cup G_2$ is a cograph; and
(3) if G_1 and G_2 are cographs, then the join $G_1 \vee G_2$ is a cograph.

The main result of this section is the following theorem.

Theorem 4. *The independent feedback vertex set problem can be solved in linear time for cographs.*

In this section, we give such an algorithm as a proof of Theorem 4.

5.1   Cotree

From the definition of cographs, we can naturally represent a cograph G by a binary tree, called a cotree of G, defined as follows (see Fig. 5 as an example): a *cotree* $T_c = (V_c, E_c)$ of a cograph G is a binary tree such that each leaf of T_c corresponds to a single vertex in G, and each internal node of T_c has exactly two children and is labeled with either union $\cup$ or join $\vee$. Such a cotree of a given cograph G can be constructed in linear time [13].

Each node $i \in V_c$ corresponds to a subgraph G_i of G which is induced by all vertices corresponding to the leaves of T_c that are the descendants of i in T_c. Clearly, $G = G_0$ for the root 0 of T_c.

5.2   Idea and Definitions

Let i be an internal node of a cotree T_c with two children l and r. Suppose that G_l and G_r have independent feedback vertex sets F_l and F_r, respectively. If i is a union node and hence $E(G_i) = E(G_l) \cup E(G_r)$, then the vertex subset $F_l \cup F_r$ is clearly an independent feedback vertex set of G_i. On the other hand, if i is a join node and hence $E(G_i) = E(G_l) \cup E(G_r) \cup \{(v, w) : v \in V(G_l), w \in V(G_r)\}$, then the vertex subset $F_l \cup F_r$ is not always an independent feedback vertex set of G_i. Indeed, $F_l = \emptyset$ or $F_r = \emptyset$ must hold to keep the independence. Furthermore, newly added edges may make cycles in $G_i \setminus (F_l \cup F_r)$. However, such a case can be avoided by the following lemma.

Lemma 4. *Let H_l and H_r be forests. Then, the graph $H_l \vee H_r$ is a forest if and only if at least one of the following (i)–(iv) holds: (i) $|V(H_l)| = 0$; (ii) $|V(H_r)| = 0$; (iii) $|V(H_l)| = 1$ and $|E(H_r)| = 0$; and (iv) $|E(H_l)| = 0$ and $|V(H_r)| = 1$.*

Note that $|V(H_r)| \geq 1$ may hold in the case (iii) above. Similarly, $|V(H_l)| \geq 1$ may hold in the case (iv) above.

According to Lemma 4, we thus characterize an independent feedback vertex set F of a graph G_i, $i \in V_c$, by the number of vertices in $G_i \setminus F$ and the number of edges in $G_i \setminus F$. Formally, for a node $i \in V_c$ and two integers $p \in \{0, 1, 2\}$ and $q \in \{0, 1\}$, a (p, q)-*set* of G_i is an independent feedback vertex set F of G_i satisfying the following (i) and (ii):

(i) $G_i \setminus F$ has no vertex (resp., exactly one vertex) if $p = 0$ (resp., $p = 1$); and
     $G_i \setminus F$ has at least two vertices if $p = 2$; and
(ii) $G_i \setminus F$ has no edge if $q = 0$; and $G_i \setminus F$ has at least one edge if $q = 1$.

Then, for a node $i \in V_c$ and two integers $p \in \{0, 1, 2\}$ and $q \in \{0, 1\}$, we define the value $g(G_i; p, q)$, as follow:

$$g(G_i; p, q) = \min\{|F| : F \text{ is a } (p, q)\text{-set of } G_i\}.$$

If G_i has no (p, q)-set, then we let $g(G_i; p, q) = +\infty$.

Our algorithm computes the values $g(G_i; p, q)$ for each node $i \in V_c$ and all integers $p \in \{0, 1, 2\}$ and $q \in \{0, 1\}$, from the leaves of T_c to the root 0 of T_c, by means of dynamic programming. Since $G_0 = G$ for the root 0 of T_c, we can compute the optimal value $\mathsf{OPT}(G)$ from the values of $g(G_0; p, q)$, as follows:

$$\mathsf{OPT}(G) = \min\{g(G_0; p, q) : p \in \{0, 1, 2\}, \; q \in \{0, 1\}\}.$$

Since our algorithm computes only six types of (p, q)-sets for each node $i \in V_c$, it runs in linear time.

6   Concluding Remark

As shown in Fig. 2, it remains open to clarify the complexity status for even-hole-free graphs. However, interestingly, the complexity status for even-hole-free graphs remains open even for the (original) feedback vertex set problem.

Acknowledgments. We are grateful to Saket Saurabh for fruitful discussions with him. This work is partially supported by JSPS KAKENHI Grant Numbers 25106504 and 25330003.

References

1. Bafna, V., Berman, P., Fujito, T.: A 2-approximation algorithm for the undirected feedback vertex set problem. SIAM J. Discrete Math. **12**, 289–297 (1999)
2. Bar-Yehuda, R., Geiger, D., Naor, J., Roth, R.M.: Approximation algorithms for the feedback vertex set problem with applications to constraint satisfaction and Bayesian inference. SIAM J. Comput. **27**, 942–959 (1998)
3. Bodlaender, H.L.: A linear-time algorithm for finding tree-decompositions of small treewidth. SIAM J. Comput. **25**, 1305–1317 (1996)
4. Bodlaender, H.L., Cygan, M., Kratsch, S., Nederlof, J.: Deterministic single exponential time algorithms for connectivity problems parameterized by treewidth. In: Fomin, F.V., Freivalds, R., Kwiatkowska, M., Peleg, D. (eds.) ICALP 2013, Part I. LNCS, vol. 7965, pp. 196–207. Springer, Heidelberg (2013)
5. Bodlaender, H.L., Drange, P.G., Dregi, M.S., Fomin, F.V., Lokshtanov, D., Pilipczuk, M.: An $O(c^k n)$ 5-approximation algorithm for treewidth. In: Proceedings of FOCS 2013, pp. 499–508 (2013)
6. Brandstädt, A., Le, V.B., Spinrad, J.P.: Graph Classes: A Survey. SIAM, Philadelphia (1999)
7. Courcelle, B.: Graph rewriting: an algebraic and logic approach. In: van Leeuwen, J. (ed.) Handbook of Theoretical Computer Science, vol. B, pp. 193–242. MIT Press, Cambridge (1990)
8. Dorn, F., Telle, J.A.: Two birds with one stone: the best of branchwidth and treewidth with one algorithm. In: Correa, J.R., Hevia, A., Kiwi, M. (eds.) LATIN 2006. LNCS, vol. 3887, pp. 386–397. Springer, Heidelberg (2006)
9. Festa, P., Pardalos, P.M., Resende, M.G.C.: Feedback set problems. In: Du, D.-Z., Pardalos, P.M. (eds.) Handbook of Combinatorial Optimization, pp. 209–258. Kluwer Academic Publishers, Dordrecht (1999)
10. Fomin, F.V., Thilikos, D.M.: Dominating sets in planar graphs: branch-width and exponential speed-up. SIAM J. Comput. **36**, 281–309 (2006)

11. Kloks, T., Lee, C.M., Liu, J.: New algorithms for k-face cover, k-feedback vertex set, and k-disjoint cycles on plane and planar graphs. In: Kučera, L. (ed.) WG 2002. LNCS, vol. 2573, pp. 282–295. Springer, Heidelberg (2002)
12. Lampis, M.: Algorithmic meta-theorems for restrictions of treewidth. Algorithmica **64**, 19–37 (2012)
13. McConnell, R.M., Spinrad, J.P.: Linear-time modular decomposition of directed graphs. Discrete Appl. Math. **145**, 198–209 (2005)
14. Misra, N., Philip, G., Raman, V., Saurabh, S.: On parameterized independent feedback vertex set. Theoret. Comput. Sci. **461**, 65–75 (2012)
15. Robertson, N., Seymour, P.D.: Graph minors. II. Algorithmic aspects of tree-width. J. Algorithms **7**, 309–322 (1986)
16. Song, Y.: An improved parameterized algorithm for the independent feedback vertex set problem. Theoret. Comput. Sci. **535**, 25–30 (2014)
17. Speckenmeyer, E.: On feedback vertex sets and nonseparating independent sets in cubic graphs. J. Graph Theory **12**, 405–412 (1988)

Lossless Seeds for Searching Short Patterns with High Error Rates

Christophe Vroland[1,2,3]([⊠]), Mikaël Salson[1,2], and Hélène Touzet[1,2]

[1] LIFL, UMR CNRS 8022, Université Lille 1, Villeneuve D'ascq, France
{christophe.vroland,mikael.salson,helene.touzet}@lifl.fr
[2] Inria Lille Nord-Europe, Villeneuve D'ascq, France
[3] Laboratoire Génétique et Evolution des Populations Végétales, UMR CNRS 8198, Université Lille1, Villeneuve D'ascq, France

Abstract. We address the problem of approximate pattern matching using the Levenshtein distance. Given a text T and a pattern P, find all locations in T that differ by at most k errors from P. For that purpose, we propose a filtration algorithm that is based on a novel type of seeds, combining exact parts and parts with a fixed number of errors. Experimental tests show that the method is specifically well-suited for short patterns with a large number of errors.

1 Introduction

We consider the approximate pattern matching problem where a pattern P is searched in a text T with a given number of errors k. An error can be defined in several ways. Here we consider an error as defined by the Levenshtein distance: either a substitution, an insertion, or a deletion. The problem is to find all the locations where the pattern matches the text with at most k errors.

Navarro *et al.* distinguish three main approaches [17]. The first one, *neighborhood generation*, consists in generating all the strings within the number of errors of the queried string. Then the generated strings are searched exactly in the text. This generation is exponential in the number of errors. It is often done with dynamic programming, bit-parallelism, or finite automata (for instance [27]).

A second approach, *partitioning approach*, consists in filtering regions of interest. Such regions are found by using pattern substrings, called *seeds*. Once the seeds are found in the text, their occurrences must be extended to check if the pattern occurs within k errors. The pigeonhole principle is used, where the pattern is split in $k + \epsilon$ non-overlapping parts, $\epsilon > 0$. Then ϵ parts are searched exactly, usually $\epsilon = 1$. However the more errors we allow, the shorter the parts will be and therefore the more potential occurrences we may have. Thus the filtration efficiency becomes lower with higher values of k. A recent example of this method, using a modified Burrows-Wheeler transform is shown in [18].

Finally, the third approach a hybrid of the two previous established approaches. The pattern is split in non-overlapping parts that can be searched with a given

This work was partly supported by the Mastodons project (CNRS).

number of errors. For instance, Navarro and Baeza-Yates [16] designed a *hybrid method* which consists of splitting the pattern in $j = (m+k)/\log_\sigma n$ parts, where σ is the alphabet size, and searching these parts with $\lfloor \frac{k}{j} \rfloor$ errors. This approach has also been used with a LZ-index or a FM-index [19].

Each of these search strategies can be directly applied to a string and will be linear in time depending on the size of the text. For a survey of online algorithms, refer to [15]. Using a text index, it is also possible to reduce time consumption at the expense of space consumption. Two main families of indexes are used: q-gram indexes and full-text indexes. The former allows to efficiently recover occurrences of a fixed-length word, while the latter allows to search for any pattern of any length. This generally allows one to backtrack so that a word can be searched with some errors. A third family of indexes consists of indexes specifically designed for approximate search [2,4,5,13]. However, these indexes are not compressed indexes, (*i.e.* whose space consumption is proportional to the empirical entropy of the text) and, to the best of our knowledge, no implementation of the proposed solutions exists.

Despite all these methods, we think there is a need for algorithms dedicated to searching short patterns (<50 letters) on a small alphabet (*e.g.* DNA alphabet) with a medium to high error-rate (7%–15%). This can be used in several applications in computational biology, such as predicting targets of non-protein coding small RNAs [25] and analysing spacers in CRISPR for potential transfers from viruses or plasmids [22,24], to cite a few. More generally, introducing some errors would improve the sensitivity in the presence of sequencing errors or variants. Since we combine a high error rate with a small alphabet, we need to design a method with a good filtration efficiency. This is necessary to limit the number of false positives, thus the number of unnecessary verifications. In this paper, we present a new hybrid method where the pattern will be split in non-overlapping parts, some of them being searched without error, while others are searched with a limited number of errors.

In Sect. 2, we detail the novel type of seeds we will use and show some properties of those seeds. We also show that in practice those seeds are efficient filters. In Sect. 3, we explain how we make use of those seeds in our algorithm and how they are searched in a compressed index. In Sect. 4 we show some experimental results on DNA with random sequences and real data.

2   Approximate Seeds for the Levenshtein Edit Distance

Let A be a finite alphabet. Given two strings u and v of A^*, define $lev(u, v)$ to be the Levenshtein distance between u and v. This is the minimum number of operations needed to transform u into v, where the only allowed operations are substitution of a single character and deletion or insertion of a single character. Each such operation is also called an *error*. From now on, we assume that a given natural number k corresponds to a maximum number of errors.

Let P be a pattern of length m over A. Using the pigeonhole principle, it is well-known that if P is partitioned into $k + 1$ parts, then every string U, such that $lev(P, U) \leq k$, contains at least one of these parts. Similarly, if P is partitioned into $k + 2$ parts, denoted $P_1, \ldots, P_{k+2}$, then U should contain at least two disjoint parts of P. The parts do not need to be of the same length. The following lemma allows to push the analysis further. It is indeed possible to request that these two parts be separated by parts with exactly one error.

Lemma 1. *Let U be a string of A^* such that $lev(P, U) \leq k$. Then there exists i, j, $1 \leq i < j \leq k + 2$, and $U_1, \ldots, U_{j-i-1}$ of A^* such that*

1. *$P_i U_1 \ldots U_{j-i-1} P_j$ is a substring of U, and*
2. *When $j > i + 1$, for each ℓ, $1 \leq \ell \leq j - i - 1$, $lev(P_{i+\ell-1}, U_\ell) = 1$.*

Example 1. Assume $k = 3$. Given the pattern $P = $ AACGTGAGGTAGGTTCCATG of length 20, we partition it into five parts, of equal length: $P_1 = $ AACG, $P_2 = $ TGAG, $P_3 = $ GTAG, $P_4 = $ GTTC, and $P_5 = $ CATG. Consider three strings whose Levenshtein distance with P is 3: AACGGAGGTAAGTTCTCATG, AACGTAGGCAAGTTCCATG and ATCGTGACGTAGGGTCCATG. For each string, we show in Fig. 1 the parts that fulfil the conditions of Lemma 1.

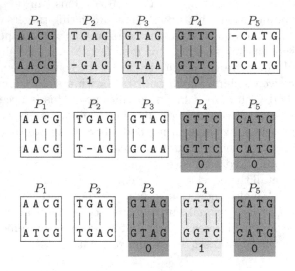

Fig. 1. Application of Lemma 1 for sequences of Example 1. The pattern AACGTGAGGTAGGTTCCATG is on the top of each alignment. The two parts with no error (written P_i and P_j in the Lemma) are highlighted in dark grey, and the parts with 1 error surrounded by P_i and P_j are highlighted in light grey.

We rephrase Lemma 1 as a pure counting problem and establish its proof.

Lemma 2. *Let k be a natural number. Assume you have $k + 2$ containers numbered from 1 to $k + 2$, and y tokens with $0 \leq y \leq k$. Then there exists two containers i and j, $1 \leq i < j \leq k + 2$, such that*

1. *containers i and j are empty, and*
2. *for each ℓ, $i < \ell < j$, the container ℓ contains exactly one token.*

Proof. The proof is by recurrence on k.

- If $k = 0$. In this case, we have two containers that are both empty: $i = 1$ and $j = 2$ is a solution for our problem.
- If $k > 0$:

Case 1: The first container is not empty, or the total amount of tokens y is strictly inferior to k. In this case, containers 2 to $k+2$ contain at most $k-1$ tokens. The recurrence hypothesis yields that there exists i and j, with $2 \leq i < j \leq k+2$, which solves our problem.

Case 2: The first container is empty, and the total amount of tokens is exactly k. In general, we know that at least one of the containers among $2, \ldots, k+2$ is empty. Let $x \geq 2$ be the number of empty containers and $i_1, i_2, \ldots, i_x$ be their indices. We must have at least one pair (i_p, i_{p+1}) with $1 \leq p < x$, such that each intervening container $\ell, i_p < \ell < i_{p+1}$ contains a single token. Therefore, we have $x - 1$ candidates.

We have $k + 2 - x$ non-empty containers. All of them have at least one token, so we have $k - (k+2-x) = x-2$ tokens left. The $x - 2$ remaining tokens must be distributed in the non-empty containers. In the worst case scenario, the tokens will be distributed in $x - 2$ distinct containers, each one being in-between a different candidate pair. Since we have $x - 1$ candidate pairs, there remains at least one candidate pair whose intervening containers contain exactly one token. $\qquad\square$

As a consequence of Lemma 1, we can design a seeding framework for lossless filtering for the approximate pattern matching problem with k errors. To this end, we introduce some terminology that will be used in the remaining of the paper.

Definition 1. *Let $P = P_1 \ldots P_{k+2}$ be a pattern divided into $k + 2$ parts. Then the 01^*0 seed for P and k is the regular expression*

$$\cup_{i=1}^{k+1} \cup_{j=i+1}^{k+2} P_i \, lev^1(P_{i+1}) \ldots lev^1(P_{j-1}) \, P_j$$

where $lev^1(u)$ denotes the set of strings whose Levenshtein distance with u is 1. A subseed is the regular expression associated to a given pair (i, j):

$$P_i \, lev^1(P_{i+1}) \ldots lev^1(P_{j-1}) \, P_j.$$

An instance of a seed, or of a subseed, is a string u of A^* which is recognized by the seed, or the subseed. Given a text T on A^*, an occurrence of the seed for P is a substring of T which is an instance of the seed. Therefore, an occurrence is characterized by its start position and its end position in T.

The filtration efficiency is the primary criterion used to evaluate the performance of a seed. To estimate it, we generated an independent and identically distributed random sequence of length 10^8 over the DNA alphabet $\{A, C, G, T\}$ as well as 100 patterns of length 20. We then searched for our 01*0 seed for $k = 3$. For each pattern, we counted the total number of occurrences of the seed in the text, including overlapping occurrences. The distribution is plotted in Fig. 2-(a). The average number of observed occurrences per pattern is 6,665. To compare with exact seeds, we report analogous results obtained with filtration based on q-grams in the same text as well as the same collection of patterns. First, we divide the pattern in $k + 1 = 4$ parts, leading to q-grams of length 5, which guarantees lossless filtration (Fig. 2-(b)). We also divide the pattern in three parts, of lengths 6, 7, and 7 (Fig. 2-(c)). This seed is less sensitive since it allows for some false negatives. In the first case, the average number of occurrences is 390,635, and in the latter case, it is 36,644. Both distributions are shown in Fig. 2-(b,c). These empirical measurements show that the 01*0 seed is significantly more selective than exact seeds, such as q-grams. Of course, this higher selectivity comes at the price of some additional work to locate seeds in the text. However, the fact that errors are not randomly distributed within the seed drastically reduces the combinatorics.

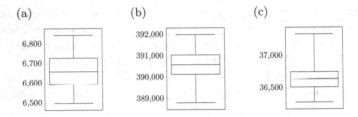

Fig. 2. Distribution of the number of occurrences of three different seeds for 100 patterns of size 20 in a random sequence of length 10^8. This filtration is done (a) with the 01*0 seed for $k = 3$ and $m = 20$, (b) by dividing the pattern in 4 parts of length 5, (c) by dividing the pattern in 3 parts of lengths 6, 7, and 7, respectively. For each box plot, the bottom and top of the box are the first and third quartiles. The band inside the box is the median, and the ends of the whiskers represent the minimum and maximum of all of the data. There is on average 26.85 occurrences of the whole pattern within 3 errors.

To conclude this section, we introduce a new Lemma that is a direct consequence of Lemma 2, and that will be useful in the forthcoming section.

Lemma 3. *Let k be a natural number. Assume you have $k + 2$ containers numbered from 1 to $k+2$, and y tokens with $0 \leq y \leq k$. Assume j to be the container furthest to the right, such that there exists some i, such that the pair (i, j) fulfils the conditions of Lemma 2. Then the total number of tokens present in containers from $j + 1$ to $k + 2$ is at least $k - j + 2$.*

Proof. Consider the containers numbered from $i+1$ to $k+2$, which makes $k-i+2$ containers. By hypothesis, these containers do not contain a 01*0 seed. So the contraposition of Lemma 2 ensures that these containers contain strictly more than $k - i$ tokens. Furthermore, by definition of i and j, there is exactly one token in each container ranging from $i+1$ to $j-1$, thus making $j - i - 1$ tokens, while the container j is empty. As a consequence, the total number of tokens in containers $j+1, \ldots, k+2$ is strictly greater than $k - i - (j - i - 1) = k - j + 1$. So, it is at least $k - j + 2$.   □

3   Algorithm

Let T be a text over the alphabet A^*. The problem we consider now is that of finding matches of P with at most k errors in T. For this we devise an efficient filtration algorithm based on the seeding framework introduced in Sect. 2. It is necessary to keep in mind that we want to search small patterns (several dozens of letters) in large texts (millions or billions of letters) with small alphabets (e.g. DNA). We first justify our choice of using a FM-index. Then we explain how seeds are searched for and how they are extended when necessary.

3.1   Choice of Index

As we have biological applications in mind (e.g. searching small DNA sequences on large genomes), we are in the situation where the text is known in advance. Moreover, we may have millions of short sequences to be queried in the text. This situation is particularly suitable for text indexes.

Since patterns do not have fixed sizes, full-text indexes are more appropriate. Furthermore, to limit space consumption, compressed indexes appear to be the indexes of choice. Among compressed indexes, FM-indexes [8] have an optimal time complexity for counting the occurrences of a pattern, while pattern search is more complex and counting is more time consuming with LZ-indexes [7].

3.2   Seed Filtration

Given a pattern P, we enumerate all possible subseeds for the pattern. Each subseed for P is characterized by two parts P_i and P_j, $1 \leq i < j \leq k + 2$, that occur exactly in the text. According to Lemma 2, all the intervening parts between P_i and P_j must be searched with exactly one error. We recall that in the FM-index, patterns are searched backwards, therefore, we first start by searching any part P_ℓ, with $1 < \ell \leq k + 2$, assuming it is P_j. This is an exact search in the index. Then the parts preceding P_ℓ are searched with at most one error (by backtracking as in BWA for instance [12]). When a part is found exactly, we know that P_i has been reached. Starting with P_ℓ, we can have several parts that fulfill our requirements; on reaching different parts $P_{i_1}, \ldots, P_{i_q}$ each of them matching exactly at different locations in the text. All the possible solutions are searched. If P_ℓ cannot be found exactly or if a part cannot be found with at most one error, this P_ℓ is skipped and we move on the next one. Therefore, at most we will have considered the $\frac{(k+1)(k+2)}{2}$ possible pairs (i, j).

Example 2. Let us continue with Example 1, also shown in Fig. 1: $k = 3$ and $P =$ AACG TGAG GTAG GTTC CATG, which is partitioned into 5 parts of equal length. Assume that this text is the concatenation of the three strings at distance 3 from P:

$T =$ AACGGAGGTAAGTTCTCATGAACGTAGGCAAGTTCCATGATCGTGACGTAGGGTCCATG.

- The algorithm first tries $j = 5$. $P_5 =$ CATG is found with no error in the FM-index. So, it has some exact occurrences in the text. Therefore, we continue to go through the FM-index to extend P_5 to the left and find all possible values for i. We find $i = 4$ (P_4 occurs exactly), $i = 3$ (P_4 occurs with one error and P_3 exactly) and $i = 1$ (P_4, P_3 and P_2 occur with one error and P_1 exactly). This gives three different seed instances, leading to three seed occurrences.

- With $j = 4$, GTTC occurs exactly in the FM-index, and correspond to two occurrences in T. By extending P_4 to the left, we keep just one instance since the second one cannot be extended to $P_3 =$ GTAG with at most one error.

Note that in this particular case, the first occurrence of P in T is covered by two overlapping 01*0 seeds, characterized by $i = 1$ and $j = 5$, and $i = 1$ and $j = 4$, respectively. This redundancy is solved with the extension and verification step, which is described in the next subsection.

- With $j = 3$, we have two occurrences of GTAG in the text. The first one cannot be extended to the left with $P_2 =$ TGAG. As for the second occurrence, P_2 is found with one error, but $P_1 =$ AACG does not exactly match. So, the occurrence is discarded.

- With $j = 2$, there is no exact occurrence of the part TGAG in the text.

At this point, all the seed *instances* occurring in the text are identified. We then proceed to the elongation and verification step.

3.3 Elongation and Verification

To perform the elongation of an instance of the seed, we first need to have a deeper look at the error distribution along the pattern. We know that the subseed instance has a Levenshtein distance of $j - i - 1$ with $P_i \ldots P_j$, which makes $j - i - 1$ errors. Via Lemma 3, we know that there are at least $k - j + 2$ errors in $P_{j+1} \ldots P_{k+2}$.

So, since the total number of errors should not exceed k, there should be at most $i - 1$ errors in $P_1 \ldots P_{i-1}$. As a consequence, each seed instance is first extended to the left, to find $P_1 \ldots P_{i-1}$ with at most $i - 1$ errors. To gain more efficiency, this extension is directly carried out in the FM-index to filter out candidates. Indeed, the retrieval of the positions of occurrences is the most time consuming part in an FM-index (in $O(\log^{1+\epsilon} n)$ per occurrence [9]). Once this extension is performed, the occurrences of $P_1 \ldots P_j$ are retrieved. Then the extension to the right is performed in the text using a banded dynamic programming algorithm. The starting point of the extension is the ending position of the occurrence of $P_1 \ldots P_j$ in the text. Let us assume that an instance of a given prefix $P_1 \ldots P_j$ has been found with e errors in the FM-index. Thus, $P_{j+1} \ldots P_{k+2}$ must be searched with at most $k - e$ errors in the text. Therefore, the bandwidth is $2 \times (k - e) + 1$ in the dynamic programming algorithm. Note that the extension to the right could also have been performed in the index using a bidirectional Burrows-Wheeler transform [3, 21]. That would, however, increase the memory footprint and provide only a moderate speed up, since many false positive seed instances have been removed at this step.

3.4   Implementation

Our algorithm was implemented in a software called Bwolo. Bwolo is written in C++, with the help of SeqAn library and the FM-Index implemented within [6]. It is open source and can be downloaded from http://bioinfo.lifl.fr/bwolo. In this implementation, patterns are divided into parts whose length differ by at most one character.

4   Experimental Results

In this section, we present some experimental results in order to measure the performance of our algorithm. We compare Bwolo to a selection of tools that were chosen for their complementarity. Widely utilized in bioinformatics, Exonerate is a generic tool for pair-wise sequence alignment, which uses exact sparse dynamic programming to perform the search. [23]. We use it as a standard for an on-line exact algorithm for our problem. RazerS3 is a read mapping program based on counting q-grams [26]. It performs the verification via an implementation of the improved Myers bit-vector algorithm proposed by Hyyr [10]. RazerS3 works without a precomputed index for the text. So, we also selected Bowtie2 [11], that, like our tools, indexes the text with an FM-index. It then uses backtracking for handling errors and dynamic programming to build the full alignment. Lastly, we used an in-house implementation for approximate search in an FM-index written with the SeqAn library. It is based on a simple breadth-first search method with no prior filtration. Unfortunately, we were not able to include hybrid methods described in [19] in our benchmark, since the implementation is not available.

All these tools were configured to be full sensitive and output all occurrences of the pattern: option `--exhaustive` for Exonerate, `--filter pigeonhole`

`--percent-identity` [*Id*] `--recognition-rate` 100 such that [*Id*] = 100 ×
$(1 - \frac{k}{m})$ for RazerS3 and `-a -L` [*Seeds*] `-i C,`[*Seeds*]`,0` such that [*Seeds*] = $\frac{m}{k+1}$
for Bowtie2. Moreover, for each tool the score system is based on the unit score,
which computes the Levenshtein distance.

The tests were run on a single thread of a server equipped with two Intel(R)
Xeon(R) CPU E5-2420 and 205 GB of RAM. The CPU time and the memory
consumption were measured using the GNU `time` command.

4.1   Randomly Generated Sequences

This first test uses independent and identically distributed sequences on the
DNA alphabet. The size n of the sequences ranges from 10^4 to 10^9. We also
generated 100 patterns of 20 nt at random and measured the computation time
of each tool for $k = 2$ and $k = 3$. Results are shown in Fig. 3.

In both cases, Bwolo is the fastest tool for long sequences, from 10^6 nt. As
expected, the added-value of Bwolo is even more obvious when $k = 3$. Tools
with no filtration, Exonerate and the exact search in the FM-index, are slow.
Bowtie2 operates slowly compared to all the other tools, especially with larger
values of n. This confirms that Bowtie2's heuristics, which have been designed
for long patterns (at least 50 nt) and few errors, is not well adapted to shorter
patterns with higher error rate. Unfortunately, there is not yet a specialized tool
for this type of problem. In our benchmark, Bowtie2 is obliged to use a seed
with low filtering power that lets too many occurrences happen. This dramatically increases the verification effort due to the cost of retrieving text positions
from the FM-index. Interestingly enough, RazerS3, which uses the same seed,
functions well on this data. This is consistent with the fact that a linear method
can, in certain conditions for large k and n, be more efficient than a method
based on a text index [16]. However, Bwolo is still five times faster than RazerS3
for sequences of length 10^9. Indeed, the number of seed occurrences is an order
of magnitude less with Bwolo, which offsets the additional time needed to query
the FM-index in the verification step.

For $k = 2$, we can observe that there are fewer differences in the CPU time
between FM-index and Bwolo on larger texts. The former takes 18.4 s on the
1 GB sequence while the latter takes 13.8 s. This small difference is actually
misleading. Loading the index from disk (which is the same in both cases) and
unserializing the data structures takes 12 s on that same sequence. Ignoring the
loading of the index leads to a three-fold speedup using the 01*0 seeds compared
to the breadth-first approach. With a higher error rate ($k = 3$) we have a seventy-
five-fold speedup on the 1 GB sequence. For the sake of comprehensiveness, we
should mention that RazerS3 takes 8 s to load the 1 GB sequence from disk.

All tools have a reasonable memory consumption, independent of the value
of k, which grows linearly with the size of the text. For example, it is 27 MB
for Bwolo, 99 MB for Bowtie2, 25 MB for RazerS3, and 31 MB for Exonerate
for $n = 10^7$. The memory consumption of Bwolo and Bowtie2 is dominated by
the size of the FM-index. It is larger for Bowtie2 because it also deals with the
inverted text and uses a different implementation. It is quite surprising that

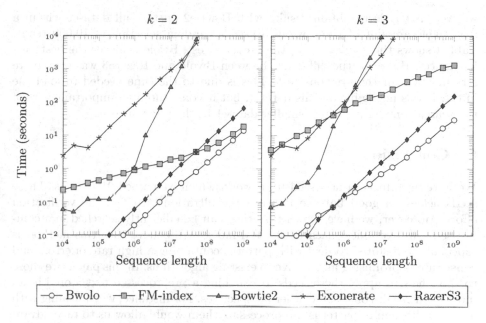

Fig. 3. Running time for 100 randomly generated sequences. Both axes are in logarithmic scale. Bwolo is our algorithm. FM-index refers to the breadth-first search implementation in a FM-index.

Table 1. Running time on the Human genome benchmark. All times are in seconds, and the memory in Mo. NA: non available.

	Index construction		$10{,}000$ reads		10^7 reads	
	Time	Memory	Time	Memory	Time	Memory
Bwolo	$7{,}594$	$9{,}584$	97	$6{,}522$	$55{,}493$	$9{,}054$
RazerS3	0	0	502	$6{,}469$	$467{,}413$	$152{,}045$
Bowtie2	$10{,}584$	$5{,}379$	$156{,}164$	$8{,}260$	NA	NA

RazerS3 and Exonerate have a memory peak in the same order. It may be possible that they load both all the text and keep all results in memory.

4.2   Reads from the Human Genome

In order to test our algorithm with an external dataset made of short sequences we relied on the work done by Schbath *et al.* [20]. Their $\mathcal{H}_3$ dataset contains 10 millions of reads of length 40nt that have been generated from the Human genome (assembly 37.1 from the NCBI, 25 chromosomes for 2.7 Gbp) with exactly three mismatches. Compared to the previous test, it allows us to evaluate the performance of the software with longer patterns, hence longer seeds. The maximum number k of errors is 3 (including indels, not only substitutions). We ran Bwolo, RazerS3, and Bowtie2 on the full set of reads (10^7 reads). Since we were not able to obtain results with Bowtie2 on the full dataset within a reasonable amount of time, we also used a random selection of 10,000 reads. Table 1 shows the results. As in the previous test, Bwolo achieves the best performances. However, the difference between Bwolo and RazerS3 was even more striking than in the previous test. This is due to the time needed to load the index. It was negligible on this dataset, but it constituted an important part of the search time with a much smaller dataset in the previous test.

5   Conclusion

We have introduced a new seed framework, which we named 01*0 seeds. These seeds achieve a good balance between the filtration step and the verification effort. Moreover, we have shown that they can be efficiently searched in a compressed full-text index, such as the FM-index. We believe that this method is especially well-suited to deal with patterns containing a high rate of errors and constitutes a promising alternative to existing algorithms. In this paper, we chose to show how to apply these seeds to searching a preprocessed text stored in an index. Our results offer some other perspectives. For instance, when dealing with a large collection of patterns, preprocessing them would allow us to take advantage of the parts that are shared among several patterns in order to speed up the algorithm. The filtration algorithm could also be applied online. Identifying the 01*0 seeds requires us to identify an exact part first, which we then extend to the 1* parts. This can be performed efficiently using bit-wise operations. Once the seeds are identified, we can compute the left and right extensions using a bit-parallel algorithm [14].

The generalisation of 01*0 seeds to $(01^*)^\varepsilon 0$ could also be promising in further studies. This would not be as straightforward as one would think, since splitting the pattern in $k + 1 + \varepsilon$ parts is not sufficient.

Finally, albeit having been beyond the scope of this paper, an important aspect to thoroughly analyze would be the average case of our algorithm, as Baeza-Yates and Perleberg did in [1].

References

1. Baeza-Yates, R.A., Perleberg, C.H.: Fast and practical approximate string matching. Inf. Process. Lett. **59**(1), 21–27 (1996)
2. Belazzougui, D.: Improved space-time tradeoffs for approximate full-text indexing with one edit error. Algorithmica, pp. 1–27 (2014)
3. Belazzougui, D., Cunial, F., Kärkkäinen, J., Mäkinen, V.: Versatile succinct representations of the bidirectional burrows-wheeler transform. In: Bodlaender, H.L., Italiano, G.F. (eds.) ESA 2013. LNCS, vol. 8125, pp. 133–144. Springer, Heidelberg (2013)
4. Chan, H.L., Lam, T.W., Sung, W.K., Tam, S.L., Wong, S.S.: A linear size index for approximate pattern matching. J. Discrete Algorithms **9**(4), 358–364 (2011)
5. Chávez, E., Navarro, G.: A metric index for approximate string matching. In: Rajsbaum, S. (ed.) LATIN 2002. LNCS, vol. 2286, pp. 181–195. Springer, Heidelberg (2002)

6. Döring, A., Weese, D., Rausch, T., Reinert, K.: SeqAn an efficient, generic C++ library for sequence analysis. BMC Bioinformatics 9(1), 11–19 (2008)
7. Ferragina, P., González, R., Navarro, G., Venturini, R.: Compressed text indexes: from theory to practice. J. Exp. Algorithmics (JEA) 13, 12 (2009)
8. Ferragina, P., Manzini, G.: Indexing compressed text. J. ACM (JACM) 52(4), 552–581 (2005)
9. Ferragina, P., Manzini, G., Mäkinen, V., Navarro, G.: Compressed representations of sequences and full-text indexes. ACM Trans. Alg. (TALG) 3(2) (2007)
10. Hyyrö, H.: A bit-vector algorithm for computing levenshtein and damerau edit distances. Nord. J. Comput. 10(1), 29–39 (2003)
11. Langmead, B., Salzberg, S.L.: Fast gapped-read alignment with Bowtie 2. Nat. Meth. 9(4), 357–359 (2012)
12. Li, H., Durbin, R.: Fast and accurate short read alignment with burrows-wheeler transform. bioinformatics 25(14), 1754–1760 (2009). (Oxford, England)
13. Maaß, M.G., Nowak, J.: Text indexing with errors. J. Discrete Algorithms 5(4), 662–681 (2007)
14. Myers, G.: A fast bit-vector algorithm for approximate string matching based on dynamic programming. J. ACM 46(3), 395–415 (1999)
15. Navarro, G.: A guided tour to approximate string matching. ACM comput. surv. (CSUR) 33(1), 31–88 (2001)
16. Navarro, G., Baeza-Yates, R.: A hybrid indexing method for approximate string matching. J. Discrete Algorithms 1, 19–27 (2001)
17. Navarro, G., Sutinen, E., Tanninen, J., Tarhio, J.: Indexing text with approximate q-grams. In: Giancarlo, R., Sankoff, D. (eds.) CPM 2000. LNCS, vol. 1848, pp. 350–363. Springer, Heidelberg (2000)
18. Petri, M., Culpepper, J.S.: Efficient indexing algorithms for approximate pattern matching in text. In: Proceedings of the Seventeenth Australasian Document Computing Symposium, ADCS 2012, pp. 9–16. ACM, New York (2012)
19. Russo, L., Navarro, G., Oliveira, A.L., Morales, P.: Approximate string matching with compressed indexes. Algorithms 2(3), 1105–1136 (2009)
20. Schbath, S., Martin, V., Zytnicki, M., Fayolle, J., Loux, V., Gibrat, J.F.: Mapping reads on a genomic sequence: an algorithmic overview and a practical comparative analysis. J. Comput. Biol. 19(6), 796–813 (2012)
21. Schnattinger, T., Ohlebusch, E., Gog, S.: Bidirectional search in a string with wavelet trees. In: Amir, A., Parida, L. (eds.) CPM 2010. LNCS, vol. 6129, pp. 40–50. Springer, Heidelberg (2010)
22. Shah, S.A., Hansen, N.R., Garrett, R.A.: Distribution of CRISPR spacer matches in viruses and plasmids of crenarchaeal acidothermophiles and implications for their inhibitory mechanism. Biochem. Soc. Trans. 37(1), 23 (2009)
23. Slater, G.S.C., Birney, E.: Automated generation of heuristics for biological sequence comparison. BMC Bioinformatics 6, 1–11 (2005)
24. Stern, A., Keren, L., Wurtzel, O., Amitai, G., Sorek, R.: Self-targeting by CRISPR: gene regulation or autoimmunity? Trends Genet. 26(8), 335–340 (2010)
25. Storz, G., Altuvia, S., Wassarman, K.M.: An abundance of RNA regulators. Annu. Rev. Biochem. 74, 199–217 (2005)
26. Weese, D., Holtgrewe, M., Reinert, K.: RazerS 3: faster, fully sensitive read mapping. Bioinformatics 28(20), 2592–2599 (2012)
27. Wu, S., Manber, U.: Fast text searching: allowing errors. Commun. ACM 35(10), 83–91 (1992)

Author Index

Abu-Khzam, Faisal N.   1
Adamaszek, Anna   13
Alfalayleh, Mousa   24
Apollonio, Nicola   37
Barton, Carl   49
Blanchet-Sadri, Francine   62, 74, 86
Blin, Guillaume   13
Bonnet, Édouard   1
Brankovic, Ljiljana   24
Caramia, Massimiliano   37
Choi, Ilkyoo   98
Chung, Fan   110
Cichacz, Sylwia   122
Colbourn, Charles J.   274
Collister, Daniel   250
Cordier, Michelle   62
Djelloul, Selma   128
Durocher, Stephane   140
Ekstein, Jan   98
El-Zanati, Saad I.   153
Eynden, Charles Vanden   153
Franciosa, Paolo Giulio   37
Holub, Přemysl   98
Iliopoulos, Costas S.   49
Irving, Robert W.   213
Ito, Takehiro   164, 351
Jäger, Gerold   188
Jayapaul, Varunkumar   176
Jo, Seungbum   176
Jongthawonwuth, Uthoomporn   153
Jordon, Heather   153
King, Valerie   307
Kirsch, Rachel   62
Kriege, Nils   200
Kurpicz, Florian   200
Kwanashie, Augustine   213
Laforest, Christian   262
Larjomaa, Tommi   226
Lidický, Bernard   98
Lindzey, Nathan   238
Liu, Daphne Der-Fen   250
Lohr, Andrew   74

Manlove, David F.   213
Manuel, Paul   298
Mehrabi, Saeed   140
Momège, Benjamin   262
Murray, Patrick C.   274
Mutzel, Petra   200
Nikkel, Jordan   86
Nooka, Hiroyuki   164
Palios, Leonidas   286
Parthiban, N.   298
Peczarski, Marcin   188
Pissis, Solon P.   49
Popa, Alexandru   13, 226
Quigley, J.D.   86
Rahmati, Zahed   307
Rajalaxmi, T.M.   298
Rajan, R. Sundara   298
Rajasingh, Indra   298
Raman, Venkatesh   176
Rizzi, Romeo   318
Sacomoto, Gustavo   318
Sagot, Marie-France   318
Salson, Mikaël   364
Saputro, Suhadi Wido   330
Satti, Srinivasa Rao   176
Sikora, Florian   1
Simanjuntak, Rinovia   330
Simpson, Olivia   110
Smyth, William F.   49
Sng, Colin T.S.   213
Starikovskaya, Tatiana   338
Tamura, Yuma   351
Touzet, Hélène   364
Tzimas, Petros   286
Uttunggadewa, Saladin   330
Vildhøj, Hjalte Wedel   338
Vroland, Christophe   364
Whitesides, Sue   307
Zhang, Xufan   86
Zhou, Xiao   164, 351

Printed in the United States
By Bookmasters